R.Q. HUANG and I

STELLAR ASTROPHYSICS

Springer

Professor R Q Huang
Yunnan Observatory
Academia Sinica
Kunming 65001
China

Dr K N Yu
Department of Physics and Materials Science
City University of Hong Kong
Tat Chee Avenue
Kowloon
Hong Kong

Huang , R. Q. , 1933 -
        Stellar Astrophysics / R. Q. Huang & K. N. Yu
                p.   cm.
        Includes bibliographical references and index.
        ISBN 9813083360
        1. Stars. 2. Astrophysics /   I. Yu, K. N., 1963-   . II. Title.
QB801.H73   1998
523.8 -- dc21

ISBN 981-3083-36-0

© Springer-Verlag Singapore Pte. Ltd. 1998
Printed in Singapore

Typesetting: Expo Holdings Sdn Bhd
SPIN 10642820   5 4 3 2 1 0

STELLAR ASTROPHYSICS

Springer
Singapore
Berlin
Heidelberg
New York
Barcelona
Budapest
Hong Kong
London
Milan
Paris
Tokyo

PREFACE

Stellar astrophysics is an extremely important component in contemporary astrophysics. It includes the studies of various physical processes, internal structure and evolution of different stellar objects, stellar oscillations, and radiative transfer in the stellar atmosphere etc. After the 1960s, research of stellar astrophysics has made rapid progress, and has led to relatively complete and systematic theories on stellar structure and evolution, stellar oscillations and stellar atmosphere. After the 1980s, along with the continual broadening of the wavelengths for astronomical observations and the continuous modernization of observational techniques, many new stellar objects and many new astronomical phenomena not known in the past have been discovered. Within these, many discoveries have posed serious problems on existing theories and concepts, which have led to a large amount of research outputs in different regimes of stellar astrophysics in the recent ten to twenty years, and to many new theories and concepts. These have also inspired the authors to write a book on stellar astrophysics. On one hand, we try to incorporate in a single book several important areas in stellar astrophysics, including stellar structure and evolution, stellar oscillations and stellar atmosphere. On the other hand, we try our best to summarize the great developments. Furthermore, we present detailed descriptions of the good models on stellar structure and evolution, stellar oscillations and stellar atmosphere in recent years.

Although we have tried every effort to avoid mistakes, these might still exist due to the limited knowledge of the authors. Comments from readers are most welcome.

R.Q. Huang
Yunnan Observatory, Yunnan
People's Republic of China

K.N. Yu
City University of Hong Kong, Hong Kong
People's Republic of China

FOREWORD

Stellar astrophysics is a fundamental subject in contemporary astrophysics. The majority of information in the universe comes either directly or indirectly from the light emitted from stars. Until now, about 10^9 astronomical objects have been collected by optical surveys, while about 10^5 objects have been collected in radio, infrared, and X-ray wavelengths etc., which may increase to 10^6 in ten years. Most of the optical objects are stars, while the rest are mainly galaxies. Since the light from galaxies are in fact contributed by billions of stars, their physical explanations also rely on stellar astrophysics.

Based on their scales and physical characteristics, astronomical objects are contemporarily categorized into regimes of planets, stars and galaxies. On one hand, each regime has its own developments. On the other hand, stars are constituents of galaxies and at the same time are main parts of planetary systems. Therefore, these three different regimes are interrelated with each other, and stellar astrophysics can be viewed as a "traditional starting point" of astrophysical research.

Astrophysical research employs physics theories in explaining astronomical observations, and physics theories are inspired and tested by astronomical observations. Therefore, the external factors in driving the development of astrophysics include the technological advancements at a certain time, which enhance the astronomical observation techniques, as well as the progress in physics research, which has diverse interrelationship with astrophysics. Of course, these phenomena also apply to stellar astrophysics. At the same time, there are two internal factors in driving the development of stellar astrophysics. First, the research in the structure, origin and evolution of stars are fundamental to astrophysical research. Second, the stellar objects on record amount to only one-hundredth of the total number in the Galaxy, and within these recorded ones, only about one in ten thousand have been studied. Therefore, even under our present observational limits, there are many more targets and dimmer targets to explore, and there are much more and much finer topics for these already known and to-be-known targets. This treasure in the world of stellar astrophysics has provided never-ending passion for astrophysics researchers.

Stellar astrophysics has evolved to a mature subject within only a century. Classical texts on stellar atmosphere and internal structure emerged in the 1920s to 1930s. During the half century after the establishment of the theoretical framework, we have witnessed immense developments in astronomical observation techniques and the data processing and analytical power. These have brought about a series of new developments, research and concepts, which enabled the perfection, development and refinement of different aspects in stellar astrophysics. In this way, stellar astrophysics texts should be modified from time to time to include new materials and to introduce new methods; these have been done by astrophysicists in different generations. This book is written based on this need. One of the authors of this book have also published Chinese astrophysics textbooks "Stellar Structure and Evolution" and "Theory of Stellar Atmosphere" in 1986, and "Theory of Stellar Oscillations" in 1990. This is based on these texts, and also on results of more recent research. The structure and the contents of Stellar Astrophysics have been outlined in the preface and can be readily known from the content pages. A remark is that the stress on calculation methods of different models of stellar astrophysics has made this book a valuable reference for research in this field.

Professor Shou-Guan Wang
Beijing
December 1996

CONTENTS

CHAPTER ONE

INTRODUCTION

1.1 SUBJECTS AND METHODOLOGY IN STELLAR ASTROPHYSICS

There are many areas related to stellar astrophysics. The main areas to be discussed in this book are briefly described in the following sections.

Theory of stellar structure and evolution This theory studies the various physical processes inside stars, and the inner structure and evolution patterns determined by these physical processes, such as the distribution and evolution patterns of temperature, pressure, density, chemical composition etc. The inside of the stars cannot be directly observed. Their structure and evolution patterns can be revealed as follows: Based on some simplified assumptions on the geometrical structure and the physical state of stars, one can establish a set of basic equations which describe the hydrostatic equilibrium, distribution of mass, generation and transfer of energy inside the stars. On solving this set of equations, the distribution of various physical and chemical parameters inside the star at different times, or in other words, a stellar model, can be obtained. The theoretically obtained stellar evolution patterns are then compared with observations. The former will be taken to be true if the comparisons show satisfactory agreements.

Theory of stellar oscillations   This theory studies the physical nature of pulsations in different pulsation variables, and properties of the oscillations, such as oscillation modes, propagating regions, and oscillation periods etc. Traditional theories of stellar oscillations employ a set of equations which describe the conservation of mass and momentum and the thermodynamic equilibrium inside stars. Through linearizing these equations, a set of oscillation equations can be set up. The numerical solution together with analytical discussions of the oscillation equations can tell whether a star will have oscillations, oscillation modes, propagation regions inside the star, and oscillation periods. Of course, these theoretical oscillation properties are further required to commensurate with observational properties.

Theory of stellar atmosphere   It studies radiative transfer in the stellar atmosphere. Based on simplified assumptions on the geometrical structure and physical conditions of the stellar atmosphere, astrophysicists have set up a set of basic equations which describe the radiative transfer, conservation of particles and statistical equilibrium at different levels, i.e., they have set up a model atmosphere. On solving this equation set, properties of stellar spectrum (such as the distribution of fluxes in the continuum or profiles of the lines) and the structure of the stellar atmosphere (such as distribution of temperature, pressure, density, etc.) can be obtained. By further comparing the theoretical spectral properties with the observational information, relevant parameters in the model atmosphere or line model are tuned for agreement between theoretical and observational properties, and the theoretical model is then assumed to reflect the true physical and chemical structure of the stellar atmosphere.

1.2 IMPORTANT PATTERNS OBTAINED FROM ASTRONOMICAL OBSERVATIONS

There is abundant observational information accumulated for different types of stellar objects, some of which are essential for driving and testing the development of stellar astrophysics. The Hertzsprung–Russel diagram (H–R diagram) is a wonderful tool in assisting astrophysicists to reveal the patterns of stellar evolution, and at the same time provides the most powerful test for theories on stellar structure and evolution, and on stellar oscillations. The spectral features have helped develop and provide a powerful test on the theory of stellar atmosphere.

1.2.1 Hertzsprung–Russel Diagram (H–R diagram)

Through astronomical observations, the luminosity (or absolute magnitude) and effective temperature (or spectral type, color index) can be obtained for a star. A plot of the luminosity L (L increasing upwards for the y-axis) against the effective temperature T_{eff} (T_{eff} increasing towards the left for the x-axis) is called the Hertzsprung–Russel diagram (H–R diagram) (see Fig. 1.1). According to the luminosity L and the effective temperature, each star should correspond to a point in the H–R diagram. Since a cool star is red while a hot star is white or blue, red stars are located on the right while white or blue stars are on the left of the H–R diagram. Stars with large luminosities are located at the top, while stars with small luminosities are at the bottom. Therefore stars located in the upper right-hand corner of the H–R diagram are cool stars with large luminosities. The very small energy radiated from a unit surface area from these cool stars, together with the very large amount of total radiated energy from the star inferred from their large luminosities imply a very large surface area for the stars or imply that the stars have very large radii; these stars are called red giants or red supergiants. Conversely, stars at the lower left corner should be hot stars with small radii and are called white dwarfs.

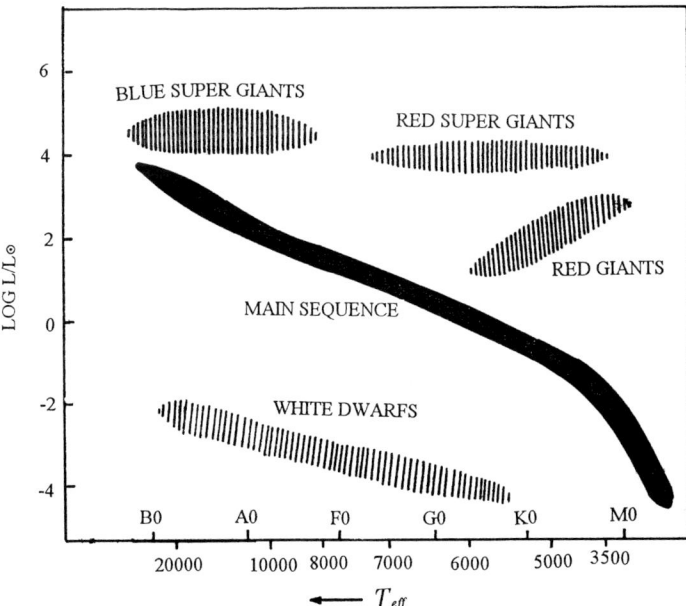

Fig. 1.1 H–R diagram showing the distribution of different types of stars.

If the several thousand stars near the Sun are plotted in the H–R diagram, very interesting patterns can be found: more than 90% of the stars are located in a narrow strip extending from the upper left to the lower right, which is thus known as the main sequence strip. The Sun is also in the main sequence strip. When descending from the top to the bottom of the main sequence strip, the stars will have their mass decreasing, colors changing from white to blue to red, and temperatures from high to low. Towards the top of the main sequence strip, there is a small number of blue supergiants. To the right of these blue supergiants, there is a small number of red supergiants. Between the main sequence strip and the red supergiants is the red giant branch. To the lower left of the main sequence strip is the white dwarf strip. There is a gap between the main sequence strip, red supergiants and red giants in which no stars are located (called Hertzsprung gap).

The luminosities for stars in the H–R diagram have a large range of values which can differ by a factor of 10^8. Therefore, the luminosities are divided into six classes, named the luminosity classes for stars. The classes are: I = supergiants; II = bright giants; III = giants; IV = subgiants; V = main sequence stars. Figure 1.2 gives the distribution of stars with different luminosity classes in the H–R diagram.

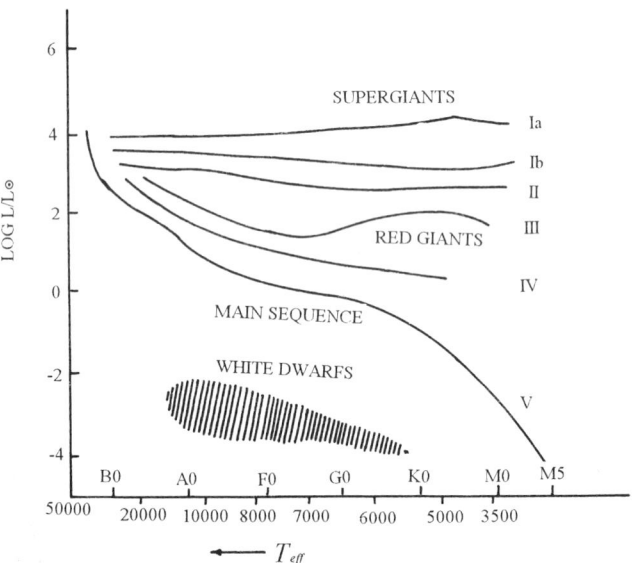

Fig. 1.2 The distribution of stars with different luminosity classes in the H–R diagram.

Figure 1.3 shows the distribution of pulsation variables in the H–R diagram. It can be seen that most of the pulsation variables are distributed in a narrow strip, called the Cepheid pulsation strip, extending from the upper right to the lower left, which is almost perpendicular to the main sequence strip. When going from top to

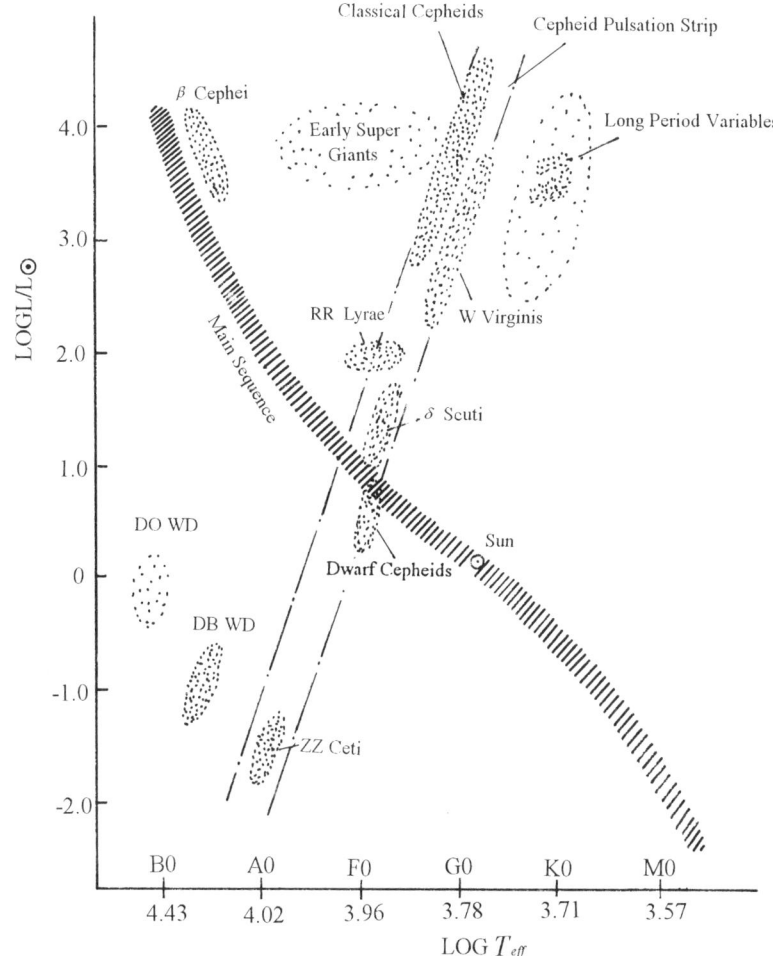

Fig. 1.3 The distribution of pulsation variables in the H–R diagram.

bottom of the Cepheid pulsation strip, we can find the variables including classical Cepheids, W Virginis, RR Lyrae, δ Scuti, dwarf Cepheid and the ZZ Ceti variables. On the left of Fig. 1.3 near the main sequence strip, we have the β Cephei variables, and to the right of it, we have early supergiant variables with quasi-periodic oscillations. To the right of the Cepheid pulsation strip are found long period variables which are then surrounded by red semi-regular variables. To the lower left of the main sequence strip and above the ZZ Ceti variables, we have DB type variables, and at an even higher location, we have DO type variables. These three types of variables are situated in the white dwarf branch and are white dwarf variables. The Sun is situated at the lower half of the main sequence strip and is not far away from the zero-age main sequence. Abundant pulsations have also been observed on the surface of the Sun. When going from top to bottom along the Cepheid pulsation strip, observations have revealed that the oscillation tends to change from longitudinal to transverse mode, from single to multiple modes, and from large amplitude to small amplitude.

According to the characteristics in the chemical composition, spatial distribution and motion, the stars can be further divided into different populations. Baad (1944) first divided the stars into two large populations: population I and population II. Regarding the chemical composition, the differences between the two populations are

$$\text{population I:} \quad X_H = 0.6\text{--}0.7, \quad X_Z = 0.01\text{--}0.04$$
$$\text{population II:} \quad X_H = 0.9, \quad\quad X_Z = 0.001$$

where X_H is the abundance of hydrogen and X_Z is the abundance of heavy metals. From these, we can see that population I stars have rich contents of heavy metals while population II stars are deficient in heavy metals. Most population I stars are situated on the galactic plane, especially in the spiral arms; while population II stars are mainly situated near the center of the galaxy.

In general, stars have a clustering phenomenon. Stellar clusters can be divided into globular clusters and open clusters. Globular clusters are distributed in a vast globular sphere whose center coincides with the center of the Galaxy. The stars inside globular clusters are in general population II stars. Open clusters are normally situated on the galactic plane. These two different types of stellar clusters have completely different properties on the H–R diagram. Figure 1.4 shows some open cluster stars on the H–R diagram. It can be seen that for each stellar cluster, most stars are concentrated on the main sequence. However, the brighter stars will deviate to the right of the main sequence starting at a certain point of the main sequence (turn off point). Some stellar clusters, such as the $h + \chi$ Persei can be observed to have a red giant branch. Others such as M11 have a giant branch. Figure 1.5 gives the H–R diagram for some globular cluster stars. It can be seen that the turn off points and the distribution for different globular cluster stars are quite similar. Another feature in the H–R diagram for globular clusters is the presence of a horizontal branch which is absent for open clusters.

1.2.2 Mass–Luminosity and the Mass–Radius Relations for Main Sequence Stars

There is a certain relation between the mass and the luminosity of stars which is known as the mass–luminosity relation. The mass–luminosity relation of main sequence stars is shown in Fig. 1.6. According to Harris *et al.* (1962), the mass–luminosity relation of main sequence stars can be expressed as

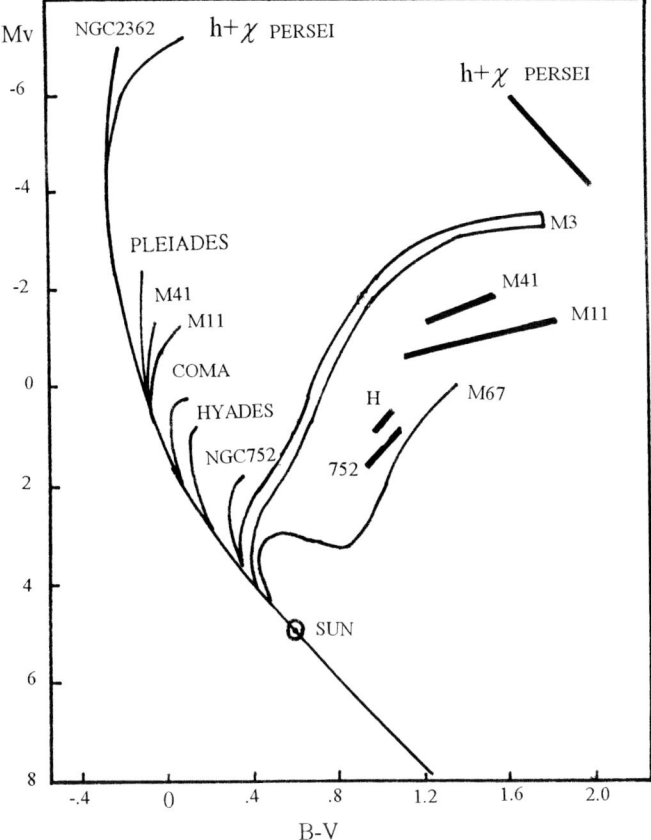

Fig. 1.4 The H–R diagram for some open clusters. Some globular clusters have been included for comparison (according to Sandage 1962).

$$L/L_\odot = 1.2(M/M_\odot)^{4.0}$$

or

$$M_b = 4.6 - 10.0\log(M/M_\odot)$$

$$\left.\right\} 0 \leq M_b \leq +7.5 \qquad (1.1)$$

$$L/L_\odot = 0.67(M/M_\odot)^{2.76}$$

or

$$M_b = 5.2 - 6.9\log(M/M_\odot)$$

$$\left.\right\} 7.5 \leq M_b \leq +11 \qquad (1.2)$$

Similarly, there is also a relation between the mass and the radius for main sequence stars, which is known as the mass–radius relation. Figure 1.7 shows this relation for main sequence stars. From Fig. 1.7, it is known that the mass–radius relation for main sequence stars with masses greater than the solar mass can be represented as

$$R \sim M^{0.6} \qquad (1.3)$$

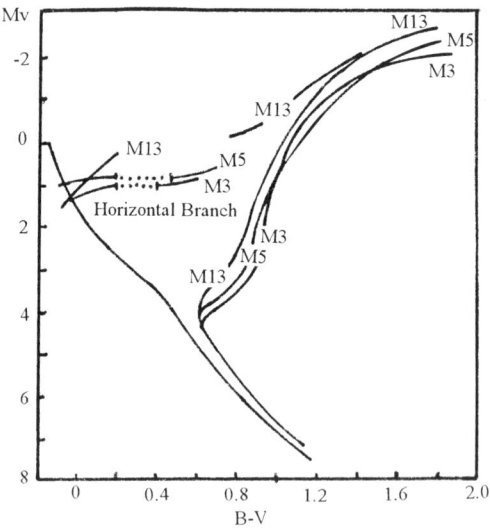

Fig. 1.5 The H–R diagram for some globular clusters (according to Sandage 1962).

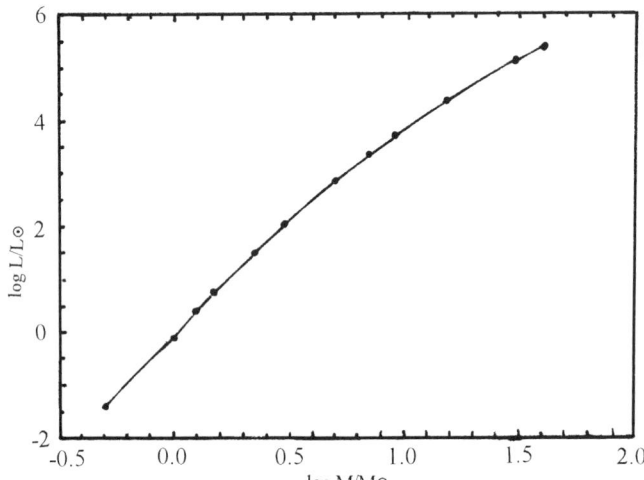

Fig. 1.6 The mass–luminosity relation for the main sequence stars ($X_H = 0.70$, $X_Z = 0.02$) (according to Iben 1965, Brunish and Truran 1982).

For masses smaller than the solar mass, the curve in Fig. 1.7 is visibly bent, so it is difficult to have a simple approximation for the relation.

1.2.3 Stellar Spectra

Stellar spectra are constituted by continuous spectra and the many superimposed absorption and emission lines. The profile of the continuous spectrum is related to the effective temperature of the star; the energy flux peak moves towards the short wavelengths as the effective temperature increases. The intensity of a spectral line depends on the

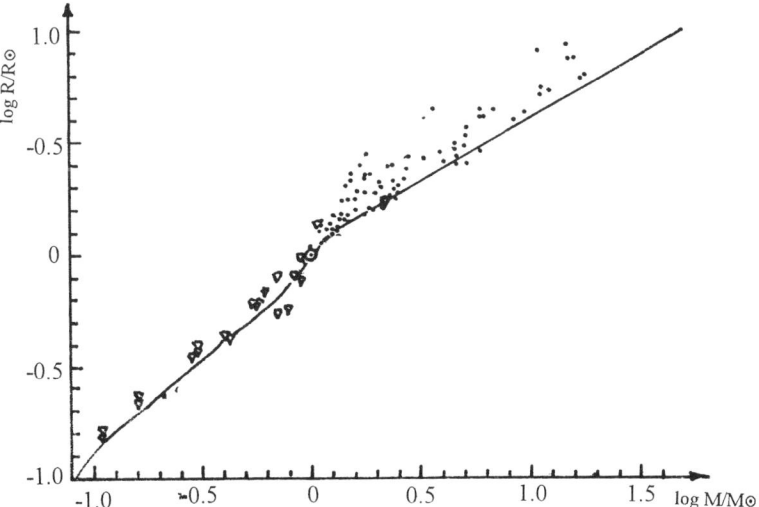

Fig. 1.7 The mass–radius relation for main sequence stars. *Black dots*: observational values for detached binaries; *Triangles*: observational values for visual binaries (according to Popper 1980); *Solid line*: theoretical calculations ($X_H = 0.70, X_Z = 0.02$).

atomic occupation number of the energy level corresponding to the spectral line, which is in turn associated with the effective temperature and pressure (or gravitational acceleration) of the stellar atmosphere.

In order to differentiate among spectra for different stars, the concept for spectral type is introduced which is based on the energy flux profile of the continuous spectra or based on features of spectral lines. For very faint stars, the color indices can also be employed for determining their spectral types. More than 99% of stellar spectra can be classified into the following spectral types

$$O, B, A, F, G, K, M$$

and each spectral type can be further divided into ten sub-classes denoted by numbers 0, 1, 2, ..., 9, e.g., B0, B1, ..., B9 and A0, A1, ..., A9 and so on. The adoption of these alphabets for the spectral types are due to historical habits. Usually, the O, B and A type stars are referred to as "early" type stars, F and G type stars as "intermediate" type stars, and K and M type stars as "late" type stars.

When the spectral type transits from O to M, the effective temperature of the stars changes from high to low, and the color of the stars changes from white and blue to red and yellow.

In the following, three methods to determine the spectral types of stars are described:

(1) Determination of spectral types according to spectral lines. For a fixed chemical composition of the stellar atmosphere, the intensity of a

spectral line is only dependent on the effective temperature and gravitational acceleration of the star, and the gravitational acceleration is directly related to the luminosity class of the star. In this way, for stars with the same chemical composition and luminosity class, such as main sequence stars, the spectral type is mainly determined by the effective temperature. This is a one-dimensional classification. The Havard classification is a one-dimensional classification based on the difference in temperature. Of course, a better classification is a two-dimensional classification based on both the effective temperature and the luminosity class. The commonly used M-K classification is a two-dimensional classification.

● *Havard classification.* In the following, the ionization state of an atom is represented by I, II, III, ... after the symbol of the element, with I representing the neutral atom, II the singly ionized atom and III the doubly ionized atom, etc. For example, neutral hydrogen atoms are written as HI while singly ionized hydrogen atoms are written as HII. Different spectral types can be determined from the following spectral features.

O type: For O0 to O4, HeII absorption lines are the most important. For O5 to O9, HeII absorption lines are replaced by HeI lines, and there are NII, OII, SiII absorption lines and relatively weak HI absorption lines (Balmer lines). The energy flux peaks in the uv region of the continuous spectrum.

B type: For B0 to B4, there are HeI and hydrogen absorption lines (H_β, $H\gamma$, H_δ etc.). For B5 to B9, HeI lines are very weak but the Balmer lines are stronger.

A type: For A0 to A4, Balmer lines are the most important and have reached their maximum intensities. For A5 to A9, Balmer lines become weaker, and there are some ionized metal lines such as FeII and CaII.

F type: The intensities for CaII, FeII and TiII lines become stronger, and there are neutral metal lines such as FeI. The Balmer lines gradually disappear.

G type: The intensity of the CaII line reaches the maximum. The neutral metal lines such as FeI and TiI become more intense and their ionization lines become weaker. The energy flux peaks in the continuous spectra move back to the visible range.

K type: The neutral metal lines become stronger, especially the CaI line. For K4 to K9, lines from simple molecules such as CH and CN emerge.

M type: Almost no lines other than neutral metal lines and molecular lines. In particular, only absorption lines due to TiO are rapidly strengthened.

Figure 1.8 shows the relationship between the relative intensity and the effective temperature of different spectral lines for different spectral types. It can be seen that for the same element, the spectral line corresponding to the more ionized element will emerge in an earlier type of spectrum.

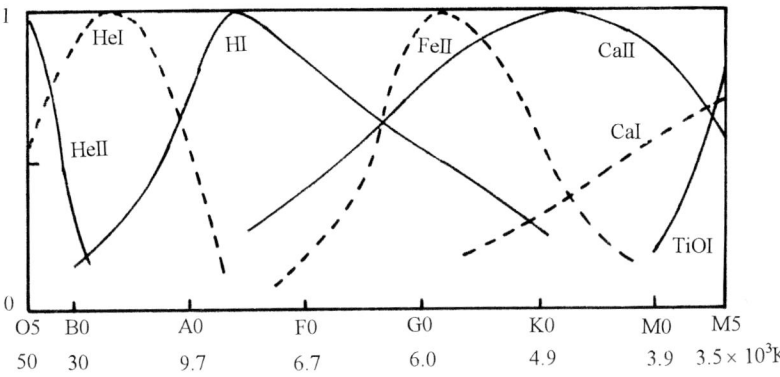

Fig. 1.8 Relationship between the relative intensity and the effective temperature of different spectral lines for different spectral types.

- *M-K classification.* One-dimensional classification cannot precisely determine the spectral types of stars. For example, if the classification is only based on the species and the intensities of spectral lines, the spectra of two stars can be taken to be very similar thus belonging to the same spectral type. However, it is possible that the lines in one spectrum are very sharp and clear while those in the other are broad and vague, which are not differentiated with each other in the one-dimensional classification scheme based solely on the effective temperature. However, for a fixed chemical composition of the stellar atmosphere, the spectral properties are also dependent on the gravitational acceleration on the stellar surface as well as the effective temperature. The luminosity L of a star is closely related to the gravitational acceleration g. According to the formula $L = R^2(T_{eff}/5787)^4$, L is proportional to the square of the radius R when the effective temperature T_{eff} is kept constant. On the other hand, g is inversely proportional to R and is thus inversely proportional to L. Furthermore, the spectral line width is also associated with L. For stars with the same T_{eff}, the spectral line width decreases and the line becomes clearer when L increases. Therefore, besides relying on the effective temperature in the determination of spectral types, the luminosity class of stars should also be taken into consideration. This is the two-dimensional classification. And M-K classification belongs to such a two-dimensional classification.

In Subsec. 1.2.1, it has been pointed out that the luminosity classes can be represented using Romans I, II, ... In the spectral type classification, the luminosity classes are represented by the Romans attached to the end of the spectral type symbol, e.g., AoI, AoII, ... Within each luminosity class, three sub-classes can be further identified as a, ab and b. The case for white dwarfs is special, and their luminosity classes are written as DA and DB. Some examples of spectral types in the M-K classification are shown in the following

$$
\begin{array}{ll}
\text{Sun:} & \text{G2V} \\
\alpha \text{ Cyg:} & \text{A2Ia} \\
\alpha \text{ Boo:} & \text{K2III} \\
\alpha \text{ Ori:} & \text{M2Iab}
\end{array}
$$

(2) Determination of spectral types according to boundary jumps for line series. In the continuous spectra of stars, abrupt changes in the energy flux at certain positions are observed. These are known as jumps. These phenomena occur at boundaries of strong absorption line series, e.g., the Balmer series, which correspond to the change from bound-bound transition to bound-free transition. The most typical example is the Balmer line series of hydrogen: $H_\alpha(\lambda = 656.3\text{nm})$ $H_\beta(\lambda = 486.1\text{nm})$, $H_\gamma(\lambda = 434.0\text{nm})$, ... When the wavelength gets smaller, both line intensities and separation between lines become smaller. Finally, the lines converge at the boundary value $\lambda_\infty = 3646\text{Å}$. We go into the continuous spectrum beyond λ_∞, the energy flux of which is always much lower than that of the Balmer series. Therefore, a conspicuous change in the energy flux will be observed near λ_∞ which is known as the Balmer jump and represented by D in Fig. 1.9(b). In a realistic spectrum, the boundary of the Balmer series is very vague, and is artificially defined to be $\lambda_\infty = 3650\text{Å}$. In this way, the Balmer jump is

$$D \equiv 2.5\log[F_\nu(\lambda 3650^+)/F_\nu(\lambda 3650^-)]$$

The size of the Balmer jump D is dependent on the effective temperature and thus the spectral type of the star. Besides, the slope of the drop in the

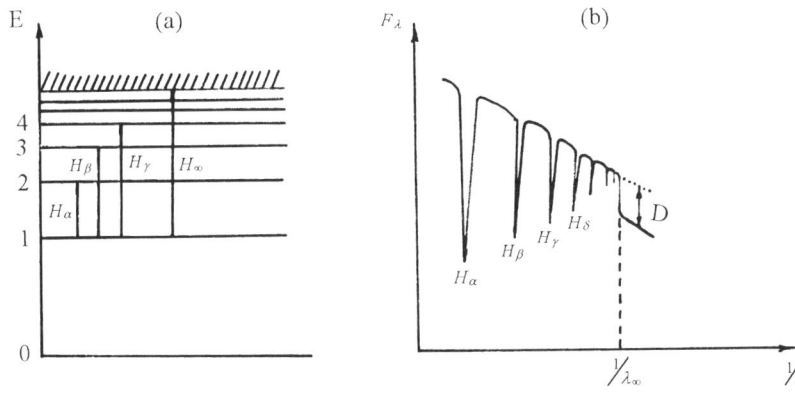

Fig. 1.9 Schematic diagrams for (a) energy levels of a hydrogen atom and (b) the Balmer jump.

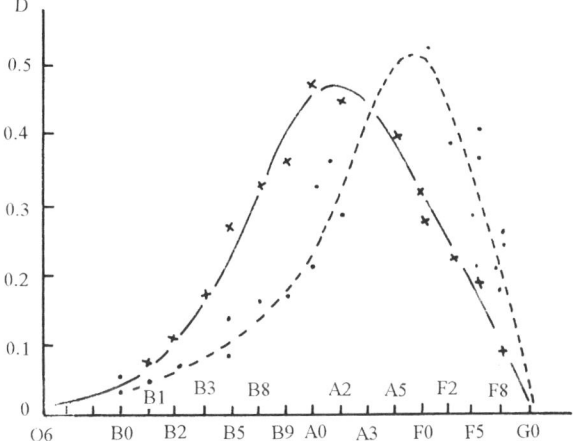

Fig. 1.10 The relationship between the Balmer jump D and the spectral type for main sequence stars and supergiants. X indicates the average value for main sequence stars, • indicates the average value for supergiants.

energy flux at the boundary of the Balmer series is dependent on the luminosity class of the star. Therefore the spectral type can be determined by D. Figure 1.10 gives the relationship between the value of D and the spectral type for main sequence stars and supergiants.

(3) Determination of spectral types according to color indices. The color indices of stars are differences in the brightness (in units of visual magnitude) measured in different spectral regions (different color regions). The color indices are associated with the energy flux profile in the spectra, so the color indices are related to the spectral types of stars and thus the spectral types can be determined by color indices. For a very faint star, it will be difficult to obtain its spectrum, but the color index can still be obtained through photo-electric measurements. In such circumstances, the color index method will be very important for determination of spectral types. The common color index system is the UBV system, with U representing

the uv region, B the blue region and V the visible region. Figure 1.11 shows the relationship between the color index FI and the spectral type for main sequence stars.

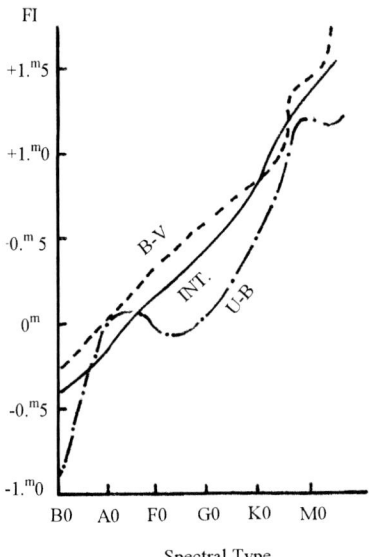

Fig. 1.11 The relationship between the color index FI and the spectral type for main sequence stars.

CHAPTER TWO

THEORY OF RADIATION

The gaseous materials and radiation fields inside a star have mutual interactions. The action of gaseous materials on the radiation can be reflected through absorption, emission and scattering. This action can change the intensities, propagation direction and frequency of the radiation. Conversely, the radiation can affect the state of thermal motion of the gaseous materials. The mutual interactions between the gaseous materials and radiation fields have important effects on the structure and evolution of stars. In this chapter, macroscopic description of radiation field and questions on radiative transfer will be discussed. In-depth discussions on the interaction between radiation and stellar matter will be presented in Chapter 5.

2.1 MACROSCOPIC DESCRIPTION OF RADIATION FIELD

For the macroscopic description of features of a radiation field, the physical quantities of radiation intensity, radiation flux, radiation field energy density and radiation pressure should be introduced, which will be defined in the following.

2.1.1 Radiation Intensity

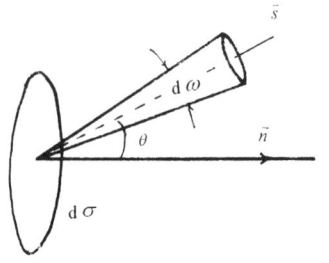

Fig. 2.1 The geometry for calculation of the radiation intensity.

To analyze a radiation field, a test area element $d\sigma$ is placed at an arbitrary position $\vec{r}$ in the radiation field with the normal $\vec{n}$ (see Fig. 2.1). The energy dE_ν passing through $d\sigma$ along the direction $\vec{s}$ in a time interval dt, in a solid angle $d\omega$ with a frequency in the range $(\nu, \nu + d\nu)$ is represented as

$$dE_\nu = I_\nu \cos\theta d\sigma d\nu d\omega dt \tag{2.1}$$

where $d\sigma \cos\theta$ is the projection of the area element $d\sigma$ perpendicular to the axis of the radiation beam, i.e., the effective area. The physical meaning of the proportionality constant I_ν can be visualized as radiation energy passing through a unit area perpendicular to the axis of the radiation beam, in a unit solid angle along this axis of the radiation beam in a unit time and a unit frequency interval. I_ν is called the specific intensity or the radiation intensity. Usually, I_ν is related to the test position and the direction, i.e., $I_\nu = I_\nu(\vec{r}, \vec{s})$. If the radiation field is time dependent, I_ν should also depend on the time t.

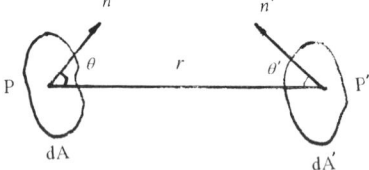

Fig. 2.2 The geometry for illustrating that the radiation intensity is independent of the distance from the radiation source.

The radiation intensity I_ν is independent of the distance from the radiation source. To illustrate this, we have to prove that I_ν is a constant (i.e., $dI_\nu/ds = 0$) when there is no radiation source in the light path. Suppose a light beam passes through area elements dA and dA' (see Fig. 2.2) and there are no radiation sources between the area elements. When viewing from point P, the solid angle subtended by dA' is $d\omega = dA'\cos\theta'/r^2$ when viewing from point P', the solid angle subtended by dA is $d\omega' = dA\cos\theta/r^2$. Denote the specific intensity at P as I_ν and that at P' as I'_ν. Since the light beam passes through dA as well as dA', and there are no radiation sources between them, for the same radiation beam, we have

$$dE_\nu = I_\nu dA\cos\theta \cdot \frac{dA'\cos\theta}{r^2} \cdot d\nu dt$$

$$= I'_\nu dA'\cos\theta' \frac{dA\cos\theta}{r^2} d\nu dt$$

from which we can obtain

$$I_\nu = I'_\nu \quad \text{or} \quad \frac{dI_\nu}{ds} = 0 \tag{2.2}$$

According to the definition of I_ν, we also know that the solid angle $d\omega$ subtended by a stellar object should be known if an observer on the ground wants to directly measure I_ν for this object. This is only possible if the circular area of the surface of the object can be resolved. Up to now, only the I_ν value for the Sun can be measured. For other stars, the I_ν values cannot be directly measured because their circular areas cannot be resolved.

If I_ν is independent of position and direction, the radiation field is homogeneous and isotropic. Averaging I_ν over all directions, the mean radiation intensity can be obtained as

$$J_\nu(\vec{r}) = \frac{1}{\oint d\omega} \cdot \oint I_\nu(\vec{r}, \vec{s}) d\omega = \frac{1}{4\pi} \int_0^{2\pi} \int_0^{\pi} I_\nu \sin\theta d\theta d\varphi \tag{2.3}$$

which is of course independent of direction, and will be frequently employed for studies of properties or problems of the radiation field which are independent of directions, e.g., in the studies of energy densities of radiation fields or isotropic scattering. For an isotropic radiation field, it is obvious that $J_\nu = I_\nu$. Integrating I_ν over all frequencies, we obtain the integrated radiation intensity I as

$$I = \int_0^\infty I_\nu d\nu \tag{2.4}$$

2.1.2 Radiation Flux

In Eq. (2.1), dE_ν represents the radiation energy passing through the area element $d\sigma$ along $\vec{s}$. Summation of the radiation energy passing through $d\sigma$ from all directions, we have

$$dE_\nu^* = d\nu d\sigma dt \cdot \oint I_\nu \cos\theta d\omega \tag{2.5}$$

When $0 \le \theta \le \pi/2$, $\cos\theta > 0$, so the summation of radiation energy is the energy flowing out from the area element $d\sigma$ in a unit time dt and a unit frequency interval dv, which is denoted as dE_v^+ and expressed as

$$dE_v^+ = dvd\sigma dt \int\limits_{\varphi=0}^{2\pi} \int\limits_{\theta=0}^{\pi/2} I_v \cos\theta \sin\theta d\theta d\varphi \tag{2.6}$$

Similarly, when $\pi/2 \le \theta \le \pi$, $\cos\theta < 0$, so the summation of radiation energy is the energy flowing into the area element $d\sigma$ in a unit time dt and a unit frequency interval dv, which is denoted as dE_v^- and expressed as

$$dE_v^- = dvd\sigma dt \int\limits_{\varphi=0}^{2\pi} \int\limits_{\theta=\pi}^{\pi/2} I_v \cos\theta \sin\theta d\theta d\varphi \tag{2.7}$$

The total energy dE_v^* passing through $d\sigma$ can also be written as

$$dE_v^* = dE_v^+ - dE_v^- \tag{2.8}$$

which represents the net energy passing through the area element $d\sigma$ in a unit time dt and a unit frequency interval dv. On dividing dE_v^* by $dtd\sigma dv$, we obtain the net radiation flux or simply the radiation flux, which is denoted as πF_v (π is a normalization factor introduced for convenience in calculations in future) and expressed as

$$\pi F_v \equiv \oint I_v \cos\theta d\omega$$
$$= \int\limits_{\varphi=0}^{2\pi} \int\limits_{\theta=0}^{\pi} I_v \cos\theta \sin\theta d\theta d\varphi \tag{2.9}$$

For an isotropic radiation field, it is obvious that $\pi F_v = 0$. Furthermore, it is worth noting that the radiation flux πF_v is a physical quantity related to the direction of the normal $\vec{n}$ to the test area element $d\sigma$. Equation (2.9) can be written as

$$\pi F_v \equiv \oint I_v \cos\theta d\omega = \oint I_v \cdot (\vec{n} \cdot \vec{s}) d\omega$$
$$= \vec{n} \cdot \oint I_v \vec{s} d\omega = \vec{n} \cdot \vec{f_v}$$

where

$$\vec{f}_\nu = \oint I_\nu \vec{s} d\omega$$

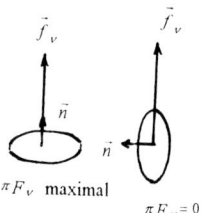

Fig. 2.3 The geometry for showing that the radiation flux is a physical quantity related to the direction of the normal to the test area element.

When $\vec{f}_\nu$ and $\vec{n}$ have the same direction, πF_ν has a maximum value; when $\vec{f}_\nu$ and $\vec{n}$ are perpendicular to each other, $\pi F_\nu = 0$ (see Fig. 2.3).

According to the definition of πF_ν, which represents the net energy flowing through a unit area in a unit time and a unit frequency interval, it is inversely proportional to the distance from the radiation source.

We now proceed to investigate whether an observer on the ground can directly measure the πF_ν value of a star. Suppose the distance between the star and the observer is l which is far greater than the radius R of the star, so that the radiation from all positions on the star surface are parallel (see Fig. 2.4). The radiation flux detected by the observer from an annular area dA on the star surface is

$$df_\nu = I_\nu(Z_0, \mu)d\omega$$

where $d\omega$ is the solid angle subtended by the annular area dA at the position of the observer, $I_\nu(Z_0, \mu)$ is the radiation intensity on the star surface and $\mu = \cos\theta$. Since $r = R\sin\theta$, $dA = 2\pi r dr = -2\pi R^2 \mu d\mu$, $d\omega = -2\pi(R/l)^2 \mu d\mu$, so the total flux received by the observer from the entire star surface is

$$f_\nu = 2\left(\frac{R}{l}\right)^2 \pi \int_0^1 I_\nu(Z_0, \mu)\mu d\mu = \left(\frac{R}{l}\right)^2 \pi F_\nu(0) \qquad (2.10)$$

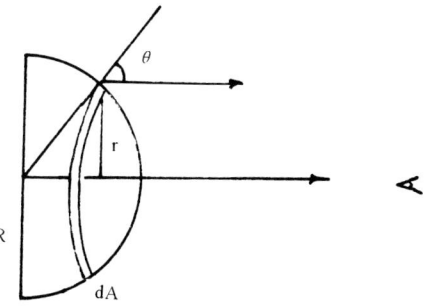

Fig. 2.4 The geometry for showing the radiation flux detected by the observer from an annular area dA on the star surface.

From Eq. (2.10), it can be seen that an observer can directly measure the $\pi F_\nu(0)$ value of the star surface [differed by a dimensionless factor $(R/l)^2$]. In the derivation of Eq. (2.10), we have employed the condition that the radiation flowing into the star surface is zero, i.e., $\pi F_\nu^-(0) = 0$. Integrating Eq. (2.9) over all frequencies, the total radiation flux can be obtained as

$$\pi F = \int_0^\infty \pi F_\nu d\nu = \int_{4\pi} I \cos\theta d\omega \qquad (2.11)$$

2.1.3 Energy Density

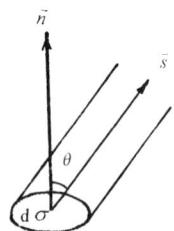

Fig. 2.5 The geometry for calculation of the energy density of a radiation field.

The energy density of a radiation field is the energy of the radiation with a frequency ν in a unit volume (e.g., 1 cm³). The radiation energy passing through $d\sigma$ along the direction $\vec{s}$ in a solid angle $d\omega$ in a time interval dt and in a frequency interval $d\nu$ (see Fig. 2.5) is

$$dE_\nu = I_\nu(\theta)\cos\theta d\sigma d\omega d\nu dt$$

This energy will fill the volume $dV = d\sigma \cos\theta \cdot c \cdot dt$ (where c is the speed of light). On dividing dE_ν by dV, we have the contribution of radiation energy in a solid angle $d\omega$ with a frequency in the range $(\nu, \nu + d\nu)$ towards the energy density as

$$\frac{dE_\nu}{dV} = \frac{I_\nu(\theta)}{c} d\omega d\nu$$

Integrating the above equation over all directions, and then dividing by $d\nu$, we obtain the energy density for the frequency ν, which is denoted as u_ν, as

$$u_\nu = \frac{1}{c}\oint I_\nu(\theta)d\omega \qquad (2.12)$$

Substituting Eq. (2.3), we have

$$u_\nu = \frac{4\pi}{c}J_\nu \qquad (2.13)$$

Integrating u_ν over all frequencies, the integrated energy density u for the radiation field is obtained as

$$u = \int_0^\infty u_\nu d\nu = \frac{4\pi}{c}\int_0^\infty J_\nu d\nu = \frac{4\pi}{c}J \qquad (2.14)$$

For an isotropic radiation field, we have

$$u_v = \frac{4\pi}{c} I_v \quad \text{and} \quad u = \frac{4\pi}{c} I \tag{2.15}$$

2.1.4 Radiation Pressure

When radiation propagates, there is energy flow as well as momentum flow, the latter leading to radiation pressure. The radiation pressure can be defined as the momentum passing perpendicularly through a unit area in a unit time. Denoting the energy passing through an area element $d\sigma$ along the direction $\vec{s}$ in a time interval dt in a solid angle $d\omega$ with a frequency in the range $(v, v + dv)$ as dE_v, the momentum carried by the radiation will be dE_v/c. Further considering its component perpendicular to the direction of $d\sigma$, the momentum perpendicularly passing through a unit area in a unit time is

$$\frac{1}{c} \frac{dE_v \cos\theta}{dt d\sigma} = \frac{I_v}{c} \cos^2\theta dv d\omega$$

which is carried in a solid angle $d\omega$ with a frequency in the range $(v, v + dv)$. Integrating the above equation over all solid angles and dividing by dv, the momentum flow at the frequency v, or the radiation pressure at the frequency v is

$$P_R(v) = \oint \frac{I_v}{c} \cos^2\theta d\omega \tag{2.16}$$

Further defining a quantity K_v

$$K_v \equiv \frac{1}{4\pi} \oint I_v \cos^2\theta d\omega \tag{2.17}$$

Equation (2.16) can be written as

$$P_R(v) = \frac{4\pi}{c} K_v \tag{2.18}$$

For an isotropic radiation field, since $I_v(\mu) = I_v$, $(\mu = \cos\theta)$, we have

$$P_R(v) = \frac{4\pi}{c} \cdot \frac{I_v}{3} \tag{2.19}$$

Integrating $P_R(v)$ over all frequencies, the total radiation pressure P_R is obtained as

$$P_R = \frac{1}{c} \int_{4\pi} I\cos^2\theta d\omega = \frac{4\pi}{3c} \int_0^\infty I_v dv \qquad (2.20)$$

Comparing Eqs. (2.19) with (2.15), we get the following relationships for an isotropic radiation field

$$P_R(v) = \frac{1}{3} u_v \qquad (2.21)$$

$$J_v = 3K_v \qquad (2.22)$$

2.1.5 Semi-Infinite Plane Radiation Field

When studying the radiation field in the stellar atmosphere, if the thickness of the atmosphere is far smaller than the stellar radius, the atmosphere can be treated as a semi-infinite plane. We now proceed to discuss various physical quantities describing radiation field properties in a semi-infinite parallel plane.

The coordinate system (Z, θ, ϕ) will be used (see Fig. 2.6). Since it has been assumed that the physical properties are uniform in every layer in the atmosphere, the radiation along the direction $\vec{s}$ is only dependent on θ and not ϕ. Defining $\mu = \cos\theta = \vec{n} \cdot \vec{s}$, we have $d\omega = \sin\theta d\theta d\phi$,

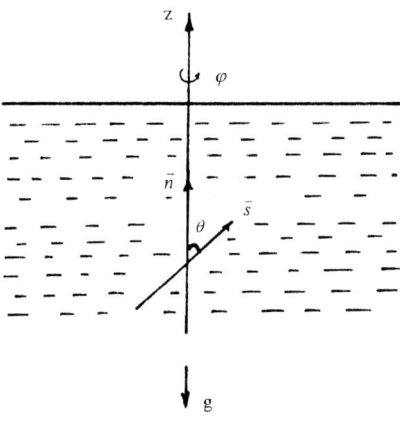

Fig. 2.6 The geometry to show a semi-infinite plane radiation field.

and the integration of an arbitrary function f with respect to $d\omega$ can be written as

$$\oint f d\omega = \int_{\varphi=0}^{2\pi} \int_{\theta=0}^{\pi} f \sin\theta d\theta d\varphi = 2\pi \int_{\mu=-1}^{+1} f d\mu$$

Using this formula, the radiation intensity, mean radiation intensity and radiation flux can be written as

radiation intensity $\qquad I_\nu = I_\nu(Z, \mu)$

mean radiation intensity $\qquad J_\nu = \dfrac{1}{4\pi} \oint I_\nu d\omega = \dfrac{1}{2} \int_{-1}^{+1} I_\nu d\mu$

radiation flux $\qquad \pi F_\nu = \oint I_\nu \cos\theta d\omega = 2\pi \int_{-1}^{+1} I_\nu \mu d\mu$

In summarizing the above expressions, we can introduce the "radiative moment" as a general form for individual expressions, for which the common form is

$$\frac{1}{2} \int_{-1}^{+1} I_\nu \mu^n d\mu \qquad (2.23)$$

where n is called the order of the moment, e.g., zeroth-order radiative moment:

$$J_\nu = \frac{1}{2} \int_{-1}^{+1} I_\nu d\mu \qquad (2.24)$$

(i.e., mean radiation intensity)
   first-order radiative moment:

$$H_\nu = \frac{1}{2} \int_{-1}^{+1} I_\nu \mu d\mu \left(= \frac{1}{4} F_\nu\right) \qquad (2.25)$$

(known as Eddington radiation flux)

second-order radiative moment:

$$K_\nu = \frac{1}{2} \int_{-1}^{+1} I_\nu \mu^2 d\mu \qquad (2.26)$$

(known as K-integral)

2.2 ABSORPTION COEFFICIENT, EMISSION COEFFICIENT AND SCATTERING COEFFICIENT

Radiation interacts with the gaseous matter inside a star. Therefore, besides introducing physical quantities describing properties of the radiation field, physical quantities describing interactions between radiation and gaseous matter should also be introduced. Macroscopically speaking, the interactions can be viewed as absorption, emission and scattering of radiation by the gaseous matter. Therefore, for macroscopic description of interactions, the absorption coefficient, emission coefficient and scattering coefficient, as well as related physical quantities such as the optical depth and source function should be introduced.

To better describe the interactions mentioned above, some microscopic processes for the interactions are first outlined here. However, further discussions and calculations of contributions from individual microscopic processes towards the absorption, emission and scattering coefficients will be presented in Chapter 5.

2.2.1 Microscopic Interactions Between Radiation and Stellar Matter

There are a lot of microscopic interaction processes between radiation and stellar matter, but they can be basically divided into two categories. The first category is called scattering in which photons will not disappear after interactions although they can suffer changes in their directions or even frequencies. The second category is called true absorption (the inverse process is called true emission) in which photons disappear after interactions with all the photon energy transferred into the thermal energy of particles in the stellar matter (the inverse process involves the conversion of thermal energy of particles in the stellar matter into new photons). Examples of microscopic processes in these two categories are discussed in the following.

(A) Microscopic Processes Belonging to Scattering

(1) Scattering of radiation by free electrons

When a radiation with frequency v is incident on an electron, the electron will have forced oscillation with the same frequency under the effect of the incident electromagnetic wave, and will at the same time emit secondary waves with that frequency in all directions, which is called Thompson scattering of radiation by an electron. Due to this scattering, part of the original radiation will be diverted to other directions and the radiation will be weakened in the original direction.

If the energy of the incident radiation is very high, the scattering will be qualitatively different. At this time, we cannot view the electron as having forced oscillation under the effect of the incident electro-magnetic wave, and have to treat the process as an electron scattering of the photon, which is called Compton scattering. In this process, the frequency of the photon can have minor changes.

(2) Scattering of radiation by molecules

Scattering of radiation by molecules can occur in the stellar atmosphere of late type cool stars, which is called Rayleigh scattering.

(3) Other scattering process

On absorbing a photon, an atom can be excited from a bound state i to a higher bound state j (bound-bound transition). It later de-excites to state i with the emission of a photon. Since the energy levels for bound states have certain widths themselves, the emitted and absorbed photons have a minute difference in frequency. This process can be viewed as a scattering process.

(B) Microscopic Processes Belonging to True Absorption or True Emission

(1) Photoionization

On absorbing the energy of a photon, the atom ionizes and a bound electron becomes a free electron (bound-free transition). The excessive energy of the incident photon (after ionizing the atom) is converted into kinetic energy of the free electron. The inverse process of photoionization is the conversion of a free electron into a bound electron with the emission of a photon in a random direction.

(2) Photoexcitation

On absorbing a photon, an atom can be excited from a bound state i to a higher bound state j. If another particle is now excited through an inelastic collision with this atom, the energy of the incident photon will be converted into thermal energies for these two colliding particles. The inverse process involves the collision of two particles with the emission of a single photon.

(3) Free-free transition

When an electron orbiting an ion absorbs the energy of a photon, its energy increases and its orbit can change to another hyperbolic orbit but it still orbits around the ion. This process is called free-free absorption. Its inverse process is the free-free emission.

(4) Other true absorption process

After photoexcitation, an atom can have collision ionization. The inverse process is the collision recombination.

2.2.2 Absorption Coefficient and Optical Depth

For macroscopic description of the interaction between radiation and stellar matter, we introduce the absorption coefficient κ_ν, or the opacity. Consider a radiation beam with an intensity $I_\nu(0)$ which passes perpendicularly through a matter layer with thickness ds and density ρ (see Fig. 2.7). Due to absorption of radiation by matter in the layer, the intensity of the radiation beam is reduced by dI_ν. We first study the relative change in the radiation intensity (denoted as $d\tau_\nu$):

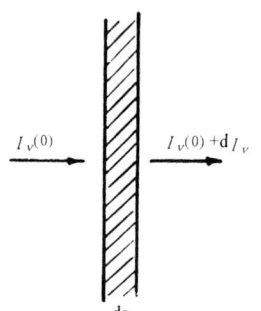

$$d\tau_\nu = -\frac{dI_\nu}{I_\nu} \tag{2.27}$$

Fig. 2.7 A radiation beam with an intensity $I_\nu(0)$ passes perpendicularly through a matter layer with thickness ds and density ρ.

where the minus sign signifies the reduction in radiation intensity. The relative change in radiation intensity is obviously dependent on properties of the absorption medium and the geometrical thickness of the absorption layer. The properties of the absorption medium can be represented by κ_ν. From Eq. (2.27), we know that the relative change $d\tau_\nu$ is dimensionless, which is related to κ_ν and the thickness of the matter layer ds through the following two equations.

(1) $d\tau_\nu = \kappa_\nu \rho ds \tag{2.28}$

In this case, since $d\tau_\nu$ is dimensionless, the unit for κ_ν is [cm²/g].

(2) $d\tau_v = \kappa_v ds$ (2.29)

In this case, the unit for κ_v is $[\text{cm}^2/\text{cm}^3] = [\text{cm}^{-1}]$.

From the above two equations and the corresponding unit for κ_v, it is common to define the dimension for absorption coefficient as a normalized cross section, i.e.,

$$[\kappa_v] = [\text{cm}^2/\text{normalized units}]$$ (2.30)

According to the normalized unit, we have different units for the absorption coefficient and relationships for $d\tau_v$, κ_v and ds. In the first case above, the normalized unit is $[g]$ and we call the corresponding κ_v the mass absorption coefficient. In the second case above, the normalized unit is $[\text{cm}^3]$ and we call the corresponding κ_v the volume absorption coefficient. However, these two absorption coefficients with different dimensions have the same symbol κ_v in the literature, so care should be exercised in choosing their dimensions and their relationships with $d\tau_v$, κ_v and ds.

Integrating Eq. (2.27), we have

$$I_v = I_v(0)e^{-\tau_v}$$ (2.31)

where $I_v(0)$ is the intensity before absorption. Equation (2.31) shows that the intensity decreases exponentially in the medium, and the index

$$\tau_v = \int d\tau_v$$ (2.32)

is called optical depth for the absorption medium. According to Eq. (2.27), $d\tau_v$ is the relative change in the intensity after passing through the layer with a thickness ds.

From Eq. (2.31), when $\tau_v \gg 1$, the intensity after absorption by the layer is $I_v \approx 0$, i.e., the radiation almost cannot pass through the layer. Therefore, a medium with $\tau_v \gg 1$ is referred as optically thick. Conversely, when $\tau_v \ll 1$, the intensity after absorption by the layer is $I_v \approx I_v(0)$, i.e., the radiation almost entirely passes through the layer. Therefore, a medium with $\tau_v \ll 1$ is referred as optically thin.

The gaseous matter inside a star is composed of atoms, ions and electrons. The absorption of radiation by the gaseous matter is the manifestation of microscopic processes of interaction between the radiation and the atoms, ions and electrons, and each microscopic process has a certain contribution towards the total absorption. Therefore, the total absorption coefficient κ_v should be composed of the absorption coefficient of individual processes. For the description of absorption characteristics of microscopic processes, we introduce absorption coefficients or atomic cross sections for microscopic particles denoted by a_ν. Its relationship to the total absorption coefficient is

$$\kappa_\nu = \begin{cases} n \cdot \frac{1}{\rho} a_\nu & (\kappa_\nu \text{ is the mass absorption coefficient}) \\ n \cdot a_\nu & (\kappa_\nu \text{ is the volume absorption coefficient}) \end{cases} \tag{2.33}$$

where n is the particle number per cm³, and the dimension for a_ν is [cm²]. In Chapter 5, we will further discuss the microscopic absorption processes, and present the formulae for a_ν for individual processes.

Previously, we have pointed out that the interaction between the gaseous matter and radiation can be intrinsically divided into scattering and true absorption processes. Therefore, the absorption coefficient κ_ν can also be represented as the combination of two different absorption coefficients:

$$\kappa_\nu = \kappa_\nu^T + \sigma_\nu \tag{2.34}$$

where κ_ν^T is the true absorption coefficient and σ_ν is the scattering absorption coefficient.

2.2.3 Emission Coefficient and Source Function

Emission is the inverse process for absorption. Similar to the absorption coefficient, the emission coefficient can be divided into two representations according to its dimension. The first one is the mass emission coefficient which is defined as the radiation energy emitted in a solid angle in a unit time interval and in a unit frequency interval per unit mass, i.e.,

$$dE_\nu = \eta_\nu \rho \, d\omega \, d\nu \, dt \, dA \, ds \tag{2.35}$$

where ρ is the matter density, $dAds$ is a volume element with a cross section area dA and a length ds. In terms of radiation intensity, this can be written as

$$dI_\nu = \eta_\nu \rho \, ds \tag{2.36}$$

where η_ν is the mass emission coefficient with the dimension [erg.g^{-1}s^{-1}Hz^{-1}sr^{-1}]. The second emission coefficient is the volume emission coefficient which is defined as the radiation energy emitted in a solid angle in a unit time interval and in a unit frequency interval per unit volume, i.e.,

$$dE_\nu = \eta_\nu d\omega d\nu dt dA ds \qquad (2.37)$$

or in terms of radiation intensity,

$$dI_\nu = \eta_\nu ds \qquad (2.38)$$

where η_ν is the volume emission coefficient with the dimension [erg.cm^{-3}s^{-1}Hz^{-1}sr^{-1}]. Similarly, the emission coefficient η_ν can also be represented as the combination of two different emission coefficients:

$$\eta_\nu = \eta_\nu^T + \eta_\nu^s \qquad (2.39)$$

where η_ν^T is the true emission coefficient and η_ν^S is the scattering emission coefficient.

When an atom of a gas absorbs energy of an incident photon, it will undergo a transition from a lower energy level i to a higher energy level j. Subsequently, two different recombination transitions may take place. One of them is the spontaneous transition for which the emitted photons are isotropic. Another one is the induced transition in which the excited atom undergoes recombination under the induced action of incident radiation and in which the emitted and incident photons have the same direction. The true emission coefficient η_ν^T represents the isotropic spontaneous emission coefficient. Since the emitted and incident photons for an induced transition have the same direction, the induced emission coefficient is normally treated as a "negative absorption coefficient".

We further define a source function which is related to the absorption and emission coefficients through

$$S_\nu \equiv \eta_\nu / \kappa_\nu \qquad (2.40)$$

The source function S_ν is a very useful physical quantity when studying radiative transfer problems.

2.2.4 General Discussions on the Scattering Coefficient

Scattering has contributions to both the absorption coefficient and the emission coefficient, namely, the scattering absorption coefficient and the scattering emission coefficient. We will present general discussions on these scattering coefficients in the following.

During scattering, there are changes in the directions as well as frequencies of the radiation. In view of these, a distribution function $W(\vec{s}', \nu', \vec{s}, \nu)$ is introduced for description of the change in direction and frequency of the radiation, which gives the probability of a radiation travelling along the direction $\vec{s}'$ in a solid angle $d\omega'$ with a frequency in the range $(\nu', \nu' + d\nu')$ being scattered to travel along the direction $\vec{s}$ in a solid angle $d\omega$ with a frequency in the range $(\nu, \nu + d\nu)$, and should satisfy the normalization condition

$$\oint_{\omega'} \oint_{\omega} \int_{\nu'=0}^{\infty} \int_{\nu=0}^{\infty} W(\vec{s}', \nu', \vec{s}, \nu) d\nu' d\nu \frac{d\omega'}{4\pi} \frac{d\omega}{4\pi} = 1 \qquad (2.41)$$

For simplicity, in studying the general representation of the scattering emission coefficient η_ν^s and the scattering absorption coefficient σ_ν, we only discuss the volume scattering emission coefficient and the volume scattering absorption coefficient.

Scattering Emission Coefficient η_ν^s

If the emission coefficient is employed for description, the energy of radiation emitted from a unit volume of matter to $\{\vec{s}, \nu\}$ is given by

$$\varepsilon_\nu = \eta_\nu^s d\nu d\omega \qquad (2.42)$$

On the other hand, ε_ν can be described by the distribution function W. Apparently, ε_ν should be proportional to $d\nu$, $d\omega$ and the total incident energy, i.e.,

$$\varepsilon_\nu = \sigma d\nu d\omega \oint_{\omega'} \oint_{\nu'=0}^{\infty} I_{\nu'} W(\vec{s}', \nu', \vec{s}, \nu) d\nu' \frac{d\omega'}{4\pi} \qquad (2.43)$$

where the proportionality constant σ is called the total scattering cross section. Comparing Eqs. (2.42) and (2.43), we have

$$\eta_\nu^s = \sigma \oint_{\omega'} \int_{\nu'=0}^{\infty} I_{\nu'} W(\vec{s}', \nu', \vec{s}, \nu) d\nu' \frac{d\omega'}{4\pi} \qquad (2.44)$$

Scattering Absorption Coefficient σ_ν

If the scattering absorption coefficient is employed for description, the energy of radiation in a unit volume of matter scattered from $\{\vec{s}, \nu\}$ is given by

$$\varepsilon_{\nu'} = I_{\nu'}\sigma_{\nu'}d\nu'd\omega' \tag{2.45}$$

On the other hand, $\varepsilon_{\nu'}$ can also be described by the distribution function W. Apparently, $\varepsilon_{\nu'}$ should be proportional to $d\nu'$, $d\omega'$ and the total scattered energy, i.e.,

$$\varepsilon_{\nu'} = \sigma d\nu'd\omega' \oint_{\omega'} \int_{\nu'=0}^{\infty} I_{\nu'} W(\vec{s}', \nu', \vec{s}, \nu)d\nu\frac{d\omega}{4\pi} \tag{2.46}$$

where $I_{\nu'}$ can be taken out of the integration sign because scattering is only dependent on the angle between $\vec{s}$ and $\vec{s}'$, and is independent of the direction of $\vec{s}$ alone. If a scattering profile is further defined as

$$\varphi(\nu') \equiv \oint_{\omega'} \oint_{\omega} \int_{\nu=0}^{\infty} W(\vec{s}', \nu', \vec{s}, \nu)d\nu\frac{d\omega\,d\omega'}{4\pi\ 4\pi} \tag{2.47}$$

Equation (2.46) can be written as

$$\varepsilon_{\nu'} = I_{\nu'}\sigma\varphi(\nu')d\nu'd\omega' \tag{2.48}$$

Comparing Eqs. (2.45) and (2.48), we have

$$\sigma_{\nu'} = \sigma\varphi(\nu') \tag{2.49}$$

The above only gives the general representation of the scattering emission coefficient and the scattering absorption coefficient. We now proceed to discuss two realistic scattering processes inside a star, and study the representations of the scattering emission coefficient and the scattering absorption coefficient.

(1) Coherent scattering

If the frequency does not change (i.e., $\nu = \nu'$) in a scattering, the process is called coherent scattering, for which the distribution function can be simplified as

$$W(\vec{s}', \nu', \vec{s}, \nu) = g(\vec{s}', \vec{s})\varphi(\nu')\delta(\nu - \nu') \tag{2.50}$$

Substituting Eq. (2.50) into Eq. (2.44), and considering Eq. (2.49), we have

$$
\begin{aligned}
\eta_\nu^s &= \sigma \oint_{\omega'} \int_{\nu'=0}^{\infty} I_{\nu'} g(\vec{s}', \vec{s})\varphi(\nu')\delta(\nu - \nu')d\nu' \frac{d\omega'}{4\pi} \\
&= \sigma \oint_{\omega'} I_\nu g\varphi(\nu) \frac{d\omega'}{4\pi} = \sigma\varphi(\nu) \oint_{w'} I_\nu g \frac{d\omega'}{4\pi} \\
&= \sigma_\nu \oint_{\omega'} I_\nu g \frac{d\omega'}{4\pi}
\end{aligned}
\tag{2.51}
$$

(2) Isotropic coherent scattering

In the isotropic case, $g(\vec{s}', \vec{s}) = 1$, so the distribution function becomes

$$W = \varphi(\nu')\delta(\nu - \nu') \tag{2.52}$$

Substituting Eq. (2.52) into Eq. (2.51), we have

$$\eta_\nu^s = \sigma_\nu \oint_{\omega} I_\nu \frac{d\omega}{4\pi} = \sigma_\nu J_\nu \tag{2.53}$$

That is, for an isotropic coherent scattering, the scattering emission coefficient η_ν^s is the product of the scattering absorption coefficient σ_ν and the mean radiation intensity J_ν. Usually, a continuous scattering can be treated as an isotropic coherent scattering. For a monochromatic light, although both the Thompson scattering by an electron and the Rayleigh scattering by an atom have a dipole feature, i.e., $g = (3/4)[1 + (\vec{s}' \cdot \vec{s})^2]$, if the incident light is isotropic, the scattered light will also be isotropic due to interference.

2.3 BLACK BODY AND ITS RADIATION

We now proceed to discuss the most important radiation field for stars, which is the isotropic radiation field under thermodynamic equili-

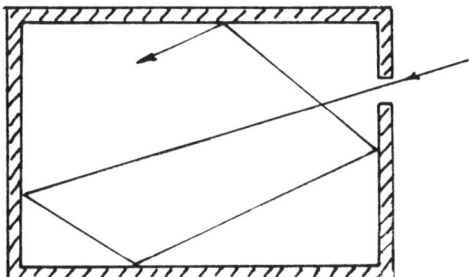

Fig. 2.8 An empty cavity can be treated as a totally absorbing black body.

brium. To illustrate the concepts of a black body and thermodynamic equilibrium, we devise a cavity totally isolated from the surroundings (see Fig. 2.8), except that there is a very tiny hole on one of the walls. The probability for any radiation which has got into the cavity through the hole to come out again is extremely small. Therefore this cavity is equivalent to a totally absorbing black body. As long as sufficient time has elapsed, hydrostatic equilibrium and thermodynamic equilibrium can be established for such a black body cavity, and every physical process that takes place inside the body will be in equilibrium with its inverse process. At equilibrium, all particles (atoms, molecules, electrons and photons) will satisfy certain statistical distributions irrespective of their initial states or physical processes they have gone through before reaching equilibrium. Temperature is the only parameter for the equilibrium state. We refer to this equilibrium state as the thermodynamic equilibrium state and denote it as "TE". Apparently, the radiation field is isotropic under thermodynamic equilibrium, and is only a function of frequency ν and temperature T. We have pointed out in §2.1 that the following relationships apply for an isotropic radiation field:

$$\left.\begin{array}{l} J_\nu = I_\nu, J = I; \\[4pt] u_\nu = \dfrac{4\pi}{c} I_\nu, u = \dfrac{4\pi}{c} I; \\[4pt] P_R(\nu) = \dfrac{1}{3} u_\nu = \dfrac{4\pi}{3c} I_\nu, P_R = \dfrac{1}{3} u = \dfrac{4\pi}{3c} I; \\[4pt] K_\nu = J_\nu/3 = I_\nu/3, K = I/3 \end{array}\right\} \qquad (2.54)$$

The black body radiation satisfies the following experimental laws:

(1) Planck radiation law

Under thermodynamic equilibrium, the relationship among radiation intensity I_ν, frequency ν and temperature T is

$$I_\nu = \frac{2h\nu^3}{c^2} \frac{1}{e^{h\nu/kT} - 1} \equiv B_\nu(T) \qquad (2.55)$$

and is called the Planck radiation law, in which $B_\nu(T)$ is called the Planck's function, and $h = 6.626 \times 10^{-27}$ erg.s^{-1} is called the Planck's constant.

(2) Wein displacement law

The wavelength λ_{max} corresponding to the radiation intensity peak will displace towards shorter wavelengths for higher temperatures according to the relationship

$$\lambda_{max} = 0.28978/T \quad (\text{cm K}^{-1}) \qquad (2.56)$$

which is called the Wien displacement law. Figure 2.9 shows the displacement of the intensity peak under different temperatures.

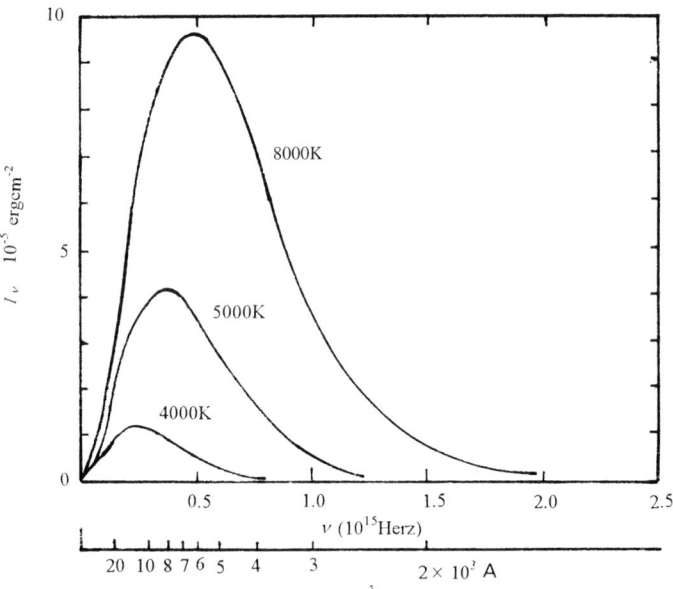

Fig. 2.9 The relationship between radiation intensity and wavelength for different temperatures.

(3) Stefan–Boltzmann law

The total radiation flux (erg.cm^{2}s^{-1}) can be obtained by summing the flux values for all wavelengths. According to the Stefan–Boltzmann law, the total radiation flux is proportional to the fourth power of temperature, i.e.,

$$\pi F = \pi B = \sigma T^4 \tag{2.57}$$

where $B = \int_0^\infty B_\nu(T)d\nu$, and the constant $\sigma = 5.67 \times 0^{-5}$ (erg.cm^{-2} s^{-1}K^{-4}) is called the Stefan–Boltzmann constant.

(4) Kirchhoff law

Under thermodynamic equilibrium, the ratio for the true emission coefficient η_ν^T to the true absorption coefficient κ_ν^T is equal to the Planck's function, i.e.,

$$\eta_\nu^T/\kappa_\nu^T = I_\nu = B_\nu(T) \tag{2.58}$$

We know that the absorption coefficient κ_ν^T and the emission coefficient η_ν^T are medium dependent quantities. However, their ratio is equal to the radiation intensity I_ν under thermodynamic equilibrium or the Planck's function $B_\nu(T)$ which are medium independent and are only functions of frequency ν and temperature T. Equation (2.58) is called the Kirchhoff law.

2.4 EQUATION OF RADIATIVE TRANSFER

We now discuss the change in the intensity of a radiation passing through a matter layer with both emission and absorption, which can be given by the equation of radiative transfer, obtained from the law of conservation of energy. When discussing radiative transfer, different coordinate systems are employed according to the geometrical structure of the matter layer.

2.4.1 Equation of Radiative Transfer in the Plane-rectangular Frame

In the discussion of radiative transfer in a stellar atmosphere, if the thickness of the atmosphere is far less than the stellar radius, the atmosphere can be treated as a semi-infinite plane and the radiative transfer can be discussed in a plane-rectangular frame.

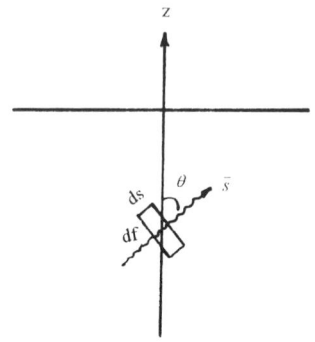

Fig. 2.10 Radiative transfer in a stellar atmosphere described in the plane-rectangular frame.

Consider a small volume element with bottom area *df* and thickness *ds* at a distance *z* from the center of the star (Fig. 2.10). The radiation passes through this volume element along the direction $\vec{s}$. Suppose the change in intensity of the radiation after passing through the volume element is given by $-dI_\nu(z, \mu)/ds$ (the minus sign refers to a reduction of the intensity; $\mu = \cos\theta$). The absorption of radiation energy in the volume element is $-\kappa_\nu I_\nu(z, \mu)ds$ (here κ_ν is the volume absorption coefficient), and the emitted radiation is $+\eta_\nu ds$ (here η_ν is the volume emission coefficient). According to the law of conservation of energy, the sum of the above terms should be zero. Further considering that $ds = dz/\mu$, we obtain the equation of radiative transfer as

$$\mu\frac{\partial I_\nu(z, \mu)}{\partial z} = -\kappa_\nu(z, \mu)I_\nu(z, \mu) + \eta_\nu(z, \mu)$$
$$= \kappa_\nu(z, \mu)[S_\nu(z, \mu) - I_\nu(z, \mu)] \quad (2.59)$$

where $S_\nu(t, \nu)$ is the source function. In practice, the variable dz is usually changed to $d\tau_\nu = -\kappa_\nu I_\nu(z, \mu)dz$ for convenience. In this way, Eq. (2.59) can be written as

$$\mu\frac{\partial I_\nu(\tau_\nu, \mu)}{\partial \tau_\nu} = I_\nu(\tau_\nu, \mu) - S(\tau_\nu, \mu) \quad (2.60)$$

2.4.2 Equation of Radiative Transfer in the Plane-polar Frame

When discussing radiative transfer inside a star, or radiative transfer in stellar atmospheres of giants or supergiants, the atmosphere can no longer be treated as a plane structure but should be treated as a spherically symmetric structure. The corresponding equation of radiative transfer can be discussed in the plane-polar frame. Suppose the direction of radiative transfer at a distance *r* from the centre of the star is determined by the angle θ between the radiation beam and the normal line to the spherical surface (Fig. 2.11). However, after traversing a small distance *ds* in the direction of the radiation, there is a change in the angle θ which causes a decrease in the angle to $\theta - d\theta$, i.e., an increase in $\mu = \cos\theta$. Since

$$\frac{d\mu}{ds} = \frac{d\mu}{d\theta}\cdot\frac{d\theta}{ds} = -\sin\theta\frac{d\theta}{ds}$$
$$= \frac{\sin^2\theta}{r} = \frac{1 - \mu^2}{r} \quad (2.61)$$

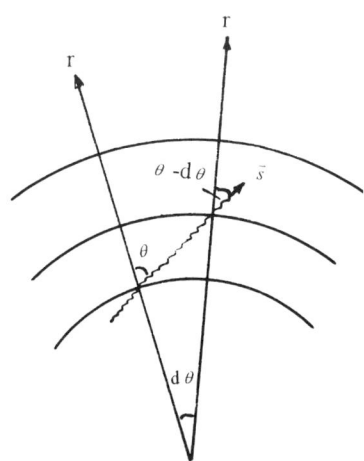

Fig. 2.11 Radiative transfer in a stellar atmosphere described in the plane-polar frame.

we have

$$
\begin{aligned}
\frac{\partial I_v(r, \mu)}{\partial s} &= \frac{\partial I_v(r, \mu)}{\partial r} \cdot \frac{dr}{ds} + \frac{\partial I_v(r, \mu)}{\partial \mu} \frac{d\mu}{ds} \\
&= \mu \frac{\partial I_v(r, \mu)}{\partial r} + \frac{1 - \mu^2}{r} \frac{\partial I_v(r, \mu)}{\partial \mu}
\end{aligned}
\tag{2.62}
$$

Therefore the equation of radiative transfer is obtained as

$$
\begin{aligned}
\mu \frac{\partial I_v(r, \mu)}{\partial r} + \frac{1 - \mu^2}{r} \frac{\partial I_v(r, \mu)}{\partial \mu} &= \kappa_v(r, \mu)[S_v(r, \mu) - I_v(r, \mu)] \\
&= -\kappa_v(r, \mu)I_v(r, \mu) + \eta_v(r, \mu)
\end{aligned}
\tag{2.63}
$$

2.4.3 Equation of Radiative Transfer in the Plane-cylindrical Frame

When the structure of the atmosphere is spherical, the plane-cylindrical frame (p, z) can be employed (Fig. 2.12) for the equation of radiative transfer, which is more convenient for studying radiative transfer problems in stellar atmospheres of giants and supergiants. Due to the spherically symmetric structure, any direction can be taken as the z-axis. We choose this direction as the one pointing to the observer, and study radiative transfer problems along this z-axis. Since

$$
r^2 = p^2 + z^2
\tag{2.64}
$$

$$
\mu = \cos\theta = z/r
\tag{2.65}
$$

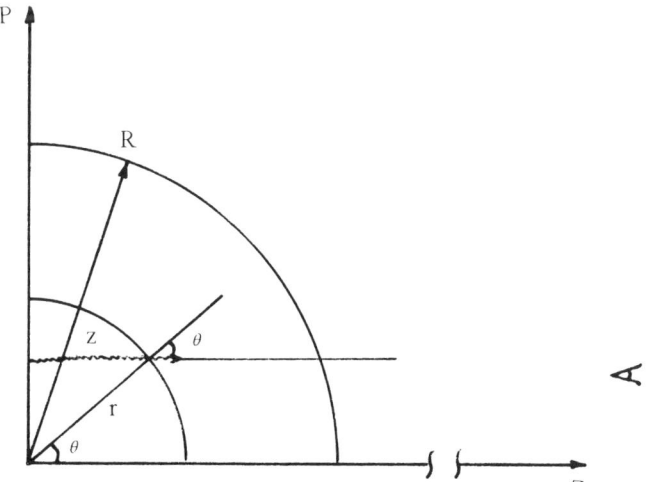

Fig. 2.12 Radiative transfer in a stellar atmosphere described in the plane-cylindrical frame.

and the direction of radiative transfer is same as the z-axis, so $(\partial I/\partial s) = (\partial I/\partial z)$ and the equation for radiative transfer is

$$\pm \left[\frac{\partial I_\nu^\pm(z, p)}{\partial z} \right] = \kappa_\nu(z, p) \cdot [S_\nu(z, p) - I_\nu^\pm(z, p)]$$

$$= -\kappa_\nu(z, p)I_\nu^\pm(z, p) + \eta_\nu(z, p) \tag{2.66}$$

where

$$\begin{aligned} I^+ &= I \quad \text{when } \mu \geq 0 \\ I^- &= I \quad \text{when } \mu < 0 \end{aligned} \tag{2.67}$$

2.4.4 General Solution to the Equation of Radiative Transfer

The radiative transfer equation (2.60) is a constant-coefficient first-order differential equation, which has an integration factor $\exp(-\tau_\nu/\mu)$. Multiplying Eq. (2.60) by $\exp(-\tau_\nu/\mu)$, we have

$$\frac{d}{d\tau_\nu}(I_\nu e^{-\tau_\nu/\mu}) = -\frac{1}{\mu}S_\nu e^{-\tau_\nu/\mu} \tag{2.68}$$

Integrating the above equation, we have

$$I_\nu e^{-\tau_\nu/\mu} \Big|_{\tau_{1\nu}}^{\tau_{2\nu}} = \int\limits_{\tau_{1\nu}}^{\tau_{2\nu}} S_\nu(t_\nu, \mu)e^{-t_\nu/\mu} \frac{dt_\nu}{\mu} \tag{2.69}$$

which can be expressed as

$$I_\nu(\tau_{1\nu}, \mu) = I_\nu(\tau_{2\nu}, \mu)e^{-(\tau_{2\nu}-\tau_{1\nu})/\mu} + \int\limits_{\tau_{1\nu}}^{\tau_{2\nu}} S_\nu(t_\nu, \mu)e^{-(t_\nu-\tau_{1\nu})/\mu} \frac{dt_\nu}{\mu} \tag{2.70}$$

Equation (2.70) is the general solution to the equation of radiative transfer. It is not difficult to observe that the first term on the right of Eq. (2.70) represents the contribution from the incident radiation (multiplied by the factor $\exp[-(\tau_{2\nu} - \tau_{1\nu})/\mu]$ due to absorption in the matter), and the second term on the right represents the contribution from the emission from the matter layer itself.

As an important application of the general solution to the equation of radiative transfer, we can consider a matter layer without incident radiation, i.e., $I_\nu(\tau_{2\nu}, \mu) = 0$, and with a thickness of L. Writing $\tau_{0\nu} = (\tau_{2\nu} - \tau_{1\nu})/\mu) = \int_0^L \kappa_\nu ds$ and denoting the outward radiation intensity $I_\nu(\tau_{1\nu}, \mu)$ as I_ν^{out}, Eq. (2.70) becomes

$$I_\nu^{out} = \int_0^{\tau_{0\nu}} S_\nu(\tau_\nu) e^{-\tau_\nu} d\tau_\nu \qquad (2.71)$$

If the matter layer is under thermodynamic equilibrium, we have $S_\nu = \eta_\nu/\kappa_\nu = I_\nu = B_\nu(T)$, where B_ν is the Planck's function. Therefore, Eq. (2.71) becomes

$$I_\nu^{out} = \int_0^{\tau_{0\nu}} B_\nu(T) e^{-\tau_\nu} d\tau_\nu \qquad (2.72)$$

If we further assume that the matter layer is homogeneous, i.e., the temperature and κ_ν are constants, the above equation becomes

$$I_\nu^{out} = B_\nu(T)(1 - e^{-\kappa_\nu L}) \qquad (2.73)$$

From this equation, for a homogeneous matter layer under thermodynamic equilibrium at a temperature T, if the optical depth $\tau_\nu = \kappa_\nu L \geq 1$ for all ν (i.e., optically thick), we have

$$I_\nu^{out} \cong B_\nu(T) \qquad (2.74)$$

This shows that the blackbody spectrum will be obtained for a matter layer which is optically thick over all frequencies. This also shows that only under the optically thick condition can we have full interactions with matter under thermodynamic equilibrium to reach the same thermodynamic equilibrium state of the matter.

When $\tau_\nu << 1$, i.e., the optically thin case, we have

$$I_\nu^{out} = \kappa_\nu L B_\nu = \eta_\nu L \qquad (2.75)$$

Therefore, the emitted radiation does not have the black body spectrum for the optically thin case.

Through the general solution of the transfer equation shown in Eq. (2.70), we can discuss the expression for radiation intensity

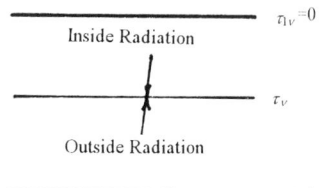

Inside Radiation

$\tau_{1\nu} = 0$

τ_ν

Outside Radiation

$\tau_{2\nu} = \infty$

Fig. 2.13 The geometry for derivation of the expression of radiation intensity at a point in the semi-infinite stellar atmosphere.

$I_\nu(\tau_\nu, \mu)$ at a point in the semi-infinite stellar atmosphere. In this case, $\tau_{1\nu} = 0$, $\tau_{2\nu} \to \infty$ (see Fig. 2.13). At the same time, we can use the following boundary conditions: at $\tau_\nu = 0$, there is no inward radiation, i.e.,

$$\tau_\nu = 0, I_\nu(0, \mu) = 0, (-1 \le \mu \le 0) \qquad (2.76)$$

and the intensity of radiation coming from the lowest layer of the atmosphere is also zero due to absorption, so

$$\lim_{\tau_{2\nu} \to \infty} I_\nu(\tau_{2\nu}, \mu) = 0 \qquad (0 \le \mu \le 1) \qquad (2.77)$$

We first study the outward radiation at τ_ν, which is the case for $\mu \ge 0$. According to Eq. (2.70) we have

$$I_\nu(\tau_\nu, \mu) = \lim_{\tau_{2\nu} \to \infty} I_\nu(\tau_{2\nu}, \mu)e^{-(\tau_{2\nu}-\tau_\nu)/\mu} + \int_\tau^\infty S_\nu(t_\nu)e^{-(t_\nu-\tau_\nu)/\mu}\frac{dt_\nu}{\mu}$$

Using the boundary condition in Eq. (2.77), the equation becomes

$$I_\nu(\tau_\nu, \mu) = \int_{\tau_\nu}^\infty S_\nu(t_\nu)e^{-(t_\nu-\tau_\nu)/\mu}\frac{dt_\nu}{\mu}, (0 \le \mu \le 1) \qquad (2.78)$$

We then study the inward radiation which is the case for $\mu \le 0$. At this time, we take $\tau_{2\nu} = 0$, so the general solution in Eq. (2.70) becomes

$$I_\nu(\tau_\nu, \mu) = I_\nu(0, \mu)e^{\tau_\nu/\mu} + \int_{\tau_\nu}^0 S_\nu(t_\nu)e^{-(t_\nu-\tau_\nu)/\mu}\frac{dt_\nu}{\mu}$$

Using the boundary condition in Eq. (2.76), the previous equation becomes

$$I_\nu(\tau_\nu, \mu) = -\int_0^\tau S_\nu(t_\nu)e^{-(t_\nu-\tau_\nu)/\mu}\frac{dt_\nu}{\mu}, (-1 \le \mu \le 0) \qquad (2.79)$$

If the above result is applied at the stellar surface, and considering only the outward radiation, we have $\tau_\nu = \tau_{1\nu} = 0$ and $\mu \ge 0$, so

$$I_v(0, \mu) = \int\limits_0^\infty S_v(t_v)e^{-t_v/\mu}\frac{dt_v}{\mu} \qquad (2.80)$$

2.4.5 Mean intensity $J_v(\tau_v)$, Radiation Flux $F_v(\tau_v)$ and K-integral

Through the general solution to the equation of radiative transfer, we can obtain expressions for the mean intensity $J_v(\tau_v)$, radiation flux $F_v(\tau_v)$ and K-integral. For simplicity, we will drop the subscript v in our derivations.

(1) Expression for $J_v(\tau_v)$
According to Eq. (2.24), we have

$$J(\tau) = \frac{1}{2}\int\limits_{-1}^{+1} I(\tau, \mu)d\mu$$

Substituting Eqs. (2.78) and (2.79) into the above, we have

$$J(\tau) = \frac{1}{2}\int\limits_{\mu=0}^{1} d\mu \int\limits_{t=\tau}^{\infty} S(t)e^{-(t-\tau)/\mu}\frac{dt}{\mu} - \frac{1}{2}\int\limits_{\mu=-1}^{0} d\mu \int\limits_{t=0}^{\tau} S(t)e^{-(t-\tau)/\mu}\frac{dt}{\mu}$$

$$(2.81)$$

Reversing the order of integrations, and introducing variables $w = 1/\mu$ or $-d\mu/\mu = dw/w$ for the first integral on the right, and $u = -1/\mu$ or $-d\mu/\mu = du/u$ for the second integral on the right, Eq. (2.81) can be written as

$$J(\tau) = \frac{1}{2}\int\limits_{\tau}^{\infty} S(t)dt \int\limits_{w=1}^{\infty} e^{-w(t-\tau)}\frac{dw}{w} + \frac{1}{2}\int\limits_0^{\tau} S(t)dt \int\limits_{u=1}^{\infty} e^{-u(\tau-t)}\frac{du}{u} \qquad (2.82)$$

Defining a function

$$E_n(x) \equiv \int\limits_1^\infty e^{-xt}\frac{dt}{t^n} = x^{n-1}\int\limits_x^\infty \frac{e^{-t}dt}{t^n} \qquad (2.83)$$

known as the nth order exponential integral, we can see that the two integrals on the right of Eq. (2.82) are both first-order exponential integrals, so Eq. (2.82) can be written as

$$J(\tau) = \frac{1}{2} \int_{\tau}^{\infty} S(t)E_1(t - \tau)dt + \frac{1}{2} \int_{0}^{\tau} S(t)E_1(\tau - t)dt \qquad (2.84)$$

Replacing the subscript ν, we have

$$J_\nu(\tau_\nu) = \frac{1}{2} \int_{0}^{\infty} S_\nu(t_\nu)E_1|t_\nu - \tau_\nu|dt_\nu \qquad (2.85)$$

(2) Expression for $F_\nu(\tau_\nu)$
From Eq. (2.25), $F(\tau)$ can be written as

$$F(\tau) = 2 \int_{-1}^{+1} I(\tau, \mu)\mu d\mu$$

Substituting the equation of radiative transfer, and using the method described above, we have

$$F(\tau) = 2 \int_{0}^{\infty} S(t)E_2(t - \tau)dt - 2 \int_{0}^{\tau} S(t)E_2(\tau - t)dt \qquad (2.86)$$

where $E_2(t - \tau)$ is a second-order exponential integral. Replacing the subscripts, the above becomes

$$F_\nu(\tau_\nu) = 2 \int_{\tau_\nu}^{\infty} S(t_\nu)E_2(t_\nu - \tau_\nu)dt_\nu - 2 \int_{0}^{\tau_\nu} S_\nu(t_\nu)E_2(\tau_\nu - t_\nu)dt_\nu \qquad (2.87)$$

(3) Expression for the K-integral
According to Eq. (2.26), the K-integral can be written as

$$K(\tau) = \frac{1}{2} \int_{-1}^{+1} I(\tau, \mu)\mu^2 d\mu$$

Substituting the equation of radiative transfer, and using the method described above, we have

$$K(\tau) = \frac{1}{2} \int_0^\infty S(t) E_3 |t - \tau| dt \tag{2.88}$$

where $E_3(t - \tau)$ is a third-order exponential integral. Replacing the subscripts, the above becomes

$$K_\nu(\tau_\nu) = \frac{1}{2} \int_0^\infty S(t_\nu) E_3 |t_\nu - \tau_\nu| dt_\nu \tag{2.89}$$

In all the above equations, we have used the exponential integrals in Eq. (2.83). Some important features for the exponential integrals will be given in the following:

(i) when $x = 0$, we have

$$E_n(0) = \int_1^\infty \frac{dt}{t^n} = \frac{1}{n - 1} \tag{2.90}$$

(ii) on differentiating the exponential integral, we have

$$\frac{dE_n(x)}{dx} = \int_1^\infty \frac{1}{t^n} \frac{d}{dx} e^{-xt} dt = - \int_1^\infty \frac{e^{-xt}}{t^{n-1}} dt$$

which can also be written as

$$\frac{dE_n(x)}{dx} = -E_{n-1}(x) \tag{2.91}$$

(iii) on partially integrating the exponential integral, we have

$$nE_{n+1}(x) = e^{-x} - xE_n(x) \tag{2.92}$$

Equation (2.92) is a very useful relationship, which gives us $E_{n+1}(x)$ when $E_n(x)$ is known.

2.4.6 Asymptotic Solution to the Equation of Radiative Transfer for Large Optical Depths

We now investigate how the equation of radiative transfer changes or what the asymptotic solution is for very large optical depths $\tau \gg 1$, e.g., when we go deeply into the lowest layers of the stellar atmosphere. We have known that the physical state is close to thermodynamic equilibrium under such a condition, so $S_\nu = B_\nu$ and the radiation is isotropic.

Expanding I_ν at a certain optical depth τ_ν,

$$I_\nu(\mu) = I_{\nu 0} + I_{\nu 1} \cdot \mu + I_{\nu 2} \cdot \mu^2 + \ldots \qquad (2.93)$$

substituting the above into the equation for radiative transfer, and considering $S_\nu = B_\nu$, we have

$$\sum_{n=0}^{\infty} \mu^{n+1} \frac{dI_{\nu n}}{d\tau_\nu} = \sum_{n=0}^{\infty} I_{\nu n} \mu^n - B_\nu(\tau_\nu) \qquad (2.94)$$

This equation should hold for all values of μ, so the coefficients in the terms with the same power in μ should be equal. We therefore arrive at the following solutions:

$$\left.\begin{aligned}
I_{\nu 0} &= B_\nu, \\
I_{\nu 1} &= \frac{dI_{\nu 0}}{d\tau_\nu} = \frac{dB_\nu}{d\tau_\nu}, \\
I_{\nu 2} &= \frac{d^2 B_\nu}{d\tau_\nu^2}, \\
&\vdots \\
I_{\nu n} &= \frac{d^n B_\nu}{d\tau_\nu^n}
\end{aligned}\right\} \qquad (2.95)$$

Substituting Eq. (2.95) into Eq. (2.93), we have

$$I_\nu(\mu) = B_\nu + \mu \frac{dB_\nu}{d\tau_\nu} + \mu^2 \frac{d^2 B_\nu}{d\tau_\nu^2} + \ldots = \sum_{n=0}^{\infty} \mu^n \frac{d^n B_\nu}{d\tau_\nu^n} \qquad (2.96)$$

which is convergent because a general term $(d^n B_\nu/d\tau_\nu^n)$ has the property $|d^n B_\nu/d\tau_\nu^n| \sim B_\nu/d\tau_\nu^n$ according to properties of B_ν, which decreases when $\tau_\nu \gg 1$.

Through Eq. (2.96), the expressions for J_ν, H_ν and K_ν can be obtained for $\tau \gg 1$.

$$J_\nu = \frac{1}{2} \int_{-1}^{+1} I_\nu d\mu = \frac{1}{2} \sum_{n=0}^{\infty} \frac{d^n B_\nu}{d\tau_\nu^n} \int_{-1}^{+1} \mu^n d\mu$$

The integral on the right will have non-zero value only when n is even, so

$$J_v = \sum_{n=0}^{\infty} \frac{1}{2^n + 1} \frac{d^{2n} B_v}{d\tau_v^{2n}} = B_v + \frac{1}{3} \frac{d^2 B_v}{d\tau_v^2} + \ldots \qquad (2.97)$$

Similarly, we have

$$H_v = \frac{1}{2} \int_{-1}^{+1} I_v \mu \, d\mu = \frac{1}{2} \sum_{n=0}^{\infty} \frac{d^n B_v}{d\tau_v^n} \int_{-1}^{+1} \mu^{n+1} d\mu$$

$$= \sum_{n=0}^{\infty} \frac{1}{2^n + 3} \frac{d^{2n+1} B_v}{d\tau_v^{2n+1}} = \frac{1}{3} \frac{dB_v}{d\tau_v} + \frac{1}{5} \frac{d^3 B_v}{d\tau_v^3} + \ldots \qquad (2.98)$$

$$K_v = \frac{1}{2} \int_{-1}^{+1} I_v \mu^2 d\mu = \sum_{n=0}^{\infty} \frac{1}{2^n + 3} \frac{d^{2n} B_v}{d\tau_v^{2n}} = \frac{1}{3} B_v + \frac{1}{5} \frac{d^2 B_v}{d\tau_v^2} + \ldots \qquad (2.99)$$

When $\tau \gg 1$, the following asymptotic solutions can be obtained from Eqs. (2.97) to (2.99):

$$I_v(\tau_v, \mu) \approx B_v + \mu \frac{dB_v}{d\tau_v} \qquad (2.100)$$

$$J_v(\tau_v) \approx B_v(\tau_v) \qquad (2.101)$$

$$H_v(\tau_v) \approx \frac{1}{3} \frac{dB_v}{d\tau_v} \qquad (2.102)$$

$$K_v(\tau_v) \approx \frac{1}{3} B_v(\tau_v) \qquad (2.103)$$

Furthermore, from Eqs. (2.101) and (2.103), we have

$$\lim_{\tau_v \to \infty} [K_v(\tau_v)/J_v(\tau_v)] = \frac{1}{3} \qquad (2.104)$$

Moreover, since $H_v = (1/4)F_v$, $\frac{d}{d\tau_v} = -\frac{1}{\kappa_v} \frac{d}{dz}$ and $B_v(\tau_v) = B_v(T(\tau_v))$, from Eq. (2.102), we have

$$F_v(\tau_v) = -\frac{4}{3} \left(\frac{1}{\kappa_v} \frac{dB_v}{dT} \right) \frac{dT}{dz} \qquad (2.105)$$

This is the asymptotic solution to the equation for radiative transfer for $\tau \gg 1$.

CHAPTER THREE

CONVECTION

Convection is a common phenomenon in fluids. For a fluid under a macroscopically stable equilibrium state, the physical quantities (such as temperature, density and pressure etc.) of each fluid element have in fact small fluctuations around their mean values. If the physical state of the entire fluid or a localized region undergoes changes thereby forming unstable conditions, the fluid elements can deviate from their equilibrium positions and have macroscopic motions. For example, the fluid elements hotter than the surroundings can move upwards while those cooler than the surroundings can move downwards, which lead to convection.

Convection is present inside stars and has significant effects on the stellar structure and evolution, since it can homogenize chemical compositions in certain regions inside the star by mixing and transferring heavy elements from the core to the surface of the star. Furthermore, convection can be a major form of energy transfer in certain regions, and the energy transfer has important impact on the temperature distribution inside a star.

In this section, we first discuss the convective instability, and then introduce local and non-local mixing length theories commonly employed in stellar theories.

3.1 CRITERION FOR THE PRODUCTION OF CONVECTIVE INSTABILITY

Inside a star, some regions have homogeneous chemical compositions while others have inhomogeneous chemical compositions. The criteria for the production of convective instabilities in these two sorts of regions are different, which will be discussed separately.

3.1.1 Schwarzchild Criterion

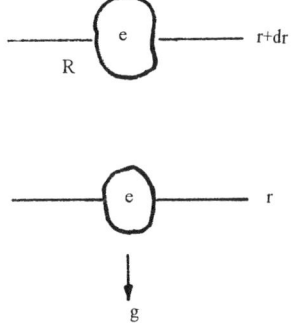

Fig. 3.1 The geometry for calculation of the Schwarzchild criterion.

We first study the criterion for production of convective instability in a region with a homogeneous chemical composition. We consider a region of homogeneous chemical composition under radiative equilibrium (which means that the major energy transfer forms in this region are radiation and conduction). Inside this region, there is a fluid element with "e" representing its interior and "R" representing its surroundings which is the radiative equilibrium region (Fig. 3.1). Suppose a temperature perturbation occurs inside the element at a certain time, so that $\Delta T = T_e - T_R > 0$ or the interior temperature is higher than the surrounding temperature. Together with the temperature perturbation, there may also be a pressure perturbation inside the element but the pressure can quickly attain equilibrium with the surroundings so $\Delta P = P_e - P_R = 0$ since the equalization of pressure takes place with the speed of sound. Comparatively, the equalization of temperature is much slower.

According to the equation of state, $\Delta T > 0$ and $\Delta P = 0$ will lead to a density perturbation $\Delta \rho = \rho_e - \rho_R < 0$, i.e., the density inside the element is smaller than that of the surroundings. Therefore, the element experiences a buoyancy $-g\Delta\rho$ and moves upwards (i.e., a motion opposite to the direction of the gravitational force). After the element has travelled an upward distance dr (dr is positive) under the buoyancy, the difference between the density inside the element and that of the surroundings becomes

$$\Delta\rho(r + dr) = \Delta\rho(r) + dr \cdot \frac{d}{dr}(\Delta\rho)$$

where $\Delta\rho(r) < 0$ and $dr > 0$. Therefore, if $\frac{d}{dr}(\Delta\rho) \leq 0$, $\Delta\rho(r + dr) < 0$, i.e., the density inside the element is still lower than that of the surroundings after travelling a distance dr, the fluid element can keep going upwards under the buoyancy. Conversely, if $\frac{d}{dr}(\Delta\rho) > 0$, the fluid element will experience a resistance force and stop moving.

Therefore, it can be seen that the criterion for production of convective instability is

$$\frac{d}{dr}(\Delta \rho) \leq 0 \tag{3.1}$$

Since the region under study has been assumed to have a homogeneous chemical composition, the equation of state can be written as

$$\rho = \rho(P, T) \tag{3.2}$$

so

$$d\ln \rho = \alpha \, d\ln P - \delta \, d\ln T \tag{3.3}$$

where

$$\alpha = \left(\frac{\partial \ln \rho}{\partial \ln P}\right)_T, \, \delta = -\left(\frac{\partial \ln \rho}{\partial \ln T}\right)_P \tag{3.4}$$

For an ideal gas, $\alpha = \delta = 1$, while for a non-ideal gas, α and δ are not equal to 1. Applying Eq. (3.3) for the interior and the surroundings of the fluid element, we have

$$\frac{d\ln \rho_e}{dr} = \alpha \frac{d\ln P_e}{dr} - \delta \frac{d\ln T_e}{dr} \qquad \text{(for the interior)} \tag{3.5}$$

$$\frac{d\ln \rho_R}{dr} = \alpha \frac{d\ln P_R}{dr} - \delta \frac{d\ln T_R}{dr} \qquad \text{(for the surroundings)} \tag{3.6}$$

Subtracting Eq. (3.6) from Eq. (3.5), we obtain

$$\frac{d}{dr}(\Delta \ln \rho) = \frac{d\ln \rho_e}{dr} - \frac{d\ln \rho_R}{dr} = \delta\left(\frac{d\ln T_R}{dr} - \frac{d\ln T_e}{dr}\right) \tag{3.7}$$

To create convective instability, we have to fulfil Eq. (3.1) which is equivalent to the requirement that the above equation should be less than zero, or

$$\frac{d\ln T_R}{dr} \leq \frac{d\ln T_e}{dr} \tag{3.8}$$

We introduce a quantity H_P as

$$H_P \equiv -\frac{dr}{d\ln P} = -P\frac{dr}{dP} \tag{3.9}$$

known as the pressure scale height. On one hand, the pressure scale height has the dimension of length, and on the other hand, it reflects the pressure change in the radial direction. When P decreases with an increase in r, $H_P > 0$. Multiplying Eq. (3.8) by H_P, we have

$$\left(\frac{d \ln T}{d \ln P}\right)_R \geq \left(\frac{d \ln T}{d \ln P}\right)_e \qquad (3.10)$$

We further define two symbols:

$$\nabla_R \equiv \left(\frac{d \ln T}{d \ln P}\right)_R \qquad (3.11)$$

$$\nabla_e \equiv \left(\frac{d \ln T}{d \ln P}\right)_e \qquad (3.12)$$

The quantity ∇_R is the radiation temperature gradient, since we have assumed that the surroundings of the fluid element is under radiative equilibrium. Furthermore, we assume that the fluid element does not exchange energy with the surroundings during its motion, i.e., the fluid element is approximated to be adiabatic. In this way, the temperature gradient ∇_e in Eq. (3.12) can be rewritten as ∇_{ad}, i.e., the adiabatic temperature gradient. Therefore the criterion for production of convective instability in Eq. (3.10) can be expressed as

$$\nabla_R \geq \nabla_{ad} \qquad (3.13)$$

which is called the Schwarzschild criterion for convection. In a certain region inside a star, the luminosity L or the mass absorption coefficient κ can be very large so that the radiation temperature gradient ∇_R becomes very large and the Schwarzschild criterion is fulfilled. In this case, convective instability and thus convection occurs in this region. Under such a convection, the fluid element is approximately adiabatic whilst in motion. This kind of instability is also called the dynamic instability. There are other forms of instabilities inside a star. For example, an instability can occur inside a star which makes the fluid move but the fluid element is not adiabatic during the motion and exchanges energy with its surroundings. This is a pulsational instability and leads to stellar oscillations.

3.1.2 Ledoux Criterion

The Schwarzchild criterion for production of convective instability is for regions having homogeneous chemical compositions. For regions having inhomogeneous chemical compositions, we need a new criterion. For a region with an inhomogeneous chemical composition, the equation of state can be written as

$$\rho = \rho(P, T, \mu) \qquad (3.14)$$

where μ is the average molecular weight and the changes in the chemical composition is reflected through the changes in μ ($d\mu \neq 0$). From Eq. (3.14), we obtain

$$d \ln \rho = \alpha d \ln P - \delta \ln T + \varphi d \ln \mu \qquad (3.15)$$

where

$$\varphi \equiv \left(\frac{\partial \ln \rho}{\partial \ln \mu} \right)_{P.T} \qquad (3.16)$$

For an ideal gas,

$$\alpha = \delta = \varphi = 1 \qquad (3.17)$$

We apply Eq. (3.15) in the interior and the surrounding of the fluid element in Fig. 3.1, respectively. For the surrounding "R" region, there is no doubt that Eq. (3.15) should be used. However, for the interior "e" region, since the fluid element is treated as closed, $d\ln\mu_e = 0$ during its motion, so we have

$$\frac{d \ln \rho_e}{dr} = \alpha \frac{d \ln P_e}{dr} - \delta \frac{d \ln T_e}{dr} \qquad \text{(for the interior)} \qquad (3.18)$$

$$\frac{d \ln \rho_R}{dr} = \alpha \frac{d \ln P_R}{dr} - \delta \frac{d \ln T_R}{dr} + \varphi \frac{d \ln \mu_R}{dr} \qquad \text{(for the surroundings)}$$

$$(3.19)$$

The subsequent derivations are similar to those for the above derivations of the Schwarzchild criterion. Finally we obtain the criterion for production of convective instability in regions having inhomogeneous chemical compositions as

$$\nabla_R \geq \nabla_{ad} + \nabla_\varphi \qquad (3.20)$$

where

$$\nabla_{\varphi} = \frac{\varphi}{\delta} \frac{d \ln \mu_R}{dr} \frac{dr}{d \ln P} \qquad (3.21)$$

Equation (3.20) is called the Ledoux criterion.

3.2 TEMPERATURE GRADIENT AND ROSSELAND MEAN OPACITY

From the above, the production of convective instabilities in a certain region inside a star is controlled by whether the radiation temperature gradient ∇_R and the adiabatic temperature gradient ∇_{ad} fulfil Eq. (3.13) (for regions with homogeneous chemical compositions) or Eq. (3.20) (for regions with inhomogeneous chemical compositions). We now investigate the relationship between ∇_R and ∇_{ad} with other physical quantities in the region such as temperature, pressure, density and radiation flux, i.e., to derive the expressions for ∇_R and ∇_{ad}. We treat the case of ∇_R first. For a certain region inside a star, from the first law of thermodynamics, we have

$$dQ = du - \frac{P}{\rho^2} d\rho = \frac{\partial u}{\partial T} dT + \left(\frac{\partial u}{\partial \rho} - \frac{P}{\rho^2} \right) d\rho \qquad (3.22)$$

where $u = u(\rho, T)$ is the internal energy of the system. To compute the expression inside the brackets on the right of the above equation, we need to study the total differential of the entropy $s = s(\rho, T)$

$$ds = \frac{dQ}{T} = \frac{1}{T} \frac{\partial u}{\partial T} dT + \left(\frac{1}{T} \frac{\partial u}{\partial \rho} - \frac{P}{T\rho^2} \right) d\rho \qquad (3.23)$$

From Eq. (3.23), we get the following relationship

$$\frac{1}{T} \frac{\partial^2 u}{\partial T \partial \rho} = -\frac{1}{T^2} \frac{\partial u}{\partial \rho} + \frac{1}{T} \frac{\partial^2 u}{\partial \rho \partial T} - \frac{1}{T\rho^2} \frac{\partial P}{\partial T} + \frac{P}{T^2 \rho^2}$$

from which we can obtain

$$\frac{\partial u}{\partial \rho} - \frac{P}{\rho^2} = -\frac{T}{\rho^2} \frac{\partial P}{\partial T} \qquad (3.24)$$

Substituting Eq. (3.24) into Eq. (3.22), and using the definition of specific heat at constant volume $C_V \equiv \left(\frac{\partial Q}{\partial T}\right)_\rho = \left(\frac{\partial u}{\partial T}\right)_\rho$, we have

$$
\begin{aligned}
dQ &= C_V dT - \frac{T}{\rho^2}\frac{\partial P}{\partial T}d\rho \\
&= \left(C_V - \frac{T}{\rho^2}\frac{\partial P}{\partial T}\frac{\partial \rho}{\partial T}\right)dT - \frac{T}{\rho^2}\frac{\partial P}{\partial T}\frac{\partial \rho}{\partial T}dP
\end{aligned}
\tag{3.25}
$$

Using the definition of specific heat at constant pressure $C_P \equiv \left(\frac{\partial Q}{\partial T}\right)_P$, Eq. (3.25) can be written as

$$
dQ = C_P dT - \frac{T}{\rho^2}\frac{\partial P}{\partial T}\frac{\partial \rho}{\partial P}dP
\tag{3.26}
$$

However,

$$
\frac{\partial P}{\partial T} = -\frac{\partial \rho/\partial T}{\partial \rho/\partial P}
\tag{3.27}
$$

so Eq. (3.26) can be written as

$$
dQ = C_P dT + \frac{T}{\rho^2}\frac{\partial \rho}{\partial T}dP
\tag{3.28}
$$

From the equation of state, we have

$$
\frac{d\rho}{\rho} = \alpha\frac{dP}{P} - \delta\frac{dT}{T}
\tag{3.29}
$$

where

$$
\alpha = \left(\frac{\partial \ln \rho}{\partial \ln P}\right)_T = \frac{P}{\rho}\left(\frac{\partial \rho}{\partial P}\right)_T
\tag{3.30}
$$

$$
\delta = -\left(\frac{\partial \ln \rho}{\partial \ln T}\right)_P = -\frac{T}{\rho}\left(\frac{\partial \rho}{\partial T}\right)_P
\tag{3.31}
$$

so Eq. (3.28) can be written as

$$
dQ = C_P dT - \frac{\delta}{\rho}dP
\tag{3.32}
$$

If the system is adiabatic, $dQ = 0$ and we can obtain from Eq. (3.32) that

$$
C_P dT_{ad} - \frac{\delta}{\rho}dP_{ad} = 0
\tag{3.33}
$$

from which we get

$$\nabla_{ad} \equiv \left(\frac{d \ln T}{d \ln P}\right)_{ad} = \frac{\delta P}{C_P T \rho} \tag{3.34}$$

In order to obtain the expression for the radiative temperature gradient ∇_R, we have to use the equation for radiative transfer and the hydrostatic equilibrium equation. According to the definition of temperature gradient ∇, we have

$$\nabla \equiv \frac{d \ln T}{d \ln P} = \frac{P}{T}\frac{dT}{dP} \tag{3.35}$$

Substituting the hydrostatic equilibrium equation in Eq. (7.7) into Eq. (3.35), we have

$$\frac{dT}{dr} = -\frac{GM_r}{r^2}\frac{\rho T}{P} \cdot \nabla \tag{3.36}$$

Multiplying the equation for radiative transfer in Eq. (2.63) by μ/c, and integrating over all solid angles, we get

$$\frac{dP_R(\nu)}{dr} + \frac{1}{r}(3P_R(\nu) - u_\nu) + \frac{\kappa_\nu \rho}{c}\pi F_\nu = 0 \tag{3.37}$$

in which we have employed Eqs. (2.9), (2.12) and (2.16).

Integrating Eq. (3.37) with respect to ν, and using the relationship $P_R(\nu) = (1/3)u_\nu$, we have

$$\begin{aligned}
\pi F &= -\frac{c}{3\rho}\int_0^\infty \frac{1}{\kappa_\nu}\frac{du_\nu}{dr}\,d\nu \\
&= -\frac{c}{3\rho}\frac{du}{dr}\frac{\int_0^\infty \frac{1}{\kappa_\nu}\frac{du_\nu}{dr}\,d\nu}{\int_0^\infty \frac{du_\nu}{dr}\,d\nu}
\end{aligned} \tag{3.38}$$

We introduce a mean value $\bar{\kappa}$ for the absorption coefficient κ_ν given by

$$\frac{1}{\bar{\kappa}} \equiv \frac{\int_0^\infty \frac{1}{\kappa_\nu}\cdot\frac{du_\nu}{dr}\,d\nu}{\int_0^\infty \frac{du_\nu}{dr}\,d\nu} = \frac{\int_0^\infty \frac{1}{\kappa_\nu}\cdot\frac{du_\nu}{dT}\frac{dT}{dr}\,d\nu}{\int_0^\infty \frac{du_\nu}{dT}\frac{dT}{dr}\,d\nu} \tag{3.39}$$

Since we have assumed that the region which we are studying is under thermodynamic equilibrium, the radiation field is an isotropic and homogeneous field, so we have

$$u_\nu = \frac{4\pi}{c} B_\nu(T) = \frac{8\pi h}{c^3} \frac{\nu^3}{e^{h\nu/kT} - 1}$$

and, Eq. (3.39) can be written as

$$\frac{1}{\bar{\kappa}} = \frac{\int_0^\infty \frac{1}{\kappa_\nu} \cdot \frac{dB_\nu(T)}{dT} d\nu}{\int_0^\infty \frac{dB_\nu(T)}{dT} d\nu} \qquad (3.40)$$

The mean value of κ_ν weighted by the differentials of the Planck's function with respect to temperature T shown in Eq. (3.40) is called the Rosseland mean opacity.

After introducing $\bar{\kappa}$, Eq. (3.38) can be written as

$$\pi F = -\frac{c}{3\bar{\kappa}\rho} \frac{du}{dr} \qquad (3.41)$$

Further considering $L_r = 4\pi r^2 \cdot \pi F$ and $u = aT^4$, Eq. (3.41) becomes

$$L_r = -4\pi r^2 \frac{4acT^3}{3\bar{\kappa}\rho} \cdot \frac{dT}{dr} \qquad (3.42)$$

which gives the relationship between the radiation flux and the temperature gradient when energy is transferred by radiation. This equation can also be written in the form

$$L_r = -4\pi r^2 K_r \frac{dT}{dr} \qquad (3.43)$$

where

$$K_r = \frac{4ac}{3} \frac{T^3}{\bar{\kappa}\rho} \qquad (3.44)$$

Eq. (3.43) is entirely similar to the equation for heat conduction, so K_r is also called the "radiative conduction coefficient". Due to exact resemblance of the two forms, under both radiative and conductive energy transfers, the relationship between the energy flux and the temperature gradient can be expressed as

$$L_r = -4\pi r^2 K \frac{dT}{dr} \tag{3.45}$$

where $K = K_r + K_L$, and K_L is the "conduction coefficient".

Usually, the conductive energy transfer is very small in the interior of a star when compared to the radiative energy transfer and can be neglected. However, if the density of the interior of the star is very high and degeneracy occurs, conductive heat transfer can no longer be neglected. In practice, Eq. (3.42) is used to express the temperature gradient inside a radiative equilibrium region. When degeneracy occurs, we need only to include the electron conduction when calculating $\bar{\kappa}_\nu$.

Applying Eq. (3.36) in the radiative equilibrium region, we have

$$\frac{dT}{dr} = -\frac{GM_r}{r^2} \frac{\rho T}{P} \cdot \nabla_R \tag{3.46}$$

Comparing Eqs. (3.46) and (3.42), we get

$$\nabla_R = \frac{3}{16\pi acG} \frac{\bar{\kappa} L_r P}{M_r T^4} \tag{3.47}$$

From the expression of ∇_R in Eq. (3.47), large values in L_r (for example, when there are' violent thermonuclear reactions and the energy generation rate is high) or $\bar{\kappa}$ (for example, in the envelope and certain elements are partially ionized) can increase ∇_R in that region. When ∇_R is greater than ∇_{ad}, convective instability will occur.

3.3 MIXING LENGTH THEORY

The mixing length theory proposed by Böhn–Vitense (1958) is the earliest and commonest convection theory used in stellar astrophysics, which handles the turbulence problem using the molecular motion theory. The theory introduces a length l similar to the molecular mean free path, and assumes that a fluid blob (or a fluid element) will not mix with other fluid blobs and change its own properties within a distance of l during the turbulent motion. The length l is called the mixing length and the theory is called the mixing length theory. The mixing length theory proposed by Böhn–Vitense is also called the local mixing length theory because the momentum of a blob at a certain position is determined by local physical quantities for this theory. The non-local mixing length theory developed in recent years has suggested that the momentum of the blob at a certain position should be determined by physical quantities at different places during the entire journey. We will present the local mixing length theory first in the following section.

3.3.1 Basic Equations

Equation (3.42) is obtained from the equation of radiative transfer, so it should be satisfied when energy transfers through radiation (including conduction). Since $\pi F = L_r/4\pi r^2$ and $H_P = -P(dr/dP)$, the relationship can also written as

$$\pi F = \frac{4ac}{3\bar{\kappa}\rho} \frac{T^4}{H_P} \cdot \nabla \qquad (3.48)$$

Applying this equation in a radiative equilibrium region, we have

$$\pi F_R = \frac{4ac}{3\bar{\kappa}\rho} \frac{T^4}{H_P} \cdot \nabla_R \qquad (3.49)$$

Since all energy in the radiative equilibrium region is transferred through radiation, πF_R is equal to the total radiation flux πF, i.e.,

$$\pi F = \frac{4ac}{3\bar{\kappa}\rho} \frac{T^4}{H_P} \cdot \nabla_R \qquad (3.50)$$

Applying Eq. (3.48) in a convective region, we can obtain the radiative energy transfer. Since the temperature gradient inside the convective region is ∇_{con}, we have

$$\pi F_R = \frac{4ac}{3\bar{\kappa}\rho} \frac{T^4}{H_P} \cdot \nabla_{con} \qquad (3.51)$$

Inside the convective region, the total energy flux πF is the summation of radiative and convective energy transfer, i.e.,

$$\pi F = \pi F_R + \pi F_{con} \qquad (3.52)$$

Based on conservation of energy, the total energy flux transferred by radiation and convection is equal to the total energy flux transferred entirely by radiation if no convection occurs in this region. Therefore, from Eqs. (3.52) and (3.50), we have

$$\pi F = \pi F_R + \pi F_{con} = \frac{4ac}{3\bar{\kappa}\rho}\frac{T^4}{H_P}\nabla_R \qquad (3.53)$$

We present the derivation of the expression for πF_{con} in the following. Consider an upward moving fluid element at a distance r from the center of a star (Fig. 3.2). Suppose the element has an upward velocity of v and a density ρ. The energy flux passing through a plane at a distance r from the center is

$$\pi F'_{con} = \rho v (C_P \Delta T) \qquad (3.54)$$

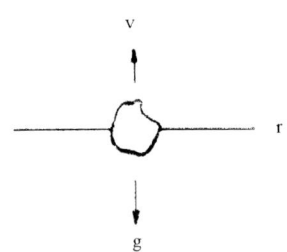

Fig. 3.2 The geometry for derivation of the expression for the convective energy flux πF_{con}.

where $C_P \Delta T$ is the energy carried by one element. Since the fluid element is in pressure equilibrium with its surroundings, the specific heat at constant pressure C_P is employed. Only one fluid element has been considered in the above discussions. To compute the mean energy flux πF_{con} passing through that surface, we have to average over all $\pi F'_{con}$. Denoting the mean values of v and ΔT to be $\bar{v}$ and $\overline{\Delta T}$, respectively, we have

$$\pi F_{con} = C_P \rho \bar{v} \overline{\Delta T} \qquad (3.55)$$

in which we have assumed that all fluid elements have the same values of C_P and ρ. In the above discussions, only the upward moving elements have been considered. The case for downward moving elements is similar. To obtain $\bar{v}$ and $\overline{\Delta T}$, a further assumption should be made. The mixing length theory assumes that an average fluid element will completely disappear after traversing a path length l, where l is called the mixing length. The values of v and ΔT for this average fluid element at the position r are the values of $\bar{v}$ and $\overline{\Delta T}$ that we need. At the same time, we assume this fluid element to start its upward journey from a distance $l/2$ perpendicularly below the position designated as r, the values v_0 and ΔT_0 are zero (so $\Delta \rho_0$ is also zero), and the changes of C_P, T and ρ are all very small within a distance of l. From these, we have

$$\overline{\Delta T} = \frac{d}{dr}(\Delta T) \cdot \frac{l}{2} \qquad (3.56)$$

where ΔT is the difference between the temperature T_e inside the element and the surrounding temperature. The temperature of the surroundings is denoted by T_{con}. In this way, Eq. (3.56) can be written as

$$\overline{\Delta T} = \left(\frac{dT_e}{dr} - \frac{dT_{con}}{dr}\right) \cdot \frac{l}{2} = T(\nabla_{con} - \nabla_e)\frac{l}{2H_P} \qquad (3.57)$$

The buoyancy on the average fluid element when it starts its upward journey from a distance $l/2$ perpendicularly below the position designated as r is $K_o = -g\Delta\rho_o = 0$, but after traversing a distance of $l/2$ and on reaching the position r, the buoyancy becomes

$$K = -g\Delta\rho = -g\frac{d}{dr}(\Delta\rho) \cdot \frac{l}{2} \qquad (3.58)$$

Similarly, we have assumed that there is negligible change in the density ρ over the distance l and the initial value is $\Delta\rho_o = 0$. Equation (3.58) can be rewritten as

$$\begin{aligned} K &= -g\rho\frac{\partial \ln \rho}{\partial \ln T}\left(\frac{d\ln T_e}{dr} - \frac{d\ln T_{con}}{dr}\right) \cdot \frac{l}{2} \\ &= -g\rho\delta\left(\frac{d\ln T_e}{d\ln p} - \frac{d\ln T_{con}}{d\ln p}\right)\frac{l/2}{H_P} \\ &= -g\rho\delta(\nabla_e - \nabla_{con})\frac{l}{2H_P} \end{aligned} \qquad (3.59)$$

After the average fluid element has traversed a distance of $l/2$, the work done by buoyancy is

$$\begin{aligned} A &= -g\rho\delta(\nabla_e - \nabla_{con})\frac{1}{4H_P} \cdot \frac{l}{2} \\ &= -g\rho\delta(\nabla_e - \nabla_{con})\frac{l^2}{8H_P} \end{aligned} \qquad (3.60)$$

Part of the work done by buoyancy has been converted to kinetic energy $\rho\bar{v}^2/2$ of the average fluid element at position r, while part of it has been converted to kinetic energy $\rho\bar{v}^2/2$ of the fluid elements forced apart by the average fluid element, so the total work done by buoyancy is the sum of the kinetic energies, i.e., $\rho\bar{v}^2$, so

$$g\rho\delta(\nabla_{con} - \nabla_e)\frac{l^2}{8H_P} = \rho\bar{v}^2 \qquad (3.61)$$

from which we have

$$\bar{v}^2 = g\delta(\nabla_{con} - \nabla_e)\frac{l^2}{8H_P} \qquad (3.62)$$

Substituting Eqs. (3.62) and (3.57) into Eq. (3.55), we obtain the expression of πF_{con} as

$$\pi F_{con} = \rho C_P T(g\delta)^{1/2}\frac{l^2}{4\sqrt{2}}H_P^{-3/2}(\nabla_{con} - \nabla_e)^{3/2} \qquad (3.63)$$

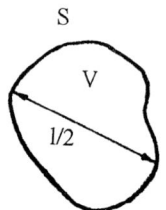

We go on to investigate the energy loss of the average fluid element due to radiation in traversing the distance l. Suppose the linear dimension of the average element is $l/2$ (Fig. 3.3), so that the temperature gradient from the fluid center to the surroundings is approximately $\Delta T/(l/4)$. Further supposing that the surface area of the fluid element is S and the volume is V, from Eq. (3.42), the radiated energy $\pi F'$ from the element surface per cm^2 per second is

Fig. 3.3 An average fluid element with volume V, surface area S and linear dimension $l/2$.

$$\pi F' = \frac{4ac}{3}\frac{T^3}{\bar{\kappa}\rho}\frac{\Delta T}{l/4} \qquad (3.64)$$

Therefore the energy loss per second per unit volume is

$$\frac{S \cdot \pi F'}{V} = \frac{4ac}{3}\frac{T^3}{\bar{\kappa}\rho}\frac{\Delta T}{l/4} \cdot \frac{S}{V} \qquad (3.65)$$

The time taken by the average fluid element to go through the distance $l/2$ is $\tau = (l/2)/\bar{v}$. Within this time, the energy loss per unit volume is

$$\frac{S \cdot \pi F'}{V} \cdot \tau = \frac{4ac}{3}\frac{T^3}{\bar{\kappa}\rho}\frac{\Delta T}{l/4} \cdot \frac{S}{V}\frac{l/2}{\bar{v}} \qquad (3.66)$$

Dividing the above equation by ρ and then substituting Eq. (3.57), we get the energy loss per gram of mass as

$$\frac{S \cdot \pi F'}{V\rho} \cdot \tau = \frac{4ac}{3}\frac{T^4}{\bar{\kappa}\rho^2}\frac{l}{H_P\bar{v}}(\nabla_{con} - \nabla_e)\frac{S}{V} = -dQ \qquad (3.67)$$

On the other hand, from the first law of thermodynamics [see Eq. (3.32)],

$$dQ = C_P dT_e - \frac{\delta}{\rho} dP = C_P T_e d\ln T_e - \frac{\delta}{\rho} P d\ln P \qquad (3.68)$$

Using Eqs. (3.34) and (3.9), Eq. (3.68) can also be written as

$$dQ = C_P T d\ln P(\nabla_e - \nabla_{ad})$$

$$= -C_P T(\nabla_e - \nabla_{ad})\frac{l}{H_P} \qquad (3.69)$$

From Eqs. (3.67) and (3.69), we have

$$\frac{4ac}{3}\frac{T^4}{\bar{\kappa}\rho^2}\frac{1}{H_P \bar{v}}(\nabla_{con} - \nabla_e)\frac{S}{V} = -C_P T(\nabla_e - \nabla_{ad})\frac{l}{H_P}$$

from which we can obtain

$$\frac{(\nabla_e - \nabla_{ad})}{(\nabla_{con} - \nabla_e)} = \frac{4ac}{3}\frac{T^3}{\bar{\kappa}\rho^2 C_P \bar{v}}\cdot\frac{S}{V} \qquad (3.70)$$

If we assume the fluid element to be spherical,

$$\frac{S}{V} = \frac{12}{l}$$

However, Böhn–Vitense (1958) obtained $\frac{S}{V} = \frac{9}{2}\frac{1}{l}$, so

$$\frac{(\nabla_e - \nabla_{ad})}{(\nabla_{con} - \nabla_e)} = \frac{6acT^3}{\bar{\kappa}\rho^2 C_P}\frac{1}{l\bar{v}} \qquad (3.71)$$

3.3.2 Solution of the Basic Equations

The following is a summary of the basic equations obtained above for the mixing length theory:

(1) $\pi F_R = \dfrac{4ac}{3\bar{\kappa}\rho}\dfrac{T^4}{H_P}\cdot\nabla_{con}$

(2) $\pi F_R + \pi F_{con} = \dfrac{4ac}{3\bar{\kappa}\rho}\dfrac{T^4}{H_P}\nabla_R$

(3) $\pi F_{con} = \rho C_P T(g\delta)^{1/2}\dfrac{l^2}{4\sqrt{2}}H_P^{-3/2}(\nabla_{con} - \nabla_e)^{3/2}$

$$(4) \quad \bar{v}^2 = g\delta(\nabla_{con} - \nabla_e)\frac{l^2}{8H_P}$$

$$(5) \quad \frac{(\nabla_e - \nabla_{ad})}{(\nabla_{con} - \nabla_e)} = \frac{6acT^3}{\bar{\kappa}\rho^2 C_P}\frac{1}{l\bar{v}}$$

There are five unknowns for these five equations, namely, $\nabla_{con}, \nabla_e, \bar{v}$ $\pi F_{con}, \pi F_R$. Among the five unknowns, the most important one is ∇_{con}. On knowing ∇_{con}, it can be substituted into Eq. (3.36) to compute the temperature gradient dT/dr in the convective region, where Eq. (3.36) is one of the basic equations for the stellar structure. Moreover, based on the basic equations (1) and (2), the convective energy transfer can be calculated after knowing ∇_{con}. In the following, we will describe how the unknown ∇_{con} can be obtained from these basic equations, but the detailed derivations will not be presented. Introducing a quantity U

$$U = \frac{12\sigma T^3}{C_P \rho^2 l\bar{\kappa}}\left(\frac{8H_P}{gl^2\delta}\right)^{1/2} \tag{3.72}$$

the above five basic equations can be combined to form a single equation

$$\nabla_{con} - \nabla_R + \frac{9}{8U}\left[(\nabla_{con} - \nabla_{ad} + U^2)^{1/2} - U\right]^3 = 0 \tag{3.73}$$

At a certain position in a star, when the values of L_r, P and T are known, ∇_R and ∇_{ad} can be computed using Eqs. (3.47) and (3.34). If l is further given, U can also be calculated. Therefore, Eq. (3.73) is a cubic equation in ∇_{con}. Hofmeister *et al.* (1964) and Baker (1963) gave a solution to Eq. (3.73) as

$$\nabla_{con} = \nabla_{ad} + \frac{1}{W^2}\left(W^2 + \frac{19}{27}WU - \frac{E}{3}\right)^2 - U^2 \tag{3.74}$$

where

$$W \equiv \left(\frac{1}{2}A + \sqrt{D}\right)^{1/3};$$

$$\frac{1}{2}A \equiv \left[\frac{4}{9}(\nabla_R - \nabla_{ad}) + \left(\frac{19^3}{27^3} - \frac{1}{9}\right)U^2\right]U;$$

$$D \equiv \left(\frac{A}{2}\right)^2 + \left(\frac{E}{3}\right)^3 \tag{3.75}$$

$$\frac{E}{3} \equiv \frac{368}{729}U^2;$$

$$H_P = \frac{\mathfrak{R}T}{g\mu\beta};$$

$$g = \frac{GM_r}{r^2};$$

$$\delta = -\left(\frac{\partial \ln \rho}{\partial \ln T}\right)_P.$$

In Eq. (3.74), when $U \to 0$, $\nabla_{con} \to \nabla_{ad}$. Generally speaking, when we go deeper into the star, U becomes smaller so $\nabla_{con} \approx \nabla_{ad}$. To understand the very small value of U inside the star, we can estimate its value at the center of the star.

Substituting the parameters of the ideal gas into Eq. (3.72), and neglecting effects of radiation pressure, i.e., taking $\delta = 1$, $C_P = \frac{5}{2}\frac{\Re}{\mu}$, $P = \frac{\Re}{\mu}\rho T$, $H_P = \frac{P}{g\rho}$ and $l = H_P$ we obtain

$$U = \frac{12}{5\sqrt{2}} \frac{acG}{\left(\frac{\Re}{\mu}\right)^{5/2}} \frac{M_r T^{3/2}}{r^2 \bar{\kappa} \rho^2} \tag{3.76}$$

If we adopt $M_r = 10^{33}$, $T = 10^7$, $r = 10^{10.5}$, $\bar{\kappa} = 10$, $\rho = 10^2$, we have $U = 10^{-13}$ which is negligible when compared to the value of ∇_{ad} (generally equals 0.4) inside the star. Therefore, $\nabla_{con} = \nabla_{ad}$ can be assumed in the convective region inside the star. Only for the convective region in the envelope of the star should we use Eq. (3.74) to compute ∇_{con}.

3.4 CONVECTIVE OVERSHOOT

The size of the convective region inside a star and the efficiency of convective energy transfer are important issues because they have significant effects on the stellar structure and evolution. In particular, the size of the convective region will directly influence the amount of matter inside a star which can have thermonuclear reactions to generate energy. The mixing length theory introduced in the last section is a local theory. The boundary of the convective region determined using such a theory will be underestimated. Astronomical observations for massive stars have shown that evolutionary features calculated using the local mixing length theory have considerable discrepancies, which have led to the development of the non-local mixing length theory in recent years. The difference between local and non-local mixing length theories will be presented in the following section.

According to the Schwarzschild criterion, the condition for convective instability is $\nabla_R \geq \nabla_{ad}$ while the condition for no convective instability is $\nabla_R < \nabla_{ad}$. Therefore, the boundary for the convective region should be determined by the condition $\nabla_R = \nabla_{ad}$. However, the realistic boundary of the convective region should be determined by the condition that the velocity of the upward (or downward) moving fluid element becomes zero, i.e., $v = 0$. For an upward moving fluid element, the entire upward journey is subjected to effects of buoyancy. Therefore, when the boundary determined by $\nabla_R = \nabla_{ad}$ (called the Schwarzschild boundary) is reached, its velocity $v \neq 0$, and the inertia will keep it rising (see Fig. 3.4). However, when it has crossed the Schwarzschild boundary, the condition has become $\nabla_R < \nabla_{ad}$, so $\Delta T = T_e - T_R < 0$. Here, the fluid element experiences a force in a direction opposite to the direction of its motion, and the velocity v will decrease until $v = 0$. Therefore, $v = 0$ should be the realistic boundary condition for the convective region. Astrophysicists refer the region between the boundaries determined by $\nabla_R = \nabla_{ad}$ and $v = 0$ to the convective overshoot region.

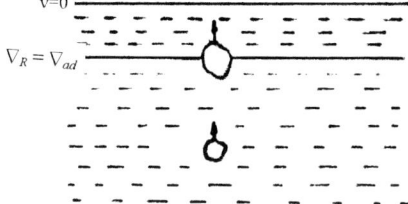

Fig. 3.4 The region between the boundaries determined by $\nabla_R = \nabla_{ad}$ and $v = 0$ is referred to as the convective overshoot region.

One can calculate the size of the convective overshoot region using local mixing length theory, but the obtained length of the convective overshoot region is only 10^{-5} of the mixing length l (Saslaw and Schwarzschild 1965), so the convective overshoot region can be neglected. Therefore, $\nabla_R = \nabla_{ad}$ can be taken as the boundary condition for the convective region using the local mixing length theory. However, Shavir and Salpeter (1973) and Maeder (1975) have pointed out that the size of the convective overshoot region is underestimated by the local mixing length theory. The reason behind this is that we have only used the physical parameters at the Schwarzschild boundary to determine the velocity v of the fluid element at the boundary through Eq. (3.62). Realistically, the velocity should be the result of buoyancy acceleration at different places during the entire upward journey, and should be determined by physical parameters at different points in the journey. Besides this factor, the local mixing length theory also supposes the temperature gradient inside the convective overshoot region to be determined only by physical parameters inside the region. Realistically, the large amount of fluid elements entering the region and releasing the carried energy will affect the temperature gradient inside the region, which can further decrease the velocity of the fluid element inside the region. The above two factors will enlarge the convective overshoot region calculated by the non-local mixing length theory.

Until now, there are a number of methods using the non-local mixing length theory to compute the convective overshoot region. However, the size of the convective overshoot regions obtained by various methods are very different. In the following, we will describe the method of Bressan *et al.* (1981) to determine the convective overshoot region boundary.

Inside the convective region, the total energy flux is

$$\pi F = \pi F_R + \pi F_{con} = \frac{4ac}{3\bar{\kappa}\rho} \frac{T^4}{H_P} \nabla_R \tag{3.77}$$

in which the radiative energy flux is

$$\pi F_R = \frac{4ac}{3\bar{\kappa}\rho} \frac{T^4}{H_P} \cdot \nabla_{con} \tag{3.78}$$

According to the non-local mixing length theory, πF_{con} can be written as

$$\pi F_{con} = KC_P \rho v_r \int_{r_1}^{r} \left[\left(\frac{\partial T}{\partial r'} \right)_e - \left(\frac{\partial T}{\partial r'} \right)_R \right] dr' \tag{3.79}$$

in which the coefficient K accounts for the contribution from a rising convective element or a descending convective element to πF_{con}. For a rising convective element, we can take $K = 0.5$. $(\partial T / \partial r')_e$ represents the temperature change inside the fluid element while $(\partial T / \partial r')_R$ represents the temperature change surrounding the fluid element. The integration boundary is $r - r_1 \le l/2$.

From the law of conservation of energy, the work done by buoyancy on the fluid element equals the change in the kinetic energy of the fluid element, i.e.,

$$v_r \frac{\partial v_r}{\partial r} = -g \frac{\Delta \rho}{\rho} \tag{3.80}$$

where v_r is the velocity of the fluid element at position r, $\Delta \rho$ is the difference between the density inside the fluid element and that of the surroundings at r. For the non-local mixing length theory, $\Delta\rho$ should be written as

$$\Delta\rho = \int_{r_1}^{r}\left[\left(\frac{\partial\rho}{\partial r'}\right)_e - \left(\frac{\partial\rho}{\partial r'}\right)_R\right]dr' \tag{3.81}$$

Since

$$dP = \left(\frac{\partial P}{\partial\rho}\right)_{T,\mu}d\rho + \left(\frac{\partial P}{\partial T}\right)_{\rho,\mu}dT + \left(\frac{\partial P}{\partial\mu}\right)_{\rho,T}d\mu = 0$$

we obtain

$$\begin{aligned}
d\rho &= -\frac{\left(\frac{\partial P}{\partial T}\right)_{\rho,\mu}}{\left(\frac{\partial P}{\partial\rho}\right)_{T,\mu}}dT - \frac{\left(\frac{\partial P}{\partial\mu}\right)_{\rho,T}}{\left(\frac{\partial P}{\partial\rho}\right)_{T,\mu}}d\mu \\
&= -\frac{\rho}{T}\frac{\chi_T}{\chi_\rho}dT - \frac{\rho}{\mu}\frac{\chi_\mu}{\chi_\rho}d\mu
\end{aligned} \tag{3.82}$$

where

$$\left.\begin{aligned}
\chi_T &\equiv \left(\frac{\partial\ln P}{\partial\ln T}\right)_{\rho,\mu} \\[2mm]
\chi_\rho &\equiv \left(\frac{\partial\ln P}{\partial\ln\rho}\right)_{T,\mu} \\[2mm]
\chi_\mu &\equiv \left(\frac{\partial\ln P}{\partial\ln\mu}\right)_{\rho,T}
\end{aligned}\right\} \tag{3.83}$$

Substituting Eq. (3.82) into Eq. (3.81), we have

$$\Delta\rho = -\int_{r_1}^{r}\frac{\rho}{T}\frac{\chi_T}{\chi_\rho}\left[\left(\frac{\partial T}{\partial r'}\right)_e - \left(\frac{\partial T}{\partial r'}\right)_R\right]dr' - \int_{r_1}^{r}\frac{\rho}{\mu}\frac{\chi_\mu}{\chi_\rho}\frac{\partial\mu}{\partial r'}dr' \tag{3.84}$$

We assume that the values of $(\rho/T)\cdot(\chi_T/\chi_\rho)$ and $(\rho/\mu)\cdot(\chi_\mu/\chi_\rho)$ do not change noticeably within the integration region, and can be treated as constants which can be taken out of the integration sign in Eq. (3.84). Using Eq. (3.79), the first integration term on the right-hand side of Eq. (3.84) can be written as πF_{con}. In this way, when Eq. (3.84) is substituted into Eq. (3.80), we obtain the following differential equation:

$$\frac{1}{3}\frac{\partial v_r^3}{\partial r} = \frac{1}{K}\frac{g}{T}\frac{\chi_r}{\chi_\rho}\frac{\pi F_{con}}{C_P\rho} - \frac{g}{\mu}\frac{\chi_\mu}{\chi_\rho}\Delta\mu v_r \tag{3.85}$$

which gives v_r on integration, and the position corresponding to $v_r = 0$ is the boundary of the convective region.

3.5 SEMI-CONVECTION

From §3.1, we know the criterion for convection in a region with homogeneous chemical composition in a star is $\nabla_{con} \geq \nabla_{ad}$, and that in a region with inhomogeneous chemical composition is $\nabla_{con} \geq \nabla_{ad} + \nabla_\varphi$. We now proceed to study the results when the temperature gradient in a region with inhomogeneous chemical composition fulfils

$$\nabla_{ad} < \nabla_R < \nabla_{ad} + \nabla_\varphi \tag{3.86}$$

Since $\nabla_R \leq \nabla_{ad} + \nabla_\varphi$, the Ledoux criterion for convective instability for a region with inhomogeneous chemical compositions is not satisfied, so no convection occurs and the fluid element will only oscillate around its mean position. Nevertheless, under the condition $\nabla_R \leq \nabla_{ad} + \nabla_\varphi$, the oscillation of the fluid element has peculiar properties. To illustrate this, we study the motion of a fluid element in a radiative equilibrium region with inhomogeneous chemical composition. When a fluid element deviates from its equilibrium position by a distance Δr, the difference between the internal and external densities for the fluid element is $\Delta\rho$, so by subtracting Eqs. (3.18) and (3.19) and further using Eqs. (3.9), (3.11) and (3.12), we obtain

$$\Delta\rho = \frac{\rho\delta}{H_P}(\nabla_e - \nabla_R + \nabla_\varphi)\Delta r \tag{3.87}$$

Since the fluid element has a density difference, it gets a buoyancy $K_r = -g\Delta\rho$. Under this buoyancy, the acceleration of the fluid element is

$$\frac{\partial^2(\Delta r)}{\partial t^2} = -\frac{g\delta}{H_P}(\nabla_e - \nabla_R + \nabla_\varphi)\Delta r \tag{3.88}$$

From Eq. (3.88), we know that the fluid element will experience an acceleration in the opposite direction when it deviates from its equilibrium position by Δr to bring it back to the equilibrium position. Eq. (3.88) is an oscillation equation. Langer *et al.* (1985) pointed out that the oscillation of the fluid element around its equilibrium position is unstable and grows with time. Therefore, for a region with inhomogeneous chemical composition, the condition of Eq. (3.86) will lead to oscillation instability in the region. Due to this oscillation, the chemical composition will become homogenized, so this region is called the semi-convective region.

Inside the semi-convective region, there is a gradual change in the chemical composition caused by oscillations of fluid elements with amplitudes growing with time. In practice, therefore, the change in the chemical composition inside a semi-convective region is normally computed by diffusion methods. In diffusion methods, careful choice of diffusion coefficients should be made. A too large coefficient will be equivalent to convection while a too small coefficient will lead to an extremely small change in the chemical composition.

CHAPTER FOUR

EQUATION OF STATE OF STELLAR MATTER

The pressure, temperature and chemical composition of stellar matter are closely related to the density, and the relationship is called the equation of state, which should be known for computations of the stellar structure and evolution models.

Since there are different temperatures, densities and chemical compositions inside a star, there may be different equations of state. For example, in the inner part of a star, all matter has been ionized. If the density is not very high, the interactions among all particles can be neglected. At this time, the stellar matter can be treated as an ideal gas and the equation of state is that for an ideal gas. This is the simplest case. If the density of the interior of the star is very high and there is degeneracy, the equation of state is no longer the simplest one mentioned above. In some regions in the envelope of the star, the temperature is relatively low and some of the elements are only partially ionized. In these partially ionized regions, the equation of state is also different from those for the totally ionized regions. The stellar atmosphere can also be divided into regions close to the interior and close to the surface. In regions close to the interior, the stellar matter is close to the thermal equilibrium state and the regions are called local thermal equilibrium regions. In regions close to the surface, the matter has apparently deviated from the thermal equilibrium state and the regions are called non-local thermal equilibrium regions. The equations of state for matter in these two regions are also different.

The interior of a star is approximately under thermal equilibrium. Therefore, in computing the equation of state for the interior of a star, we need to apply some statistical distributions of particles under thermal equilibrium, such as the Boltzmann equation and the Saha equation. However, if the density of the interior of the star is very high, interaction among particles cannot be neglected and the Boltzmann and the Saha equations become invalid. Similarly, in non-local thermal equilibrium regions in the stellar atmosphere, since the density is very low and thermal equilibrium is not satisfied, the Boltzmann equation and the Saha equation are also invalid.

4.1 STATISTICAL LAWS UNDER THERMAL EQUILIBRIUM

In §2.3, we introduced the concept of thermal equilibrium with help of the black body cavity. When the particles inside the black body cavity reach thermal equilibrium, hydrostatic equilibrium as well as thermal equilibrium are established inside the cavity. When the radiation field inside the cavity is isotropic and homogeneous, the black body radiation law is valid. Under thermal equilibrium, the particles inside the cavity will obey some statistical laws irrespective of their initial states. The interior of a star is approximately under thermal equilibrium so the statistical laws for particles under thermal equilibrium can be employed for the equation of state here. The major statistical laws for particles under thermal equilibrium include the following.

4.1.1 Maxwell Velocity Equation

From classical statistical physics, the velocity distribution of particles under thermal equilibrium should obey the Maxwell velocity equation:

$$\frac{dn}{n} = 4\pi v^2 \left(\frac{m}{2\pi kT} \right)^{3/2} e^{-mv^2/2kT} dv \tag{4.1}$$

where n is the number density of the particles per cm³ with the velocity range $(v, v + dv)$, k is the Boltzmann constant 1.380662×10^{-16} erg.K^{-1}.

4.1.2 Boltzmann Equation

Figure 4.1 shows the energy levels of an atom. The excitation energy χ_s is the energy difference between a bound state and the ground state, while

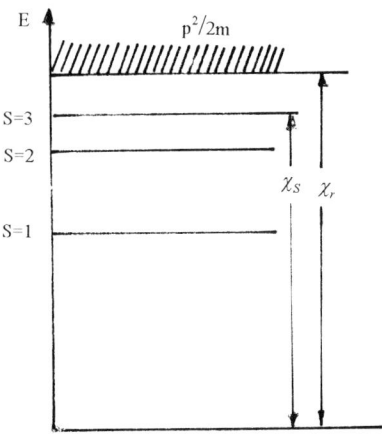

Fig. 4.1 The energy states of an atom. χ_s is the excitation energy while χ_r is the ionization energy. The kinetic energy of a free electron is $p^2/2m$.

the ionization energy χ_r is the energy difference between the continuous state and the ground state. The number of atoms per unit volume at an atomic energy level s is represented by n_s and called the occupation number of the energy level s. Under thermal equilibrium, the occupation number of an atomic energy level is proportional to the statistical weighting factor g_s of that level and the factor $\exp(-\chi_s/kT)$, i.e.,

$$n_s = \text{const} \cdot g_s e^{-\chi_s/kT} \qquad (4.2)$$

which is the Boltzmann equation. The statistical weighting factor g_s is also called the degree of degeneracy, which represents the number of quantum states for the same energy level. For hydrogen atoms, $g_s = 2n^2$ where n is the principal quantum number. For a multi-electron atom, $g_s = 2J + 1$ where J is the inner quantum number.

From Eq. (4.2), we can obtain the relationship among the occupation numbers of different excitation states of the atom as

$$\frac{n_s}{n_i} = \frac{g_s}{g_i} e^{-\Delta_\chi/kT} , \qquad (4.3)$$

where $\Delta_\chi = \chi_s - \chi_i$. From Eq. (4.2), we can also obtain the ratio between the occupation number of a certain excitation state of an atom and the total number of atoms per unit volume as

$$\frac{n_s}{n} = \frac{g_s e^{-\chi_s/kT}}{g_1 + g_2 e^{-\chi_2/kT} + g_3 e^{-\chi_3/kT} + \ldots} \qquad (4.4)$$
$$= \frac{g_s}{U(T)} e^{-\chi_s/kT}$$

where

$$U(T) \equiv \sum_i g_i e^{-\chi_i/kT} \tag{4.5}$$

is called the sum of states.

4.1.3 Saha Equation

We proceed to discuss the distribution of occupation numbers for atoms under different degrees of ionization. We denote $n_{r,o}$ as the number of atoms of a particular element in 1 cm^3 which are at the rth ionized ground state, and $g_{r,o}$ as the corresponding statistical weighting; $dn_{r,o}$ as the number of atoms of the same element in 1 cm^3 which are at the $(r+1)$th ionized ground state and of which the electrons after ionization have a kinetic energy of $p_e^2/2m_e$, and $dg(p_e)$ as the statistical weighting of the corresponding system (i.e., the atoms are at the $(r+1)$th ionized ground state and the electrons have a kinetic energy of $p_e^2/2m_e$). According to the Boltzmann equation, we have

$$\frac{dn_{r+1,0}}{n_{r,0}} = \frac{dg(p_e)}{g_{r,s}} e^{-(\chi_r + p_e^2/2m_e)/kT} \tag{4.6}$$

where $dg(p_e)$ is equal to the product of the statistical weighting $g_{r+1,o}$ for atoms at the $(r+1)$th ionized ground state and the statistical weighting of electrons having a kinetic energy $p_e^2/2m_e$, which is $(\Delta V \cdot \Delta^3 p_e \cdot 2/h^3)$, or

$$dg(p_e) = g_{r+1,0} \left(\frac{\Delta V \cdot \Delta^3 p_e}{h^3} \cdot 2 \right) \tag{4.7}$$

The statistical weighting of electrons is obtained by the number of phase cells in the phase space occupied by electrons and the Pauli exclusion principle which allows maximally two electrons with different spins in the same phase cell.

Substituting Eq. (4.7) into Eq. (4.6), integrating over all possible momenta p_e of electrons, and considering $\Delta V = 1/n_e$ and $\Delta^3 p_e = 4\pi p_e^2 dp_e$, we have

$$\frac{n_{r+1,0}}{n_{r,0}} = \frac{g_{r+1,0}}{g_{r,0}} \cdot \frac{2}{n_e h^3} e^{-\chi_r/kT} \int_0^\infty 4\pi p_e^2 e^{-p_e^2/2m_e kT} dp_e \tag{4.8}$$

where n_e is the number of electrons in 1 cm^3. Using the relationship

$$\int_0^\infty x^2 e^{-a^2 x^2} dx = \frac{\sqrt{\pi}}{4a^3} \qquad (a > 0)$$

Eq. (4.8) can be written as

$$\frac{n_{r+1,0}}{n_{r,0}} n_e = \frac{g_{r+1,0}}{g_{r,0}} \cdot 2 \cdot \frac{(2\pi m_e kT)^{3/2}}{h^3} e^{-\chi_r/kT} \qquad (4.9)$$

which gives the relationship between occupation numbers of two adjacent ionized ground states. From Eq. (4.9), the distribution of occupation numbers of two adjacent ionized states can also be obtained. Summing over all excited states, we have

$$\frac{n_{r+1}}{n_r} n_e = \frac{U_{r+1}}{U_r} \cdot 2 \cdot \frac{(2\pi m_e kT)^{3/2}}{h^3} e^{-\chi_r/kT} \qquad (4.10)$$

where U_{r+1} and U_r represent the summation of states for the $(r+1)$th and the rth ionization. In practice, the electron pressure $P_e(= n_e kT)$ sometimes replaces n_e for convenience, so Eq. (4.10) can also be written as

$$\frac{n_{r+1}}{n_r} \cdot P_e = 2 \frac{U_{r+1}}{U_r} \frac{(2\pi m_e)^{3/2}}{h^3} (kT)^{5/2} e^{-\chi_r/kT} \qquad (4.11)$$

or in logarithmic form as

$$\lg \frac{n_{r+1}}{n_r} = -\lg P_e - \frac{5040}{T} \chi_r + 2.5\lg T + \lg\left(\frac{U_{r+1}}{U_r} 2\right) - 0.48 \qquad (4.12)$$

All the Eqs. (4.10), (4.11) and (4.12) represent the relationship between occupation numbers of atoms in two adjacent ionized states, and are called the Saha equation.

In practice, sometimes we also need to know the ratio between the occupation number of ions at an arbitrary ionized state j and the total number of atoms:

$$\eta_j \equiv \frac{n_j}{n} = \frac{(n_j/n_{j-1})(n_{j-1}/n_{j-2})(n_{j-2}/n_{j-3})...(n_1/n_0)}{1+(n_1/n_0)+(n_2/n_1)(n_1/n_0)+...(n_n/n_{n-1})(n_{n-1}/n_{n-2})...(n_1/n_0)} \qquad (4.13)$$

To compute η_j, we need to substitute the Saha equation in all terms in the above equation.

4.2 LIMITATION OF THE SAHA EQUATION

We should be aware that the Saha equation is only valid within a certain density range. For example, the external part of the stellar atmosphere has an extremely low density and is a region which has deviated from thermal equilibrium (called a non-local thermal equilibrium region). Since the Saha equation is obtained based on thermal equilibrium conditions, the Saha equation is no longer valid here. Another example is found in the deep interior of a star. Consider the most abundant hydrogen in the Sun. According to the Saha equation, $(n_{H+}/n_{Ho}) \approx 1$ at $T \sim 10^4$ K, i.e., about half of the hydrogen atoms have been ionized. In the central region of the Sun, the Saha equation gives $(n_{H+}/n_{Ho}) \approx 3.17$, i.e., about 24% of the hydrogen atoms are still neutral atoms. In fact, however, all hydrogen atoms have already been ionized at this time, and (n_{H+}/n_{Ho}) should be infinite. To understand why the Saha equation is invalid in the very dense region, one should be aware that it has assumed no interactions among particles. When the density is larger than a certain limit, interactions among atoms can no longer be neglected so the Saha equation is also invalid. Mutual interactions among atoms can reduce the occupation probability of bound states of atoms, which causes more atoms to stay at ionized states. To illustrate this, we consider a gas made up of pure hydrogen. The average distance between hydrogen atoms in this gas is d, and the orbital radius for electrons in hydrogen atoms is a, so

$$d \approx \left(\frac{3}{4\pi n_H} \right)^{1/3}, \quad a = a_0 s^2 \qquad (4.14)$$

where n_H is the number of hydrogen atoms in 1 cm^3, $a_0 = 5.3 \times 10^{-9}$ cm is the Bohr radius of the hydrogen atom and s is a quantum number.

If the density of gas increases so that d is comparable to a, bound states with quantum numbers larger than that (s) corresponding to a should not exist. We can use $a < d/2$ as a crude condition to calculate the quantum number of the highest bound state. This condition can be written as

$$s^2 < \left(\frac{3}{4\pi n_H} \right)^{1/3} \frac{1}{2a_0} \qquad (4.15)$$

From the density at the center of the Sun given by $\rho_c \approx 170$ g/cm^3, we obtain $n_H = \rho_c/m_p \sim 10^{26}$cm^{-3}, so from Eq. (4.15), $s^2 < 0.13$. This shows that the ground state for hydrogen atoms does not exist, so all hydrogen atoms have been ionized.

4.3 EQUATION OF STATE FOR THE MATTER IN THE INTERIOR OF STARS

4.3.1 Equation of State for Completely Ionized Ideal Gas

(A) Total pressure

The total pressure at a position inside a star is made up of the gas pressure P_g and the radiation pressure P_R at that position. The gas pressure can be further divided into the ion pressure P_I and the electron pressure P_e:

$$P = P_g + P_R = P_I + P_e + P_R \qquad (4.16)$$

The radiation pressure can be derived as follows. For an isotropic homogeneous radiation field, the entropy can be written as

$$ds = \frac{dQ}{T} = \frac{1}{T}(dU + p_R dV)$$
$$= \frac{1}{T}\left[d(V \cdot u) + \frac{1}{3}u dV \right] \qquad (4.17)$$

where U and u are the internal energy and the energy density of the radiation field, respectively. From Eq. (2.54), $P_R = u/3$ for an isotropic and homogeneous radiation field. From Eq. (4.17), we have

$$ds = \frac{V}{T}\frac{du}{dT}dT + \frac{4}{3}\frac{u}{T}dV \qquad (4.18)$$

Since entropy is a total differential, we obtain

$$\frac{1}{T}\frac{du}{dT} = \frac{4}{3}\left(\frac{du}{dT}\cdot\frac{1}{T} - \frac{u}{T^2} \right)$$

or

$$\frac{du}{u} = 4\frac{dT}{T}$$

so

$$u = aT^4 \tag{4.19}$$

From Eq. (2.54), the relationship between the total radiation intensity I and the energy density u for an isotropic and homogeneous radiation field is $I = cu/4\pi$, so we get from Eq. (4.19) $I = acT^4/4\pi$, and the total radiation flux can be written as

$$\pi F = \int_0^{\pi/2} I\cos\theta 2\pi\sin\theta d\theta = \frac{ac}{4}T^4 \tag{4.20}$$

According to the Stefan–Boltzmann law in Eq. (2.57), $\pi F = \sigma T^4$, so we evaluate the coefficient $a = 4\sigma/c$. The radiation pressure P_R is expressed as

$$P_R = \frac{1}{3}u = \frac{1}{3}aT^4 \tag{4.21}$$

The gas pressure, ion pressure and electron pressure for a completely ionized ideal gas can be written as

$$P_g = n_g kT = \frac{\Re}{\mu}\rho T \tag{4.22}$$

$$P_I = n_I kT = \frac{\Re}{\mu_I}\rho T \tag{4.23}$$

$$P_e = n_e kT = \frac{\Re}{\mu_e}\rho T \tag{4.24}$$

where n_g, n_I and n_e represent the numbers of particles, ions and electrons in a unit volume, respectively; μ, μ_I and μ_e the mean molecular weights for particles, ions and electrons respectively; $\rho = n\mu m_p$, $m_p = 1.6735 \times 10^{-24}$ g is the proton mass, $\Re = k/m_p = 8.314 \times 10^7 \mathrm{ergK^{-1}g^{-1}}$ is the universal gas constant. Therefore, Eq. (4.16) can be written as

$$P = \frac{\Re}{\mu}\rho T + \frac{a}{3}T^4 = \frac{\Re}{\mu_I}\rho T + \frac{\Re}{\mu_e}\rho T + \frac{a}{3}T^4 \tag{4.25}$$

which is the equation of state for a completely ionized ideal gas, and gives the relationship among the density ρ, pressure P, temperature T and chemical composition. For convenience in calculations, we introduce a quantity β as

$$\beta = \frac{P_g}{P} \tag{4.26}$$

so from Eq. (4.25) we have

$$1 - \beta = \frac{aT^4}{3P} \tag{4.27}$$

The above discussions are valid for a completely ionized ideal gas. If the gaseous matter is not completely ionized, or when degeneracy occurs, the expression for the electron pressure in Eq. (4.25) should be changed.

(B) Mean molecular weight

To obtain expressions for the three mean molecular weights, μ, μ_I and μ_e, which appeared in the above equation for pressure, we analyze a mixed gas made of i types of different elements. We denote the molecular weight, total electron number, partial density and abundance for the ith element as μ_i, Z_i, ρ_i and X_i, respectively. Apparently, $X_i = \rho_i/\rho$. For complete ionization, the nucleon number per unit volume of the ith element is

$$n_{Ki} = \frac{\rho_i}{\mu_i m_P} = \frac{X_i \rho}{\mu_i m_P} \tag{4.28}$$

The number of electrons released from atoms of the ith element is

$$n_{ei} = n_{Ki} \cdot Z_i \tag{4.29}$$

so the total number of ions, electrons and particles in a unit volume of the mixed gas are, respectively,

$$n_I = \sum_i n_{Ki} = \sum_i \frac{X_i \rho}{\mu_i m_P} \tag{4.30}$$

$$n_e = \sum_i n_{ei} = \sum_i \frac{X_i Z_i \rho}{\mu_i m_P} \tag{4.31}$$

$$n_g = n_I + n_e = \sum_i (1 + Z_i) \frac{X_i \rho}{\mu_i m_p} \tag{4.32}$$

Substituting Eq. (4.32) into Eq. (4.22), we have

$$P_g = n_g kT = \Re \sum_i \frac{X_i(1 + Z_i)}{\mu_i} \rho T \tag{4.33}$$

from which we can obtain the mean molecular weight for particles as

$$\mu = \frac{1}{\sum_i \frac{(1 + Z_i)}{\mu_i} X_i} \tag{4.34}$$

Similarly, substituting Eq. (4.30) into Eq. (4.23), we can obtain the mean molecular weight for ions as

$$\mu_I = \frac{1}{\sum_i \frac{X_i}{\mu_i}} \tag{4.35}$$

Substituting Eq. (4.31) into (4.24), we can obtain the mean molecular weight for electrons as

$$\mu_e = \frac{1}{\sum_i \frac{X_i Z_i}{\mu_i}} \tag{4.36}$$

We consider a star mainly composed of hydrogen, helium and a small amount of heavy elements, of which the abundances are represented by X_H, X_{He} and X_Z. For heavy elements, most of the atomic nuclei are constituted by the same number of protons and neutrons, and the electron number equals the proton number, so on using Eqs. (4.34) to (4.36), we have

$$\mu = \frac{1}{2X_H + \frac{3}{4}X_{He} + \frac{1}{2}X_z} = \frac{1}{0.5 + 1.5X_H + 0.25X_{He}} \tag{4.37}$$

$$\mu_e = \frac{1}{X_H + \frac{1}{2}X_{He} + \frac{1}{2}X_z} = \frac{2}{1 + X_H} \tag{4.38}$$

$$\mu_I = \frac{1}{X_H + \dfrac{1}{4}X_{He} + \dfrac{X_z}{\mu_z}} \tag{4.39}$$

where we have made use of the relationship $X_H + X_{He} + X_Z = 1$, and μ_Z represents the mean molecular weight of heavy elements.

(C) Expressions of some thermodynamic quantities

In discussing the interior structure, evolution and oscillation etc., of stars, we normally need to know the expression for some thermo-dynamic quantities such as the internal energy U, specific heat at constant pressure C_P and the adiabatic temperature gradient ∇_{ad}, which can be obtained through the equation of state. For a completely ionized ideal gas, from the equation of state in Eq. (4.25), we have

$$\rho = \frac{\mu}{\Re}\frac{1}{T}\left(P - \frac{aT^4}{3}\right) = \frac{\mu P}{\Re T}\beta \tag{4.40}$$

so

$$d\ln\rho = d\ln\beta + d\ln P - d\ln T + d\ln\mu$$
$$= \left[1 + \left(\frac{\partial\ln\beta}{\partial\ln P}\right)_T\right]d\ln P - \left[1 - \left(\frac{\partial\ln\beta}{\partial\ln T}\right)_P\right]d\ln T + d\ln\mu \tag{4.41}$$

where

$$\left(\frac{\partial\ln\beta}{\partial\ln P}\right)_T = -\frac{P}{\beta}\frac{\partial(1-\beta)}{\partial P} = \frac{1-\beta}{\beta} \tag{4.42}$$

$$\left(\frac{\partial\ln\beta}{\partial\ln T}\right)_P = -\frac{T}{\beta}\frac{\partial(1-\beta)}{\partial T} = \frac{4(\beta-1)}{\beta} \tag{4.43}$$

Substituting Eqs. (4.42) and (4.43) into Eq. (4.41), we have

$$d\ln\rho = \frac{1}{\beta}d\ln P - \frac{4-3\beta}{\beta}d\ln T + d\ln\mu \tag{4.44}$$

On the other hand, $d\ln\rho$ can be written as Eq. (3.15). Comparing Eqs. (4.44) and (3.15), we have

$$\alpha = \frac{1}{\beta} \ , \quad \delta = \frac{4 - 3\beta}{\beta} \ , \quad \varphi = 1 \tag{4.45}$$

In fact, if the radiation pressure can be neglected, i.e., $\beta = 1$, we can obtain $\alpha = \delta = 1$, which is the result for a monatomic ideal gas. If the gas is monatomic, the internal energy per unit mass is composed of the kinetic energy, radiation energy and ionisation energy. It can be written as

$$u = \frac{3}{2} kT \frac{n_g}{\rho} + \frac{aT^4}{\rho} + \frac{1}{k} [\nu_H \chi_H + \nu_{He} \chi_{He} + \nu_{He} \chi_{He^+}]$$

$$= \frac{\Re T}{\mu} \left[\frac{3}{2} + \frac{3(1 - \beta)}{\beta} \right] + \frac{1}{k} [\nu_H \chi_H + \nu_{He} \chi_{He} + \nu_{He} \chi_{He^+}] \tag{4.46}$$

where the third term on the right side of Eq. (4.47) is the ionization energy (see Eq. (4.67)) and we have used the relationship

$$\frac{aT^4}{\rho} = \frac{3P_R}{\rho} = \frac{3(1 - \beta)P}{\rho} = \frac{3(1 - \beta)}{\beta} \frac{P_g}{\rho} = \frac{3(1 - \beta)}{\beta} \frac{\Re T}{\mu}$$

From Eqs. (4.46) and (4.43), we have

$$\left(\frac{\partial u}{\partial T} \right)_P = \frac{\Re}{\mu} \left[\frac{3}{2} + \frac{3(4 + \beta)(1 - \beta)}{\beta^2} \right] \tag{4.47}$$

From the first law of thermodynamics,

$$dQ = du - \frac{P}{\rho^2} d\rho \tag{4.48}$$

Considering $C_P = (\partial Q/\partial T)_P$, from Eq. (4.48),

$$C_P = \left(\frac{\partial Q}{\partial T} \right)_P - \frac{P}{\rho^2} \left(\frac{\partial \rho}{\partial T} \right)_P = \left(\frac{\partial u}{\partial T} \right)_P - \frac{P}{\rho T} \left(\frac{\partial \ln \rho}{\partial \ln T} \right)_P \tag{4.49}$$

Substituting $\delta = -\left(\frac{\partial \ln \rho}{\partial \ln T} \right)_P = \frac{4 - 3\beta}{\beta}$ of Eqs. (4.45) and (4.46) into Eq. (4.50), we have

$$C_P = \frac{\Re}{\mu} \left[\frac{3}{2} + \frac{3(4 + \beta)(1 - \beta)}{\beta^2} + \frac{4 - 3\beta}{\beta^2} \right] \tag{4.50}$$

Furthermore, from Eq. (3.34), the adiabatic temperature gradient can be written as

$$\nabla_{ad} = \frac{\delta P}{C_P T \rho} = \frac{\Re \delta}{\beta \mu C_P} \qquad (4.51)$$

Substituting Eqs. (4.46) and (4.51) into Eq. (4.52), we obtain

$$\nabla_{ad} = \frac{1 + (1 - \beta)(4 + \beta)/\beta^2}{\frac{5}{2} + 4(1 - \beta)(4 + \beta)/\beta^2} \qquad (4.52)$$

When the radiation pressure is extremely small, i.e., $\beta \to 1$, Eqs. (4.50) and (4.52) give $C_P = 5\Re/2\mu$ and $\Delta_{ad} = 0.4$, which are the values for an monatomic ideal gas.

4.3.2 Equation of State for a Partially Ionized Ideal Gas

(A) Equation of state

The temperature of the envelope of a star is relatively low, and some of the elements are only partially ionized. The case for partial ionization is different from the case for complete ionization. The number of electrons in the gas particles is now smaller and hence the electron pressure is smaller, but the partial pressure of ions and the radiation pressure do not change. If the number of electrons released by an average atom is E' in a partial ionization, the total pressure can be written as

$$P = \frac{\Re}{\mu_I} \rho T + \frac{\Re E'}{\mu_I} \rho T + \frac{a}{3} T^4 \qquad (4.53)$$

which is the equation of state for a partially ionized ideal gas. From this, we know that E' is critical in setting up the equation of state. We now discuss how we can use the Saha equation to calculate E' for a gas mixture made of hydrogen and helium.

We consider a gas mixture in which the abundances of hydrogen and helium are X_H and X_{He}. We denote η_H as the degree of ionization for hydrogen, and η_{He} and η_{He^+} for the primary and secondary ionized helium, respectively, i.e.,

$$\eta_H = \frac{n_{H^+}}{n_H}, \quad \eta_{He} = \frac{n_{He^+}}{n_{He}}, \quad \eta_{He^+} = \frac{n_{He^{++}}}{n_{He}} \qquad (4.54)$$

where $n_H = n_{H^o} + n_{H^+}$, $n_{He} = n_{He^o} + n_{He^+} + n_{He^{++}}$. Furthermore, the ratios of hydrogen and helium atoms to the total number of atoms in the gas mixture are denoted by ν_H and ν_{He}. From Eq. (4.30), we get

$$\nu_H = \frac{n_H}{n_H + n_{He}} = \frac{4X_H}{4X_H + X_{He}} \tag{4.55}$$

$$\nu_{He} = \frac{n_{He}}{n_H + n_{He}} = \frac{X_{He}}{4X_H + X_{He}} \tag{4.56}$$

The number of electrons E' released by an atom in the gas mixture can then be written as

$$E' = \nu_H \eta_H + \nu_{He}(\eta_{He} + \eta_{He^+}) + \nu_{He}\eta_{He^+} \tag{4.57}$$

We apply the Saha equation for hydrogen and helium. For hydrogen, since

$$\frac{n_{H^+}}{n_{H^\circ}} = \frac{n_{H^+}}{n_H - n_{H^+}} = \frac{n_{H^+}/n_H}{1 - n_{H^+}/n_H} = \frac{\eta_H}{1 - \eta_H} \tag{4.58}$$

and since $P_e = E'P_I$, $P_g = P_I + P_e = (1 + E')P_I$,

$$P_e = \frac{E'}{1 + E'}P_g = \frac{E'\beta}{1 + E'}P \tag{4.59}$$

Further defining

$$K_H = \frac{2}{\beta P} \frac{(2\pi m_e)^{3/2}(kT)^{5/2}}{h^3} e^{-\chi_H/kT} \tag{4.60}$$

substituting Eqs. (4.58) and (4.60) into the Saha equation in Eq. (4.11), and considering $\frac{U_{H^+}}{U_{H^\circ}} = \frac{1}{2}$, we have

$$\frac{\eta_H}{1 - \eta_H} \cdot \frac{E'}{1 + E'} = \frac{1}{2}K_H \tag{4.61}$$

Similarly, for helium,

$$\frac{n_{He^+}}{n_{He^\circ}} = \frac{n_{He^+}/n_{He}}{1 - n_{He^+}/n_{He} - n_{He^{++}}/n_{He}} = \frac{\eta_{He}}{1 - \eta_{He} - \eta_{He^+}} \tag{4.62}$$

$$\frac{n_{He^{++}}}{n_{He^+}} = \frac{n_{He^{++}}/n_{He}}{n_{He^+}/n_{He}} = \frac{\eta_{He^+}}{\eta_{He}} \tag{4.63}$$

Substituting Eqs. (4.62) and (4.63) into the Saha equation in Eq. (4.11), and considering $\frac{U_{He^+}}{U_{He^o}} = 2$, $\frac{U_{H^{++}}}{U_{H^+}} = \frac{1}{2}$, we have

$$K_{He} = \frac{2}{\beta P} \frac{(2\pi m_e)^{3/2}(kT)^{5/2}}{h^3} e^{-\chi_{He}/kT}$$

$$K_{He^+} = \frac{2}{\beta P} \frac{(2\pi m_e)^{3/2}(kT)^{5/2}}{h^3} e^{-\chi_{He^+}/kT}$$

and

$$\frac{\eta_{He}}{1 - \eta_{He} - \eta_{He^+}} \frac{E'}{1 + E'} = 2K_{He} \qquad (4.64)$$

$$\frac{\eta_{He^+}}{\eta_{He}} \frac{E'}{1 + E'} = \frac{1}{2} K_{He^+} \qquad (4.65)$$

In the above, there are four equations (4.57), (4.61), (4.64) and (4.65) consisting of four unknowns η_H, η_{He}, η_{He^+} and E'. At a certain position inside a star, when P and T are known, K_H, K_{He} and K_{He^+} can be computed. To solve the four equations, we first give a seed value of E', denoted by $E' = E^{(1)}$. Substituting $E^{(1)}$ into Eqs. (4.62), (4.65) and (4.66), we obtain a set of η_H, η_{He}, η_{He^+} values. Substituting these η_H, η_{He}, η_{He^+} values back into Eq. (4.57), we can get a new value of E', denoted as $E^{(2)}$. If $E^{(2)}$ is not equal to $E^{(1)}$, the initial value of $E^{(1)}$ is not accurate enough, so we put $E^{(2)}$ into Eqs. (4.61), (4.64) and (4.65) and repeat the above computations. Iterations will go on until the E' evaluated from Eq. (4.57) differs from the one just substituted into Eqs. (4.61), (4.64) and (4.65) by an acceptably small precision, and the corresponding E' is the value we require.

(B) Expressions for some thermodynamic quantities for partial ionization

The internal energy is composed by the kinetic energy, ionization energy and radiation energy. For a partial ionization of the gas, it can be written as

$$u = \frac{1}{\mu m_p} \cdot \frac{3}{2} kT + \frac{1}{\mu_I m_p} \left\{ \nu_H \eta_H \chi_H + \nu_{He} \left[\eta_{He} \chi_{He} + \eta_{He^+} \cdot (\chi_{He} + \chi_{He^+}) \right] \right\} + \frac{aT^4}{\rho}$$

$$= \frac{3}{2} \frac{1}{\mu} \Re T + \frac{\Re}{\mu_I k} \left[\nu_H \eta_H \chi_H + \nu_{He}(\eta_{He} + \eta_{He^+})\chi_{He} + \nu_{He}\eta_{He^+} \chi_{He^+} \right]$$

$$+ \frac{3(1-\beta)}{\beta} \frac{\Re}{\mu} T$$

$$= \frac{\mathfrak{R}}{\mu_I} \left\{ \left[\frac{3}{2} + \frac{3(1-\beta)}{\beta} \right] (1+E')T + \frac{1}{k} \left[\nu_H \eta_H \chi_H + \nu_{He}(\eta_{He} + \eta_{He^+}) \chi_{He} \right. \right.$$
$$\left. \left. + \nu_{He} \eta_{He^+} \chi_{He^+} \right] \right\} \tag{4.66}$$

From Eq. (4.49), we have

$$C_P = \left(\frac{\partial u}{\partial T} \right)_P + \frac{P}{\rho T} \delta \tag{4.67}$$

where $(\partial U / \partial T)_P$ can be evaluated through Eq. (4.66) as

$$\left(\frac{\partial u}{\partial T} \right)_P = \frac{\mathfrak{R}}{\mu_I} \left\{ \left[\frac{3}{2} + \frac{3(1-\beta)(4+\beta)}{\beta^2} \right] (1+E') + \left[\frac{3}{2} + \frac{3(1-\beta)}{\beta} \right] T \left(\frac{\partial E'}{\partial T} \right)_P \right.$$
$$\left. + \frac{1}{k} \left[\nu_H \chi_H \left(\frac{\partial \eta_H}{\partial T} \right)_P + \nu_{He} \chi_{He} \left(\frac{\partial (\eta_{He} + \eta_{He^+})}{\partial T} \right)_P + \nu_{He} \chi_{He} \left(\frac{\partial \eta_{He^+}}{\partial T} \right)_P \right] \right\} \tag{4.68}$$

To obtain an expression for δ in Eq. (4.67), we have to make use of the following relationships:

$$P = \frac{1}{\beta} \frac{\mathfrak{R}}{\mu} \rho T \tag{4.69}$$

$$\mu = \frac{\mu_I}{1 + E'} \tag{4.70}$$

From Eq. (4.70), we obtain

$$\mu_T = \left(\frac{\partial \ln \mu}{\partial \ln T} \right)_P = -\frac{T}{1+E'} \left(\frac{\partial E'}{\partial T} \right)_P$$
$$\mu_P = \left(\frac{\partial \ln \mu}{\partial \ln P} \right)_T = -\frac{P}{1+E'} \left(\frac{\partial E'}{\partial P} \right)_T \tag{4.71}$$

From Eq. (4.70) and substituting Eqs. (4.42) and (4.43), we get

$$\delta = -\left(\frac{\partial \ln \rho}{\partial \ln T} \right)_P = 1 - \mu_T - \left(\frac{\partial \ln \beta}{\partial \ln T} \right)_P$$
$$= \frac{4 - 3\beta}{\beta} + \left(\frac{\partial E'}{\partial T} \right)_P \frac{T}{1+E'} \tag{4.72}$$

$$\alpha = \left(\frac{\partial \ln \rho}{\partial \ln P}\right)_T = 1 + \mu_P + \left(\frac{\partial \ln \beta}{\partial \ln P}\right)_T$$

$$= \frac{1}{\beta} - \frac{P}{1+E'}\left(\frac{\partial E'}{\partial P}\right)_T \qquad (4.73)$$

where $(\partial E'/\partial T)_P$ and $(\partial E'/\partial P)_T$ can be determined through Eq. (4.57) as

$$\left(\frac{\partial E'}{\partial T}\right)_P = v_H\left(\frac{\partial \eta_H}{\partial T}\right)_P + v_{He}\left[\left(\frac{\partial \eta_{He}}{\partial T}\right)_P + 2\left(\frac{\partial \eta_{He^+}}{\partial T}\right)_P\right] \qquad (4.74)$$

$$\left(\frac{\partial E'}{\partial P}\right)_T = v_H\left(\frac{\partial \eta_H}{\partial P}\right)_T + v_{He}\left[\left(\frac{\partial \eta_{He}}{\partial P}\right)_T + 2\left(\frac{\partial \eta_{He^+}}{\partial P}\right)_T\right] \qquad (4.75)$$

where $(\partial \eta_H/\partial T)_P$, $(\partial \eta_H/\partial P)_T$, $(\partial \eta_{He}/\partial T)_P$, $(\partial \eta_{He}/\partial P)_T$, $(\partial \eta_{He^+}/\partial T)_P$ and $(\partial \eta_{He^+}/\partial P)_T$ can be obtained from differentiation of Eqs. (4.61), (4.64) and (4.65). In this way, we finally obtain equations for C_P, δ and α as

$$C_P = \frac{\Re}{\mu_I}\left\{\left[\frac{5}{2} + \frac{4(1-\beta)(4+\beta)}{\beta^2}\right](1+E') + \frac{v_H}{G_H}\phi_H^2 + \frac{v_{He}}{G_{He}}\phi_{He}^2 + \frac{v_{He}}{G_{He^+}}\phi_{He^+}^2\right\} \qquad (4.76)$$

$$\delta = \frac{4-3\beta}{\beta} + \frac{1}{1+E'}\left[\frac{v_H}{G_H}\phi_H + \frac{v_{He}}{G_{He}}\phi_{He} + \frac{v_{He}}{G_{He^+}}\phi_{He^+}\right] \qquad (4.77)$$

$$\alpha = \frac{1}{\beta}\left[1 + \frac{1}{1+E'}\left(\frac{v_H}{G_H} + \frac{v_{He}}{G_{He}} + \frac{v_{He}}{G_{He^+}}\right)\right] \qquad (4.78)$$

where

$$\phi_H = \frac{5}{2} + \frac{4(1-\beta)}{\beta} + \frac{\chi_H}{kT} \qquad (4.79)$$

$$\phi_{He} = \frac{5}{2} + \frac{4(1-\beta)}{\beta} + \frac{\chi_{He}}{kT} \qquad (4.80)$$

$$\phi_{He^+} = \frac{5}{2} + \frac{4(1-\beta)}{\beta} + \frac{\chi_{He^+}}{kT} \qquad (4.81)$$

$$G_H = \frac{1}{\eta_H(1-\eta_H)} + \frac{v_H}{E'(1+E')} \qquad (4.82)$$

$$G_{He} = \frac{1}{(\eta_{He} + \eta_{He^+})(1 - \eta_{He} - \eta_{He^+})} + \frac{v_{He}}{E'(1+E')} \qquad (4.83)$$

$$G_{He^+} = \frac{1}{\eta_{He^+}(1-\eta_{He^+})} + \frac{\nu_{He}}{E'(1+E')} \tag{4.84}$$

Substituting Eqs. (4.76) and (4.77) into Eq. (4.51), we get

$$\nabla_{ad} = \frac{1+E'}{\beta}\frac{\delta}{C_P\frac{\mu_I}{\Re}} = \frac{\left[1+\frac{(1-\beta)(4+\beta)}{\beta^2}\right](1+E')+\frac{1}{\beta}\left[\frac{\nu_H\phi_H}{G_H}+\frac{\nu_{He}\phi_{He}}{G_{He}}+\frac{\nu_{He}\phi_{He^+}}{G_{He^+}}\right]}{\left[\frac{5}{2}+\frac{4(1-\beta)(4+\beta)}{\beta^2}\right](1+E')+\frac{\nu_H\phi_H^2}{G_H}+\frac{\nu_{He}\phi_{He}^2}{G_{He}}+\frac{\nu_{He}\phi_{He^+}^2}{G_{He^+}}}$$

$$\tag{4.85}$$

4.3.3 Equation of State for Degenerate Gas

(A) Degenerate gas, Fermi distribution

From classical statistical physics, the momenta of electrons follow the Maxwell distribution

$$dN_m = N\frac{4\pi p^2}{(2\pi m_e kT)^{3/2}}e^{-p^2/2m_e kT}dp \tag{4.86}$$

where p is the electron momentum, dN_m is the electron number in a unit volume having a momentum in the range $(p, p + dp)$, N is the total electron number in the volume V. However, the classical Maxwell distribution is only valid for certain density and temperature ranges. If the density increases to a certain level, the momenta of electrons no longer follow the Maxwell distribution, but follow the quantum statistical Fermi distribution, i.e.,

$$dN_F = V\frac{8\pi p^2}{h^3}\frac{1}{e^{-\psi}e^{\varepsilon_l/kT}+1}dp \tag{4.87}$$

where $\varepsilon_l = p^2/2m_e$ and p is the electron momentum, dN_F is the number of electrons in the volume V having a momentum in the range $(p, p + dp)$.

A fundamental difference between the quantum statistics and the classical statistics is that the former should strictly obey the Pauli's exclusion principle which only allows the occupation number of a microscopic particle in every quantum state to be one or zero. For electrons, the Pauli exclusion principle only allows each phase cell in the phase space to hold maximally two electrons with different spins.

When the density is small, the average number of electrons in a phase cell is far smaller than 2, so the Pauli exclusion principle does not have any effects in practice, and the momenta of electrons can follow the classical Maxwell distribution. We refer to these electrons as non-degenerate. However, when the density increases to a certain extent such that almost every phase cell is occupied by two electrons, the momenta of electrons should follow the Fermi distribution. We refer to these electrons as degenerate.

The magnitude of the constant ψ in the Fermi distribution in Eq. (4.87) reflects the degree of degeneracy and is called the degeneracy parameter. When $\psi << 0$, $e^{-\psi}e^{\varepsilon_l/kT} >> 1$, so the Fermi distribution in Eq. (4.87) becomes the classical Maxwell distribution in Eq. (4.86). At this time, there is no or very weak degeneracy, and the electron occupation number in each phase cell is far smaller than two. When $\psi >> 0$, $e^{-\psi}e^{\varepsilon_l/kT} << 1$, so Eq. (4.87) shows that the electron occupation number in each phase cell is close to two and there is strong degeneracy.

We now discuss the general non-relativistic degenerate case, in which the electron speeds are far smaller than the speed of light. Integrating Eq. (4.87), we have

$$N = V \cdot \frac{8\pi}{h^3} \int\limits_0^\infty \frac{p^2}{e^{-\psi}e^{p^2/2m_ekT} + 1} dp \qquad (4.88)$$

Introducing the variable

$$x = \frac{P}{(2m_ekT)^{1/2}} \qquad (4.89)$$

Eq. (4.88) can be written as

$$N = V \cdot \frac{8\pi}{h^3}(2m_ekT)^{(3/2)} \int\limits_0^\infty \frac{x^2}{e^{-\psi}e^{x^2} + 1} dx \qquad (4.90)$$

The integration on the right should have a finite value, since there is only a finite number N on the left, and $T \neq 0$, $V \neq 0$. Writing

$$f(\psi) = \int\limits_0^\infty \frac{x^2}{e^{-\psi}e^{x^2} + 1} dx \qquad (4.91)$$

and considering

$$\frac{N}{V}\mu_e \cdot m_P = \rho \qquad (4.92)$$

Where m_P is the ion mass, we obtain from Eq. (4.90)

$$f(\psi) = \frac{h^3}{8\pi\mu_e m_P(2m_e k)^{3/2}} \cdot \frac{\rho}{T^{3/2}} \qquad (4.93)$$

from which we obtain the following results.

(i) When $\psi = $ constant, $f(\psi) = $ constant, so

$$\frac{\rho}{T^{3/2}} = \text{const} \qquad (4.94)$$

i.e., $\psi = $ constant represents a set of straight lines with a slope of 2/3 on the $\lg T - \lg \rho$ plane. Figure 4.2 shows the family of straight lines on the $\lg T - \lg\rho$ plane.

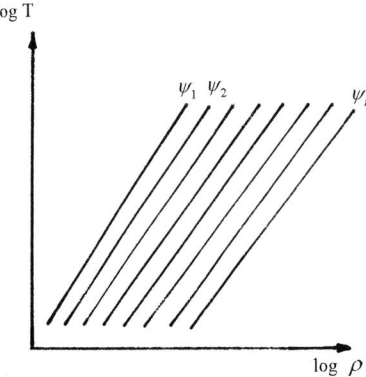

Fig. 4.2 The family of straight lines with a slope of 2/3 on the lgT-lgρ plane.

(ii) When the temperature T decreases or when the density ρ increases, ψ or the degree of degeneracy will increase.

(iii) For ions, Eq. (4.93) is still valid with μ_e and m_e replaced by μ_I and m_P for the ions. It is much more difficult for ions than electrons to become degenerate, which can be seen from the following consideration. For no degeneracy, the kinetic energy of the ions should be equal to that of electrons, so the momentum ratio should be $p_I/p_e = (m_P/m_e)^{1/2}$. On the other hand, the ion mass is larger than the electron mass by a factor of 10^3, so in the phase space, the number of phase cells for ions is larger than that for electrons by a factor of 10^5. Therefore, the density required for the ions to become degenerate is larger than that for electrons also by a factor of 10^5. Because of this, normally only electron degeneracy will occur deep inside the stars. Only in special stars with very large densities, such as neutron stars, will ion degeneracy occur.

(B) Equations for density ρ, electron pressure P_e and internal energy U_e under degeneracy

From Eqs. (4.80) and (4.92), the equation for the density is

$$\frac{\rho}{\mu_e} = \frac{8\pi m_P}{h^3} \int_0^\infty \frac{p^2}{e^{-\psi}e^{\varepsilon_l/kT} + 1}\, dp \qquad (4.95)$$

where ε_l is the kinetic energy of the electron. From the special theory of relativity, an electron with speed v has momentum p and energy ε_{tot} given by

$$p = \frac{m_e v}{\sqrt{1 - \dfrac{v^2}{c^2}}} \qquad (4.96)$$

$$\varepsilon_{tot} = \frac{m_e c^2}{\sqrt{1 - \dfrac{v^2}{c^2}}} = m_e c^2 \sqrt{1 + \frac{p^2}{m_e^2 c^2}} \qquad (4.97)$$

so the kinetic energy ε_l of the electron is

$$\varepsilon_l = \varepsilon_{tot} - m_e c^2 = m_e c^2 \left[\left(1 + \frac{p^2}{m_e^2 c^2}\right)^{1/2} - 1 \right] \qquad (4.98)$$

The equation for electron pressure can be obtained from the momentum flux flowing across 1 cm² per second:

$$P_e = \frac{1}{3} \int_0^\infty pv \frac{dN}{V} = \frac{8\pi}{3h^3} \int_0^\infty \frac{p^3 \dfrac{d\varepsilon_l}{dp}\, dp}{e^{-\psi}e^{\varepsilon_l/kT} + 1} \qquad (4.99)$$

The integrations on the right of Eqs. (4.95) and (4.99) are related to the value of the degeneracy parameter ψ.

Table 4.1 The constants a_i, ε_i and α_i for equations (4.100) and (4.101)

i	a_i	ε_i	α_i
1	7.28905×10^{-2}	0.11896	8.87844×10^{-1}
2	2.04648×10^{-1}	0.47652	6.20940×10^{-1}
3	2.54685×10^{-1}	1.07476	3.41379×10^{-1}
4	1.96668×10^{-1}	1.91722	1.47015×10^{-1}
5	1.04294×10^{-1}	3.00898	4.93418×10^{-2}
6	3.95716×10^{-2}	4.35703	1.28164×10^{-2}
7	1.09024×10^{-2}	5.96967	2.55507×10^{-3}
8	2.21504×10^{-3}	7.86147	3.85307×10^{-4}
9	3.14295×10^{-4}	10.02706	4.41879×10^{-5}
10	3.64498×10^{-5}	12.57936	3.44234×10^{-6}
11	2.42533×10^{-6}	15.19876	2.050763×10^{-7}
12	1.49830×10^{-7}	18.87814	6.32888×10^{-9}

If we write $e^{-\psi} \equiv 1/\Lambda$, according to Kippenhahn and Thomas (1964), when $0.001 < \Lambda < e^7$, Eqs. (4.95) and (4.99) can be approximated as

$$\frac{\rho}{\mu_e} = c_1 \gamma^{3/2} \sum_{i=1}^{12} a_i \frac{\left(1 + \frac{1}{2}\gamma\varepsilon_i\right)^{1/2}(1 + \gamma\varepsilon_i)}{\left(\frac{1}{\Lambda}\right) + \alpha_i} \tag{4.100}$$

$$P_e = c_2 \gamma^{5/2} \sum_{i=1}^{12} a_i \frac{\varepsilon_i\left(1 + \frac{1}{2}\gamma\varepsilon_i\right)^{3/2}}{\left(\frac{1}{\Lambda}\right) + \alpha_i} \tag{4.101}$$

where

$$\left.\begin{aligned} c_1 &= 8\pi\sqrt{2} \cdot m_P \left(\frac{m_e}{h}\right)^3 \\ c_2 &= \frac{16\pi}{3}\sqrt{2} \cdot m_e c^2 \left(\frac{m_e c}{h}\right)^3 \\ \gamma &= \frac{kT}{m_e c^2} \end{aligned}\right\} \tag{4.102}$$

and the constants a_i, ε_i and α_i can be taken from Table 4.1.

When $\Lambda > e^7$, according to Chandrasekhar (1939), Eqs. (4.95) and (4.99) can be approximated as

$$\frac{\rho}{\mu_e} = \frac{c_1}{\sqrt{2}} \left[\frac{1}{3} y^3 + \frac{\pi^2}{6} \frac{\gamma^2}{y} (1 + 2y^2) + \cdots \right] \qquad (4.103)$$

$$P_e = \frac{c_2}{2\sqrt{2}} \left[\frac{1}{8} f(y) + \frac{\pi^2}{6} \frac{\gamma^2}{y^3} f_2(y) + \cdots \right] \qquad (4.104)$$

where

$$\left. \begin{array}{c} y = [(1 + \gamma \ln \Lambda)^2 - 1]^{1/2} \\ f(y) = y(2y^2 - 3)(1 + y^2)^{1/2} - 3\ln((1 + y^2)^{1/2} - y) \\ f_2(y) = 3y^4 (1 + y^2)^{1/2} \end{array} \right\} \qquad (4.105)$$

We now consider the internal energy of electrons in a unit volume. Apparently, we have

$$\frac{U_e}{V} = \frac{8\pi}{h^3} \int_0^\infty \frac{p^2 \varepsilon_l dp}{e^{-\psi} e^{\varepsilon_l/kT} + 1} \qquad (4.106)$$

Similar to computations for density ρ and electron pressure P_e, our considerations are also based on the magnitude of Λ:

When $0.001 < \Lambda < e^7$, we have

$$\frac{\rho}{\mu_e} U_e = c_3 \gamma^{5/2} \sum_{i=1}^{12} a_i \frac{\varepsilon_i \left(1 + \frac{1}{2} \gamma \varepsilon_i \right)^{1/2} (1 + \gamma \varepsilon_i)}{\left(\frac{1}{\Lambda} \right) + \alpha_i} \qquad (4.107)$$

When $\Lambda > e^7$, we have

$$\frac{\rho}{\mu_e} U_e = \frac{c_3}{\sqrt{2}} \left[\frac{1}{8} g(y) + \frac{\pi^2}{6} \frac{\gamma^2}{y^3} g_2(y) + \cdots \right] \qquad (4.108)$$

where

$$\left. \begin{array}{c} c_3 = 8\pi\sqrt{2} \cdot m_e c \left(\frac{m_e c}{h} \right)^3 m_P/m_e, \\ 3 \cdot g(y) = 8y^3[(1 + y^2)^{1/2} - 1] - f(y), \\ g_2(y) = (3y^4 + y^2)(1 + y^2)^{1/2} - 2y^4 - y^2 \end{array} \right\} \qquad (4.109)$$

(C) Equation of state for complete degeneracy

For complete degeneracy, $\psi \gg 0$, and there exists a ψ which can be written as

$$\psi = \varepsilon_0/kT \tag{4.110}$$

so that

$$\frac{1}{\varepsilon^{-\psi}e^{\varepsilon_l/kT}+1} = \frac{1}{e^{\psi(\varepsilon_l/\varepsilon_0-1)}+1} \approx \begin{cases} 1, & \text{when } \varepsilon_l < \varepsilon_0 \\ 0, & \text{when } \varepsilon_l > \varepsilon_0 \end{cases} \tag{4.111}$$

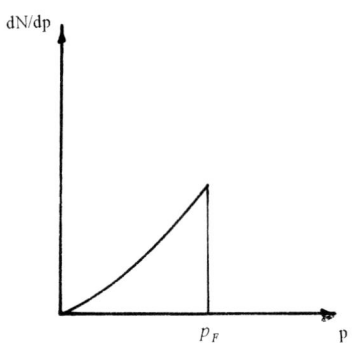

Employing the relationship between energy and momentum in Eq. (4.98), we can obtain the p value corresponding to ε_0, which is known as the Fermi momentum and denoted as p_F. From Eq. (4.87), we have

$$\frac{dN}{dp} = V \cdot \frac{8\pi p^2}{h^3} \frac{1}{e^{\psi(\varepsilon_l/\varepsilon_0-1)}+1} \tag{4.112}$$

Figure 4.3 gives the relationship between (dN/dp) and p. It can be observed that $(dN/dp) = 0$ at $p > p_F$, so we need only to integrate to p_F when integrating Eq. (4.87), i.e.,

Fig. 4.3 The relationship between dN/dp and p.

$$N = \int_0^{P_F} dN = V \cdot \frac{8\pi}{h^3} \int_0^{P_F} p^2 dp = V \frac{8\pi p_F^3}{3h^3} \tag{4.113}$$

According to $\rho = (N/V)\mu_e m_P$, we get

$$\frac{\rho}{\mu_e} = \frac{8\pi m_P}{3h^3} p_F^3 \tag{4.114}$$

For electron pressure P_e, we have

$$P_e = \frac{1}{3}\int_0^{P_F} pv\frac{dN}{N} = \frac{8\pi}{3h^3}\int_0^{P_F} p^3 v dp \tag{4.115}$$

(i) Non-relativistic degeneracy ($v \ll c$)
   Here $v = p/m_e$, so Eq. (4.115) becomes

$$P_e = \frac{8\pi}{15m_e h^3} p_F^5 \tag{4.116}$$

Substituting Eq. (4.114), we get

$$P_e = K_1 \left(\frac{\rho}{\mu_e} \right)^{5/3} \tag{4.117}$$

where

$$K_1 = \frac{h^2}{20 m_e m_P} \left(\frac{3}{\pi m_P} \right)^{2/3} = 9.91 \times 10^{12} [\text{c.g.s.}] \tag{4.118}$$

From Eq. (4.117), we see that electron pressure is only dependent on density ρ and is independent of temperature T.

It has been pointed out that ions are more difficult to become degenerate than electrons. Under normal circumstances, when electron degeneracy occurs in stars, the ions still behave as an ideal gas. Furthermore, Eq. (4.117) gives an electron pressure much larger than the ion pressure so that under normal circumstances for stars, when the electrons are completely degenerate, the total pressure is given by $P = P_e$. Thus, from Eq. (4.117),

$$P = K_1 \left(\frac{\rho}{\mu_e} \right)^{5/3} \tag{4.119}$$

which is the equation of state for complete non-relativistic degeneracy.

(ii) Relativistic degeneracy $(v \sim c)$

Here,

$$v = \frac{p}{m_e} = \frac{p}{m_0} \sqrt{1 - \frac{v^2}{c^2}} \tag{4.120}$$

from which we get

$$1 - \frac{v^2}{c^2} = \frac{1}{1 + \dfrac{p^2}{m_0^2 c^2}} \tag{4.121}$$

so the integration of Eq. (4.115) gives

$$P_e = \frac{8\pi}{3 m_0 h^3} \int_0^{p_F} \frac{p^4 dp}{\left(1 + \dfrac{p^2}{m_0^2 c^2} \right)^{1/2}} \tag{4.122}$$

Introducing the parameter

$$\xi = p_F/m_0 c \tag{4.123}$$

known as the "relativistic degeneracy parameter", Eq. (4.122) becomes

$$P_e = \frac{\pi m_0^4 c^5}{3h^3} f(\xi) = 10^{22.8} \cdot f(\xi) \tag{4.124}$$

where

$$f(\xi) = \xi(2\xi^2 - 3)(1 + \xi^2)^{1/2} + 3\sinh^{-1}\xi \tag{4.125}$$

When the electron speed equals the speed of light, we have

$$P_e = \frac{2\pi c p_F^4}{3h^3} \tag{4.126}$$

Substituting Eq. (4.114), we get

$$P_e = K_2 \left(\frac{\rho}{\mu_e} \right)^{4/3} \tag{4.127}$$

where

$$K_2 = \frac{hc}{8m_P} \left(\frac{3}{\pi m_P} \right)^{1/3} = 1.244 \times 10^{15} [\text{c.g.s.}] \tag{4.128}$$

Similarly, the equation of state for a complete relativistic degeneracy is obtained as

$$P = K_2 \left(\frac{\rho}{\mu_e} \right)^{4/3} \tag{4.129}$$

(D) Expressions for some thermodynamic properties for degenerate conditions

From Eq. (4.49), we have

$$C_P = \left(\frac{\partial u}{\partial T} \right)_P + \frac{P\delta}{\rho T} \tag{4.130}$$

From Eq. (4.51), we have

$$\nabla_{ad} = \frac{P}{\rho T} \cdot \frac{\delta}{C_P}$$

(4.131)

The u in Eq. (4.130) represents the internal energy of 1 gram of stellar matter, which is composed of kinetic energy and radiation energy of electrons and ions, i.e.,

$$u = \frac{3}{2} \frac{\Re}{\mu_I} T + u_e(P \cdot T) + \frac{aT^4}{\rho}$$

(4.132)

Partial differentiating with respect to u, we get

$$\left(\frac{\partial u}{\partial T} \right)_P = \frac{3}{2} \frac{\Re}{\mu_I} + \left(\frac{\partial u_e}{\partial T} \right)_P + \frac{4aT^3}{\rho} - \frac{aT^4}{\rho^2} \left(\frac{\partial \rho}{\partial T} \right)_P$$

(4.133)

Substituting Eq. (4.133) into (4.130), and using Eq. (4.26), we have

$$C_P = \frac{3}{2} \frac{\Re}{\mu_I} + \left(\frac{\partial u_e}{\partial T} \right)_P + \frac{P}{\rho T} \left[\delta + \frac{4aT^4}{P} - \frac{aT^4}{P} \cdot \frac{T}{\rho} \left(\frac{\partial \rho}{\partial T} \right)_P \right]$$

$$= \frac{3}{2} \frac{\Re}{\mu_I} + \left(\frac{\partial u_e}{\partial T} \right)_P + \frac{P}{\rho T} [(4 - 3\beta)\delta + 12(1 - \beta)]$$

(4.134)

We further introduce two quantities

$$\eta = \frac{\varepsilon}{kT} \ , \ \zeta = \frac{kT}{m_e c^2} \eta = \frac{\varepsilon}{m_e c^2}$$

(4.135)

From Eqs. (4.106) and (4.99), and considering that U_e/V in Eq. (4.106) represents the electron kinetic energy per unit volume while u_e in Eq. (4.134) is the kinetic energy of 1 gram of electron gas so their relationship is $u_e = U_e/V\rho$, we have

$$u_e = 3 \frac{I_1}{I_2} \frac{P_e}{\rho}$$

(4.136)

where I_1 and I_2 are

$$I_1 = \int_0^\infty \frac{(2\zeta + \zeta^2)^{1/2}(\zeta + \zeta^2)}{e^{\eta - \psi} + 1} d\zeta$$

(4.137)

$$I_2 = \int_0^\infty \frac{(2\zeta + \zeta^2)^{3/2}}{e^{\eta - \psi} + 1} d\zeta \tag{4.138}$$

For normal stars, the non-relativistic approximation is adopted, i.e., $\zeta << 1$ (valid for $\rho << 10^6$ g/cm^3, $T << 10^9$ K). Thus, the terms with ζ^2 in Eqs. (4.137) and (4.138) can be neglected, so

$$I_1 = \frac{1}{2} I_2 \tag{4.139}$$

$$u_e = \frac{3}{2} \frac{P_e}{\rho} \tag{4.140}$$

Since

$$P = \frac{\Re}{\mu_I} \rho T + P_e + \frac{a}{3} T^4 \text{ and } 1 - \beta = \frac{a}{3} \frac{T^4}{P}$$

we have

$$u_e = \frac{3}{2} \beta \frac{P}{\rho} - \frac{3}{2} \frac{\Re}{\mu_I} T \tag{4.141}$$

Partial differentiating with respect to u_e, we get

$$\left(\frac{\partial u_e}{\partial T} \right)_P = -\frac{3}{2} \frac{\Re}{\mu_I} + \frac{P}{\rho T} \left[\frac{3}{2} \beta \delta - 6(1 - \beta) \right] \tag{4.142}$$

Substituting into Eq. (4.134), we obtain

$$C_P = \frac{P}{\rho T} \left[\left(4 - \frac{3}{2} \beta \right) \delta + 6(1 - \beta) \right] \tag{4.143}$$

Substituting C_P into the expression of ∇_{ad} in Eq. (4.131), we have

$$\nabla_{ad} = \frac{\delta}{\left(4 - \frac{3}{2} \beta \right) \delta + 6(1 - \beta)} \tag{4.144}$$

from which we observe that, in the non-relativistic case, if the radiation pressure can be neglected ($\beta = 1$), $\nabla_{ad} = 0.4$ is independent of the degree of degeneracy.

4.3.4 Equation of State Considering Turbulent Pressure

In the above discussions of the equation of state for matter in the interior of the stars, turbulent pressure has been neglected. In other words, at a certain position inside a star, the total pressure has been assumed to arise only from the gas pressure and the radiation pressure. When intermediate-massive or low mass stars evolve to giant stars or AGB stars, the envelope is convective, and the energy transferred per second through a unit area through convection in the convective region is far greater than the energy transferred by radiation. Thus, the momentum transferred per second through a unit area through convection, i.e., the turbulent pressure, will be far greater than the momentum transferred by radiation, or the radiation pressure. When the turbulent pressure cannot be neglected, the total pressure should arise from the gas pressure, radiation pressure and turbulent pressure, i.e.,

$$P = P_g + P_R + P_t \qquad (4.145)$$

where the turbulent pressure P_t can be expressed as

$$P_t = Q\rho v_t^2 \qquad (4.146)$$

Q is a coefficient with a value of unity (De Jager 1980, Henyey *et al.* 1965), and v_t is the turbulent velocity to be determined from turbulent convection theories. There being no rigorous time dependent turbulent convection theories at the moment, Jiang and Huang (1997a) employed the basic equation set (see §3.3.2) of the mixed length convection theory and equated v_t as the convective velocity v to solve for an expression for v_t as

$$v_t^2 = \frac{g\delta\alpha^2 H_P}{8}\left(W - \frac{8}{27}U - \frac{E}{3W}\right)^2 \qquad (4.147)$$

where α, δ, g, H_p, W, U, E and A are the same as those in Eq. (3.75).

(A) Equation of state for completely ionized ideal gas and some thermodynamic quantities

When the gas is a completely ionized ideal gas, the total pressure can be written as

$$P = P_g + P_R + P_t = \frac{R}{\mu}\rho T + \frac{a}{3}T^4 + \rho v_t^2 \tag{4.148}$$

Introducing two dimensionless quantities

$$C_t = \frac{P_t}{P} \tag{4.149}$$

$$C = \frac{P_g + P_t}{P} \tag{4.150}$$

so from Eqs. (4.147) to (4.151), we obtain

$$C_t = \frac{\alpha^2 \delta}{8}\left(W - \frac{8}{27}U - \frac{E}{3W}\right)^2 \tag{4.151}$$

$$C = 1 - \frac{aT^4}{3P} \tag{4.152}$$

and the equation of state for a completely ionized ideal gas is obtained as

$$\rho = \frac{\mu}{\Re}(C - C_t)\frac{P}{T} \tag{4.153}$$

where μ is the mean molecular weight of a free gas particle [Eq. (4.34)], and $\Re$ is the gas constant.

Jiang and Huang (1997a) gave the expressions of some thermodynamic quantities such as C_P and ∇_{ad} under such conditions as

$$C_P = \frac{P}{\rho T}\left[\left(4 - \frac{3}{2}C + \frac{3}{2}C_t\right)\delta + 6(1 - C) + \frac{3}{2}\left(\frac{\partial C_t}{\partial \ln T}\right)_P\right] \tag{4.154}$$

$$\nabla_{ad} = \frac{\delta}{\left(4 - \frac{3}{2}C + \frac{3}{2}C_t\right)\delta + 6(1 - C) + \frac{3}{2}\left(\frac{\partial C_t}{\partial \ln T}\right)_P} \tag{4.155}$$

where

$$\alpha \equiv \left(\frac{\partial \ln \rho}{\partial \ln P}\right)_T = \frac{1 - C_t - \left(\frac{\partial C_t}{\partial \ln P}\right)_T}{C - C_t} \tag{4.156}$$

$$\delta \equiv -\left(\frac{\partial \ln \rho}{\partial \ln T}\right)_P = \frac{4 - 3C - 3C_t - \left(\frac{\partial C_t}{\partial \ln T}\right)_P}{C - C_t} \tag{4.157}$$

$$\left(\frac{\partial C_t}{\partial \ln T}\right)_P = C_t \left[a_1 \left(\frac{\partial \ln \rho}{\partial \ln T}\right)_P + a_2 \left(\frac{\partial \ln \kappa}{\partial \ln T}\right)_P + a_3 \left(\frac{\partial \nabla_{ad}}{\partial \ln T}\right)_P \right.$$
$$\left. + \frac{a_4}{\delta}\left(\frac{\partial \delta}{\partial \ln T}\right)_P + a_5 \right] \tag{4.158}$$

$$\left(\frac{\partial C_t}{\partial \ln P}\right)_T = C_t \left[a_1 \left(\frac{\partial \ln \rho}{\partial \ln P}\right)_T + a_2 \left(\partial \frac{\ln \kappa}{\partial \ln P}\right)_T + a_3 \left(\frac{\partial \nabla_{ad}}{\partial \ln P}\right)_T \right.$$
$$\left. + \frac{a_4}{\delta}\left(\frac{\partial \delta}{\partial \ln P}\right)_T + a_6 \right] \tag{4.159}$$

$$\left(\frac{\partial \nabla_{ad}}{\partial \ln T}\right)_P = \frac{\nabla_{ad}^2}{\delta^2}\left\{ -6(1-C)(4+\delta)\delta + \left[6(1-C)+\frac{3}{2}\left(\frac{\partial C_t}{\partial \ln T}\right)_P\right]\left(\frac{\partial \delta}{\partial \ln T}\right)_P \right.$$
$$\left. - \frac{3}{2}\delta^2 \left(\frac{\partial C_t}{\partial \ln T}\right)_P\right\} \tag{4.160}$$

$$\left(\frac{\partial \nabla_{ad}}{\partial \ln P}\right)_T = \frac{\nabla_{ad}^2}{\delta^2}\left\{ \frac{3}{2}(1-C)(4+\delta)\delta + \left[6(1-C)+\frac{3}{2}\left(\frac{\partial C_t}{\partial \ln T}\right)_P\right]\left(\frac{\partial \delta}{\partial \ln P}\right)_T \right.$$
$$\left. - \frac{3}{2}\delta^2 \left(\frac{\partial C_t}{\partial \ln P}\right)_T\right\} \tag{4.161}$$

$$\left(\frac{\partial \delta}{\partial \ln T}\right)_P = \frac{1}{(C-C_t)^2}\left\{ 4(1-C)\left[4(1-C_t)+\left(\frac{\partial C_t}{\partial \ln T}\right)_P\right] + \left[4(1-C)+ \right.\right.$$
$$\left.\left. \left(\frac{\partial C_t}{\partial \ln T}\right)_P\right]\left(\frac{\partial C_t}{\partial \ln T}\right)_P\right\} \tag{4.162}$$

$$\left(\frac{\partial \delta}{\partial \ln P}\right)_T = \frac{1}{(C-C_t)^2}\left\{ -(1-C)\left[4(1-C_t)+\left(\frac{\partial C_t}{\partial \ln T}\right)_P\right] + \left[4(1-C) \right.\right.$$
$$\left.\left. + \left(\frac{\partial C_t}{\partial \ln T}\right)_P\right]\left(\frac{\partial C_t}{\partial \ln P}\right)_T\right\} \tag{4.163}$$

$$\left.\begin{aligned}
&a_1 = \frac{1}{b_1}(-b_2 + b_3), \quad a_2 = -2a_1 + b_4 \nabla_R, \\
&a_3 = \frac{2a_1}{\nabla_{ad}} - b_4, \quad a_4 = 1 - 3a_1, \\
&a_5 = 8a_1 - 4b_4 \nabla_R, \quad a_6 = -5a_1 + b_4 \nabla_R. \\
&b_1 = W - \frac{8}{27}U - \frac{1}{W}\frac{E}{3}, \quad b_2 = \frac{8}{27}U + \frac{2}{W}\frac{E}{3}, \\
&b_3 = \frac{W}{3\sqrt{D}}\left(1 + \frac{1}{W^2}\frac{E}{3}\right)\left[\frac{3}{W^2}\left(\frac{E}{3}\right)^2 + \frac{A}{2} + 2U^3\left(\frac{19^3}{27^3} - \frac{1}{9}\right)\right], \\
&b_4 = \frac{8}{27}\frac{UW}{b_1 \sqrt{D}}\left(1 + \frac{1}{W^2}\frac{E}{3}\right).
\end{aligned}\right\} \tag{4.164}$$

(B) Equation of state for partially ionized ideal gas and some thermodynamic quantities

When the gas is a partially ionized ideal gas, the total pressure can be written as [see Eq. (4.54)]

$$P = P_g + P_R + P_t = \frac{(1+E')}{\mu_I} \Re \rho T + \frac{a}{3} T^4 + \rho v_t^2 \tag{4.165}$$

where μ_I is the mean molecular weight for ions [see Eq. (4.35)] and E' is the number of electrons released by an atom in the gas mixture [see Eq. (4.57)].

From Eqs. (4.165) and (4.146) to (4.150), the equation of state for a partially ionized ideal gas is obtained as

$$\rho = \frac{\mu'}{\Re}(C - C_t)\frac{P}{T} \tag{4.166}$$

where

$$\mu' = \frac{\mu_I}{(1+E')}$$

Jiang and Huang (1997a) gave expressions of some thermodynamic quantities for a partially ionized gas mixture composed of hydrogen and helium as

$$C_P = \frac{P}{\rho T}\left\{\left(4 - \frac{3}{2}C + \frac{3}{2}C_t\right)(\delta + \mu_T') + 6(1-C) - 4\mu_T'C_t + \left(\frac{3}{2} + \mu_T'\right)\left(\frac{\partial C_t}{\partial \ln T}\right)_P\right.$$
$$\left. + \frac{C - C_t}{1 + E'}\left(\frac{v_H}{G_H}\phi_H^2 + \frac{v_{He}}{G_{He}}\phi_{He}^2 + \frac{v_{He}}{G_{He^+}}\phi_{He^+}^2\right)\right\} \tag{4.167}$$

$$\nabla_{ad} =$$
$$\frac{\delta}{\left(4 - \frac{3}{2}C + \frac{3}{2}C_t\right)(\delta + \mu_T') + 6(1-C) - 4\mu_T'C_t + \left(\frac{3}{2} + \mu_T'\right)\left(\frac{\partial C_t}{\partial \ln T}\right)_P + \frac{C - C_t}{1 + E'}\left(\frac{v_H}{G_H}\phi_H^2 + \frac{v_{He}}{G_{He}}\phi_{He}^2 + \frac{v_{He}}{G_{He^+}}\phi_{He^+}^2\right)} \tag{4.168}$$

$$\alpha = \mu_P' + \frac{1 + C_t - \left(\frac{\partial C_t}{\partial \ln P}\right)_T}{C - C_t} \quad , \quad \delta = -\mu_T' + \frac{4 - 3C - C_t + \left(\frac{\partial C_t}{\partial \ln T}\right)_P}{C - C_t} \tag{4.169}$$

$$\mu_P' \equiv \left(\frac{\partial \ln \mu'}{\partial \ln P}\right)_T = \frac{1 - C_t - \left(\frac{\partial C_t}{\partial \ln P}\right)_T}{(1 + E')(C - C_t)}\left(\frac{v_H}{G_H} + \frac{v_{He}}{G_{He}} + \frac{v_{He}}{G_{He^+}}\right) \tag{4.170}$$

$$\mu'_T \equiv \left(\frac{\partial \ln \mu'}{\partial \ln T}\right)_P = \frac{-1}{(1+E')}\left(\frac{v_H}{G_H}\phi_H + \frac{v_{He}}{G_{He}}\phi_{He} + \frac{v_{He}}{G_{He^+}}\phi_{He^+}\right) \quad (4.171)$$

$$\frac{v_H}{G_H} = \frac{v_H \eta_H (1-\eta_H)}{1+\dfrac{v_H \eta_H (1-\eta_H)}{E'(1+E')}} \quad (4.172)$$

$$\frac{v_{He}}{G_{He}} = \frac{v_{He}(\eta_{He}+\eta_{He^+})[1-(\eta_{He}+\eta_{He^+})]}{1+\dfrac{v_{He}(\eta_{He}+\eta_{He^+})[1-(\eta_{He}+\eta_{He^+})]}{E'(1+E')}} \quad (4.173)$$

$$\frac{v_{He}}{G_{He^+}} = \frac{v_{He}\eta_{He^+}(1-\eta_{He^+})}{1+\dfrac{v_{He}\eta_{He^+}(1-\eta_{He^+})}{E'(1+E')}} \quad (4.174)$$

$$\left.\begin{aligned}
\phi_H &= \frac{5}{2} + \frac{1}{C-C_t}\left[4(1-C)+\left(\frac{\partial C_t}{\partial \ln T}\right)_P\right] + \frac{\chi_H}{kT}\\[6pt]
\phi_{He} &= \frac{5}{2} + \frac{1}{C-C_t}\left[4(1-C)+\left(\frac{\partial C_t}{\partial \ln T}\right)_P\right] + \frac{\chi_{He}}{kT}\\[6pt]
\phi_{He^+} &= \frac{5}{2} + \frac{1}{C-C_t}\left[4(1-C)+\left(\frac{\partial C_t}{\partial \ln T}\right)_P\right] + \frac{\chi_{He^+}}{kT}
\end{aligned}\right\} \quad (4.175)$$

$$\left.\begin{aligned}
v_H &= \frac{4X_H}{4X_H+X_{He}}, \quad v_{He} = \frac{X_{He}}{4X_H+X_{He}}\\[6pt]
\eta_H &= \frac{n_{H^+}}{n_H}, \quad \eta_{He} = \frac{n_{He^+}}{n_{He}}, \quad \eta_{He^+} = \frac{n_{He^{++}}}{n_{He}}
\end{aligned}\right\} \quad (4.176)$$

$$\left(\frac{\partial \nabla_{ad}}{\partial \ln T}\right)_P = \frac{\nabla_{ad}^2}{\delta^2}\left\{-6(1-C)(4+\delta)\delta + \left[6(1-C)+\frac{3}{2}\left(\frac{\partial C_t}{\partial \ln T}\right)_P\right]\left(\frac{\partial \delta}{\partial \ln T}\right)_P\right.$$
$$\left. -\frac{3}{2}\delta^2\left(\frac{\partial C_t}{\partial \ln T}\right)_P + A_1\right\} \quad (4.177)$$

$$\left(\frac{\partial \nabla_{ad}}{\partial \ln P}\right)_T = \frac{\nabla_{ad}^2}{\delta^2}\left\{\frac{3}{2}(1-C)(4+\delta)\delta + \left[6(1-C)+\frac{3}{2}\left(\frac{\partial C_t}{\partial \ln T}\right)_P\right]\left(\frac{\partial \delta}{\partial \ln P}\right)_T\right.$$
$$\left. -\frac{3}{2}\delta^2\left(\frac{\partial C_t}{\partial \ln P}\right)_T + A_2\right\} \quad (4.178)$$

$$\left(\frac{\partial \delta}{\partial \ln T}\right)_P = \frac{1}{(C-C_t)^2}\left\{4(1-C)\left[4(1-C_t)+\left(\frac{\partial C_t}{\partial \ln T}\right)_P\right] + \left[4(1-C)\right.\right.$$
$$\left.\left. +\left(\frac{\partial C_t}{\partial \ln T}\right)_P\right]\left(\frac{\partial C_t}{\partial \ln T}\right)_P\right\} + A_3 \quad (4.179)$$

$$\left(\frac{\partial \delta}{\partial \ln P}\right)_T = \frac{1}{(C - C_t)^2}\left\{-(1-C)\left[4(1-C_t) + \left(\frac{\partial C_t}{\partial \ln T}\right)_P\right] + \left[4(1-C)\right.\right.$$

$$\left.\left. + \left(\frac{\partial C_t}{\partial \ln T}\right)_P\right]\left(\frac{\partial C_t}{\partial \ln P}\right)_T\right\} + A_4 \tag{4.180}$$

$$A_1 = a_3'\mu_T'[a_4' + (2 - a_2'\mu_T')\delta] + \delta\left(a_3' + \frac{a_6'}{a_1'}\right)\left(\frac{a_3'}{a_5}\mu_P' + \mu_T'\right)$$

$$+ \frac{a_1'}{1+E'}\left[\delta(2 + a_2'\mu_T') + a_4'\right]\left(\frac{\nu_H}{G_H}\phi_H^2 + \frac{\nu_{He}}{G_{He}}\phi_{He}^2 + \frac{\nu_{He}}{G_{He^+}}\phi_{He^+}^2\right)$$

$$+ \frac{\delta}{1+E'}\left\{(a_3' - a_1'\phi_H)\left[a_2'\mu_T' + \frac{(1 - 2\eta_H)\phi_H}{G_H\eta_H(1 - \eta_H)}\right]\frac{\nu_H\phi_H}{G_H^2\eta_H(1 - \eta_H)}\right.$$

$$+ (a_3' - a_1'\phi_{He})\left[a_2'\mu_T' + \frac{(1 - 2\eta_{He} - 2\eta_{He^+})\phi_{He}}{G_{He}(\eta_{He} + \eta_{He^+})(1 - \eta_{He} - \eta_{He^+})}\right]$$

$$\frac{\nu_{He}\phi_{He}}{G_{He}^2(\eta_{He} + \eta_{He^+})(1 - \eta_{He} - \eta_{He^+})}$$

$$\left. + (a_3' - a_1'\phi_{He^+})\left[a_2'\mu_T' + \frac{(1 - 2\eta_{He^+})\phi_{He^+}}{G_{He}\eta_{He^+}(1 - \eta_{He^+})}\right]\frac{\nu_{He}\phi_{He^+}}{G_{He^+}^2\eta_{He^+}(1 - \eta_{He^+})}\right\} \tag{4.181}$$

$$A_2 = a_3'\mu_T'[a_4' + (a_8' + a_2'\mu_P')\delta] + \frac{a_7'}{a_1'}\left(\frac{a_3'}{a_5'}\mu_P' + \mu_T'\right)\delta$$

$$+ \frac{a_1'}{1+E'}(a_2'\mu_P'\delta + a_8)\left(\frac{\nu_H}{G_H}\phi_H^2 + \frac{\nu_{He}}{G_{He}}\phi_{He}^2 + \frac{\nu_{He}}{G_{He^+}}\phi_{He^+}^2\right)$$

$$+ \frac{\delta}{1+E'}\left\{(a_3' - a_1'\phi_H)\left[a_2'\mu_P' - \frac{a_5'}{a_1'}\frac{(1 - 2\eta_H)}{G_H\eta_H(1 - \eta_H)}\right]\frac{\nu_H\phi_H}{G_H^2\eta_H(1 - \eta_H)}\right.$$

$$+ (a_3' - a_1'\phi_{He})\left[a_2'\mu_P' - \frac{a_5'}{a_1'}\frac{(1 - 2\eta_{He} - 2\eta_{He^+})}{G_{He}(\eta_{He} + \eta_{He^+})(1 - \eta_{He} - \eta_{He^+})}\right]$$

$$\frac{\nu_{He}\phi_{He}}{G_{He}^2(\eta_{He} + \eta_{He^+})(1 - \eta_{He} - \eta_{He^+})}$$

$$+ (a_3' - a_1'\phi_{He^+})\left[a_2'\mu_P' - \frac{a_5'}{a_1'} + \frac{(1 - 2\eta_{He^+})}{G_{He^+}\eta_{He^+}(1 - \eta_{He^+})}\right]$$

$$\left. \frac{\nu_{He}\phi_{He^+}}{G_{He^+}^2\eta_{He^+}(1 - \eta_{He^+})}\right\} \tag{4.182}$$

$$A_3 = \mu_T'[1 + \mu_T'(a_2' - 1)] + \frac{\mu_P'}{a_5'}\left[\frac{5}{2}a_1' + \frac{a_6'}{a_1'} + 4(1 - C) + \left(\frac{\partial C_t}{\partial \ln T}\right)_P\right]$$

$$
+ \frac{1}{1+E'} \left\{ \left[a_2' \mu_T' + \frac{(1-2\eta_H)\phi_H}{G_H \eta_H (1-\eta_H)} \right] \frac{\nu_H \phi_H}{G_H^2 \eta_H (1-\eta_H)} \right.
$$

$$
+ \left[a_2' \mu_T' + \frac{(1-2\eta_{He} - 2\eta_{He^+})\phi_{He}}{G_{He}(\eta_{He} + \eta_{He^+})(1 - \eta_{He} - \eta_{He^+})} \right]
$$

$$
\frac{\nu_{He}\phi_{He}}{G_{He}^2 (\eta_{He} + \eta_{He^+})(1 - \eta_{He} - \eta_{He^+})}
$$

$$
+ \left[a_2' \mu_T' + \frac{(1-2\eta_{He^+})\phi_{He^+}}{G_{He^+} \eta_{He^+}(1 - \eta_{He^+})} \right] \frac{\nu_{He}\phi_{He^+}}{G_{He^+}^2 \eta_{He^+}(1 - \eta_{He^+})} \qquad (4.183)
$$

$$
A_4 = \mu_P' \left[\mu_T'(a_2' - 1) + \frac{a_7'}{a_1' \cdot a_5'} \right] + \frac{1}{1+E'} \left\{ \left[a_2' \mu_P' - \frac{a_5'}{a_1'} \frac{(1-2\eta_H)\phi_H}{G_H \eta_H (1-\eta_H)} \right] \right.
$$

$$
\frac{\nu_H \phi_H}{G_H^2 \eta_H (1-\eta_H)}
$$

$$
+ \left[a_2' \mu_P' - \frac{a_5'}{a_1'} \frac{(1-2\eta_{He} - 2\eta_{He^+})\phi_{He}}{G_{He}(\eta_{He} + \eta_{He^+})(1 - \eta_{He} - \eta_{He^+})} \right]
$$

$$
\frac{\nu_{He}\phi_{He}}{G_{He}^2 (\eta_{He} + \eta_{He^+})(1 - \eta_{He} - \eta_{He^+})}
$$

$$
+ \left[a_2' \mu_P' - \frac{a_5'}{a_1'} \frac{(1-2\eta_{He^+})\phi_{He^+}}{G_{He^+} \eta_{He^+}(1 - \eta_{He^+})} \right] \frac{\nu_{He}\phi_{He^+}}{G_{He^+}^2 \eta_{He^+}(1 - \eta_{He^+})} \qquad (4.184)
$$

$$
\mu_P' = \frac{a_5'}{a_1'(1+E')} \left(\frac{\nu_H}{G_H} + \frac{\nu_{He}}{G_{He}} + \frac{\nu_{He}}{G_{He^+}} \right) \qquad (4.185)
$$

$$
\left. a_1' = C - C_t \ , \quad a_2' = \left(2 + \frac{1}{E'} \right) \right.
$$

$$
a_3' = 4 - \frac{3}{2}C - \frac{5}{2}C_t + \left(\frac{\partial C_t}{\partial \ln T} \right)_P
$$

$$
a_4' = \delta^2 + \left(\frac{\partial \delta}{\partial \ln T} \right)_P , \quad a_5' = 1 - C_t - \left(\frac{\partial C_t}{\partial \ln P} \right)_T
$$

$$
a_6' = 4(1-C) \left[4(1-C_t) + \left(\frac{\partial C_t}{\partial \ln T} \right)_P \right] \times \left[4(1-C) + \left(\frac{\partial C_t}{\partial \ln T} \right)_P \right] \left(\frac{\partial C_t}{\partial \ln T} \right)_P
$$

$$
a_7' = -(1-C) \left[4(1-C_t) + \left(\frac{\partial C_t}{\partial \ln T} \right)_P \right] \times \left[4(1-C) + \left(\frac{\partial C_t}{\partial \ln T} \right)_P \right] \left(\frac{\partial C_t}{\partial \ln P} \right)_T
$$

$$
a_8' = (1-\alpha)\delta + \left(\frac{\partial \delta}{\partial \ln P} \right)_T
$$

$$
(4.186)
$$

where

$$\left(\frac{\partial C_t}{\partial \ln T}\right)_P, \left(\frac{\partial C_t}{\partial \ln P}\right)_T$$

are the same as those in Eqs. (4.158) and (4.159).

4.3.5 Equation of State for Non-ideal Gas

In §4.2, we have pointed out that when the density of the interior of the star increases to a certain extent, interactions among gas particles can no longer be neglected. At the same time, the Boltzmann equation and the Saha equation for determination of occupation number of particles at different energy states have also become invalid. We refer to a gas with interactions among particles as a non-ideal gas.

In recent years, two methods have been proposed to establish the equation of state and the related thermodynamic quantities for a non-ideal gas, namely, the minimum free energy method proposed by Mihalas *et al.* (1988), Hummer and Mihalas (1988), and Däppen *et al.* (1988), and the grand canonical partition function method proposed by Luo (1994a,b,c). The treatments of interactions among particles in both methods are the same.

To illustrate the interactions among gas particles, we first introduce the concept of "atomic configuration". An atom or ion consisting of m bound electrons, the total energy of which is E, is called an atomic configuration and represented by the symbol $|mE\rangle$. The number of bound electrons m in an atomic configuration is related to the degree of ionization of the atom. For example, an atom with nuclear charge number Z has an atomic configuration with a degree of ionization $Z - m$. The total energy of the bound electrons E is determined by the ionization energies and excitation energies of each energy level of the atom. For example, the total energy of bound electrons for a hydrogen atom of the nth ionization state is given by $E = -\chi_{HI} + 13.6 \text{ eV}/n^2$ where $\chi_{HI} = 13.6$ eV is the ionization energy of the hydrogen atom.

Hummer and Mihalas (1988) used a method similar to the perturbation method to describe the mutual interactions among gas particles. When an atomic configuration is subjected to perturbation effects of surrounding particles, the statistical weightings of its energy states will be reduced, and the reduction rate w_i is called the occupation probability of the atomic configuration which can be expressed as

$$\ln W_{is} = -\left(\frac{4\pi}{3V}\right)\left\{\sum_v N_v(r_{is}+r_{1v})^3 + 16\left[\frac{(Z_s+1)e^2}{K_{is}^{1/2}\chi_{is}}\right]^3\sum_{\alpha\neq 2}N_\alpha Z_\alpha^{3/2}\right\} \quad (4.187)$$

where the first term on the right is the perturbation term for neutral particles (atoms) so that the summation $\sum_v$ is performed over all neutral atoms, r_{is} is the atomic radius of the sth element at the ith energy level and r_{1v} is the atomic radius of the vth element at the ground state, the second term on the right is the perturbation term for charged particles (ions), Z_s and χ_{is} are nuclear charge number and ionization energy of the atoms, N_α and Z_α are number density and nuclear charge number of the αth ions, K_{is} is the quantum mechanical correction factor. From Eq. (4.187), we can see that the perturbation term for charged particles is related to the minimum distance between atoms, i.e., $R_s = 2(Z_s+1)^{1/2}/\chi_{is}$.

The detailed derivation for the equation of state for a non-ideal gas through the minimum free energy method or the grand canonical partition function method are very complicated, so only the main points for the two methods will be presented in the following.

(1) Minimum free energy method
We consider a gas mixture composed of hydrogen and helium with temperature T and volume V. Suppose the total number of hydrogen nuclei is N_H, the total number of helium nuclei is N_{He}, the total number of electrons is N_e, and the numbers of particles in different energy states are

$$\{N_s\} = \left\{N_{H_2}, N_{HI}, N_{HII}, N_{HeI}, N_{HeII}, N_{HeIII}, N_e\right\}$$

The free energy of the gas $F(T, V, \{N_s\})$ is divided into four parts:

(i) the free energy of molecules, atoms and ions, which can be written as

$$F_1 = -kT\sum_{s\neq e} N_s\left(\frac{3}{2}\ln T + \ln\left(\frac{V}{N_s}\right) + \ln G_s + 1\right) \quad (4.188)$$

where

$$G_s = \left(\frac{2\pi m_s k}{h^2}\right)^{3/2} \quad (4.189)$$

(ii) the free energy of excited states and ionized states inside molecules, atoms and ions, which can be written as

$$F_2 = \sum_{s \neq e} N_s(E_{1s} - kT\ln 2s) \tag{4.190}$$

where E_{1s} is the ground state energy of the sth particles,

$$2s = \sum_i w_{is} g_{is} \exp[-(E_{is} - E_{1s})/kT] \tag{4.191}$$

where w_{is}, g_{is} and E_{is} are occupation probability, statistical weighting and excitation energy of the ith energy state of the sth particles, respectively.

(iii) the free energy of electrons, which can be written as

$$F_3 = -kTN_e \left[\frac{2}{3} \frac{F_{3/2}(\psi)}{F_{1/2}(\psi)} - \psi \right] \tag{4.192}$$

where ψ is the degeneracy parameter, $F_{3/2}(\psi)$ and $F_{1/2}(\psi)$ are the Fermi–Dirac integrations defined as

$$F_n(\psi) = \int_0^\infty \frac{x^n dx}{e^{x-\psi} + 1}, \quad (n = 3/2, \ 1/2) \tag{4.193}$$

(iv) the free energy of Coulomb interactions among charged particles, which can be written as

$$F_4 = -\left(\frac{2\pi^{1/2} e^3}{3k^{1/2}} \right) \frac{1}{V^{1/2} T^{1/2}} \left(\sum_\alpha N_\alpha Z_\alpha^2 \theta_\alpha \right)^{3/2} \tau(x) \tag{4.194}$$

where summation over all charged particles is carried out, Z_α is the electric charge number of the αth particles, $\theta_\alpha = 1$, $\theta_e = \frac{1}{2} F_{-1/2}(\psi)/F_{1/2}(\psi)$, $\tau(x)$ is correction for the finite size of particles.

Therefore the total free energy of the gas mixture is

$$F(T, V, \{N_s\}) = F_1 + F_2 + F_3 + F_4 \tag{4.195}$$

which is a function of T, V and $\{N_s\}$.

Applying the condition of minimum free energy under equilibrium to the ionization of particles $j \rightarrow (j+1) + e$, we have

$$\frac{\partial F}{\partial N_j} - \frac{\partial F}{\partial N_{j+1}} - \frac{\partial F}{\partial N_e} = 0 \qquad (4.196)$$

Applying the condition to the dissociation of molecules $AB \leftrightarrow A + B$, we have

$$\frac{\partial F}{\partial N_{AB}} - \frac{\partial F}{\partial N_A} - \frac{\partial F}{\partial N_B} = 0 \qquad (4.197)$$

According to Eqs. (4.196) and (4.197), we obtain the following four equations:

$$\frac{\partial F}{\partial N_{H2}} - 2\frac{\partial F}{\partial N_{HI}} = 0 \qquad (4.198)$$

$$\frac{\partial F}{\partial N_{HI}} - \frac{\partial F}{\partial N_{HII}} - \frac{\partial F}{\partial N_e} = 0 \qquad (4.199)$$

$$\frac{\partial F}{\partial N_{HeI}} - \frac{\partial F}{\partial N_{HeII}} - \frac{\partial F}{\partial N_e} = 0 \qquad (4.200)$$

$$\frac{\partial F}{\partial N_{HeII}} - \frac{\partial F}{\partial N_{HeIII}} - \frac{\partial F}{\partial N_e} = 0 \qquad (4.201)$$

Adding two more equations for conservation of particle numbers:

$$N_H = N_{HI} + N_{HII} + 2N_{H2} \qquad (4.202)$$

$$N_{He} = N_{HeI} + N_{HeII} + N_{HeIII} \qquad (4.203)$$

and one more equation for the conservation of electric charge:

$$N_e = N_{HII} + N_{HeII} + 2N_{HeIII} \qquad (4.204)$$

we have a total of seven equations, which can be solved for seven unknowns:

$$\{N_s\} = \left\{ N_{H_2}, N_{HI}, N_{HII}, N_{HeI}, N_{HeII}, N_{HeIII}, N_e \right\}$$

or the particle numbers in different energy states. When these numbers have been obtained under equilibrium, the total free energy $F(T, V,$

$\{N_s\}$) for the gas mixture can be determined, so the pressure and the internal energy can be obtained through the Maxwell equations:

$$P = -\left(\frac{\partial F}{\partial V}\right)_{T,\{N_s\}} \tag{4.205}$$

$$U = -T^2 \frac{\partial}{\partial T}\left(\frac{F}{T}\right)_{V,\{N_s\}} \tag{4.206}$$

Equation (4.205) gives the expression for the gas pressure. If the radiation pressure is also taken into account, the expression for the total pressure can be obtained, which is the equation of state. The other thermodynamic quantities can be obtained through the derivatives of P and U, i.e.,

$$\left.\begin{aligned}
C_V &= \left(\frac{\partial U}{\partial T}\right)_V, \\
C_P &= C_V + \frac{P\chi_T^2}{T\rho\chi_\rho}, \\
\nabla_{ad} &= \frac{\gamma - 1}{\gamma\chi_T}, \\
\gamma &= C_P/C_V, \\
\chi_T &= \left(\frac{\partial \ln P}{\partial \ln T}\right)_\rho, \\
\chi_\rho &= \left(\frac{\partial \ln P}{\partial \ln \rho}\right)_T.
\end{aligned}\right\} \tag{4.207}$$

(2) Grand canonical method

In the first place, the grand canonical method gives the grand canonical partition function for electrons bound to atomic configurations and free electrons under the equilibrium state. The grand canonical partition function for electrons bound to an atomic configuration can be written as

$$Z_b = \prod_i \left\{\sum_m \sum_E w_{mE} g_{mE} \exp(\psi m - \beta E)\right\}^{N_i} \tag{4.208}$$

where m and E are the number of bound electrons and total energy of the bound electrons, respectively, for the atomic configuration $|mE\rangle$;

w_{mE} and g_{mE} are the occupation probability and statistical weighting for the atomic configuration, ψ is the degeneracy parameter, $\beta = 1/kT$, and N_i is the total number of atoms for the ith element. The multiplication is performed over all elements.

The grand canonical partition function for free electrons Z_e can be written as

$$\ln Z_e = \frac{8\pi V}{h^3} \int_0^\infty p^2 \ln\left[1 + \exp(\psi - \beta\varepsilon)\right] dp \qquad (4.209)$$

where ε is the electron energy which can be written as

$$\varepsilon = p^2/2m_e - \Delta E_f \qquad (4.210)$$

Here, ΔE_f is the reduction in the total energy due to Coulomb interactions among charged particles (ions and electrons). ΔE_f can be obtained from the Debye–Hückel model (see Cox and Giuli 1968), which has taken into account effects of the finite size of ions.

According to the grand canonical partition function for electrons bound to atomic configurations and free electrons, the equation for conservation of electron number can be obtained as

$$\sum_i N_i Z_i = \frac{\partial}{\partial\psi}\left[\ln(Z_b \cdot Z_e)\right]_{\beta V}$$

$$= GV\beta^{-3/2}F_{1/2}(\psi + \beta\Delta E_f) + \sum_i N_i \sum_{im} m \sum_E \chi_{imE} \qquad (4.211)$$

and the probability of an atomic configuration $|mE\rangle$ as

$$\chi_{imE} = \frac{w_{mE}g_{mE}e^{\psi m - \beta E}}{\sum_m \sum_E w_{mE}g_{mE}e^{\psi m - \beta E}} \qquad (4.212)$$

which gives the probabiliy of the ith element having the atomic configuration $|mE\rangle$, where $G = \sqrt{2}m_e^{3/2}/\pi^2\hbar^3$, V is the volume and $F_{1/2}(\psi + \beta\Delta E_f)$ is the Fermi–Dirac integration.

In Eq. (4.211) for the conservation of electron numbers, the left-hand side gives the total electron number (including electrons bound to atomic configurations and free electrons) in the gas, which is determined by the number of nuclei N_i and the nuclear charge number Z_i. The first term on the right-hand side gives the total number of free state electrons $N_e = GV\beta^{-3/2}F_{1/2}(\psi + \beta\Delta E_f)$. The second term on the right gives the number of electrons bound to atomic configurations.

The expression of the probability for an atomic configuration χ_{imE} in Eq. (4.169) can be used directly for determining the number of particles of the ith element in the atomic configuration $|mE\rangle$, i.e., the atomic occupation number. It can be also be shown that the Boltzmann and Saha equations are only its special cases (Luo 1994a). Therefore, it is a general formula for determining occupation numbers in all energy states for all types of atoms.

When N_i, Z_i, β, V, $\{E\}$, $\{g_{mE}\}$ are fixed for a stellar gas, we can obtain $\{w_{mE}\}$ from Eq. (4.187) and ε from Eq. (4.210). In this way, from Eq. (4.211) for the conservation of electron numbers and Eq. (4.212) for the probability of atomic configurations, we can obtain the degeneracy parameter ψ and the probability of atomic configurations $\{\chi_{mE}\}$. Subsequently, we can compute the occupation number for all types of atoms in all energy states and the electron number N_e. In this way, it is easy to get the equation of state and other thermodynamic quantities. For example, the electron pressure P_e can be written as

$$P_e = \frac{2}{3}kT\frac{N_e}{V}\frac{F_{3/2}(\psi + \beta\Delta E_f)}{F_{1/2}(\psi + \beta\Delta E_f)} \tag{4.213}$$

The internal energy of electrons U_e can be written as

$$U_e = kTN_e\frac{F_{3/2}(\psi + \beta\Delta E_f)}{F_{1/2}(\psi + \beta\Delta E_f)} \tag{4.214}$$

Due to the mutual interactions among particles, the perturbation terms for electron pressure and electron internal energy are

$$P'_e = N_e\left(\frac{\partial\Delta E_f}{\partial V}\right)_{\psi\beta}, \tag{4.215}$$

$$U'_e = -N_e\Delta E_f - \beta N_e\left(\frac{\partial\Delta E_f}{\partial\beta}\right)_{\psi V} \tag{4.216}$$

The pressure of the nuclei (including atoms and ions) P_K in the gas and the internal energy U_K can be written as

$$P_K = kT\sum_i\frac{N_i}{V} \tag{4.217}$$

$$U_K = \frac{3}{2} kT \sum_i N_i + \sum_i N_i \sum_m \sum_E \chi_{imE} E \qquad (4.218)$$

The first term on the right of Eq. (4.218) is the thermal energy of atoms and ions, while the second term is the total internal energy of atoms and ions (including the excitation energy and the ionization energy). Due to mutual interactions among particles, the perturbation term for the nuclei pressure and the internal energy are

$$P'_K = kT \sum_i N_i \sum_m \sum_E \chi_{imE} \left(\frac{\partial \ln w_{imE}}{\partial V} \right)_{\psi \beta} \qquad (4.219)$$

$$U'_K = - \sum_i N_i \sum_m \sum_E \chi_{imE} \left(\frac{\partial \ln w_{imE}}{\partial \beta} \right)_{\psi V} \qquad (4.220)$$

Therefore, the total pressure can be written as

$$P = P_g + P_R = P_K + P_e + P'_K + P'_e + \frac{a}{3} T^4 \qquad (4.221)$$

and the total internal energy as

$$U = U_K + U_e + U'_K + U'_e + aT^4 V \qquad (4.222)$$

where aT^4 is the radiation energy.

Equation (4.219) is the equation of state for a non-ideal gas. After getting the expressions for the total pressure P and the internal energy U, other thermodynamic quantities such as C_P, C_V and ∇_{ad} can be obtained from Eq. (4.207).

4.4 EQUATION OF STATE FOR THE MATTER IN THE STELLAR ATMOSPHERE

Because of different densities and temperatures, the matter in the stellar atmosphere has different equations of state. For a normal star (e.g., a main sequence star for which the temperature is not very high), the part of stellar atmosphere close to the interior has a relatively higher density so, as an approximation, we can use the Boltzmann and Saha equations to calculate the occupation number in all energy states for all types of atoms and then obtain the equation of state and other thermodynamic quantities. These regions are referred to as local thermal equilibrium regions and denoted as LTE regions. However, in regions close to the surface of the stellar atmosphere, or in the major part of the atmosphere of giant stars and supergiants, the density is very low so the state is clearly deviated from thermal equilibrium and the Boltzmann and Saha equations can no longer be used to determine the occupation number of each energy state. These regions are referred to as non-local thermal equilibrium regions (non-LTE regions).

The temperature also affects the equation of state of the atmosphere. We present a simple illustration here. When the stellar atmosphere fulfills the hydrostatic equilibrium condition (see §7.2.2), the particle number density can be obtained as

$$n_g = \frac{g}{kT} m \tag{4.223}$$

where $m = \rho dx$ represents the mass of a cylinder with bottom area 1 cm² and height dx (the height being measured from the surface to the interior). Equation (4.224) is an approximation which has assumed that the radiation pressure and the temperature gradient can be negligible. We further assume that, when the temperature of the stellar atmosphere is very high, the electron number density $n_e \approx n_g/2$. Thus, it is easy to observe from Eq. (4.224) that the number density of the particles and electrons should decrease when the temperature T rises. Therefore, it is easier for a stellar atmosphere with a higher effective temperature to deviate from the LTE state. In general, for a main sequence star, when the effective temperature exceeds 25000 K, the non-LTE equation of state should be employed for computation of spectral lines; when the effective temperature exceeds 30000 K, the non-LTE equation of state should also be employed for computation of the continuous spectrum. For a cool supergiant, both spectral lines and the continuous spectrum should be treated under a non-LTE even at effective temperatures higher than 10000 K due to the extremely low atmospheric densities.

4.4.1 LTE Equation of State

Suppose a stellar atmosphere is composed of h types of elements, and the abundance of the αth type element is X_α. We denote the number (or occupation number) of the αth type element of the ith bound state and the βth ionization per unit volume as $n_i^{\alpha\beta}$. Moreover, we denote n_g, n_I and n_e as the occupation number of particles (ions and electrons), ions and electrons, respectively.

The total pressure can be expressed as

$$P = P_g + P_R + P_t \qquad (4.224)$$

where P_R is the radiation pressure, which can be simplified under LTE as

$$P_R = \frac{a}{3} T_{eff}^4 \qquad (4.225)$$

P_t is the turbulent pressure, which can be written as

$$P_t = \rho v_t^2 \qquad (4.226)$$

where v_t is the turbulent velocity. In a general stellar atmosphere, $P_t \ll P_g$ and can be neglected. Only in some exceptions where the value of v_t in the atmosphere is comparable to the speed of sound v_s will we need to consider P_t.

The gas pressure P_g can be expressed as

$$P_g = n_g kT = (n_I + n_e)kT \qquad (4.227)$$

where n_I and n_e can be written as

$$n_I = \sum_\alpha \sum_\beta \sum_i n_i^{\alpha\beta} \qquad (4.228)$$

$$n_e = \sum_\alpha \sum_\beta \beta \sum_i n_i^{\alpha\beta} \qquad (4.229)$$

and the density ρ can be written as

$$\rho = \sum_\alpha m_\alpha \sum_\beta \sum_i n_i^{\alpha\beta} \qquad (4.230)$$

From the above equations, if we know the occupation number $n_i^{\alpha\beta}$ in each energy state of different atoms, we can get the relationship among ρ, n_e, $n_i^{\alpha\beta}$, pressure P, temperature T and chemical composition, which is known as the LTE equation of state.

To obtain $n_i^{\alpha\beta}$, we need to use the Boltzmann and Saha equations:

$$\frac{n_i^{\alpha\beta}}{n_o^{\alpha\beta}} = \frac{g_i^{\alpha\beta}}{g_o^{\alpha\beta}} e^{-\varepsilon_i^{\alpha\beta}/kT} \qquad (4.231)$$

$$(\varepsilon_i^{\alpha\beta} = E_i^{\alpha\beta} - E_0^{\alpha\beta})$$

$$\frac{n_0^{\alpha\beta+1}}{n_0^{\alpha\beta}} n_e = \frac{g_0^{\alpha\beta+1}}{g_0^{\alpha\beta}} \cdot 2 \frac{(2\pi m_e kT)^{3/2}}{h^3} e^{-\chi_\beta^\alpha/kT} \qquad (4.232)$$

where $g_0^{\alpha\beta}$ and $n_0^{\alpha\beta}$, respectively represent the statistical weightings of the αth type atoms of the βth ionization and the ground state occupation number.

From Eqs. (4.231) and (4.232), we get

$$n_i^{\alpha\beta} = n_0^{\alpha\beta+1} \cdot n_e \frac{h^3}{(2\pi m_e kT)^{3/2}} \frac{g_i^{\alpha\beta}}{2g_0^{\alpha\beta+1}} e^{(\chi_\beta^\alpha - \varepsilon_i^\alpha)/kT} \qquad (4.233)$$

Summing over all excited states for the above equation, we can obtain the occupation number of the αth type atoms of the βth ionization as

$$n^{\alpha\beta} = n_0^{\alpha\beta+1} \cdot n_e \frac{h^3}{(2\pi m_e kT)^{3/2}} \frac{U^{\alpha\beta}}{2g_0^{\alpha\beta+1}} e^{\chi_\beta^\alpha/kT} \qquad (4.234)$$

where $U^{\alpha\beta} = \sum_i g_i^{\alpha\beta} e^{-\varepsilon_i^\alpha/kT}$ is the sum of states.

Furthermore, we also need the conservation of number of atoms for each element

$$n^\alpha + n^{\alpha 1} + n^{\alpha 2} + \dots + n^{\alpha\beta} + \dots = X_\alpha n_g , \quad (\alpha = 1, 2, \dots, h) \qquad (4.235)$$

The detailed computations to solve for ρ, n_e, $n_i^{\alpha\beta}$ through the equation set Eq. (4.227) to (4.235) for given values of X_α, T and n_g will be presented in Chapter 13.

4.4.2 Non-LTE Equation of State

Under non-LTE, the total pressure is still given by Eq. (4.224) but the radiation pressure P_R in the equation can be obtained by integrating Eq. (2.16) over all frequencies:

$$P_R = \int_\nu \oint_\omega \frac{I_\nu}{c} \cos^2\theta d\omega d\nu \qquad (4.236)$$

The equations for other quantities such as turbulent pressure P_t, gas pressure P_g and density ρ are the same as those in Eqs. (4.226), (4.227) and (4.230).

The computation for the occupation number $n_i^{\alpha\beta}$ in each energy state of different atoms for the non-LTE case is apparently different from that for the LTE case. For the LTE case, $n_i^{\alpha\beta}$ is determined through the Boltzmann and Saha equations, and conservation of particle number and electric charge, i.e., from local thermodynamic quantities of the volume element. On the other hand, for the non-LTE case, $n_i^{\alpha\beta}$ is dependent on the radiation field, and the radiation field inside a certain volume element is also dependent on the $n_i^{\alpha\beta}$ value of other volume elements. In other words, the physical quantities at different positions in the entire stellar atmosphere are correlated. The value of $n_i^{\alpha\beta}$ cannot be determined by thermodynamic quantities at a particular position, but is determined by statistical equilibrium equations based on statistical equilibrium laws of transitions between energy levels, together with conservation of particle number and electric charge.

(A) Statistical equilibrium equations

Based on the fact that all spectral lines are relatively stable in stellar spectra, we can suppose that the occupation numbers of all energy states in atoms within a volume element should be constants and time independent, i.e.,

$$\frac{d}{dt} n_i^{\alpha\beta} = 0 \qquad (4.237)$$

In other words, within a unit time interval, the number of atoms with transitions from the energy state i to other states should be equal to the number of atoms with transitions from other states to the energy state i. For simplicity, when discussing the αth type atoms of the βth ionization, the subscripts α and β are omitted, so Eq. (4.237) becomes

$$n_i \sum_{\substack{j \neq i}}^{K} P_{ij} = \sum_{\substack{j \neq i}}^{K} n_j P_{ji} \qquad (4.238)$$

where P_{ij}, P_{ji} represent the probability of an atom to have a transition from the energy state i to the energy state j per second and vice versa respectively, which are called the transition rates. K represents the continuous spectrum. Theoretically speaking, an atom can have an infinite number of bound states. In reality, however, an atom has only a finite number of bound states due to mutual interactions among atoms. Therefore, in the equation set (4.238), we assumed that an atom has only K bound energy levels. The transition rates P_{ij} and P_{ji} are quantities related to the radiation field, which will be discussed in the following.

The transition of atoms can arise from radiation or collision, so the transition rates P_{ij} should also have different parts arising from radiation and collision, i.e.,

$$P_{ij} = R_{ij} + C_{ij} \qquad (4.239)$$

where R_{ij} is the radiative transition rate and C_{ij} is the collisional transition rate.

(1) The radiative transition rate R_{ij}

We first study the bound-bound transition, and consider the transition from a lower energy level i to a higher energy level j. For simplicity, we assume the profile functions for levels i and j are the same and represented by φ_v. Thus, the number of transitions from level i to j in a unit time is

$$n_i B_{ij} I_v \varphi_v \frac{d\omega}{4\pi} dv \qquad (4.240)$$

where the coefficient B_{ij} represents the probability of this type of transitions per second and is called the Einstein transition coefficient (see §5.1). Equation (4.240) shows that the number of transitions from i to j is dependent on the occupation number n_i of energy level i of atoms and the radiation intensity I_v.

If the atomic absorption cross section is a_{ij}, from Eq. (5.44), we can convert Eq. (4.240) into

$$n_i B_{ij} I_v \varphi_v \frac{d\omega}{4\pi} dv = n_i \left(\frac{a_{ij}}{hv} \right) I_v d\omega dv \qquad (4.241)$$

Integrating the above equation over all frequencies and solid angles, we have

$$R_{ij} = 4\pi \int_0^\infty \frac{a_{ij}}{hv} J_v dv \qquad (4.242)$$

Further considering the transition from energy level j to level i, the number of transitions in a unit time from energy level j to level i is given by

$$n_j A_{ji} \varphi_v dv \frac{d\omega}{4\pi} + n_j B_{ji} I_v \varphi_v \frac{d\omega}{4\pi} dv \qquad (4.243)$$

The first term is the number of spontaneous transitions, which is only related to the atomic occupation number n_j of the jth energy level, and is independent of the radiation intensity. The second term is the number of induced transitions, which is dependent on both n_j and intensity I_ν of the incident radiation. The coefficients A_{ji} and B_{ji} are the transition probabilities of spontaneous and induced transitions respectively, in a unit time, and are known as the Einstein transition coefficients. Using the established Einstein relationships in Eqs. (5.17) and (5.18),

$$A_{ji} = \frac{2h\nu_{ij}^3}{c^2} B_{ji} = \frac{2h\nu_{ij}^3}{c^2} \frac{g_i}{g_j} B_{ij} \tag{4.244}$$

and the Boltzmann equation

$$\left(\frac{n_i}{n_j}\right)^* = \frac{g_i}{g_j} \exp\left(\frac{h\nu_{ij}}{kT}\right) \tag{4.245}$$

where $(n_i/n_j)^*$ represents the ratio of occupation numbers under LTE. Substituting Eqs. (4.245) and (4.246) into Eq. (4.244), and integrating over all frequencies and solid angles, we have

$$n_j R_{ji} = n_j \cdot 4\pi \left(\frac{n_i}{n_j}\right)^* \int_0^\infty \frac{a_{ij}}{h\nu} \left(\frac{2h\nu^3}{c^2} + J_\nu\right) e^{-h\nu/kT} d\nu \tag{4.246}$$

from which we get

$$R_{ji} = 4\pi \left(\frac{n_i}{n_j}\right)^* \int_0^\infty \frac{a_{ij}}{h\nu} \left(\frac{2h\nu^3}{c^2} + J_\nu\right) e^{-h\nu/kT} d\nu \tag{4.247}$$

In some circumstances, the introduction of the net transition rate Z_{ji} from energy level j to energy level i is very useful, which is defined as

$$n_j A_{ji} Z_{ji} = n_j (A_{ji} + B_{ji} \bar{J}_{ij}) - n_i B_{ij} \bar{J}_{ij} \tag{4.248}$$

so

$$\begin{aligned} Z_{ji} &= 1 - \bar{J}_{ij}(n_i B_{ij} - n_j B_{ji})/(n_j A_{ji}) \\ &= 1 - (\bar{J}_{ij}/S_{ij}) \end{aligned} \tag{4.249}$$

where S_{ij} is the source function and is independent of frequency.

For a bound-free transition, derivations similar to those for the bound-bound case can be employed, which yield

$$R_{iK} = 4\pi \int_0^\infty \frac{a_{iK}}{h\nu} J_\nu d\nu \tag{4.250}$$

where a_{iK} is the atomic absorption cross section for photo-ionization, and

$$R_{Ki} = 4\pi \left(\frac{n_i}{n_K}\right)^* \int_0^\infty \frac{a_{iK}}{h\nu} \left(\frac{2h\nu^3}{c^2} + J_\nu\right) e^{-h\nu/kT} d\nu \tag{4.251}$$

(2) Collisional transition rate

Collisions of atmospheric particles can lead to excitation, ionization and recombination of atoms. Among the collisions, those between electrons and atoms or ions are the most important because of the high velocity of electrons which is larger than the velocity of atoms by $(m_P/m_e)^{1/2} A^{1/2} \approx 43 A^{1/2}$ (A is the atomic weight) under thermal equilibrium.

We denote the cross section for collision between an electron with a velocity v and an atom as $\sigma_{ij}(v)$ which leads to a transition in the atom from energy level i to energy level j. The speeds of electrons follow the Maxwell distribution represented by $f(v)$. Thus, the number of collisions between an electron with a velocity in the range (v, $v + dv$) and an atom in a unit time is $n_e n_i \sigma_{ij}(v) f(v) v dv$, where n_i and n_e represent the number of atoms and electrons, respectively in a unit volume. Integrating over all velocities, the number of collisional transitions from i to j is

$$n_i C_{ij} = n_i n_e \int_{v_0}^\infty \sigma_{ij}(v) f(v) v dv \equiv n_i n_e \Omega_{ij}(T) \tag{4.252}$$

where v_0 is the velocity corresponding to the minimum energy E_o for this process to occur. The integral Ω_{ij} is

$$\Omega_{ij}(T) = \int_{v_0}^\infty \sigma_{ij}(v) f(v) v dv \tag{4.253}$$

We now consider the collisional transition from a higher energy level j to a lower energy level i. Under LTE, every physical process is in equilibrium with its inverse process, so we should have

$$n_i^* C_{ij} = n_j^* C_{ji} \qquad (4.254)$$

and the collisional transition rate from a higher energy level j to a lower energy level i is obtained as

$$n_j C_{ji} = n_j \left(\frac{n_i}{n_j} \right)^* C_{ij} = n_j \left(\frac{n_i}{n_j} \right)^* n_e \Omega_{ij}(T) \qquad (4.255)$$

from which we get

$$C_{ij} = n_e \Omega_{ij}(T) \qquad (4.256)$$

$$C_{ji} = n_e \left(\frac{n_i}{n_j} \right)^* \Omega_{ij}(T) \qquad (4.257)$$

Similar to Eq. (4.248), the introduction of the net collision transition rate Y_{ij} is very useful, so

$$n_i C_{ij} Y_{ij} \equiv n_i C_{ij} - n_j C_{ji} = n_i C_{ij} \left[1 - \frac{n_j \, n_i^*}{n_j^* \, n_i} \right] \qquad (4.258)$$

Using similar methods, we obtain for a bound-free transition

$$C_{iK} = n_e \Omega_{iK}(T) \qquad (4.259)$$

$$C_{Ki} = \left(\frac{n_i}{n_K} \right)^* C_{iK} \qquad (4.260)$$

(B) Equations for conservation of particles and electric charges

To determine the occupation number in each energy state of atoms, we not only need statistical equilibrium equations, but also need equations for conservation of particles and electric charges, which include

Equation for conservation of particles

$$n_g = n_e + \sum_\alpha \sum_\beta \sum_i n_i^{\alpha\beta} \tag{4.261}$$

Equation for conservation of particles for the αth element

$$n_g X_\alpha = \sum_\beta \sum_i n_i^{\alpha\beta} \tag{4.262}$$

Equation for conservation of electric charges

$$n_e = \sum_\alpha \sum_\beta \beta \sum_i n_i^{\alpha\beta} \tag{4.263}$$

The detailed computation of the occupation number in each energy state through Eqs. (4.238), and (4.261) to (4.263), and subsequent evaluation of the equation of state will be discussed in Chapter 13.

CHAPTER FIVE

OPACITY

To macroscopically describe the mutual interactions between radiation and matter, we introduced the absorption coefficient κ_ν and the emission coefficient η_ν in §2.2. κ_ν is also called the opacity of matter with the dimension [cm²/normalized unit]. We have different absorption coefficients according to different normalized units: when the normalized unit is [g], we have the mass absorption coefficient; when the normalized unit is [cm³], we have the volume absorption coefficient. We have also introduced the atomic absorption cross section a_ν with the dimension [cm²]. According to Eq. (2.33), the relationship between opacity κ_ν and atomic absorption cross section a_ν is

$$\rho\kappa_\nu = n \cdot a_\nu \qquad \text{when } \kappa_\nu \text{ is the mass absorption coefficient;}$$
$$\kappa_\nu = n \cdot a_\nu \qquad \text{when } \kappa_\nu \text{ is the volume absorption coefficient,}$$

where n is the number of atoms in 1 cm³. For the sake of convenience, we will adopt the volume absorption coefficient for our discussions in this chapter.

There are different interactions between radiation and matter. Therefore the opacity κ_ν is contributed by absorption coefficients from all the microscopic processes including

bound-bound transition:	κ_{ij}
bound-free transition:	κ_{iK}
free-free transition:	κ_{KK}
scattering:	σ_e

Therefore the opacity can be written as

$$\kappa_\nu = \sum (\kappa_{ij} + \kappa_{iK} + \kappa_{KK}) + \sigma_e \tag{5.1}$$

and the summation is performed over different elements.

Similarly, the corresponding atomic absorption cross section for microscopic processes are denoted as a_{ij}, a_{iK}, a_{KK} and a_e. In the following section, we will discuss the formula for the atomic absorption cross sections and the absorption coefficients for the above microscopic processes.

5.1 BOUND-BOUND TRANSITION

Here, we study the transition between two bound states i and j after a photon is absorbed by an atom. Suppose the statistical weighting of the bound state energy level i of the atom is g_i and that of the bound state energy level j is g_j. We further denote the number of atoms in 1 cm³ on energy levels i and j as n_i and n_j, respectively. Due to the finite time of an atom staying at a certain energy state and effects of thermal motion, the energy levels i or j cannot be sharp and should have certain widths or certain profiles. We denote φ_ν as the profile function of the energy level i and call it the absorption profile. If $n_{i\nu}$ is the number of atoms in 1 cm³ at the energy level i which can absorb radiation in the frequency range $(\nu, \nu + d\nu)$ in a transition, the relationship between $n_{i\nu}$ and n_i is

$$n_{i\nu} = n_i \varphi_\nu \tag{5.2}$$

Obviously, φ_ν should fulfil the normalization condition

$$\int_0^\infty \varphi_\nu d\nu = 1 \tag{5.3}$$

Similarly, if ψ_ν is the profile function for the energy level j which is called the emission profile, and $n_{j\nu}$ is the number of atoms in 1 cm³ at the energy level j which can emit radiation in the frequency range $(v, v + dv)$ in a transition, the relationship between n_{jv} and n_j is

$$n_{jv} = n_j \psi_v \tag{5.4}$$

Similarly, ψ_v should also fulfil the normalization condition

$$\int_0^\infty \psi_v dv = 1 \tag{5.5}$$

Under normal conditions of stars, we can treat $\psi_v = \varphi_v$, i.e., the absorption profile is the same as the emission profile, which is called the complete redistribution assumption.

   Three different transition processes between the energy levels i and j for atoms can occur:

(1) Phototransition. When an atom absorbs a photon with energy $hv_{ij} = \varepsilon_j - \varepsilon_i$, there is a transition from a lower energy level i to a higher energy level j. The number of transitions is dependent on n_{iv} and the radiation intensity I_v. We introduce the coefficient B_{ij} to represent the probability of this transition per second, so the number of phototransitions occurring in one second in one cm^3 in a unit solid angle $d\omega$ is

$$n_{iv}B_{ij}I_v dv \frac{d\omega}{4\pi} = n_i\varphi_v B_{ij}I_v dv \frac{d\omega}{4\pi} \tag{5.6}$$

(2) Spontaneous transition. When atoms at a high energy level j collide with particles (mainly with electrons), they undergo transitions to a low energy level i. The radiation produced by atoms in spontaneous transitions is isotropic. The number of transitions is dependent on n_{jv} but is independent of the radiation intensity I_v. We introduce the coefficient A_{ji} to represent the probability of this transition per second, so the number of spontaneous transitions occurring in one second in one cm^3 in a unit solid angle $d\omega$ is

$$n_{jv}A_{ji}dv \frac{d\omega}{4\pi} = n_j\psi_v A_{ji}dv \frac{d\omega}{4\pi} \tag{5.7}$$

(3) Induced transition. When atoms at a high energy level j are subjected to the induced effects of radiation, they can also undergo transitions to a low energy level i. The radiation produced by atoms in induced transitions has a direction same as that of the incident

radiation. The number of transitions is dependent on both n_{jv} and the radiation intensity I_v. We introduce the coefficient B_{ji} to represent the probability of this transition per second, so the number of induced transitions occurring in one second in one cm^3 in a unit solid angle $d\omega$ is

$$n_{jv}B_{ji}I_v dv \frac{d\omega}{4\pi} = n_j \psi_v B_{ji}I_v dv \frac{d\omega}{4\pi} \tag{5.8}$$

The coefficients B_{ij}, B_{ji} and A_{ji} are called the Einstein transition coefficients. The induced emission has a direction same as that of the induced radiation, and has only an effect opposite to that of the absorption. It can be treated as a negative absorption and can be incorporated into the absorption coefficient. All the above transitions either absorb an energy $h\nu_{ij}$ from I_v or emit an energy $h\nu_{ij}$ to I_v, so

$$h\nu_{ij}(n_i \varphi_v B_{ij}I_v - n_j \psi_v B_{ji}I_v)dv \frac{d\omega}{4\pi} = \kappa_{ij}I_v dv d\omega \tag{5.9}$$

$$h\nu_{ij}(n_j \psi_v A_{ji})dv \frac{d\omega}{4\pi} = \eta_{ij}dv d\omega \tag{5.10}$$

From these two equations, we obtain the absorption coefficient κ_{ij} and the emission coefficient η_{ij} as

$$\kappa_{ij} = h\nu_{ij}n_i B_{ij}\varphi_v \left(1 - \frac{\psi_v}{\varphi_v}\frac{n_j}{n_i}\frac{B_{ji}}{B_{ij}}\right)/4\pi \tag{5.11}$$

$$\eta_{ij} = n_j \psi_v A_{ji}h\nu_{ij}/4\pi \tag{5.12}$$

In general, there are relationships among the Einstein coefficients B_{ij}, B_{ji} and A_{ji}, which reflect properties of the atoms themselves so the relationships are the same under LTE or non-LTE conditions. Thus, the relationships can be determined by just using the simple LTE conditions. Under LTE, we have the following relationships:

(1) Planck radiation law:

$$I_v = \frac{2h\nu^3}{c^2}\frac{1}{e^{h\nu_{ij}/kT} - 1} \equiv B_v(T) \tag{5.13}$$

(2) Boltzmann distribution:

$$\frac{n_j}{n_i} = \frac{g_j}{g_i}e^{-h\nu_{ij}/kT} \tag{5.14}$$

(3) Equation of statistical equilibrium:

$$n_i B_{ij} I_\nu = n_j A_{ji} + n_j B_{ji} I_\nu \tag{5.15}$$

Note that φ_ν and ψ_ν do not appear in the above three equations. This is because the change in $B_\nu(T)$ is extremely small within the line width so this can be neglected. Replacing I_ν in Eq. (5.15) by $B_\nu(T)$, we obtain the expression of $B_\nu(T)$ as

$$B_\nu(T) = \frac{n_j A_{ji}}{n_i B_{ij} - n_j B_{ji}} = \frac{A_{ji}/B_{ji}}{\left(\frac{g_i B_{ij}}{g_j B_{ji}}\right)e^{h\nu_{ij}/kT} - 1} \tag{5.16}$$

Comparing Eqs. (5.16) and (2.55), we get the relationships

$$A_{ji} = \frac{2h\nu_{ij}^3}{c^2} B_{ji} \tag{5.17}$$

$$g_i B_{ij} = g_j B_{ji} \tag{5.18}$$

which are the relationships among the Einstein coefficients and are called the Einstein relationships. Substituting the Einstein relationships into Eqs. (5.11) and (5.12), and applying the complete redistribution assumption we can get equations for κ_{ij} and η_{ij} as

$$\kappa_{ij} = n_i B_{ij} \varphi_\nu h\nu_{ij}\left(1 - \frac{g_i}{g_j}\frac{n_j}{n_i}\right)/4\pi \tag{5.19}$$

$$\eta_{ij} = n_j A_{ji}\varphi_\nu h\nu_{ij}/4\pi \tag{5.20}$$

From these, it is noticed that we need to know the equations for the Einstein coefficients in order to know κ_{ij} and η_{ij}. The computations of the Einstein coefficients are briefly described in the following sections.

We need to compute only one of the Einstein coefficients B_{ij}, B_{ji} and A_{ji}, and the other two can then be determined through the Einstein relationships in Eqs. (5.17) and (5.18). In the following, we first derive the expression for B_{ij} by use of the classical oscillator theory. In the classical theory, the mutual interaction between radiation and the atom is viewed as the interaction between a plane electromagnetic wave and a dipole. We consider the interaction between a beam of electromagnetic wave E with a number of dipoles, with the latter representing the absorbing atoms. We choose a co-ordinate system in which the electromagnetic wave propagates in the z direction with the electric field oscillating in the x direction, so the equation of wave motion is

$$\frac{\partial^2 E}{\partial t^2} = v^2 \frac{\partial^2 E}{\partial Z^2} \tag{5.21}$$

where v is the wave velocity. The wave velocity is related to electromagnetic properties of the medium, which can be expressed as

$$v = \left(\frac{\varepsilon_0 \mu_0}{\varepsilon \mu}\right)^{1/2} \cdot c \tag{5.22}$$

where ε is the dielectric constant, μ is the permeability, ε_0 and μ_0 are respectively their values in vacuum, and c is the speed of light. Since the medium is a gas, the permeability is extremely small so $\mu = \mu_0$, and

$$v = \left(\frac{\varepsilon_0}{\varepsilon}\right)^{1/2} \cdot c \tag{5.23}$$

The solution to the wave equation (5.21) is

$$E = E_0 e^{i\omega(t - Z/v)}$$
$$= E_0 e^{i\omega\left(t - \sqrt{\frac{\varepsilon}{\varepsilon_0}} \cdot \frac{Z}{c}\right)} \tag{5.24}$$

where ω is the angular frequency. The term $\sqrt{\varepsilon/\varepsilon_0}$ can be obtained as follows. The difference between ε and ε_0 is due to the production of an additional polarized electric field in a medium in the presence of an external electric field. The total electric field D is the sum of the incident electric field E and the polarized electric field, and the ratio between the total electric field to the incident electric field is $\varepsilon/\varepsilon_0$. If we denote the number of oscillators in 1 cm^3 as n, the electric charge of an oscillator as q and the distance between two electric charges as x, we have

$$\frac{D}{E} = \frac{E + 4\pi nqx}{E} = 1 + \frac{4\pi nqx}{E} = \frac{\varepsilon}{\varepsilon_0} \tag{5.25}$$

For most stars, we can take $q = e = 4.803 \times 10^{-10}$ electrostatic unit. In the above equation, the distance x between two electric charges can be obtained as follows. Consider an oscillator at $Z = 0$, for which the oscillation equation in the electromagnetic field is

$$\frac{d^2 x}{dt^2} + \gamma \frac{dx}{dt} + \omega_0^2 x = \frac{e}{m_e} E_0 e^{i\omega t} \tag{5.26}$$

where $m_e = 9.10955 \times 10^{-28}$ g is the electron mass, and γ is the damping constant. It is not difficult to solve this equation. Choosing $x = x_0 e^{i\omega t}$, $(dx/dt) = i\omega x$, $(d^2x/dt^2) = -\omega^2 x$, and substituting into Eq. (5.26), we have

$$-\omega^2 x + i\gamma\omega x + \omega_0^2 x = \frac{e}{m_e} E_0 e^{i\omega t} \qquad (5.27)$$

Further substituting Eq. (5.27) into Eq. (5.25), we obtain

$$\frac{\varepsilon}{\varepsilon_0} = 1 + \frac{4\pi n e^2}{m_e} \cdot \frac{1}{\omega_0^2 - \omega^2 + i\gamma\omega} \qquad (5.28)$$

We know that, for a normal gas, $\varepsilon \sim \varepsilon_0$, so the second term on the right of the above equation is far less than unity. Employing the approximation $(1 + \delta)^{1/2} \approx 1 + \delta/2$ when $\delta << 1$, we have

$$\sqrt{\frac{\varepsilon}{\varepsilon_0}} \cong 1 + \frac{2\pi n e^2}{m_e} \frac{1}{\omega_0^2 - \omega^2 + i\gamma\omega}$$

which can be written as

$$\sqrt{\frac{\varepsilon}{\varepsilon_0}} \cong 1 + \frac{2\pi n e^2}{m_e} \frac{\omega_0^2 - \omega^2}{(\omega_0^2 - \omega^2)^2 + \gamma^2\omega^2} - i\frac{2\pi n e^2}{m_e} \frac{\gamma\omega}{(\omega_0^2 - \omega^2)^2 + \gamma^2\omega^2}$$

$$(5.29)$$

For the sake of convenience, we can write $\sqrt{\varepsilon/\varepsilon_0}$ in the form

$$\sqrt{\frac{\varepsilon}{\varepsilon_0}} \equiv n - iK \qquad (5.30)$$

so Eq. (5.24) becomes

$$E = E_0 e^{i\omega\left[t - (n - iK)\frac{Z}{c}\right]} \qquad (5.31)$$

or

$$E = E_0 e^{i\omega\left(t - \frac{nZ}{c}\right) - \omega K \frac{Z}{c}} \qquad (5.32)$$

Since the intensity I_v is proportional to EE*, we have

$$I_v = I_0 e^{-2K\omega Z/c} \qquad (5.33)$$

On the other hand, the reduction in radiation intensity can be expressed through the absorption coefficient as

$$I_\nu = I_o e^{-\kappa_{ij} Z} \tag{5.34}$$

Comparing Eqs. (5.33) and (5.34), and substituting the expression for K in Eq. (5.30), we get

$$\kappa_{ij} = \frac{4\pi n e^2}{m_e c} \frac{\gamma \omega^2}{(\omega_0^2 - \omega^2)^2 + \gamma^2 \omega^2} \tag{5.35}$$

which represents a sharp peak distribution centred at ω_0, so $\omega \sim \omega_0$ and $\omega_0^2 - \omega^2 \approx 2\omega(\omega_0 - \omega)$, and from Eq. (5.35), we have

$$\kappa_{ij} = \frac{n\pi e^2}{m_e c} \left[\frac{\gamma}{(\omega_0 - \omega)^2 + (\gamma/2)^2} \right] \tag{5.36}$$

Note that the absorption coefficient here does not include the negative absorption, i.e., the induced emission has not been included as the negative absorption, so

$$\kappa_{ij} = n_i B_{ij} h\nu_{ij} \varphi_\nu / 4\pi \tag{5.37}$$

Comparing Eqs. (5.36) and (5.37), we arrive at the following relationships:

$$B_{ij} = \frac{4\pi^2 e^2}{m_e c h \nu_{ij}} \tag{5.38}$$

$$\varphi_\nu = \frac{\gamma}{(\omega_0 - \omega)^2 + (\gamma/2)^2} \tag{5.39}$$

Obviously, the absorption profile φ_ν can fulfill the normalization condition because

$$\int_0^\infty \varphi_\nu d\nu = \int_0^\infty \frac{(\gamma/4\pi^2) d\nu}{(\nu - \nu_0)^2 + (\gamma/4\pi)^2}$$

$$= \frac{1}{\pi} \int_0^{+\infty} \frac{dx}{1 + x^2} = 1 \tag{5.40}$$

where $x = 4\pi(\nu - \nu_0)/\gamma$.

Equation (5.38) is the expression for B_{ij} derived from the classical oscillator theory, which is not very precise. A more precise formula can be obtained from quantum mechanics. Here, we do not want to go into details of quantum mechanical derivations and we just present the quantum mechanical result for B_{ij} as

$$B_{ij} = \frac{4\pi^2 e^2}{m_e c h \nu_{ij}} f_{ij} \qquad (5.41)$$

Comparing Eqs. (5.38) and (5.41), we notice that the classical result and the quantum mechanical result only differ by a coefficient f_{ij} which is called the oscillator strength, and can be computed by quantum mechanics and determined experimentally. For example, from the several spectral lines of hydrogen, we obtain the following values:

$$H_\alpha : f = 0.640742$$
$$H_\beta : f = 0.119321$$
$$H_\gamma : f = 0.044670$$
$$H_\delta : f = 0.022093$$
$$H_\varepsilon : f = 0.012704$$
$$H_\zeta : f = 0.008036$$

If the induced emission is treated as negative absorption and incorporated into the absorption coefficient, the expression for the absorption coefficient will be

$$\kappa_{ij} = n_i \frac{\pi e^2}{m_e c} f_{ij} \varphi_\nu \left(1 - \frac{g_i}{g_j} \frac{n_j}{n_i} \right) \qquad (5.42)$$

To derive the atomic absorption cross section, we still adopt the expression excluding the negative absorption. From Eq. (2.33), the relationship between κ_{ij} and a_{ij} for bound-bound transition is

$$\kappa_{ij} = n_i a_{ij} \qquad (5.43)$$

so the atomic cross section a_{ij} for bound-bound transition is

$$a_{ij} = \frac{\pi e^2}{m_e c} f_{ij} \varphi_\nu = \frac{B_{ij} \varphi_\nu h \nu_{ij}}{4\pi} \qquad (5.44)$$

Similarly, the emission coefficient can be obtained as

$$\eta_{ij} = \frac{1}{4\pi} n_j \varphi_v A_{ji} h v_{ij} = \frac{2h v^3}{c^2} \frac{g_i}{g_j} n_j B_{ij} \varphi_v \frac{h v_{ij}}{4\pi}$$
$$= n_j \frac{2h v_{ij}^3}{c^2} \frac{g_i}{g_j} a_{ij} \tag{5.45}$$

For a stellar atmosphere under LTE, $\frac{g_i}{g_j} \frac{n_j}{n_i} = e^{-h v_{ij}/kT}$, so we can obtain from Eq. (5.42)

$$\kappa_{ij} = n_i \frac{\pi e^2}{m_e c} f_{ij}(1 - e^{-h v_{ij}/kT}) \varphi_v \tag{5.46}$$

5.2 BOUND-FREE TRANSITION

The above treatments for bound-bound transitions can also be applied to bound-free transitions. For simplicity, we first discuss a transition between a single bound state and a free state. In 1 cm³, we denote the number of neutral atoms as n_0, the number of ions as n_1 and the number of electrons as n_e. We further suppose that the speeds of free electrons always follow the Maxwell distribution, and denote the number of electrons with velocities in the range $(v, v + dv)$ in 1 cm³ as dn_e. Moreover, similar to the Einstein coefficients introduced above, we have to introduce C_v which is the probability of having a bound-free transition per second, $G(v)$ which is the probability of having a spontaneous recombination per second and is a function of the velocity of electrons, $F(v)$ which is the probability of having an induced recombination per second and is also a function of the velocity of electrons.

Since the number of photoionization transitions from a bound state to a free state is proportional to n_e and I_v, the number of inverse processes, i.e., spontaneous recombination transitions, is proportional to n_1, dn_e and v, and the number of induced recombination transitions is proportional to n_1, dn_e, v and I_v, the employment of the method similar to that used in the above discussions will give

$$hv[n_0 C_v I_v - n_1 F(v) dn_e v I_v / dv] dv \frac{d\omega}{4\pi} = \kappa_{iK} I_v dv d\omega$$

$$hv n_1 G(v) dn_e v \frac{d\omega}{4\pi} = \eta_{iK} dv d\omega$$

or

$$\kappa_{iK} = h\nu \left[n_o C_\nu - n_1 \frac{dn_e}{d\nu} F(\nu)\nu \right] / 4\pi \tag{5.47}$$

$$\eta_{iK} = h\nu n_1 G(\nu)\nu \frac{dn_e}{d\nu} / 4\pi \tag{5.48}$$

Similarly, there are general relationships among the transition coefficients C_ν, $G(\nu)$ and $F(\nu)$ which only reflect properties of the atoms themselves and are independent of the states of the atoms. Therefore we employ special conditions of the atoms under LTE to determine these relationships. Under LTE, we have the following relationships:

(1) Planck radiation law:

$$I_\nu = \frac{2h\nu^3}{c^2} \frac{1}{e^{h\nu/kT} - 1} \equiv B_\nu(T) \tag{5.49}$$

(2) Maxwell speed distribution:

$$dn_e = n_e 4\pi v^2 \left(\frac{m_e}{2\pi kT} \right)^{3/2} e^{-m_e v^2/2kT} dv$$

since

$$h\nu = \chi + \frac{1}{2} m_e v^2$$
$$h d\nu = m_e v dv$$

so

$$dn_e = n_e \frac{4\pi h}{m_e} \left(\frac{m_e}{2\pi kT} \right)^{3/2} v e^{(\chi - h\nu)/kT} d\nu \tag{5.50}$$

(3) Equation of statistical equilibrium:

$$n_o^* C_\nu I_\nu d\nu = n_1^* dn_e \nu [G(\nu) + F(\nu)I_\nu] \tag{5.51}$$

where n_o^* and n_1^* represent the occupation numbers under the LTE condition.

(4) Saha equation:

$$\frac{n_1^* n_e}{n_o^*} = \frac{2U_1}{U_0} \left(\frac{2\pi kT}{m_e} \right)^{3/2} \frac{m_e^3}{h^3} e^{-\chi/kT} \tag{5.52}$$

From Eqs. (5.49) to (5.51), we get

$$C_\nu B_\nu(T) = \frac{n_1^* n_e}{n_0^*} \frac{4\pi h}{m_e} \left(\frac{m_e}{2\pi kT}\right)^{3/2} \nu^2 e^{(\chi - h\nu)/kT} [G(\nu) + F(\nu) B_\nu(T)]$$

Substituting the Saha equation, we have

$$C_\nu B_\nu(T) = \frac{2U_1}{U_0} \frac{4\pi m_e^2 \nu^2}{h^2} e^{-h\nu/kT} [G(\nu) + F(\nu) B_\nu(T)]$$

Solving for $B_\nu(T)$, we arrive at

$$B_\nu(T) = \frac{G(\nu)}{F(\nu)} \left[\frac{C_\nu U_0 h^2}{2U_1 4\pi m_e^2 \nu^2 F(\nu)} e^{h\nu/kT} - 1\right]^{-1} \qquad (5.53)$$

Comparing Eqs. (5.49) and (5.53), we get

$$\frac{G(\nu)}{F(\nu)} = \frac{2h\nu^3}{c^2} \qquad (5.54)$$

$$C_\nu = \frac{2U_1}{U_0} \frac{4\pi m_e^2 \nu^2}{h^2} F(\nu) = \frac{U_1}{U_0} \frac{4\pi m_e^2 \nu^2 c^2}{h^3 \nu^3} G(\nu) \qquad (5.55)$$

which give relationships among C_ν, $G(\nu)$ and $F(\nu)$ and are called Einstein–Milne relationships. Through Eqs. (5.47), (5.50), (5.52) and (5.55), the expression for bound-free transition is derived as

$$\kappa_{iK} = h\nu C_\nu (n_0 - n_0^* e^{-h\nu/kT})/4\pi \qquad (5.56)$$

which has incorporated induced emission as negative absorption in the correction term in the brackets. To obtain the atomic cross section, we should adopt the expression excluding negative absorption. From Eq. (2.33), the relationship between κ_{iK} and a_{iK} for bound-free transition is

$$\kappa_{iK} = n_0 a_{iK} \qquad (5.57)$$

so the atomic cross section is

$$a_{iK} = C_\nu h\nu/4\pi \qquad (5.58)$$

through which Eq. (5.56) can be written as

$$\kappa_{iK} = (n_o - n_o^* e^{-h\nu/kT}) a_{iK} \qquad (5.59)$$

Using similar methods, the emission coefficient for bound-free transition can be obtained through Eq. (5.48) as

$$\eta_{iK} = n_1 \frac{2h\nu^3}{c^2} \frac{g_i}{g_K} a_{iK} \qquad (5.60)$$

Under the LTE condition, the emission coefficient can be expressed as

$$\eta_{iK} = \frac{2h\nu^3}{c^2} e^{-h\nu/kT} a_{iK} n_i^* \qquad (5.61)$$

where n_i^* represents the occupation number under LTE.

From Eq. (5.58), the atomic absorption cross section a_{iK} for bound-free transition is mainly determined by the transition coefficient C_ν which can be calculated using quantum mechanics. The theoretical calculated values of a_{iK} will be presented in §5.5.

It should be noted that the above discussions are for one type of elements and have assumed that these elements have values of κ_{ij} and a_{iK} for a single bound state for simplicity. If the gas has many types of atoms, and each type of atom has many bound states, the absorption coefficient should be the sum of bound-free transition absorption coefficients for all bound states for all types of atoms, and Eq. (5.59) becomes

$$\kappa_{iK} = \sum_A \sum_n \frac{X_A \rho}{A m_P} n_{An} (1 - e^{-h\nu/kT}) a_{AnK} \qquad (5.62)$$

where ρ is the density, X_A is the abundance of the element with atomic weight A, m_P is the proton mass, n_{An} is the occupation number of the Ath type atoms on the nth energy level, a_{AnK} is the atomic absorption cross section when the Ath type atoms undergo transitions from the nth energy level to the free state.

5.3 FREE-FREE TRANSITION

After a bound electron of an atom has gone into a free state, it can still absorb photon energies and undergo transitions to another free state.

This process is referred to as the free-free transition, for which the values of κ_{KK} and a_{KK} can be obtained through methods entirely similar to those for bound-bound transition. The difference is that the number of free-free transitions is proportional to $n_1 \cdot n_e$, where n_e and n_1 are the number of electrons and ions per cm³, so the relationship between κ_{KK} and a_{KK} is

$$\kappa_{KK} = n_1 \cdot n_e \cdot a_{KK}(1 - e^{-h\nu/kT}) \qquad (5.63)$$

which has incorporated negative absorption as the correction term in the brackets. The atomic absorption cross section a_{KK} for free-free transition can be calculated using quantum mechanics. The calculations of a_{KK} for some elements will be given in §5.5. The emission coefficient for free-free transition can be written as

$$\eta_{KK} = \frac{2h\nu^3}{c^2} e^{-h\nu/kT} \cdot a_{KK} \cdot n_e n_1 \qquad (5.64)$$

For a multi-element gas, the absorption coefficient of free-free transition is given through Eq. (5.63) as

$$\kappa_{KK} = \sum_A \int_\nu \frac{X_A \rho}{A m_P} n_e(v)dv(1 - e^{-h\nu/kT}) \cdot a_{AKK} \qquad (5.65)$$

In this equation, since a_{AKK} is related to the electron speed, it is therefore first multiplied by the number of electrons $n_e(v)$ determined by the Maxwell distribution of speeds, which are then integrated over all speeds. Summation is then carried out over all types of elements.

5.4 SCATTERING

Particles (including electrons, atoms and molecules) are forced into oscillation under the action of the incident electromagnetic wave. The oscillator can emit secondary waves with the same frequency but into different directions. This phenomenon is called scattering. Under temperatures of normal stars, there is no momentum exchange between the incident photon and the particles during scattering process, so the scattering only changes the direction of the radiation and does not change its frequency. We refer to this as coherent scattering. If the scattering process occurs under a very high temperature, there is momentum exchange between the incoming photon and the particle, i.e., part of the energy of the incoming photon is transferred to the particle which leads to a change in the frequency of the incident radiation. We refer to this as Compton scattering.

(1) Thompson scattering

Under temperatures of normal stars, the scattering of radiation by an electron is called Thompson scattering. Here, the scattering cross section can be calculated using classical electrodynamics. The method is presented in §5.1. The expression is

$$a_e = \frac{8\pi e^4}{3m_e^2 c^4}\left[\frac{\omega^4}{(\omega^2 - \omega_0^2)^2 + \gamma^2\omega^2}\right] \tag{5.66}$$

where ω_0 is the characteristic frequency of the oscillator, ω is the angular frequency of the incident radiation and γ is the damping constant. For a free electron, $\omega_0 = \gamma = 0$ so Eq. (5.66) becomes

$$a_e = \frac{8\pi e^4}{3m_e^2 c^4} = 6.65 \times 10^{-25} \ [\text{cm}^2] \tag{5.67}$$

Since

$$\sigma_e = n_e a_e \tag{5.68}$$

and $n_e = \frac{\rho}{\mu_e m_p}$, $\mu_e = \frac{2}{1+X_H}$, we have

$$\sigma_e = 0.2004(1 + X_H)\rho \tag{5.69}$$

It should be noted that σ_e is the volume absorption coefficient. For the mass absorption coefficient, we have $\sigma_e = 0.2004(1 + X_H)$.

(2) Compton scattering

When the temperature is very high, there is momentum exchange between the incoming photon and the particle during the scattering, which leads to a change in the frequency of the radiation. We refer to this scattering as Compton scattering. The Compton scattering cross section is smaller than the Thompson scattering cross section, and is equal to the multiplication of the Thompson scattering cross section by a correction factor (see Hayashi *et al.* 1962):

$$\sigma_c = \sigma_e(1 + 0.35T_9^{1/2} + 0.73T_9)^{-1} \tag{5.70}$$

where $T_9 = T \cdot 10^{-9}$. When $T_9 \leq 2$, the maximum error of Eq. (5.70) is approximately $\pm$ 5%.

(3) Rayleigh scattering on neutral atoms
In general, there are neutral atoms and Rayleigh scatterings on these neutral atoms in cool stars such as G and K late type stars. Dalgarno and Willians (1962) calculated the Rayleigh scattering cross sections of neutral hydrogen atoms and neutral helium atoms, which were described as

Neutral hydrogen atoms:

$$a_H = 5.799 \times 10^{-13}/\lambda^4 + 1.422 \times 10^{-6}/\lambda^6 + 2.784/\lambda^8 \tag{5.71}$$

where $\lambda = 2.997925 \times 10^{18}/\min$ (ν, 2.922×10^{15});

Neutral helium atoms:

$$a_{He} = \frac{5.484 \times 10^{-14}}{\lambda^4}\left[1 + \frac{2.44 \times 10^5}{\lambda^2} + \frac{5.94 \times 10^{-10}}{\lambda^2(\lambda^2 - 2.90 \times 10^5)}\right]^2 \tag{5.72}$$

where $\lambda = 2.997925 \times 10^{18}/\min$ (ν, 5.15×10^{15}).
    Similarly, we have

$$\begin{aligned}\sigma_H &= n_H a_H \\ \sigma_{He} &= n_{He} a_{He}\end{aligned} \tag{5.73}$$

(4) Rayleigh scattering on hydrogen molecules
Dalgarno and Willians (1962) also gave the Rayleigh scattering cross section of hydrogen molecules as

$$a_{H_2} = 8.14 \times 10^{-13}/\lambda^4 + 1.28 \times 10^{-6}/\lambda^6 + 1.61/\lambda^8 \tag{5.74}$$

where the unit of λ is [Å]. Multiplying Eq. (5.74) by the number of hydrogen molecules per cm³, the Rayleigh scattering absorption coefficient can be obtained as

$$\sigma_{H_2} = n_{H_2} \cdot a_{H_2} \tag{5.75}$$

5.5 ATOMIC CROSS SECTIONS AND ABSORPTION COEFFICIENTS FOR SOME ATOMS

In this section, the atomic cross sections and absorption coefficients for some atoms will be calculated by quantum mechanics.

(1) Absorption coefficient for bound-free transition for hydrogen
Figure 5.1 shows the energy levels, excitation energies and ionization energies for hydrogen atoms. Table 5.1 gives the wavelengths corresponding to the principal quantum numbers for the hydrogen atom. Kramers (1923) and Gaunt (1930) gave the atomic cross section for bound-free transition for hydrogen as

$$a_n = \frac{\alpha_0 g_n' \lambda^3}{n^5} \, [\text{cm}^2] \qquad (5.76)$$

where $\alpha_0 = 1.044 \times 10^{-26}$ and g_n' is the Gaunt factor which can be calculated by

$$g_n' = 1 - 0.3456(\lambda R)^{-1/3} \left(\frac{\lambda R}{n^2} - \frac{1}{2} \right) \qquad (5.77)$$

and $R = 2\pi^2 m e^4 / h^3 c$ is the Rydberg constant.

Table 5.1 The wavelengths corresponding to the principal quantum numbers for the hydrogen atom.

n	$\lambda(\text{Å})$
1	912
2	3647
3	8206
4	14588

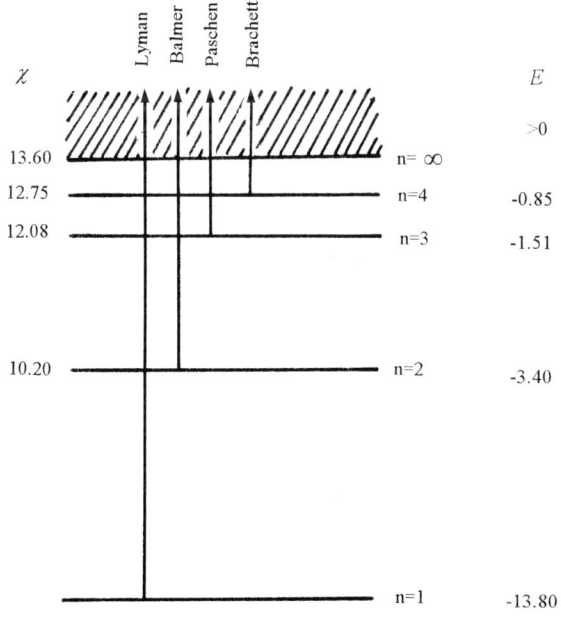

Fig. 5.1 The energy levels, excitation energies and ionization energies for hydrogen atoms.

Multiplying the atomic occupation number n_n on each energy level by a_n, and then summing over all energy levels, we can obtain the absorption coefficients for bound-free transition for the hydrogen atom as

$$\kappa_{iK}(H) = \alpha_0 \sum_n g_n' \frac{\lambda^3}{n^5} n_n \tag{5.78}$$

(2) Absorption coefficient for free-free transition for hydrogen
Kramers (1923) and Gaunt (1930) assumed that the free electrons followed the Maxwell distribution and gave the absorption cross section for free-free transition for hydrogen atoms as

$$a_f = \frac{2}{3\sqrt{3}} \frac{h^2 e^2 R}{\pi m^3} \frac{1}{\nu^3} \left(\frac{2m}{\pi kT}\right)^{1/2} g_f' \tag{5.79}$$

where the Gaunt factor g_f' was calculated by

$$g_f' = 1 + 0.3456(\lambda R)^{-1/3}\left(\frac{\lambda kT}{hc} + \frac{1}{2}\right) \tag{5.80}$$

and the absorption coefficient is proportional to the product of the electron number density n_e and the proton number density n_p. The absorption coefficient in units of cm² calculated for a neutral hydrogen atom can be written as

$$\kappa_{KK}(H) = \frac{g_f n_p n_e}{n_0} a_f \tag{5.81}$$

where n_0 is the number density of neutral hydrogen atoms and g_f is the weighting factor for the free state. Under LTE, by using the Saha equation, Eq. (5.81) can be written as

$$\kappa_{KK}(H) = a_f g_f \frac{(2\pi mkT)^{3/2}}{h^3} e^{-\chi/kT} \tag{5.82}$$

(3) Negative hydrogen ion
The negative hydrogen ion is a system consisting of a neutral hydrogen atom and an attached electron, and the formation is due to incomplete shielding of the electric field of the nucleus by electrons in the hydrogen atom. In the stellar atmosphere, there are a lot of hydrogen atoms and the electron density is also very high, so the chance of

getting negative hydrogen ions is high. A negative hydrogen ion has only one bound state with a binding energy of 0.754 eV. Due to the low binding energy, negative hydrogen ions do not exist under very high temperatures. In general, in stellar atmospheres with temperatures higher than that of the sun, negative hydrogen ions do not exist. Conversely, in atmospheres of stars with the same spectral types of the sun (G type) or with later spectral types such as the K and M types, negative hydrogen ions can be very important and constitute a major component of the continuous absorption coefficient.

The main interactions between negative hydrogen ions and radiation are bound-free and free-free transitions.

(i) Bound-free Transition

Krogdahl and Miller (1967), Doughty et al. (1966) and Geltman (1962) have calculated the atomic absorption cross section for bound-free transition for negative hydrogen ions. For different regions of λ, the atomic absorption cross sections are different which can be approximated by the following expressions:

$1500\text{Å} < \lambda < 5250\text{Å}$:
$$\lg a_{iK} = -16.20450 + 0.17280 \times 10^{-3}(\lambda - 8500) + 0.39422$$
$$\times 10^{-7}(\lambda - 8500)^2 + 0.51345 \times 10^{-11}(\lambda - 8500)^3$$

$5250\text{Å} < \lambda < 11250\text{Å}$
$$\lg a_{iK} = -16.40383 + 0.61356 \times 10^{-6}(\lambda - 8500) - 0.11095$$
$$\times 10^{-7}(\lambda - 8500)^2 + 0.44965 \times 10^{-13}(\lambda - 8500)^3$$

$11250\text{Å} < \lambda < 15000\text{Å}$
$$\lg a_{iK} = -15.95015 - 0.36067 \times 10^{-3}(\lambda - 8500) + 0.86108$$
$$\times 10^{-7}(\lambda - 8500)^2 + 0.90741 \times 10^{-11}(\lambda - 8500)^3$$

$$(5.83)$$

where a_{iK} has the unit cm²(per H⁻ per electron pressure).

The ionization of negative hydrogen ions can be obtained from the Saha equation. In this case, $U_0 = 1$ and $U_1 = 2$, so

$$\lg \frac{n(H)}{n(H^-)} = -\lg P_e - \frac{5040}{T}\chi + 2.5 \lg T + 0.1248$$

Thus the absorption coefficient for bound-free transition is

$$\kappa_{iK}(H^-) = 4.158 \times 10^{-10} a_{iK} P_e \theta^{5/2} \cdot 10^{0.754\theta} \qquad (5.84)$$

where P_e is the electron pressure, $\theta = 5040/T$ and $\kappa_{iK}(H^-)$ has a unit of cm²/(per neutral hydrogen atom). Note that there is very little H^- so they are normalized into the absorption of neutral hydrogen atoms.

(ii) Free-free Transition

In the infrared region, free-free absorption becomes the most important. John (1966), Geltman (1965), Doughty *et al.* (1966) and Stilley and Callaway (1970) have calculated the absorption coefficient for free-free transition for negative hydrogen ions. The calculated results show that the magnitude of the absorption coefficient is related to the temperature and the wavelength. For $0.5 < \theta < 2.0$ and $3038\text{Å} < \lambda < 91130\text{Å}$, the absorption coefficient for free-free transition for negative hydrogen ions can be approximated as

$$\kappa_{KK}(H^-) = P_e 10^{f_0 + f_1 \lg \theta + f_2 \lg^2 \theta} \tag{5.85}$$

where

$$f_0 = -31.3602 + 0.48735 \lg\lambda + 0.296586 \lg^2\lambda - 0.0193562 \lg^3\lambda$$
$$f_1 = 15.3126 - 9.33651 \lg\lambda + 2.000242 \lg^2\lambda - 0.1422568 \lg^3\lambda$$
$$f_2 = -2.6117 + 3.22259 \lg\lambda - 1.082785 \lg^2\lambda + 0.1072635 \lg^3\lambda$$

(4) Absorption coefficient for bound-free transition for helium

Helium is the second most abundant element in stars after hydrogen. The atomic absorption cross section for bound-free transition for helium can be calculated by

$$a_{iK}(HeI) = \sum_\lambda \kappa_\nu(i) \frac{g_i}{g_0} e^{-\chi_i/kT} \tag{5.86}$$

Table 5.2 lists some of the parameters for the above equation, and κ_ν are given as

$1^1S - \frac{1}{K} P$ $\kappa_\nu = 7.3 \times 10^{-18} \exp(1.373 - 2.311 \times 10^{-16} \cdot \nu)$

$2^3S - \frac{3}{K} P$ $\kappa_\nu = \exp(-278.3 + 14.88 \cdot Z - 0.2311 \cdot Z^2)$

$2^1S - \frac{1}{K} P$ $\kappa_\nu = \exp(-31.67 - 0.3168 \cdot Z + 0.02619 \cdot Z^2 + 0.1016$
$\qquad\qquad\qquad\qquad \times 10^{-3} \cdot Z^3 - 2.408 \times 10^{-5} \cdot Z^4 + 0.4761 \times 10^{-7} \cdot Z^5)$

$2^3P - \frac{3}{K} D$ $\kappa_\nu = \exp(-18.51 - 3.418 \cdot y + 0.9968 \cdot y^2 + 13.00 \cdot y^3$
$\qquad\qquad\qquad\qquad + 31.34 \cdot y^4 + 24.03 \cdot y^5)$

$2^1P - \frac{1}{K} D$ $\log\kappa_\nu = -3.5 \log\nu + 35.310$

Table 5.2 Parameters for calculating the atomic absorption cross section for HeI.

i	transition	$\chi_i(\text{eV})$	λ_i	$\nu_i \times 10^{-15}$	g_i
0	$1^1S - \frac{1}{K}P$	0.00	504.3	5.9447	1
1	$2^3S - \frac{3}{K}P$	19.72	2601	1.1526	3
2	$2^1S - \frac{3}{K}P$	20.51	3122	0.96026	1
3	$2^3P - \frac{3}{K}D$	20.86	3422	0.87607	9
4	$2^1P - \frac{1}{K}D$	21.11	3680	0.81465	3
5	$2^3P - \frac{3}{K}S$	20.86	3422	0.87607	9
6	$2^1P - \frac{1}{K}S$	21.11	3680	0.41465	3

$$2^3P - \frac{3}{K}S \qquad \log \kappa_\nu = -3.3 \log \nu + 31.059$$
$$2^1P - \frac{1}{K}S \qquad \log \kappa_\nu = -3.6 \log \nu + 35.487$$

where $Z = \ln\nu$, $y = \ln(\nu/R_{He})$ and R_{He} is the Rydberg constant for helium.

(5) Negative helium ions

The bound-free absorption of He$^-$ can be neglected in common stellar atmospheres since He$^-$ has only one bound state with an excitation potential of 19 eV and the number of particles at this level is too small. Nevertheless, free-free absorption of negative helium ions can be important at long wavelengths for cool stars. McDowell *et al.* (1966) gave an approximation of free-free absorption of He$^-$ as

$$\kappa_{KK}(He^-) = P_e \cdot 10^{a_0 + a_1\lambda^{1/3} + a_2\lambda^{1/2} + a_3\lambda A(He)} \cdot \left[1 + \frac{\phi(He)}{P_e}\right]^{-1} \qquad (5.87)$$

where $A(He)$ is the abundance of helium relative to that of hydrogen, P_e is the electron pressure, $\phi(He)$ can be obtained from

$$\phi(He) = 0.6665 \frac{U_1}{U_0} T^{5/2} \cdot 10^{-\frac{5040}{T}\chi}$$

and the coefficients in the index are

$$a_0 = -0.3183 \times 10^2 + 0.1358 \times 10^1 \cdot \theta - 0.1047 \cdot \theta^2 - 0.2819 \times 10^{-1} \cdot \theta^3$$
$$a_1 = 0.4332 - 0.1268 \cdot \theta - 0.3535 \times 10^{-2} \cdot \theta^2 + 0.6869 \times 10^{-2} \cdot \theta^3$$
$$a_2 = -0.4808 \times 10^{-1} + 0.1960 \times 10^{-1} \cdot \theta + 0.4305 \times 10^{-4} \cdot \theta^2$$
$$\quad - 0.9558 \times 10^{-3} \cdot \theta^3$$
$$a_3 = 0.1788 \times 10^{-4} - 0.1192 \times 10^{-4} \cdot \theta + 0.7870 \times 10^{-6} \cdot \theta^2 + 0.3936$$
$$\quad \times 10^{-6} \cdot \theta^3$$

where $\theta = 5040/T$.

5.6 APPROXIMATIONS FOR OPACITIES FOR MAJOR ABSORPTION PROCESSES

In many cases, calculated absorption coefficients should be substituted into Eq. (3.40) to evaluate the Rosseland mean values. In the following, we will present the approximations for opacities for major absorption processes after evaluating the Rosseland mean values.

5.6.1. Bound-free Transition

Kramers (1923) computed the atomic cross section for bound-free transition as

$$a_{iK} = \frac{64\pi^4}{3\sqrt{3}} \frac{Z_i^4 m_e e^{10}}{ch^6} \frac{1}{n^5} \frac{1}{v^3} S_{ni}^4 g_{iK} \quad (v > v_n) \tag{5.88}$$

where n is the principal quantum number, $Z_i e$ is the electric charge of the atomic nucleus, S_{ni} is the shielding coefficient and g_{ik} is the Gaunt factor. If the electron gas is partially degenerate, Eq. (5.88) should be further multiplied by a degeneracy factor q_{iK} given by

$$q_{iK} = \frac{1}{e^{[\psi - h v/kT + \chi_n/kT]} + 1} \tag{5.89}$$

where ψ is the degeneracy parameter and χ_n is the ionization potential. We substitute Eq. (5.88) into Eq. (5.62) to get the Rosseland mean values, and n_{An} in Eq. (5.62) can be calculated by

$$n_{An} = n_e \cdot n^2 \frac{h^3}{2(2\pi m_e kT)^{3/2}} e^{\chi_n/kT} \qquad (5.90)$$

where n_e is the number of electrons in 1 cm^3. The above equation is obtained by the Saha equation under the assumption that the interior of the star is highly ionized. Finally, an approximation to the Rosseland mean value of opacity of photoionization is

$$\bar{\kappa}_{iK} = \kappa_o \rho T^{-3.5} \qquad (5.91)$$

where

$$\kappa_o = 4.34 \times 10^{25} X_Z (1 + X_H) \frac{\bar{g}_{iK}}{t} \qquad (5.92)$$

Here X_H is the hydrogen abundance, X_Z is the heavy element abundance, $\bar{g}_{iK}$ is the mean Gaunt factor, t is the cutoff factor which can be obtained through the form given by Morse and Harisson to be

$$t \approx \rho^{0.25} \qquad (5.93)$$

5.6.2 Free-free Transition

Kramers (1923) calculated the atomic absorption cross section for free-free transition as

$$a_{KK} = \frac{4\pi e^6}{3\sqrt{3}chm_e^2} \frac{Z_i^2}{v} \frac{1}{v^3} S_{fi} \cdot g_{KK} \qquad (5.94)$$

where v is the electron speed, S_{fi} is the shielding coefficient and g_{KK} is the Gaunt factor for free-free transition. If the electron gas is partially degenerate, Eq. (5.94) should be multiplied by the degeneracy factor q_{KK}

$$q_{KK} = \frac{1}{e^{(\psi - hv/kT - \varepsilon/kT)} + 1} \qquad (5.95)$$

where ψ is the degeneracy parameter and ε is the electron energy before absorption of the photon.

Substituting Eq. (5.94) into Eq. (5.65), and evaluating the Rosseland mean value, the approximation to the Rosseland mean value of opacity for free-free transition can be obtained as

$$\bar{\kappa}_{KK} = \kappa_0 \rho T^{-3.5} \tag{5.96}$$

where

$$\kappa_0 = 3.68 \times 10^{22}(1 + X_H)(1 - X_Z)\bar{g}_{KK} \tag{5.97}$$

X_H and X_Z are the abundances of hydrogen and heavy elements, respectively, and $\bar{g}_{KK}$ is the mean value for the Gaunt factor for free-free transitions.

5.6.3 Electron Scattering

Since the scattering cross section σ_e for electron is unrelated to frequency [see Eq. (5.67)], the evaluation of the Rosseland mean value is very simple and only needs the substitution of n_e in Eq. (5.68) by [see Eq. (4.39)]

$$n_e = \frac{1}{\mu_e}\frac{\rho}{m_P}, \mu_e = \frac{2}{1 + X_H} \tag{5.98}$$

so

$$\bar{\kappa}_e = \frac{\sigma_e n_e}{\rho} = 0.2004(1 + X_H) \tag{5.99}$$

CHAPTER SIX

NUCLEAR ENERGY GENERATION

Stars are emitting vast amounts of energy continuously. So where does the energy come from? Research on this issue has a long history. Today, people know that the energy comes from the energy released from thermonuclear reactions occurring in the interior of the stars. The thermonuclear reactions not only provide energy for the star, but also lead to changes in chemical compositions and thus the internal structure of the star, which are the basic causes for stellar evolution. In this chapter, our objectives are to outline causes for thermonuclear reactions in the interior of stars, nuclear reaction rate, energy generation rate and changes in the chemical composition arising from nuclear reactions.

6.1 SYNTHETIC REACTIONS OF ATOMS AND ENERGY GENERATION

Synthetic reactions of atoms refer to those which convert several light atomic nuclei into a heavy atomic nucleus. During the synthesis, part of the mass will be converted into energy. According to Einstein's mass energy relationship

$$E = mc^2 \qquad (6.1)$$

although only a small amount of mass is converted into energy in the synthesis, the amount of the generated energy is enormous. We choose the conversion of two protons and two neutrons into a $_{2}^{4}He$ atomic nucleus as our example to illustrate this point. The mass of two protons and two neutrons is

$$2 \times 1.007825187m_u + 2 \times 1.008665197m_u = 4.032980768m_u$$

where m_u is the atomic mass unit and $1\ m_u = 1.6603 \times 10^{-24}$ g. When two protons and two neutrons fuse into a $_{2}^{4}He$ atomic nucleus, the mass of the $_{2}^{4}He$ atomic nucleus is $4.002603132\ m_u$. Therefore the total mass reduction is $0.030377647\ m_u$ after fusion which has been converted into energy. According to Eq. (6.1), the energy released will be 28.296 MeV. When this is compared to the energy released when one carbon atom combines with two oxygen atoms to form carbon dioxide during carbon burning, it can be seen that the latter only releases an energy of 4 eV which is only six-millionth of the former. [In the above computations, the atomic mass of hydrogen has been used instead of the proton mass, and the $_{2}^{4}He$ mass has also included the mass of two electrons, so the mass reduction of $0.030377647\ m_u$ is still the reduction of mass when two protons and two neutrons fuse into a $_{2}^{4}He$ atomic nucleus.]

The energy released during an atomic fusion is the binding energy of atomic nuclei. Since a mass of $1\ m_u$ is equivalent to 931.5016 MeV, for an atom with nucleon number A and total electric charge of Z, the binding energy of the nucleus is

$$E_b = 931.5016[Z \cdot m_H + (A - Z)m_n - M_i] \quad (\text{MeV}) \qquad (6.2)$$

where M_i is the atomic weight, m_H is the atomic weight of a hydrogen atom and m_n is the neutron mass.

When the binding energy E_b of an atom is divided by the nucleon number A of that atom, we obtain the mean binding energy per nucleon (proton or neutron) as

$$f = \frac{E_b}{A} \qquad (6.3)$$

Figure 6.1 gives the variation of mean binding energy f with nucleon number A. From Fig. 6.1, it can be seen that the mean binding energy is lowest for the lightest and heaviest atoms and is highest for intermediate atoms ($40 \leq A \leq 80$), with the maximum value occurring at ^{56}Fe which is 8.5 MeV. The higher the mean binding energy, the more stable is the atom, so it is more feasible to fuse lighter atoms into heavier atoms. Elements up to ^{56}Fe in a star can undergo exothermic fusion reaction. From Fig. 6.1, it can also be seen that the curve rises sharply for the lightest atoms, which shows that the largest amount of binding energy can be released when atoms in this region undergo fusion reactions. For example, when hydrogen atoms fuse into helium, the energy released is $E_{H \to He} = 10^{18.8}$(erg/g). However, when helium atoms fuse into carbon, the released energy is only one-tenth of the above value.

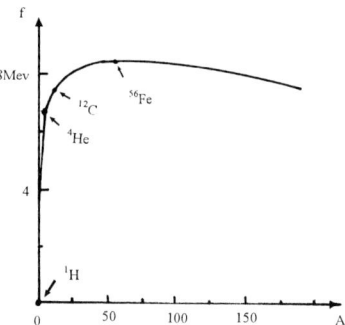

Fig. 6.1 The variation of mean binding energy f with nucleon number A.

6.2 THERMONUCLEAR REACTIONS

Before two atoms undergo fusion, they should get close to within the nuclear force range. The nuclear force binds the nucleons (protons and neutrons) into a nucleus. It is an attractive force with enormous strength, but the range is very small, only 10^{-13} cm. When the distance of two nucleons is greater than the nuclear force range, the force between them is the common Coulomb repulsive force. Figure 6.2 shows the variation in the potential energy near an atomic nucleus with electric charge number Z_1 and the energy levels of the atomic nucleus. It is noted that when the distance is 10^{-13} cm, the potential energy suddenly changes to a negative value due to the nuclear force. For two atoms with electric charge numbers Z_1 and Z_2 to fuse, their relative kinetic energies should overcome the so-called Coulomb barrier shown in Fig. 6.2. For simplicity, we consider the Coulomb barrier for two nuclei both with an electric charge number of 1, which is

$$V_{coul} = \frac{Z_1 Z_2}{R} \approx 1.6 \times 10^{-6} \cong 1 \text{ MeV} \qquad (6.4)$$

Suppose the temperature of the interior of the star is 10^7 K, so that the average kinetic energy for particles is

$$E_K = \frac{3}{2}kT \approx 10^{-4} \cdot T \cong 10^3 \text{ eV}$$

In other words, the kinetic energy is a thousand times smaller than the Coulomb barrier. It seems that nuclear reactions cannot occur under

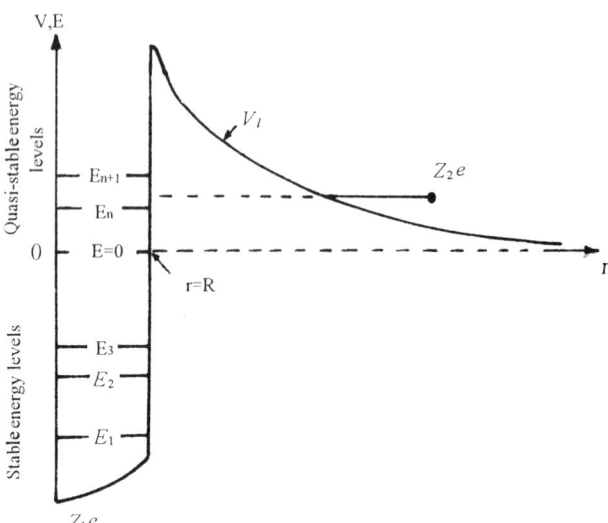

Fig. 6.2 The variation in potential energy near an atomic nucleus with electric charge number Z_1 and the energy levels of the atomic nucleus.

such a temperature. However, according to quantum mechanics, even when $E_K < V_{coul}$, the particles can still penetrate the Coulomb barrier with finite probabilities, which is referred to as the tunnel effect. The probability for tunnel effect is very small. For example, when $Z_1 \cdot Z_2 = 1$, $T = 10^7$ K, the probability for tunnel effect is about 10^{-20}. Therefore the rate of a thermonuclear reaction should be very slow, and the duration of a thermonuclear reaction maintained for a star is very long. The high temperature provides kinetic energies of particles and enables the nuclear reactions with finite probabilities, so these nuclear reactions are called thermonuclear reactions.

6.3 REACTION RATE

Consider a thermonuclear reaction in which particle a collides with nucleus X to undergo fusion reaction and form nucleus Y and particle b, which can be represented as

$$a + X \longrightarrow Y + b \tag{6.5}$$

or

$$X(a, b)Y \tag{6.6}$$

We denote $dn_a(v_a)$ and $dn_X(v_X)$ as the number of particles a and nuclei X, respectively in 1 cm³ and they are in the volume elements $d^3 v_a$ and $d^3 v_X$ in the velocity space, respectively (v_a and v_X relative to a fixed reference frame). Suppose the relative velocity for the two types of particles is $v(v = v_a - v_X)$, and the reaction cross section is $\sigma(v)$ (which is the probability of interaction between the incident particle and the target nucleus). The number of nuclear reactions in 1 cm³ per second, or the reaction rate, is

$$r_{aX} = \int\limits_{-\infty}^{+\infty} \int\limits_{-\infty}^{+\infty} \sigma(v) v \, dn_a(v) dn_X(v) \tag{6.7}$$

For conditions in a normal star, it is difficult for atoms to become degenerate (which requires a density $\rho \geq 10^{14}$ g/cm³). Under no degeneracy for atoms, we can assume the speed of particles to follow the Maxwell distribution, i.e.,

$$dn_i(v_i) = n_i \left(\frac{m_i}{2\pi kT} \right)^{3/2} e^{-m_i v_i^2 / 2kT} d^3 v_i \tag{6.8}$$

Substituting Eq. (6.8) into Eq. (6.7), we have

$$r_{aX} = n_a n_X \left(\frac{m_a}{2\pi kT} \right)^{3/2} \left(\frac{m_X}{2\pi kT} \right)^{3/2} \int\limits_{-\infty}^{+\infty} \int\limits_{-\infty}^{+\infty} e^{-\left(\frac{m_a v_a^2}{2kT} + \frac{m_X v_X^2}{2kT} \right)} \sigma(v) v \, d^3 v_a d^3 v_X \tag{6.9}$$

If U represents the velocity of the centre of mass of the two particles,

$$\left. \begin{aligned} v &= v_a - v_X \\ (m_a + m_X)U &= m_a v_a + m_X v_X \end{aligned} \right\} \tag{6.10}$$

so

$$\left. \begin{aligned} v_a &= U + \frac{m_X}{m_a + m_X} v \\ v_X &= U - \frac{m_a}{m_a + m_X} v \end{aligned} \right\} \tag{6.11}$$

and

$$\frac{1}{2} m_a v_a^2 + \frac{1}{2} m_X v_X^2 = \frac{1}{2} (m_a + m_X) U^2 + \frac{1}{2} m v^2 \tag{6.12}$$

where

$$m = \frac{m_a m_X}{m_a + m_X} \tag{6.13}$$

Substituting Eq. (6.12) into Eq. (6.9), and noting that the product $d^3 v_a \cdot d^3 v_b$ in the velocity space can be replaced by

$$d^3 U \cdot d^3 v = 4\pi U^2 dU \cdot 4\pi v^2 dv$$

we can obtain

$$r_{aX} = n_a n_X \left(\frac{m_a}{2\pi kT}\right)^{3/2} \left(\frac{m_X}{2\pi kT}\right)^{3/2} \int_0^\infty \int_0^\infty e^{-\left(\frac{(m_a + m_X)U^2}{2kT} + \frac{mv^2}{2kT}\right)}$$

$$\sigma(v) v 4\pi U^2 dU \cdot 4\pi v^2 dv \tag{6.14}$$

Integrating with respect to U, we have

$$\int_0^\infty e^{-(m_a + m_X)U^2/2kT} 4\pi U^2 dU = \left(\frac{2\pi kT}{m_a + m_X}\right)^{3/2} \tag{6.15}$$

so Eq. (6.14) becomes

$$r_{aX} = 4\pi n_a n_X \left(\frac{m}{2\pi kT}\right)^{3/2} \int_0^\infty e^{-mv^2/2kT} \sigma(v) v^3 dv \tag{6.16}$$

When using the relative kinetic energy

$$E = \frac{1}{2} m v^2 \tag{6.17}$$

Eq. (6.16) can be written as

$$r_{aX} = n_a n_X \int_0^\infty f(E)\sigma v dE = n_a n_X \langle \sigma v \rangle \tag{6.18}$$

where

$$\langle \sigma v \rangle = \int_0^\infty f(E) \sigma v dE \qquad (6.19)$$

$$f(E) = \frac{2}{\pi^{1/2}} \frac{1}{(kT)^{3/2}} e^{-E/kT} E^{1/2} \qquad (6.20)$$

The above discussion is based on the assumption that particle a is different from particle X. If they are the same, the probability of having a collision will be proportional to $n_a(n_a - 1)/2 \approx n_a^2/2$, so the general form of Eq. (6.18) can be expressed as

$$r_{aX} = \frac{1}{1 + \delta_{aX}} n_a n_X \langle \sigma v \rangle \qquad (6.21)$$

where

$$\delta_{aX} = \begin{cases} 0, & a \neq X \\ 1, & a = X \end{cases}$$

From Eq. (6.21), it can be seen that the physical meaning of $\langle \sigma v \rangle$ is the average probability of a pair of particles in 1 cm^3 having a reaction.

6.4 ENERGY RELEASED BY A NUCLEAR REACTION

Consider a nuclear reaction $X(a,b)Y$. According to conservation of energy, we have

$$E_{aX} + (\Delta M_a + \Delta M_X) = E_{bY} + (\Delta M_b + \Delta M_Y) \qquad (6.22)$$

where E_{aX} is the kinetic energy of the centre of mass of a and X, ΔM_a and ΔM_X are respectively, the binding energies of a and X, E_{bY} is the kinetic energy of the centre of mass of b and Y, and ΔM_b and ΔM_Y are respectively, the binding energies of b and Y. The binding energy of an atomic nucleus with electric charge number Z and nucleon number A is given by Eq. (6.2) as

$$\Delta M_{AZ} = 931.5016(M_{AZ} - A) \ \text{MeV} \qquad (6.23)$$

where M_{AZ} is the atomic mass (in units of m_u).

Table 6.1 lists the values of $(M_{AZ} - A)$ for atoms with different values of (Z, A) (in units of MeV), which can be used to compute the energy Q released by different nuclear reactions. For example, for the nuclear reaction $^{12}C(d, p)^{13}C$, the conservation of energy can be expressed as

$$E_{d,^{12}C} + 13.1359 + 0 = E_{p,^{13}C} + 7.2890 + 3.1246 \qquad (6.24)$$

or

$$E_{p,^{13}C} = E_{d,^{12}C} + 2.7223 \ \text{MeV} \qquad (6.25)$$

so that the energy released is calculated to be $Q_{d,^{12}C} = 2.7223$ MeV. It should be pointed out that we have not considered neutrinos (ν) in the particles produced in the above nuclear reaction. If neutrinos are produced, they can penetrate through the whole star due to their extremely small cross section, so the energy q_ν carried away by neutrinos should be subtracted from the energy Q_{ij} released in the nuclear reaction.

6.5 CHANGES IN CHEMICAL COMPOSITION AND RATE OF ENERGY GENERATION

The various reactions occurring in a certain region inside a star will definitely lead to changes in chemical composition in that region, which will in turn lead to changes in the internal structure of the star such as temperature, pressure and density. Therefore, the basic cause for stellar evolution is the internal thermonuclear reactions. The computation of changes in chemical composition due to nuclear reactions is important to model calculations of the internal structure and evolution of stars. The rate of change of particle number of the ith element produced from different nuclear reactions in 1 cm^3 is

$$\frac{dn_i}{dt} = -\sum_j r_{ij}(m) + \sum_{k,l} r_{kl}(i) \qquad (6.26)$$

Table 6.1 The values of $(M_{AZ} - A)$ for elements with electric charge number Z and nucleon number A, where M_{AZ} is the mass of that atom (in units of m_u). (From Clayton 1968)

Z	Element	A	M-A, Mev	Z	Element	A	M-A, Mev
0	n	1	8.07144	7	N	12	17.36400
1	H	1	7.28899			13	5.34520
	D	2	13.13591			14	2.86370
	T	3	14.94995			15	0.10040
	H	4	28.22000			16	5.68510
		5	31.09000			17	7.87100
2	He	3	14.93134	8	O	14	8.00800
		4	2.42475			15	2.85990
		5	11.45400			16	−4.73655
		6	17.59820			17	−0.80770
		7	26.03000			18	−0.78243
		8	32.00000			19	3.33270
3	Li	5	11.67900			20	3.79900
		6	14.08840	9	F	16	10.90400
		7	14.90730			17	1.95190
		8	20.94620			18	0.87240
		9	24.96500			19	−1.48600
4	Be	6	18.37560			20	−0.01190
		7	15.76890			21	−0.04600
		8	4.94420	10	Ne	18	5.31930
		9	11.35050			19	1.75200
		10	12.60700			20	−7.04150
		11	20.18100			21	−5.72990
5	B	7	27.99000			22	−8.02490
		8	22.92310			23	−5.14830
		9	12.41860			24	−5.94900
		10	12.05220	11	Na	20	8.28000
		11	8.66768			21	−2.18500
		12	13.37020			22	−5.18220
		13	16.56160			23	−9.52830
6	C	9	28.99000			24	−8.41840
		10	15.65800			25	−9.35600
		11	10.64840			26	−7.69000
		12	0	12	Mg	22	−0.14000
		13	3.12460			23	−5.47240
		14	3.01982			24	−13.93330
		15	9.87320			25	−13.19070

Table 6.1 *(continued).*

Z	Element	A	M-A, Mev	Z	Element	A	M-A, Mev
12	Mg	26	−16.21420	17	Cl	38	−29.8030
		27	−14.58260			39	−29.8000
		28	−15.02000			40	−27.5000
13	A1	24	0.10000	18	Ar	34	−18.3940
		25	−8.93100			35	−23.0510
		26	−12.2108			36	−30.2316
		27	−17.1961			37	−30.9509
		28	−16.8554			38	−34.7182
		29	−18.2180			39	−33.2380
		30	−17.1500			40	−35.0383
14	Si	26	−7.1320			41	−33.0674
		27	−12.3860			42	−34.4200
		28	−21.4899	19	K	36	−16.7300
		29	−21.8936			37	−24.8100
		30	−24.4394			38	−28.7860
		31	−22.9620			39	−33.8033
		32	−24.2000			40	−33.5333
15	P	28	−7.6600			41	−35.5524
		29	−16.9450			42	−35.0180
		30	−20.1970			43	−36.5790
		31	−24.4376			44	−35.3600
		32	−24.3027			45	−36.6300
		33	−26.3346			46	−35.3400
		34	−24.8300			47	−36.2500
16	S	30	−14.0900	20	Ca	38	−21.6900
		31	−18.9920			39	−27.3000
		32	−26.0127			40	−34.8476
		33	−26.5826			41	−35.1400
		34	−29.9335			42	−38.5397
		35	−28.8471			43	−38.3959
		36	−30.6550			44	−41.4596
		37	−27.0000			45	−40.8085
		38	−26.8000			46	−43.1380
17	Cl	32	−12.8100			47	−42.3470
		33	−21.0410			48	−44.2160
		34	−24.4510			49	−41.2880
		35	−29.0145	21	Sc	40	−20.900
		36	−29.5196			41	−28.6450
		37	−31.7648			42	−32.1410

Table 6.1 (continued).

Z	Element	A	M-A, Mev	Z	Element	A	M-A, Mev
21	Sc	43	−36.1740	25	Mn	52	−50.7020
		44	−37.8130			53	−54.6820
		45	−41.0606			54	−55.5520
		46	−41.7557			55	−57.7048
		47	−44.3263			56	−56.9038
		48	−44.5050			57	−57.4800
		46	−46.5490			58	−55.6500
		50	−44.9600	26	Fe	52	−48.3280
22	Ti	42	−25.1230			53	−50.6930
		43	−29.3400			54	−56.2455
		44	−37.6580			55	−57.4735
		45	−39.0020			56	−60.6054
		46	−44.1226			57	−60.1755
		47	−44.9266			58	−62.1465
		48	−48.4831			59	−60.6599
		49	−48.5577			60	−61.5110
		50	−51.4307			61	−59.1300
		51	−49.7380	27	Co	54	−47.9940
		52	−49.5400			55	−54.0140
23	V	46	−37.0600			56	−56.0310
		47	−42.0100			57	−59.3389
		48	−44.4700			58	−59.8380
		49	−47.9502			59	−62.2327
		50	−49.2158			60	−61.6513
		51	−52.1989			61	−62.9300
		52	−51.4360			62	−61.5280
		53	−52.1800			63	−61.9200
		54	−49.6300	28	Ni	56	−53.8990
24	Cr	48	−42.8130			57	−56.1040
		49	−45.3900			58	−60.2280
		50	−50.2490			59	−61.1587
		51	−51.4472			60	−60.4707
		52	−55.4107			61	−64.2200
		53	−55.2807			62	−66.7480
		54	−56.9305			63	−65.5160
		55	−55.1130			64	−67.1060
		56	−55.2900			65	−65.1370
25	Mn	50	−42.6480			66	−66.0550
		51	−48.2600	29	Cu	58	−51.6590

Table 6.1 *(continued).*

Z	Element	A	M-A, Mev	Z	Element	A	M-A, Mev
29	Cu	59	−56.3590	30	Zn	66	−68.8810
		60	−58.3460			67	−67.8630
		61	−61.9840			68	−69.9940
		62	−62.8130			69	−68.4250
		63	−65.5831			70	−69.5500
		64	−65.4276			71	−67.5200
		65	−67.2660			72	−68.1440
		66	−66.2550	31	Ga	63	−56.7200
		67	−67.2910			64	−58.9280
		68	−65.4100			65	−62.6580
30	Zn	60	−54.1860			66	−63.7060
		61	−56.5800			67	−66.8650
		62	−61.1230			68	−67.0740
		63	−62.2170			69	−69.3260
		64	−66.0003			70	−68.8970
		65	−65.9170				

where $r_{ij}(m)$ is the nuclear reaction rate, i.e., the total number of nuclear reactions per second, between the target nucleus i and the incident particle j which produces the particle m; $r_{kl}(i)$ is the nuclear reaction rate between the target nucleus k and the incident particle l which produces the particle i. The first term on the right of Eq. (6.26) represents the reduction in number density of nucleus i per second through nuclear reactions with other incident particles. The second term on the right of Eq. (6.26) represents the increase in number density of nucleus i per second through nuclear reactions with other incident particles. Besides, Eq. (6.26) is based on the assumption that only one target nucleus i disappears or emerges in a nuclear reaction. If more than one nuclei disappear or emerge in a nuclear reaction, and the same particles can react in the reaction, the general representation of Eq. (6.26) is

$$\frac{dn_i}{dt} = -\sum_j \frac{a_i}{1 + \delta_{ij}} r_{ij}(m) + \sum_{k,l} \frac{b_i}{1 + \delta_{kl}} r_{kl}(i) \qquad (6.27)$$

where

$$\delta_{ij} = \begin{cases} 0, & i \neq j \\ 1, & i = j \end{cases}, \quad \delta_{kl} = \begin{cases} 0, & k \neq l \\ 1, & k = l \end{cases}$$

a_i represents the number of particles i that have taken part in the reaction and disappear and b_i represents the number of particles i produced in the reaction. The number density n_i of particles i can be written as

$$n_i = \rho N_A \frac{X_i}{A_i} = \rho N_A Y_i \qquad (6.28)$$

where ρ is the density, N_A ($= 6.02217 \times 10^{23}$ mole^{-1}) is the Avogadro's constant, X_i, Y_i and A_i are the mass abundance, number abundance and atomic weight of element i, respectively. Substituting Eqs. (6.28) and (6.18) into Eq. (6.27), we obtain the rate of change of number density of element i to be

$$\frac{dY_i}{dt} = -\rho N_A \sum_j \frac{a_i}{1 + \delta_{ij}} Y_i Y_j \langle \sigma v \rangle_{ij} + \rho N_A \sum_{k,l} \frac{b_i}{1 + \delta_{kl}} Y_k Y_i \langle \sigma v \rangle_{kl} \quad (6.29)$$

The energy generation rate ε, i.e., the energy generated by one gram of stellar matter in one second, is written as

$$\varepsilon = \frac{1}{\rho} \sum r_{ij} Q_{ij} \qquad (erg \cdot g^{-1} s^{-1}) \qquad (6.30)$$

where r_{ij} is the nuclear reaction rate between target nuclei i and incident particles j, Q_{ij} is the energy released from that nuclear reaction, and the summation is over all nuclear reactions in the star. Substituting Eqs. (6.21) and (6.28) into Eq. (6.30), we have

$$\varepsilon = \rho N_A \sum \frac{1}{1 + \delta_{ij}} Y_i Y_j Q_{ij} N_A \langle \sigma v \rangle_{ij} \qquad (erg \cdot g^{-1} s^{-1}) \qquad (6.31)$$

If the unit of Q_{ij} is MeV, and considering $N_A = 6.02217 \times 10^{23} g^{-1}$, Eq. (6.31) becomes

$$\varepsilon = 9.649 \times 10^{17} \cdot \rho \sum \frac{1}{1 + \delta_{ij}} Y_i Y_j N_A \langle \sigma v \rangle_{ij} \qquad (6.32)$$

6.6 Reaction Probability $\langle \sigma v \rangle$

From Eqs. (6.29) and (6.32), $\langle \sigma v \rangle$ should be known for calculations of changes in chemical composition and energy generation rate. The critical step in computing $\langle \sigma v \rangle$ is to evaluate σv, which is defined as

$$\sigma v = \frac{r_{ij}}{dn_i dn_j} \tag{6.33}$$

where dn_i and dn_j are the numbers of particles for elements i and j in one cm^3 and r_{ij} is the nuclear reaction rate. Therefore, σv represents the probability of reaction between the incident particle and the target. This can be represented by the product of two probabilities, i.e.,

$$\sigma v = P_{coul} P_{nu} \tag{6.34}$$

where P_{coul} is the probability of tunnel effect, i.e., the probability of a particle going through the Coulomb barrier and entering the nuclear force range; and P_{nu} is the probability of the particle having a reaction in the nuclear force range.

Under the temperatures in the interior of a normal star, the particles have low kinetic energies. When these particles enter the nuclear force range, there are two possible reactions. The first one is called non-resonance reaction or direct reaction. Here, the incident particle hits a nucleon which then acquires sufficient energy to leave the nucleus. The second one is called resonance reaction. Here, the energy of the incident particle is distributed among all nucleons but none of the nucleons can escape from the nucleus so the incident particle forms a compound nucleus with the target nucleus. After a certain time, the compound nucleus decays to release a particle (proton, neutron or γ ray etc.). We will discuss the expressions of $\langle \sigma v \rangle$ separately for these two cases.

(A) Non-resonance reaction

For non-resonance reaction, P_{nu} is independent of E and can be treated as a constant denoted by s, so Eq. (6.34) becomes

$$\sigma v = P_{coul} \cdot s \tag{6.35}$$

The probability for tunnel effect has been established in quantum mechanics as

$$P_{coul} = \left(\frac{2}{m}\right)^{1/2} E^{-1/2} e^{-B/E^{1/2}} \qquad (6.36)$$

where

$$B = \frac{4\pi^2 Z_1 Z_2 e^2}{h} \left(\frac{m}{2}\right)^{1/2} \qquad (6.37)$$

Substituting Eq. (6.36) into Eq. (6.35), we have

$$\sigma v = s \cdot \left(\frac{2}{m}\right)^{1/2} E^{-1/2} e^{-B/E^{1/2}} \qquad (6.38)$$

From Eq. (6.38), we see that the reaction probability σv decreases quickly with increasing B (i.e., an increase of the Coulomb barrier, or in the value of $Z_1 \cdot Z_2$) and decreasing kinetic energy E of the particles (i.e., decrease in the temperature T). Therefore, under a temperature of 10^7 K in the interior of a star, hydrogen will first have nuclear reactions because of its smallest Coulomb barrier. The nuclear reactions for other elements have too small probabilities. When the temperature reaches 10^8 K, helium can have reactions. When the temperature reaches 5–10×10^8 K, C and O can have reactions.

Substituting Eq. (6.38) into Eq. (6.19), $\langle \sigma v \rangle$ can be expressed as

$$\langle \sigma v \rangle = s\left(\frac{2}{m}\right)^{1/2} \frac{2}{\pi^{1/2}} \frac{1}{(kT)^{3/2}} \int_0^\infty e^{-E/kT - B/E^{1/2}} \, dE \qquad (6.39)$$

in which the integration should be further computed. Figure 6.3 shows the integrand which has a sharp peak known as the "Gamow peak".

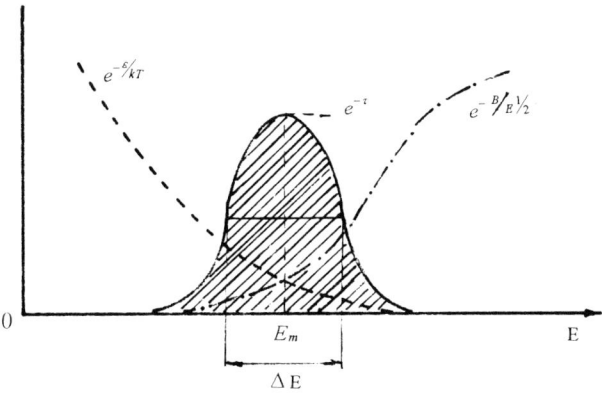

Fig. 6.3 The Gamow peak (solid curve) as the product of the Maxwell distribution (dashed) and the penetration factor (dot-dashed). The hatched area under the Gamov peak determines the reaction rate. All three curves are on different scales.

When $E = E_m$, the integrand has a maximum value $e^{-\tau}$ and the "Gamow peak" has a half-width ΔE. If these two parameters are known, we have

$$\int e^{-E/kT - B/E^{1/2}} dE \approx \Delta E \cdot e^{-\tau} \tag{6.40}$$

We now discuss the above two parameters.

(1) when $E = E_m$, the integrand has a maximum value $e^{-\tau}$
Apparently, when $\left(\frac{E}{kT} + \frac{B}{E^{1/2}}\right)$ is minimum, the integrand is maximum. From

$$\frac{d}{dE}\left(\frac{E}{kT} + \frac{B}{E^{1/2}}\right) = 0 \tag{6.41}$$

we have

$$E_m = \left(\frac{1}{2}BkT\right)^{2/3} \tag{6.42}$$

Putting E_m into the integrand, we have

$$e^{-\tau} = e^{-\left(\frac{E_m}{kT} + \frac{B}{E_m^{1/2}}\right)} = e^{-3E_m/kT} \tag{6.43}$$

$$\tau = \frac{3E_m}{kT} \tag{6.44}$$

(2) the "Gamow peak" has a half-width ΔE
Writing the integrand as

$$g(E) = e^{-t(E)}, \; t(E) = \frac{B}{E^{1/2}} + \frac{E}{kT} \tag{6.45}$$

the first and second order derivatives of $g(E)$ are

$$\frac{dg(E)}{dE} = -e^{-t}\left(-\frac{1}{2}\frac{B}{E^{3/2}} + \frac{1}{kT}\right) \tag{6.46}$$

$$\frac{d^2g(E)}{dE^2} = e^{-t}\left(-\frac{1}{2}\frac{B}{E^{3/2}} + \frac{1}{kT}\right)^2 - e^{-t}\frac{3}{4}\frac{B}{E^{5/2}} \tag{6.47}$$

At the maximum value,

$$E = E_m, \ t = \tau, \ \left.\frac{dg}{dE}\right|_{E=E_m} = 0$$

so

$$\left.\frac{d^2 g(E)}{dE^2}\right|_{E=E_m} = -\frac{3}{4} e^{-\tau} \cdot \frac{B}{E_m^{5/2}} \tag{6.48}$$

Expanding $g(E)$ at the maximum value g_m,

$$g = g_m + \left.\frac{dg}{dE}\right|_{E=E_m} \cdot \delta E + \frac{1}{2}\frac{d^2 g}{dE^2}(\delta E)^2 + \dots \tag{6.49}$$

Substituting Eq. (6.48), and considering when $\delta E = \Delta E/2$, $g = g_m/2 = e^{-\tau}/2$, we can obtain through Eq. (6.49)

$$\frac{1}{2} e^{-\tau} = e^{-\tau} - \frac{3}{8} e^{-\tau} \cdot \frac{B}{E_m^{5/2}} \cdot \Delta E^2 \tag{6.50}$$

from which we get

$$\Delta E = \frac{4}{\sqrt{3}}\frac{E_m^{5/4}}{B^{1/2}} \tag{6.51}$$

Since

$$\tau = \frac{B}{E_m^{1/2}} + \frac{E_m}{kT} = \frac{B}{E_m^{1/2}}\left(1 + \frac{E_m^{3/2}}{BkT}\right) = \frac{3}{2}\frac{B}{E_m^{1/2}} \tag{6.52}$$

we have

$$\frac{E_m^{1/4}}{B^{1/2}} = \sqrt{\frac{3}{2}}\frac{1}{\tau^{\frac{1}{2}}} \tag{6.53}$$

Substituting Eq. (6.53) into Eq. (6.51), we have

$$\Delta E = \frac{\sqrt{8}E_m}{\tau^{1/2}} \tag{6.54}$$

Substituting Eq. (6.54) into Eq. (6.40), we get

$$\int e^{-E/kT - B/E^{1/2}} dE \approx \Delta E \cdot e^{-\tau} = \frac{\sqrt{8}E_m}{\tau^{1/2}} e^{-\tau} \tag{6.55}$$

From Eq. (6.39), we have

$$\langle \sigma v \rangle = s \left(\frac{2}{m} \right)^{1/2} \frac{2}{\pi^{1/2}} \frac{1}{(kT)^{3/2}} \int e^{-E/kT - B/E^{1/2}} \, dE$$
$$= s \left(\frac{2}{m} \right)^{1/2} \frac{2}{\pi^{1/2}} \frac{1}{(kT)^{3/2}} \frac{\sqrt{8E_m}}{\tau^{1/2}} e^{-\tau} \tag{6.56}$$

Considering Eqs. (6.42) and (6.44), we obtain from Eq. (6.56)

$$\langle \sigma v \rangle \propto \frac{1}{T^{3/2}} e^{-\tau} \tag{6.57}$$

(B) Resonance reaction

Resonance reactions can be divided into two stages. In the first stage, the energy of the incident particle is distributed to all nucleons inside the nucleus but none of the nucleons can leave the nucleus and the incident particle forms a compound nucleus with the target nucleus. The compound nucleus is situated at a quasi-stationary energy level shown in Fig. 6.2, with an energy $E_n = E_{rs}$ (called the resonance energy). According to the uncertainty principle, the energy level has a certain width Γ represented by

$$\Gamma = \frac{\hbar}{t_n} \tag{6.58}$$

where t_n is the lifetime of the compound nucleus and $\hbar = 6.58 \times 10^{-16}$ eV.s.

Since the compound nucleus is only a quasi-stationary energy state, it will decay after a certain period of time to form a stable remnant nucleus and emitted particles, which is referred to as the second stage. From quantum mechanics, we obtain the reaction probability for the resonance reaction to be

$$\sigma v = \pi \lambda^2 w \frac{\Gamma_1 \Gamma_2}{(E - E_{rs})^2 + \frac{\Gamma^2}{4}} \tag{6.59}$$

where λ is the de Broglie wavelength of a particle having a momentum $p [\lambda = \hbar/p = \hbar/(2mE)^{1/2}]$, w is the statistical weighting factor

$$w = \frac{2J + 1}{(2J_i + 1)(2J_j + 1)} \tag{6.60}$$

J is the spin of the compound nucleus, J_i is the spin of the target nucleus and J_j is the spin of the incident particle, Γ_1 and Γ_2 are the energy widths of the incident channel and the reaction channel, $\Gamma(= \Gamma_1 + \Gamma_2)$ is the width of the resonance energy level, and E is the relative kinetic energy of the target nucleus and the incident particle.

From Eq. (6.59), we see that the probability has a maximum when $E = E_{rs}$, which is the reason why we call this type of nuclear reaction the resonance reaction. Substituting Eq. (6.59) into Eq. (6.19), the mean resonance reaction probability is obtained as

$$\langle \sigma v \rangle = \left(\frac{2\pi}{m}\right)^{1/2} \sigma_{rs} \Gamma_{rs} E_{rs} (kT)^{-3/2} \exp\left(\frac{-E_{rs}}{kT}\right) \tag{6.61}$$

where E_{rs} is the resonance energy, Γ_{rs} is the width of the resonance energy level, σ_{rs} is the reaction cross section when E equals the resonance energy; the value of σ_{rs} is usually determined experimentally.

6.7 ELECTRON SHIELDING

In the above discussions of $\langle \sigma v \rangle$, the target nucleus and the incoming particles are treated as naked nuclei, which is valid only for extremely low densities. In fact, a dense electron gas exists in the interior of a star, and a target nucleus tends to accrete surrounding electrons to form a negative electron cloud around itself. This electron cloud will partially shield the electric charge of the target nucleus, which is equivalent to lowering the Coulomb barrier or increasing the energy of the incident particle. As a result, the probability of the incident particle penetrating the Coulomb barrier through the tunnel effect and thus the thermonuclear reaction rate are enhanced. The effects of electron shielding will be briefly discussed in the following according to the method of Salpeter (1954).

We first study the expression for the static electric potential Φ from a system consisting of a target nucleus with electric charge $+Z_i e$ and surrounded by a negative electron cloud, which can be solved from the Poisson equation

$$\nabla^2 \Phi = -4\pi \rho_e \tag{6.62}$$

where ρ_e is the electric charge density at the point of investigation. From statistical physics, we know that the distribution of number density of type j particles with electric charge $+Z_je$ around particles with a static electric potential Φ is

$$n_j = n_{oj}\exp(-Z_je\Phi/kT) \tag{6.63}$$

where n_{oj} is the number density of type j particles at $\Phi = 0$. Equation (6.63) shows that the number density of other particles surrounding the target nucleus is reduced ($n_j < n_{oj}$), which is due to repulsion between the target nucleus and other particles. On the other hand, the number density distribution of non-degenerate electrons surrounding the target nucleus is

$$n_e = n_{oe}\exp(+e\Phi/kT) \tag{6.64}$$

where n_{oe} is the electron number density when $\Phi = 0$. Equation (6.64) shows that the number density of non-degenerate electrons surrounding the target nucleus is increased ($n_e > n_{oe}$), which is due to accretion of surrounding electrons by the target nucleus.

The exponents in Eqs. (6.63) and (6.64) can be less than unity, i.e., $|Z_je\Phi/kT| < 1$ and $|e\Phi/kT| < 1$, when the temperature is relatively high and the density is relatively low, which is referred to as the weak shielding case. At this time, Eqs. (6.63) and (6.64) can be approximated to be

$$n_j = n_{oj}\left(1 - \frac{Z_je\Phi}{kT}\right) \tag{6.65}$$

$$n_e = n_{oe}\left(1 + \frac{e\Phi}{kT}\right) \tag{6.66}$$

so the total electric charge density ρ_e of a mixed gas at the point of study can be written as

$$\rho_e = \sum_j (Z_je)n_j - en_e \tag{6.67}$$

Substituting Eqs. (6.65) and (6.66) into Eq. (6.67), we have

$$\begin{aligned} \rho_e &= \sum_j n_{oj}Z_je\left(1 - \frac{Z_je\Phi}{kT}\right) - n_{oe}e\left(1 + \frac{e\Phi}{kT}\right) \\ &= \sum_j n_{oj}Z_je - n_{oe}e - \sum_j n_{oj}\frac{Z_j^2 e^2 \Phi}{kT} - n_{oe}\frac{e^2 \Phi}{kT} \end{aligned} \tag{6.68}$$

When $\Phi = 0$, the mixed gas is a plasma and electrically neutral, so the sum of the first two terms on the right of Eq. (6.68) is zero, and

$$n_{oe} = \sum_j n_{oj} Z_j \qquad (6.69)$$

Therefore Eq. (6.68) can be written as

$$\rho_e = -\frac{\Phi e^2}{kT} \chi n_o \qquad (6.70)$$

where n_o is the total particle number density at $\Phi = 0$, or

$$n_o \equiv \sum_j n_{oj} + n_{oe} \qquad (6.71)$$

and χ in Eq. (6.70) is given by

$$\chi \equiv \sum_j \frac{n_{oj}}{n_o} Z_j(Z_j + 1) \qquad (6.72)$$

Substituting the relationships $n_o = N_A \rho / \mu$ and $n_{oj} = N_A \rho Y_j$ into Eqs. (6.71) and (6.72), we obtain from Eq. (6.70)

$$\rho_e = -\frac{\hat{Z}^2 e^2 N_A \rho}{kT} \Phi \qquad (6.73)$$

In the above equations, N_A is the Avogadro's constant, ρ is the gas density, μ is the molecular weight of particles in a mixed gas, Y_j is the number abundance of type j element and $\hat{Z}$ is the mean square value of electric charge, i.e.,

$$\hat{Z} = \left[\sum_j Z_j(Z_j + 1) Y_j \right]^{1/2} \qquad (6.74)$$

Substituting the expression for total electric charge density ρ_e in Eq. (6.73) into the Poisson equation (6.62), and assuming spherical symmetry for the static electric potential Φ, we have

$$\frac{1}{r} \frac{d^2(r\Phi)}{dr^2} = 4\pi \frac{\hat{Z}^2 e^2 N_A \rho}{kT} \Phi = \frac{\Phi}{r_D^2} \qquad (6.75)$$

where r_D is called the Debye length computed by

$$r_D = \sqrt{\frac{kT}{4\pi \hat{Z}^2 e^2 N_A \rho}} \tag{6.76}$$

Equation (6.75) can also be written as

$$\frac{1}{2}\frac{d}{d(r\Phi)}\left[\frac{d(r\Phi)}{dr}\right]^2 = \frac{r\Phi}{r_D^2} \tag{6.77}$$

Integrating Eq. (6.77), we have

$$\frac{d(r\Phi)}{dr} = \pm \frac{r\Phi}{r_D} + const \tag{6.78}$$

We take the negative sign and consider that the total electric charge density $\rho_e = 0$ when $r \to \infty$, so the integration constant in Eq. (6.78) is zero and

$$\frac{d(r\Phi)}{dr} = -\frac{r\Phi}{r_D} \tag{6.79}$$

On integrating, we have

$$\Phi = const \cdot \frac{1}{r}\exp(-r/r_D) \tag{6.80}$$

The integration constant in Eq. (6.80) should be equal to $+Z_i e$, since the static electric potential Φ should approach $Z_i e/r$ for the electric charge $+Z_i e$ for an isolated target nucleus when $r \to \infty$. Therefore,

$$\Phi = \frac{Z_i e}{r}\exp\left(-\frac{r}{r_D}\right) \tag{6.81}$$

which gives the electric potential for the system consisting of the target nucleus with electric charge $+Z_i e$ and the surrounding negative electron cloud. From Eq. (6.81), when $r \geq r_D$, $\Phi \leq (Z_i e/r)/2.718$ so the electric potential for the electric charge $+Z_i e$ has been basically shielded. In this way, r_D can be viewed as the radius of the negative electron cloud.

The shielding of target electric charge by the negative electron cloud will affect the probability of penetration of the potential barrier by the incident particle. From quantum mechanics, this probability depends on the integration

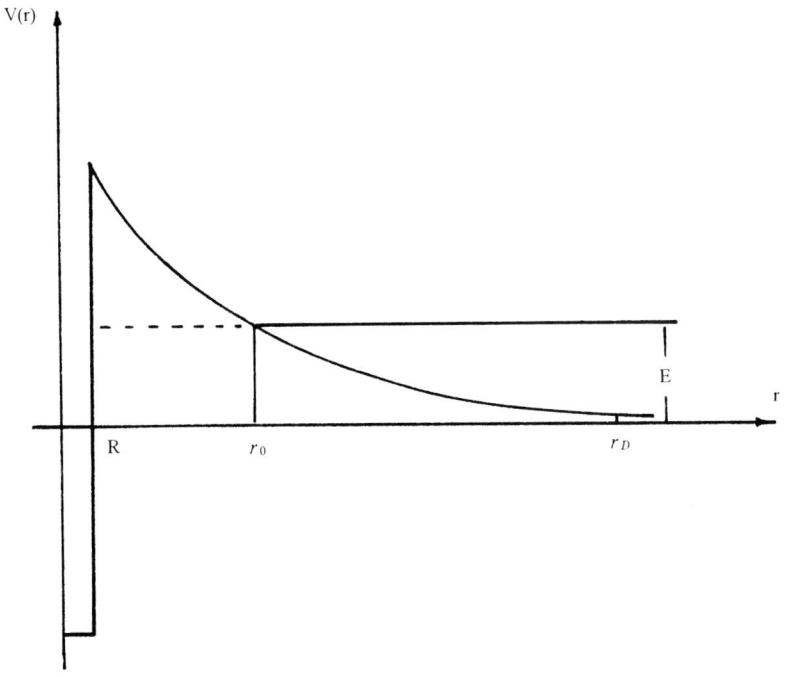

Fig. 6.4 The potential V of a nucleus. An incident particle with energy E approaches the nucleus.

$$\int\limits_{R}^{r_o} (V(r) - E)^{1/2} dr \qquad (6.82)$$

where

$$V(r) = \frac{Z_i Z_j e^2}{r}$$

R is the radius of the target nucleus, r_o is the radius at the standard turn off point (see Fig. 6.4). The range of r in Eq. (6.82) is $R \le r \le r_o$. In this range, the condition $r/r_D \ll 1$ is always fulfilled. According to Eq. (6.81), the Coulomb barrier of the target nucleus with electron shielding for the incident particle is

$$V(r) = \frac{Z_i Z_j e^2}{r} e^{r/r_D} \qquad (6.83)$$

Since $r/r_D \ll 1$, Eq. (6.83) can be approximated as

$$V(r) = \frac{Z_i Z_j e^2}{r} - \frac{Z_i Z_j e^2}{r_D} \qquad (6.84)$$

Apparently, in the above equation, since $r/r_D << 1$, the second term is small compared to the first term. According to Eq. (6.84), the integrand in Eq. (6.82) can be written as

$$V(r) - E = \frac{Z_i Z_j e^2}{r} - \left(E + \frac{Z_i Z_j e^2}{r_D} \right) \tag{6.85}$$

From Eq. (6.85), we see that the electron shielding is also equivalent to an increase in the relative kinetic energy of the incident particle from E to $E + U_o$, i.e., the interaction cross section changes from $\sigma(E)$ to $\sigma(E + U_o)$. Here,

$$U_o = \frac{Z_i Z_j e^2}{r_D} \tag{6.86}$$

Apparently, U_o is also small compared to E, since $E = Z_i Z_j e^2 / r_o$ and $r_o << r_D$.

We now continue to discuss the mean probability $\langle \sigma v \rangle$ for the nuclear reaction. When there is no electron shielding, from Eq. (6.19),

$$\langle \sigma v \rangle = \int_0^\infty f(E) \sigma(E) v(E) dE \tag{6.87}$$

where $f(E)$ is the Maxwell distribution of speeds and $f(E) \propto E^{1/2} e^{-E/kT}$, and the velocity $v(E) \propto E^{1/2}$, so Eq. (6.87) becomes

$$\langle \sigma v \rangle \propto \int_0^\infty E e^{-E/kT} \sigma(E) dE \tag{6.88}$$

When there is electron shielding, the mean probability for nuclear reactions to occur is represented as $\langle \sigma v \rangle_s$ and written as

$$\langle \sigma v \rangle_s = \int_0^\infty f(E) \sigma(E + U_o) v(E) dE \tag{6.89}$$

Writing $E + U_o$ as E', the above equation becomes

$$\langle \sigma v \rangle_s = \int_{U_o}^\infty f(E' - U_o) \sigma(E') v(E' - U_o) dE' \tag{6.90}$$

Since $f(E'-U_o) \propto (E'-U_o)^{1/2}e^{-(E'-U_o)/kT}$, and $v(E'-U_o) \propto (E'-U_o)^{1/2}$, Eq. (6.90) becomes

$$\langle \sigma v \rangle_s \propto \int_{U_o}^{\infty} (E'-U_o)e^{-E'/kT} \cdot e^{U_o/kT}\sigma(E')dE'$$

Since U_o is small compared to E', the lower limit U_o for the above integration can be replaced by zero, and the above equation can be approximated as

$$\langle \sigma v \rangle_s \propto \int_{0}^{\infty} E'e^{-E'/kT} \cdot e^{U_o/kT}\sigma(E')dE' \qquad (6.91)$$

Comparing Eqs. (6.91) and (6.88), it is easy to observe that

$$\langle \sigma v \rangle_s = \langle \sigma v \rangle \cdot f \qquad (6.92)$$

where the factor f is the electron shielding factor and represented as

$$f \equiv e^{U_o/kT} \qquad (6.93)$$

Substituting Eqs. (6.76) and (6.86) into Eq. (6.93), we have

$$f = \exp\left(1.88 \times 10^8 Z_i Z_j \hat{Z}\rho^{1/2}T^{-3/2}\right) \qquad (6.94)$$

Equation (6.94) is only valid for the weak electron shielding case, i.e.,

$$\frac{U_o}{kT} = \frac{Z_i Z_j e^2}{r_D kT} \ll 1$$

that is, the case where Eqs. (6.63) and (6.64) can be approximated by Eqs. (6.65) and (6.66). If the gas density is very large, Eqs. (6.63) and (6.64) can no longer be approximated by Eqs. (6.65) and (6.66), i.e., $U_o/kT \gg 1$, and this is the condition for strong electron shielding. Here, the computation for the electron shielding factor f is very complicated, which is given as

$$f = \exp\left\{2.05 \times 10^5 \left[(Z_i + Z_j)^{5/3} - Z_i^{5/3} - Z_j^{5/3}\right]\frac{(\rho/\mu_e)^{1/3}}{T}\right\} \qquad (6.95)$$

where μ_e is the mean molecular weight for the free electron, i.e.,

$$\mu_e = \left(\sum_j Z_j Y_j \right)^{-1} \tag{6.96}$$

6.8 RELATIONSHIP BETWEEN THE ENERGY GENERATION RATE ε AND THE TEMPERATURE T

Under the condition for non-resonance reaction, from Eqs. (6.32) and (6.56), we know

$$\varepsilon = 9.649 \times 10^{17} \rho \sum \frac{1}{1 + \delta_{ij}} Y_i Y_j Q_{ij} N_A \langle \sigma v \rangle \propto T^{-2/3} e^{-\tau} \tag{6.97}$$

Further substituting Eqs. (6.42) and (6.44), we have

$$\nu = \frac{\partial \ln \varepsilon}{\partial \ln T} = -\frac{2}{3} - \frac{\partial \tau}{\partial \ln T} = -\frac{2}{3} - \tau \frac{\partial \ln \tau}{\partial \ln T}$$

$$= -\frac{2}{3} + \frac{\tau}{3} \approx \frac{\tau}{3} \tag{6.98}$$

Moreover, from Eqs. (6.44), (6.42), (6.13) and (6.37), we get

$$\tau = \frac{3E_m}{kT} = 3 \frac{\left(\frac{1}{2} BkT \right)^{2/3}}{kT} = 3 \left(\frac{2\pi^4 Z_1^2 Z_2^2 e^4 m}{h^2 kT} \right)^{1/3}$$

$$= 4250 \left(\frac{A Z_1^2 Z_2^2}{T} \right)^{1/3} \tag{6.99}$$

where

$$A = \frac{A_1 A_2}{A_1 + A_2} \tag{6.100}$$

We consider the ^{1}H + ^{14}N reaction in the CNO cycle as our example, for which

$$Z_1 \cdot Z_2 = 7 , \quad A = \frac{14}{15} , \quad \text{therefore } \tau = \frac{15200}{T^{1/3}}$$

This thermonuclear reaction occurs at $T = 1.5 \times 10^7$ K, so $\tau = 60$. From Eq. (6.98), $\nu = 20$, so $\varepsilon \sim T^{20}$. Figure 6.5 shows the relationship

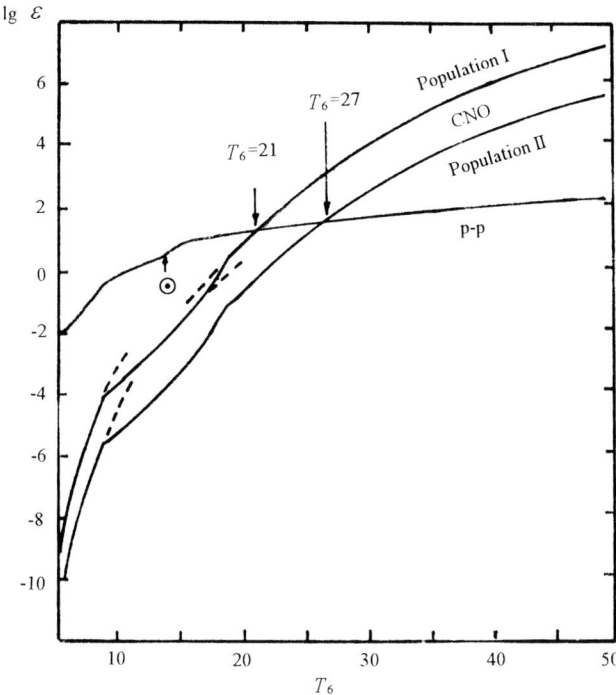

Fig. 6.5 The relationship between the energy generation rate ε and the temperature T for hydrogen burning given by Fowler (1960). $\rho = 100[\text{g}/\text{cm}^3]$, $X_H = 0.5$. Population I stars: $X_C = 0.03$, $X_N = 0.01$ and $X_O = 0.012$. Population II stars: abundance values are 1/25 of those for Population I stars.

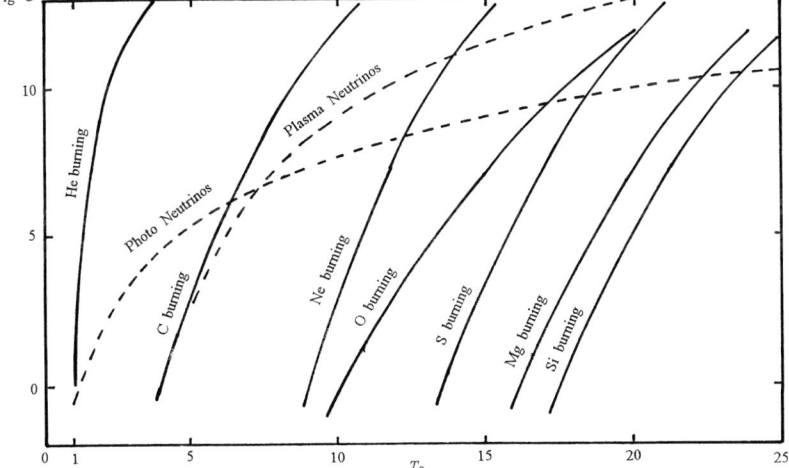

Fig. 6.6 The relationship for the burning of elements other than hydrogen given by Hayashi *et al.* (1962).

between the energy generation rate ε and the temperature T for hydrogen burning given by Fowler (1960). Figure 6.6 shows the relationship for the burning of other elements at higher temperatures given by Hayashi *et al.* (1962). For helium burning, the density is taken as $\rho = 10[\text{gcm}^{-3}]$; for other burnings, $\rho = 10^5[\text{gcm}^{-3}]$; for neutrino processes, $\rho = 10^5[\text{gcm}^{-3}]$, $\mu_e = 2$ and no degeneracy is assumed.

6.9 HYDROGEN BURNING

Hydrogen is the most abundant element in stars, which can undergo thermonuclear reactions at $T \geq 7 \times 10^6$ K. The reactions can be divided into two categories. The first is called the proton-proton chain reaction or the p-p reaction. When the temperature is relatively low (less than 15–20×10^6 K), p-p reaction dominates. The second is called the Carbon-Nitrogen-Oxygen cycle reaction or the CNO cycle reaction. When the temperature is relatively high (higher than 15–20×10^6 K), the CNO cycle reaction dominates. Both the outcomes of the above reactions are the fusion of four hydrogen atoms into one helium atom with the simultaneous release of a large amount of energy. These two types of reactions will be discussed in the following section.

(1) p-p reaction. It has three possible processes:

$$^1H(p, e^+v)^2H(p, \gamma)^3He \Big\langle \begin{array}{ll} (^3He, 2p)^4He & \text{(ppI)} \\ (\alpha, \gamma)^7Be \Big\langle \begin{array}{ll} (e^-v)^7Li(p, \gamma)^8Be(\alpha)^4He & \text{(ppII)} \\ (p, \gamma)^8B(e^+v)^8Be(\alpha)^4He & \text{(ppIII)} \end{array} \end{array}$$

(2) CNO cycle reaction

$$^{14}N(p, \gamma)^{15}O(e^+v)^{15}N \Big\langle \begin{array}{l} (p, \alpha)^{12}C(p, \gamma)^{13}N(e^+v)^{13}C(p, \gamma)^{14}N \\ (p, \gamma)^{16}O(p, \gamma)^{17}F(e^+v)^{17}O \Big\langle \begin{array}{l} (p, \alpha)^{14}N \\ (p, \gamma)^{18}F(e^+v)^{18}O \Big\langle \begin{array}{l} (p, \gamma)^{19}F(p, \alpha)^{16}O \\ (p, \alpha)^{15}N \end{array} \end{array} \end{array}$$

The energy Q_r released and the neutrino energy loss $(q_v)_r$ for all nuclear reactions during hydrogen burning are listed in Table 6.2.

According to Eq. (6.29), the differential equation set for the change in chemical composition during hydrogen burning can be written as

$$^1\text{H}: \quad \frac{dY_1}{dt} = -2R_1 - R_2 + 2R_3 - R_6 - R_7 - R_9 - R_{11} - R_{12} - R_{14}$$
$$- R_{15} - R_{16} - R_{18} - R_{19} - R_{21} - R_{22} - R_{23} \quad (6.101)$$

$$^2\text{H}: \quad \frac{dY_2}{dt} = R_1 - R_2 \quad (6.102)$$

Table 6.2 The energy Q_r released and the neutrino energy loss $(q_\nu)_r$ for all nuclear reactions during hydrogen burning. (According to Georgeanne and Fowler 1988; Harris *et al.* 1983; Fowler *et al.* 1975).

r	Reaction	Q_r(MeV)	$(q_\nu)_r$(MeV)
1	$^1\mathrm{H}(p,e^+\nu)^2\mathrm{H}$	1.442	0.265
2	$^2\mathrm{H}(p,\gamma)^3\mathrm{He}$	5.494	
3	$^3\mathrm{He}(^3\mathrm{He},2p)^4\mathrm{He}$	12.860	
4	$^3\mathrm{He}(\alpha,\gamma)^7\mathrm{Be}$	1.588	
5	$^7\mathrm{Be}(e^-\nu)^7\mathrm{Li}$	0.862	0.862
6	$^7\mathrm{Li}(p,\gamma)^8\mathrm{Be}(\alpha)^4\mathrm{He}$	17.346	
7	$^7\mathrm{Be}(p,\alpha)^8\mathrm{B}$	0.137	
8	$^8\mathrm{B}(e^+\nu)^8\mathrm{Be}(\alpha)^4\mathrm{He}$	18.072	6.710
9	$^{12}\mathrm{C}(p,\gamma)^{13}\mathrm{N}$	1.944	
10	$^{13}\mathrm{N}(e^+\nu)^{13}\mathrm{C}$	2.221	0.7067
11	$^{13}\mathrm{C}(p,\gamma)^{14}\mathrm{N}$	7.551	
12	$^{14}\mathrm{N}(p,\gamma)^{15}\mathrm{O}$	7.297	
13	$^{15}\mathrm{O}(e^+\nu)^{15}\mathrm{N}$	2.754	0.9965
14	$^{15}\mathrm{N}(p,\alpha)^{12}\mathrm{C}$	4.966	
15	$^{15}\mathrm{N}(p,\gamma)^{16}\mathrm{O}$	12.128	
16	$^{16}\mathrm{O}(p,\gamma)^{17}\mathrm{F}$	0.600	
17	$^{17}\mathrm{F}(e^+\nu)^{17}\mathrm{O}$	2.762	0.9994
18	$^{17}\mathrm{O}(p,\alpha)^{14}\mathrm{N}$	1.191	
19	$^{17}\mathrm{O}(p,\gamma)^{18}\mathrm{F}$	5.607	
20	$^{18}\mathrm{F}(e^+\nu)^{18}\mathrm{O}$	1.655	0.3965
21	$^{18}\mathrm{O}(p,\alpha)^{15}\mathrm{N}$	3.980	
22	$^{18}\mathrm{O}(p,\gamma)^{19}\mathrm{F}$	7.994	
23	$^{19}\mathrm{F}(p,\alpha)^{16}\mathrm{O}$	8.114	

$$^3\mathrm{H}: \quad \frac{dY_3}{dt} = R_2 - 2R_3 - R_4 \tag{6.103}$$

$$^4\mathrm{H}: \quad \frac{dY_4}{dt} = R_3 - R_4 + 2R_6 + 2R_8 + R_{14} + R_{18} + R_{21} + R_{23} \tag{6.104}$$

$$^7\mathrm{B}: \quad \frac{dY_7}{dt} = R_4 - R_5 - R_7 \tag{6.105}$$

$$^7\mathrm{Li}: \quad \frac{dY_{7'}}{dt} = R_5 - R_6 \tag{6.106}$$

$${}^{8}\text{B}: \quad \frac{dY_8}{dt} = R_7 - R_8 \tag{6.107}$$

$${}^{12}\text{C}: \quad \frac{dY_{12}}{dt} = -R_9 + R_{14} \tag{6.108}$$

$${}^{13}\text{C}: \quad \frac{dY_{13}}{dt} = R_{10} - R_{11} \tag{6.109}$$

$${}^{13}\text{N}: \quad \frac{dY_{13'}}{dt} = R_9 - R_{10} \tag{6.110}$$

$${}^{14}\text{N}: \quad \frac{dY_{14}}{dt} = R_{11} - R_{12} + R_{18} \tag{6.111}$$

$${}^{15}\text{N}: \quad \frac{dY_{15}}{dt} = R_{13} - R_{14} - R_{15} + R_{21} \tag{6.112}$$

$${}^{15}\text{O}: \quad \frac{dY_{15'}}{dt} = R_{12} - R_{13} \tag{6.113}$$

$${}^{16}\text{O}: \quad \frac{dY_{16}}{dt} = R_{15} - R_{16} + R_{23} \tag{6.114}$$

$${}^{17}\text{O}: \quad \frac{dY_{17}}{dt} = R_{17} - R_{18} - R_{19} \tag{6.115}$$

$${}^{17}\text{F}: \quad \frac{dY_{17'}}{dt} = R_{16} - R_{17} \tag{6.116}$$

$${}^{18}\text{O}: \quad \frac{dY_{18}}{dt} = R_{20} - R_{21} - R_{22} \tag{6.117}$$

$${}^{18}\text{F}: \quad \frac{dY_{18'}}{dt} = R_{19} - R_{20} \tag{6.118}$$

$${}^{19}\text{F}: \quad \frac{dY_{19}}{dt} = R_{22} - R_{23} \tag{6.119}$$

In the above equation set, $R_r(r = 1, 2,..., 23)$ are the reaction rates which represent the total number of reactions per second between particles a and X in the nuclear reaction $X(a, b)Y$, and are expressed as

$$R_r = \frac{1}{1 + \delta_{aX}} Y_a Y_X \rho N_A \langle \sigma v \rangle_{aX} \cdot f_{aX} = \frac{1}{1 + \delta_{aX}} Y_a Y_X \rho \lambda_{aX} f_{aX} \tag{6.120}$$

$$\delta_{aX} = \begin{cases} 0 & a \neq X \\ 1 & a = X \end{cases} \tag{6.121}$$

During hydrogen burning, some nuclear reactions are very fast. For example, the β decays for the radioactive elements ^{8}B, ^{13}N, ^{15}O, ^{17}F, and ^{18}F are very fast and these elements will be instantly changed into other elements, so $(dY_8/dt) = (dY_{13'}/dt) = (dY_{15'}/dt) = (dY_{18'}/dt) = 0$, that is,

$$R_8 = R_7, \ R_{10} = R_9, \ R_{13} = R_{12}, \ R_{20} = R_{19} \tag{6.122}$$

Similarly, since the reactions such as $^{2}\text{H}(p,\gamma)^{3}\text{He}$, $^{7}\text{Be}(e^-\nu)^{7}\text{Li}$, $^{7}\text{Li}(p,\alpha)^{4}\text{He}$, $^{19}\text{F}(p,\alpha)^{16}\text{O}$ are very fast, there are no changes in the abundance of the elements $^{2}\text{He}, ^{7}\text{Be}, ^{7}\text{Li}$ and ^{19}F, i.e., $(dY_2/dt) = (dY_{7'}/dt) = (dY_{19}/dt) = 0$, and at the same time,

$$R_2 = R_1, \ R_5 = R_6, \ R_4 = R_6 + R_7, \ R_{23} = R_{22} \tag{6.123}$$

Considering Eqs. (6.122) and (6.123), the set of differential equations for changes in chemical compositions during hydrogen burning becomes

$$^{1}\text{H}: \quad \frac{dY_1}{dt} = -3R_1 + 2R_3 - R_4 - R_9 - R_{11} - R_{12} - R_{14} - R_{15} - R_{16}$$
$$- R_{18} - R_{19} - R_{21} - R_{22} \tag{6.124}$$

$$^{3}\text{H}: \quad \frac{dY_3}{dt} = R_1 - 2R_3 - R_4 \tag{6.125}$$

$$^{4}\text{H}: \quad \frac{dY_4}{dt} = R_3 + R_4 + R_{14} + R_{18} + R_{21} + R_{22} \tag{6.126}$$

$$^{12}\text{C}: \quad \frac{dY_{12}}{dt} = -R_9 + R_{14} \tag{6.127}$$

$$^{13}\text{C}: \quad \frac{dY_{13}}{dt} = R_9 - R_{11} \tag{6.128}$$

$$^{14}\text{N}: \quad \frac{dY_{14}}{dt} = R_{11} - R_{12} + R_{18} \tag{6.129}$$

$$^{15}\text{N}: \quad \frac{dY_{15}}{dt} = R_{12} - R_{14} - R_{15} + R_{21} \tag{6.130}$$

$$^{16}\text{O}: \quad \frac{dY_{16}}{dt} = R_{15} - R_{16} + R_{22} \tag{6.131}$$

$$^{17}\text{O}: \quad \frac{dY_{17}}{dt} = R_{16} - R_{18} - R_{19} \tag{6.132}$$

$$^{18}O: \quad \frac{dY_{18}}{dt} = R_{19} - R_{21} - R_{22} \tag{6.133}$$

According to Eq. (6.32), the energy generation rate from hydrogen burning is

$$
\begin{aligned}
\varepsilon = 9.649 \times 10^{17}\{ & [Q_1 - (q_v)_1]R_1 + Q_2R_2 + Q_3R_3 + Q_4R_4 + [Q_5 - (q_v)_5]R_5 \\
& + Q_6R_6 + Q_7R_7 + [Q_8 - (q_v)_8]R_8 + Q_9R_9 + [Q_{10} - (q_v)_{10}]R_{10} \\
& + Q_{11}R_{11} + Q_{12}R_{12} + [Q_{13} - (q_v)_{13}]R_{13} + Q_{14}R_{14} + Q_{15}R_{15} \\
& + Q_{16}R_{16} + [Q_{17} - (q_v)_{17}]R_{17} + Q_{18}R_{18} + Q_{19}R_{19} \\
& + [Q_{20} - (q_v)_{20}]R_{20} + Q_{21}R_{21} + Q_{22}R_{22} + Q_{23}R_{23}\}
\end{aligned}
\tag{6.134}
$$

Substituting Eqs. (6.122) and (6.123) into Eq. (6.124), and considering $Q_5 = (q_v)_5$, we have

$$
\begin{aligned}
\varepsilon = 9.649 \times 10^{17}\{ & [Q_2 + Q_1 - (q_v)_1]R_1 + Q_3R_3 + \left[Q_4 + Q_{16} + \frac{R_7}{R_5 + R_7}\right. \\
& \left. (-Q_6 + Q_7 + Q_8 - (q_v)_8)\right]R_4 + [Q_9 + Q_{10} - (q_v)_{10}]R_9 + Q_{11}R_{11} \\
& + [Q_{12} + Q_{13} - (q_v)_{13}]R_{12} + Q_{14}R_{14} + Q_{15}R_{15} + [Q_{16} + Q_{17} \\
& - (q_v)_{17}]R_{16} + Q_{18}R_{18} + [Q_{19} + Q_{20} - (q_v)_{20}]R_{19} + Q_{21}R_{21} \\
& + (Q_{22} + Q_{23})R_{22}\}
\end{aligned}
\tag{6.135}
$$

Further substituting the energy Q_r released and the neutrino energy loss $(q_v)_r$ for the nuclear reactions into the above equation, we have

$$
\begin{aligned}
\varepsilon = 9.649 \times 10^{17}\{ & 6.671R_1 + 12.860R_3 + \left[18.933 - \frac{5.848R_7}{R_5 + R_7}\right]R_4 + 3.457R_9 \\
& + 7.551R_{11} + 9.054R_{12} + 4.966R_{14} + 12.128R_{15} + 2.364R_{16} + 1.193R_{18} \\
& + 6.867R_{19} + 3.980R_{21} + 16.108R_{22}\}[\text{erg/g} \cdot \text{s}]
\end{aligned}
\tag{6.136}
$$

The expressions for the nuclear reaction rates R_r in Eqs. (6.124) to (6.133) are

$$R_1 = \frac{1}{2}Y_1^2 \rho\lambda_{1,1}f_{1,1} \tag{6.137}$$

$$\lambda_{1,1} = 4.01 \times 10^{-15}T_9^{-2/3}(1 + 0.123T_9^{1/3} + 1.09T_9^{2/3}$$

$$+ 0.938T_9)\exp\left(-\frac{3.380}{T_9^{1/3}}\right) \tag{6.138}$$

$$f_{1,1} = \begin{cases} \exp(\psi_{1,1}), & \psi_{1,1} \leq 0.1 \\ \exp(\Phi_{1,1}), & 0.1 < \psi_{1,1} \leq 2 \end{cases} \qquad (6.139)$$

$$\psi_{1,1} = \psi \qquad (6.140)$$

$$\psi = 5.945 \times 10^{-6} \rho^{1/2} T_9^{-3/2} \left[\sum_r Z_r^2 Y_r + \frac{1}{2} \left(\frac{\partial \ln \rho}{\partial \psi} \right)_T \right]^{1/2} \qquad (6.141)$$

$$\Phi_{1,1} = 1.630 \Phi \qquad (6.142)$$

$$\Phi = 1.218 \times 10^{-5} \rho^{0.43} T_9^{-1.29} \left(\sum_r Z_r^{1.58} Y_r \right) \left(\sum_r Z_r Y_r \right)^{-0.28}$$

$$\left[\sum_r Z_r^2 Y_r + \frac{1}{2} \left(\frac{\partial \ln \rho}{\partial \psi} \right)_T \right]^{-0.29} \qquad (6.143)$$

$$R_3 = \frac{1}{2} Y_3^2 \rho \lambda_{3,3} f_{3,3} \qquad (6.144)$$

$$\lambda_{3,3} = 6.04 \times 10^{10} T_9^{-2/3} (1 + 0.034 T_9^{1/3} - 0.522 T_9^{2/3} - 0.124 T_9 + 0.353 T_9^{4/3}$$

$$+ 0.213 T_9^{5/3}) \exp\left(\frac{-12.276}{T_9^{1/3}} \right) \qquad (6.145)$$

$$f_{3,3} = \begin{cases} \exp(4\psi), & 4\psi \leq 0.1 \\ \exp(5.917\Phi), & 0.1 < 4\psi \leq 2 \end{cases} \qquad (6.146)$$

$$R_4 = Y_3 Y_4 \rho \lambda_{3,4} f_{3,4} \qquad (6.147)$$

$$\lambda_{3,4} = 5.61 \times 10^6 \frac{\exp[-12.826 T_9^{-1/3} (1 + 0.0495 T_9)^{1/3}]}{(1 + 0.0495 T_9)^{5/6} T_9^{2/3}} \qquad (6.148)$$

$$f_{3,4} = \begin{cases} \exp(4\psi), & 4\psi \leq 0.1 \\ \exp(5.917\Phi), & 0.1 < 4\psi \leq 2 \end{cases} \qquad (6.149)$$

$$R_9 = Y_1 Y_{12} \rho \lambda_{1,12} f_{1,12} \qquad (6.150)$$

$$\lambda_{1,12} = 2.04 \times 10^7 T_9^{-2/3} \exp\left[-\frac{13.690}{T_9^{1/3}} - \left(\frac{T_9}{1.5} \right)^2 \right] (1 + 0.03 T_9^{1/3}$$

$$+ 1.19 T_9^{2/3} + 0.254 T_9 + 2.06 T_9^{4/3} + 1.12 T_9^{5/3}) + 1.08 \qquad (6.151)$$

$$\times 10^5 T_9^{-3/2} \exp\left(-\frac{4.925}{T_9} \right) + 2.15 \times 10^5 T_9^{-3/2} \exp\left(-\frac{18.179}{T_9} \right)$$

$$f_{1,12} = \begin{cases} \exp(6\psi), & 6\psi \leq 0.1 \\ \exp(8.302\Phi), & 0.1 < 6\psi \leq 2 \end{cases} \tag{6.152}$$

$$R_{11} = Y_1 Y_{13} \rho \lambda_{1,13} f_{1,13} \tag{6.153}$$

$$\begin{aligned} \lambda_{1,13} = 8.01 \times 10^7 T_9^{-2/3} \exp\left[-\frac{13.717}{T_9^{1/3}} - \left(\frac{T_9}{2}\right)^2 \right] &(1 + 0.03 T_9^{1/3} \\ + 0.958 T_9^{2/3} + 0.204 T_9 + 1.39 T_9^{4/3} + 0.753 T_9^{5/3}) &+ 1.21 \\ \times 10^6 T_9^{-6/5} \exp\left(-\frac{5.701}{T_9}\right) \end{aligned} \tag{6.154}$$

$$f_{1,13} = f_{1,12} \tag{6.155}$$

$$R_{12} = Y_1 Y_{14} \rho \lambda_{1,14} f_{1,14} \tag{6.156}$$

$$\begin{aligned} \lambda_{1,14} = 4.90 \times 10^7 T_9^{-2/3} \exp\left[-\frac{15.228}{T_9^{1/3}} - \left(\frac{T_9}{3.294}\right)^2 \right] &(1 + 0.027 T_9^{1/3} \\ - 0.778 T_9^{2/3} - 0.149 T_9 + 0.261 T_9^{4/3} + 0.127 T_9^{5/3}) & \\ + 2.37 \times 10^3 T_9^{-3/2} \exp\left(-\frac{3.011}{T_9}\right) + 2.19 \times 10^4 \exp\left(-\frac{12.530}{T_9}\right) \end{aligned} \tag{6.157}$$

$$f_{1,14} = \begin{cases} \exp(7\psi), & 7\psi \leq 0.1 \\ \exp(9.52\Phi), & 0.1 < 7\psi \leq 2 \end{cases} \tag{6.158}$$

$$R_{14} = Y_1 Y_{15} \rho \lambda_{1,15} f_{1,15} \tag{6.159}$$

$$\begin{aligned} \lambda_{1,15} = 1.08 \times 10^{12} T_9^{-2/3} \exp\left[-\frac{15.251}{T_9^{1/3}} - \left(\frac{T_9}{0.522}\right)^2 \right] &(1 + 0.027 T_9^{1/3} \\ + 2.62 T_9^{2/3} + 0.501 T_9 + 5.36 T_9^{4/3} + 2.60 T_9^{5/3}) &+ 1.19 \\ \times 10^8 T_9^{-3/2} \exp\left(-\frac{3.676}{T_9}\right) + 5.41 \times 10^8 T_9^{-1/2} \exp\left(-\frac{8.926}{T_9}\right) & \\ + 4.72 \times 10^7 T_9^{-3/2} \exp\left(-\frac{7.721}{T_9}\right) + 2.20 \times 10^9 T_9^{-3/2} & \\ \exp\left(-\frac{11.418}{T_9}\right) \end{aligned} \tag{6.160}$$

$$f_{1,15} = f_{1,14} \tag{6.161}$$

$$R_{15} = Y_1 Y_{15} \rho \lambda'_{1,15} f_{1,15} \tag{6.162}$$

$$\lambda'_{1,15} = 9.78 \times 10^8 T_9^{-2/3} \exp\left[-\frac{15.251}{T_9^{1/3}} - \left(\frac{T_9}{0.450}\right)^2\right](1 + 0.027 T_9^{1/3}$$

$$+ 0.219 T_9^{2/3} + 0.042 T_9 + 6.83 T_9^{4/3} + 3.32 T_9^{5/3}) + 1.11$$

$$\times 10^4 T_9^{-3/2} \exp\left(-\frac{3.328}{T_9}\right) + 1.49 \times 10^4 T_9^{-3/2} \exp\left(-\frac{4.665}{T_9}\right)$$

$$+ 3.80 \times 10^6 T_9^{-3/2} \exp\left(-\frac{11.048}{T_9}\right) \tag{6.163}$$

$$f_{1,15} = f_{1,14} \tag{6.164}$$

$$R_{16} = Y_1 Y_{16} \rho \lambda_{1,16} f_{1,16} \tag{6.165}$$

$$\lambda_{1,16} = \frac{1.50 \times 10^8 T_9^{-2/3} \exp\left(-\frac{16.692}{T_9^{1/3}}\right)}{1 + 2.13\left[1 - \exp(-0.728 T_9^{2/3})\right]} \tag{6.166}$$

$$f_{1,16} = \begin{cases} \exp(8\psi), & 8\psi \leq 0.1 \\ \exp(10.716\Phi), & 0.1 < 8\psi \leq 2 \end{cases} \tag{6.167}$$

$$R_{18} = Y_1 Y_{17} \rho \lambda_{1,17} f_{1,16} \tag{6.168}$$

$$\lambda_{1,17} = 1.53 \times 10^7 T_9^{-2/3} \exp\left[-\frac{16.712}{T_9^{1/3}} - \left(\frac{T_9}{0.565}\right)^2\right](1 + 0.025 T_9^{1/3}$$

$$+ 5.39 T_9^{2/3} + 0.940 T_9 + 13.5 T_9^{4/3} + 5.98 T_9^{5/3}) + 2.92 \times 10^6 T_9$$

$$\exp\left(-\frac{4.247}{T_9}\right) + 4.81 \times 10^9 T_9 \exp\left[-\frac{16.712}{T_9^{1/3}} - \left(\frac{T_9}{0.04}\right)^2\right] + 5.05$$

$$\times 10^{-6} T_9^{-3/2} \exp\left(-\frac{0.723}{T_9}\right) + 1.31 T_9^{-3/2} \exp\left(-\frac{1.961}{T_9}\right) \tag{6.169}$$

$$R_{19} = Y_1 Y_{17} \rho \lambda'_{1,17} f_{1,16} \tag{6.170}$$

$$\lambda'_{1,17} = 7.97 \times 10^7 T_9^{-2/3} (1 + 2.69 T_9)^{-5/6} \exp\left[-\frac{16.712(1 + 2.69 T_9)^{1/3}}{T_9^{1/3}}\right]^{1/3}$$

$$+ 1.51 \times 10^8 T_9^{-2/3}(1 + 0.025 T_9^{1/3} - 0.051 T_9^{2/3} - 8.82$$

$$\times 10^{-3} T_9) \exp\left(-\frac{16.712}{T_9^{1/3}}\right) + 1.56 \times 10^5 T_9^{-1} \exp\left(-\frac{6.272}{T_9}\right)$$

$$+ 1.31 T_9^{-3/2} \exp\left(-\frac{1.961}{T_9}\right) \tag{6.171}$$

$$R_{21} = Y_1 Y_{18} \rho \lambda_{1,18} f_{1,16} \tag{6.172}$$

$$\lambda_{1,18} = 3.63 \times 10^{11} T_9^{-2/3} \exp\left[-\frac{16.729}{T_9^{1/3}} - \left(\frac{T_9}{1.361}\right)^2\right](1 + 0.025 T_9^{1/3}$$

$$+ 1.88 T_9^{2/3} + 0.327 T_9 + 4.66 T_9^{4/3} + 2.06 T_9^{5/3}) + 9.9$$

$$\times 10^{-14} T_9^{-3/2} \exp\left(-\frac{0.231}{T_9}\right) + 2.66 \times 10^4 T_9^{-3/2} \exp\left(-\frac{1.670}{T_9}\right)$$

$$+ 2.41 \times 10^9 T_9^{-3/2} \exp\left(-\frac{7.638}{T_9}\right) + 1.46$$

$$\times 10^9 T_9^{-1} \exp\left(-\frac{8.31}{T_9}\right) \tag{6.173}$$

$$R_{22} = Y_1 Y_{18} \rho \lambda'_{1,18} f_{1,16} \tag{6.174}$$

$$\lambda'_{1,18} = 3.45 \times 10^8 T_9^{-2/3} \exp\left[-\frac{16.729}{T_9^{1/3}} - \left(\frac{T_9}{0.139}\right)^2\right](1 + 0.025 T_9^{1/3}$$

$$+ 2.26 T_9^{2/3} + 0.394 T_9 + 30.56 T_9^{4/3} + 13.55 T_9^{5/3}) + 1.25$$

$$\times 10^{-15} T_9^{-3/2} \exp\left(-\frac{0.231}{T_9}\right) + 1.64 \times 10^2 T_9^{-3/2} \exp\left(-\frac{1.67}{T_9}\right)$$

$$+ 1.28 \times 10^4 T_9^{1/2} \exp\left(-\frac{5.098}{T_9}\right) \tag{6.175}$$

$$R_5/R_7 = \frac{Y_e \lambda_{e,7}}{Y_1 \lambda_{1,7}} \tag{6.176}$$

$$Y_e = \sum_r Z_r Y_r \tag{6.177}$$

$$\lambda_{e,7} = 1.34 \times 10^{-10} T_9^{-1/2}\left[1 - 0.537 T_9^{1/3} + 3.86 T_9^{2/3}\right.$$

$$\left. + 0.0027 T_9^{-1} \exp\left(\frac{2.515 \times 10^{-3}}{T_9}\right)\right] \tag{6.178}$$

$$\lambda_{1,7} = 3.11 \times 10^5 T_9^{-2/3}\left(-\frac{10.262}{T_9^{1/3}}\right) + 2.53 \times 10^3 T_9^{-3/2} \exp\left(-\frac{7.306}{T_9}\right) \tag{6.179}$$

6.10 HELIUM BURNING

Helium is also a relatively abundant element in a star which can undergo nuclear reactions in the temperature range 1–2 $\times$ 10^8 K. There

Table 6.3 The detailed reaction equation, energy Q_r released and neutrino energy loss $(q_v)_r$ for nuclear reactions during helium burning. (According to Georgeanne and Fowler 1988; Harris *et al.* 1983; Fowler *et al.* 1975).

r	Reaction	Q_r (MeV)	$(q_v)_r$ (MeV)
24	$^4\text{He}(2\alpha, \gamma)^{12}\text{C}$	7.275	
25	$^{12}\text{C}(\alpha, \gamma)^{16}\text{O}$	7.162	
26	$^{16}\text{O}(\alpha, \gamma)^{20}\text{Ne}$	4.734	
27	$^{13}\text{C}(\alpha, n)^{16}\text{O}$	2.216	
28	$^{14}\text{N}(\alpha, \gamma)^{18}\text{F}$	4.415	
29	$^{18}\text{F}(e^+v)18\text{O}$	1.655	0.3965
30	$^{18}\text{O}(\alpha, \gamma)^{22}\text{Ne}$	9.669	
31	$^{22}\text{Ne}(\alpha, n)^{25}\text{Mg}$	−0.481	
32	$^{22}\text{Ne}(\alpha, \gamma)^{26}\text{Mg}$	10.612	

are several reaction processes which can be summarised in a general reaction equation:

$$n \cdot {}^4\text{He} \rightarrow {}^{12}\text{C}, {}^{16}\text{O}, {}^{20}\text{Ne}$$

When $n = 3$, it is called the 3α reaction and the product is ^{12}C; when $n = 4$, the product is ^{16}O; when $n = 5$, the product is ^{20}Ne. In the interior of a normal star, the 3α reaction dominates and the last reaction has only a very small probability. The detailed reaction equation, energy Q_r released and neutrino energy loss $(q_v)_r$ for the above reactions are shown in Table 6.3.

According to Eq. (6.29), the set of differential equations for changes in chemical compositions during helium burning can be written as

neutron n: $\quad \dfrac{dY_n}{dt} = R_{27} + R_{31}$ $\qquad\qquad$ (6.180)

^4He: $\quad \dfrac{dY_4}{dt} = -3R_{24} - R_{25} - R_{26} - R_{27} - R_{28} - R_{30} - R_{31} - R_{32}$ $\quad$ (6.181)

^{12}C: $\quad \dfrac{dY_{12}}{dt} = R_{24} - R_{25}$ $\qquad\qquad$ (6.182)

^{13}C: $\quad \dfrac{dY_{13}}{dt} = -R_{27}$ $\qquad\qquad$ (6.183)

$$^{14}\text{N}: \quad \frac{dY_{14}}{dt} = -R_{28} \tag{6.184}$$

$$^{16}\text{O}: \quad \frac{dY_{16}}{dt} = R_{25} - R_{26} + R_{27} \tag{6.185}$$

$$^{18}\text{O}: \quad \frac{dY_{18}}{dt} = R_{18} - R_{30} \tag{6.186}$$

$$^{20}\text{Ne}: \quad \frac{dY_{20}}{dt} = R_{26} \tag{6.187}$$

$$^{22}\text{Ne}: \quad \frac{dY_{22}}{dt} = R_{30} - R_{31} - R_{32} \tag{6.188}$$

$$^{25}\text{Mg}: \quad \frac{dY_{25}}{dt} = R_{31} \tag{6.189}$$

$$^{26}\text{Mg}: \quad \frac{dY_{26}}{dt} = R_{32} \tag{6.190}$$

According to Eq. (6.32), the energy generation rate during helium burning is

$$\varepsilon = 9.649 \times 10^{17}[Q_{24}R_{24} + Q_{25}R_{25} + Q_{26}R_{26} + Q_{27}R_{27} + (Q_{28} + Q_{29}$$
$$- (q_v)_{29})R_{28} + Q_{30}R_{30} + Q_{31}R_{31} + Q_{32}R_{32}] \tag{6.191}$$

Substituting Q_r and $(q_v)_r$ into Eq. (6.191), we have

$$\varepsilon = 9.649 \times 10^{17}[7.275R_{24} + 7.162R_{25} + 4.730R_{26} + 2.215R_{27} + 5.675R_{28}$$
$$+ 9.668R_{30} - 0.480R_{31} + 10.613R_{32}] \tag{6.192}$$

Since the β decay for the radioactive element ^{18}F is instantaneous, the abundance of the element ^{18}F is conserved and equals zero. In Eqs. (6.180) and (6.190), we have used the relationship $R_{29} = R_{28}$.

The expressions for nuclear reaction rates R_r in Eqs. (6.180) to (6.190) are:

$$R_{24} = \frac{1}{6}Y_4^3 \rho^2 \lambda_{3,\alpha} f_{3,\alpha} \tag{6.193}$$

$$\lambda_{3,\alpha} = 2.79 \times 10^{-8}T_9^{-3}\exp\left(-\frac{4.4027}{T_9}\right) + 1.35 \times 10^{-8}T_9^{-3/2}$$
$$\exp\left(-\frac{24.811}{T_9}\right) \tag{6.194}$$

$$f_{3,\alpha} = \begin{cases} \exp(8\psi), & 8\psi \leq 0.1 \\ \exp(11.206\Phi), & 0.1 < 8\psi \leq 2 \end{cases} \tag{6.195}$$

$$R_{25} = Y_4 Y_{12} \rho \lambda_{12,4} f_{12,4} \tag{6.196}$$

$$\lambda_{12,4} = 1.04 \times 10^8 T_9^{-2} \frac{1}{(1 + 0.0489 T_9^{-2/3})^2} \exp\left[-\frac{32.120}{T_9^{1/3}} - \left(\frac{T_9}{3.496}\right)^2\right]$$
$$+ 1.76 \times 10^8 T_9^{-2} \frac{\exp\left(-\frac{32.120}{T_9^{1/3}}\right)}{(1 + 0.2654 T_9^{-2/3})^2} + 1.25 \times 10^3 T_9^{-3/2} \tag{6.197}$$
$$\exp\left(-\frac{27.499}{T_9}\right) + 1.43 \times 10^{-2} T_9^5 \exp\left(-\frac{15.541}{T_9}\right)$$

$$f_{12,4} = \begin{cases} \exp(12\psi), & 12\psi \leq 0.1 \\ \exp(16.192\Phi), & 0.1 < 12\psi \leq 2 \end{cases} \tag{6.198}$$

$$R_{26} = Y_4 Y_{16} \rho \lambda_{16,4} f_{16,4} \tag{6.199}$$

$$\lambda_{16,4} = 9.37 \times 10^9 T_9^{-2/3} \exp\left[\frac{-39.757}{T_9^{1/3}} - \left(\frac{T_9}{1.586}\right)^2\right] + 6.21$$
$$\times 10^1 T_9^{-3/2} \exp\left(-\frac{10.297}{T_9}\right) + 5.38 \times 10^2 T_9^{-3/2} \exp\left(-\frac{12.226}{T_9}\right)$$
$$+ 1.30 \times 10^1 T_9^2 \exp\left(-\frac{20.093}{T_9}\right) \tag{6.200}$$

$$f_{16,4} = \begin{cases} \exp(16\psi), & 16\psi \leq 0.1 \\ \exp(20.978\Phi), & 0.1 < 16\psi \leq 2 \end{cases} \tag{6.201}$$

$$R_{27} = Y_4 Y_{13} \rho \lambda_{13,4} f_{12,4} \tag{6.202}$$

$$\lambda_{13,4} = 6.77 \times 10^{15} T_9^{-2/3} \exp\left[\frac{-32.329}{T_9^{1/3}} - \left(\frac{T_9}{1.284}\right)^2\right](1 + 0.013 T_9^{1/3}$$
$$+ 2.04 T_9^{2/3} + 0.184 T_9) + 3.82 \times 10^5 T_9^{-3/2} \exp\left(-\frac{9.373}{T_9}\right)$$
$$+ 1.41 \times 10^6 T_9^{-3/2} \exp\left(-\frac{11.873}{T_9}\right) + 2 \times 10^9 T_9^{-3/2} \exp\left(-\frac{20.409}{T_9}\right)$$
$$+ 2.92 \times 10^9 T_9^{-3/2} \exp\left(-\frac{29.283}{T_9}\right) \tag{6.203}$$

$$R_{28} = Y_4 Y_{14} \rho \lambda_{14,4} f_{14,4} \tag{6.204}$$

$$
\begin{aligned}
\lambda_{14,4} &= 7.78 \times 10^9 T_9^{-2/3} \exp\left[-\frac{36.031}{T_9^{1/3}} - \left(\frac{T_9}{0.881}\right)^2 \right] (1 + 0.012 T_9^{1/3} \\
&\quad + 1.45 T_9^{2/3} + 0.117 T_9 + 1.97 T_9^{4/3} + 0.406 T_9^{5/3}) + 2.36 \\
&\quad \times 10^{-10} T_9^{-3/2} \exp\left(-\frac{2.798}{T_9}\right) + 2.03 T_9^{-3/2} \exp\left(-\frac{5.054}{T_9}\right) \\
&\quad + 1.15 \times 10^4 T_9^{-2/3} \exp\left(-\frac{12.310}{T_9}\right)
\end{aligned}
\tag{6.205}
$$

$$
f_{14,4} = \begin{cases} \exp(14\psi), & 14\psi \le 0.1 \\ \exp(18.606\Phi), & 0.1 < 14\psi \le 2 \end{cases}
\tag{6.206}
$$

$$R_{30} = Y_4 Y_{18} \rho \lambda_{18,4} f_{16,4} \tag{6.207}$$

$$
\begin{aligned}
\lambda_{18,4} &= 1.82 \times 10^{12} T_9^{-2/3} \exp\left[-\frac{40.057}{T_9^{1/3}} - \left(\frac{T_9}{0.343}\right)^2 \right] (1 + 0.010 T_9^{1/3} \\
&\quad + 0.988 T_9^{2/3} + 0.072 T_9 + 3.17 T_9^{4/3} + 0.586 T_9^{5/3}) \\
&\quad + 7.54 T_9^{-3/2} \exp\left(-\frac{6.228}{T_9}\right) + 3.48 \times 10^1 T_9^{-3/2} \exp\left(-\frac{7.301}{T_9}\right) \\
&\quad + 6.23 \times 10^3 T_9 \exp\left(-\frac{16.897}{T_9}\right) + 10^{-12} T_9^{-3/2} \exp\left(-\frac{1.994}{T_9}\right)
\end{aligned}
\tag{6.208}
$$

$$R_{31} = Y_4 Y_{22} \rho \lambda_{22,4} f_{22,4} \tag{6.209}$$

$$
\begin{aligned}
\lambda_{22,4} &= 4.16 \times 10^{19} T_9^{-2/3} \\
&\quad \frac{\exp\left[-47.004\left(\frac{1+0.0548 T_9}{T_9}\right)^{1/3} - 3.975 \times 10^{-4}\left(\frac{1+0.0548 T_9}{T_9}\right)^{4.82} \right]}{(1 + 0.0548 T_9)^{5/6}\left[1 + 5.0\exp\left(-\frac{14.791}{T_9}\right)\right]} \\
&\quad + \frac{1.44 \times 10^{-4} \exp\left(-\frac{5.574}{T_9}\right)}{1 + 5.0\exp\left(-\frac{14.791}{T_9}\right)}
\end{aligned}
\tag{6.210}
$$

$$
f_{22,4} = \begin{cases} \exp(20\psi), & 20\psi \le 0.1 \\ \exp(25.616\Phi), & 0.1 < 20\psi \le 2 \end{cases}
\tag{6.211}
$$

$$R_{32} = Y_4 Y_{22} \rho \lambda'_{22,4} f_{22,4} \tag{6.212}$$

$$\lambda'_{22,4} = \frac{4.16 \times 10^{19} T_9^{-2/3} \exp\left[-47.004\left(\frac{1+0.0548T_9}{T_9}\right)^{1/3}\right]}{\left[1 + 5.0 \exp\left(-\frac{14.791}{T_9}\right)\right](1+0.0548T_9)^{5/6}}$$
$$\left\{\exp\left[-24.819\left(\frac{T_9}{1+0.0548T_9}\right)^{2.31}\right]\right.$$
$$\left.+ 5.0 \times 10^{-4}\exp\left[-3.975 \times 10^{-4}\left(\frac{1+0.0548T_9}{T_9}\right)^{4.82}\right]\right\}$$

(6.213)

6.11 Carbon Burning

The temperature for nuclear reactions of carbon to occur is $T \geq 7 \times 10^8$ K. The reaction equations and the released energies are shown in Table 6.4.

Table 6.4 The reaction equation, energy Q_r released and neutrino energy loss $(q_\nu)_r$ for nuclear reactions during carbon burning (according to Georgeanne and Fowler 1988; Harris *et al.* 1983; Fowler *et al.* 1975).

r	Reaction	Q_r (MeV)	$(q_\nu)_r$ (MeV)
33	$^{12}C(^{12}C, \alpha)^{20}Ne$	4.621	
34	$^{12}C(^{12}C, p)^{23}Na$	2.242	
9	$^{12}C(p, \gamma)^{13}N$	1.944	
10	$^{13}N(e^+\nu)^{13}C$	2.221	0.710
27	$^{13}C(\alpha, n)^{16}O$	2.216	
26	$^{16}O(\alpha, \gamma)^{20}Ne$	4.734	
35	$^{20}Ne(\alpha, \gamma)^{24}Mg$	9.312	
31	$^{22}Ne(\alpha, n)^{25}Mg$	−0.481	
36	$^{24}Mg(p, \gamma)^{25}Al$	2.271	
37	$^{25}Al(e^+\nu)^{25}Mg$	4.260	1.773
38	$^{25}Mg(p, \gamma)^{26}Al$	6.306	
39	$^{26}Al(e^+\nu)^{26}Mg$	4.003	1.642
40	$^{22}Ne(p, \gamma)^{23}Na$	8.794	
41	$^{23}Na(p, \gamma)^{24}Mg$	11.691	
42	$^{23}Na(p, \alpha)^{20}Ne$	2.379	
43	$^{23}Na(\alpha, p)^{26}Mg$	1.822	
44	$^{26}Mg(p, \gamma)^{27}Al$	8.272	

The major nuclear reaction for carbon burning is the fusion of two carbon atoms to form the compound nucleus ^{24}Mg, which then quickly decays through proton (p) chain and α chain decays. The probability B_p for p chain decays and the probability B_α for α chain decays are the same, i.e., $B_p = B_\alpha = \frac{1}{2}$, so

$$R_{33} = R_{34} = \frac{1}{2}R \qquad (6.214)$$

Furthermore, considering the zero rates of abundance change for all elements with β decays during carbon burning, we have

$$R_{10} = R_9, \quad R_{37} = R_{36}, \quad R_{39} = R_{38} \qquad (6.215)$$

Therefore, the set of differential equations for changes in chemical compositions during carbon burning are

$$n: \frac{dY_n}{dt} = R_{27} + R_{31} \qquad (6.216)$$

$$p: \frac{dY_1}{dt} = -R_9 + \frac{1}{2}R - R_{36} - R_{38} - R_{40} - R_{41} - R_{42} + R_{43} - R_{44} \qquad (6.217)$$

$$\alpha: \frac{dY_4}{dt} = -R_{26} - R_{27} - R_{31} + \frac{1}{2}R - R_{35} + R_{42} - R_{43} \qquad (6.218)$$

$$^{12}\text{C}: \frac{dY_{12}}{dt} = -R_9 - 2R \qquad (6.219)$$

$$^{13}\text{C}: \frac{dY_{13}}{dt} = R_9 - R_{27} \qquad (6.220)$$

$$^{20}\text{Ne}: \frac{dY_{20}}{dt} = R_{26} + \frac{1}{2}R - R_{35} + R_{42} \qquad (6.221)$$

$$^{22}\text{Ne}: \frac{dY_{22}}{dt} = -R_{31} - R_{40} \qquad (6.222)$$

$$^{23}\text{Na}: \frac{dY_{23}}{dt} = \frac{1}{2}R + R_{40} - R_{41} - R_{42} - R_{43} \qquad (6.223)$$

$$^{24}\text{Mg}: \frac{dY_{24}}{dt} = R_{35} - R_{36} + R_{41} \qquad (6.224)$$

$$^{25}\text{Mg}: \frac{dY_{25}}{dt} = R_{31} + R_{36} - R_{38} \qquad (6.225)$$

$$^{26}\text{Mg}: \frac{dY_{26}}{dt} = R_{38} + R_{43} - R_{44} \tag{6.226}$$

$$^{27}\text{Al}: \frac{dY_{27}}{dt} = R_{44} \tag{6.227}$$

The energy generation rate during the carbon burning process is

$$\varepsilon = 9.649 \times 10^{17}\{[Q_9 + Q_{10} - (q_\nu)_{10}]R_9 + Q_{26}R_{26} + Q_{27}R_{27} + Q_{31}R_{31}$$
$$+ \frac{1}{2}(Q_{33} + Q_{34})R + Q_{35}R_{35} + [Q_{36} + Q_{37} - (q_\nu)_{37}]R_{37} + [Q_{38}$$
$$+ Q_{39} - (q_\nu)_{39}]R_{38} + Q_{40}R_{40} + Q_{41}R_{41} + Q_{42}R_{42} + Q_{43}R_{43} + Q_{44}R_{44}\} \tag{6.228}$$

Substituting the Q_r and $(q_\nu)_r$ values into the above equation, we have

$$\varepsilon = 9.649 \times 10^{17}(3.455R_9 + 4.730R_{26} + 2.215R_{27} - 0.480R_{31} + 3.428R$$
$$+ 9.315R_{35} + 4.758R_{36} + 8.667R_{38} + 8.793R_{40} + 11.692R_{41}$$
$$+ 2.377R_{42} + 1.822R_{43} + 8.271R_{44}) \tag{6.229}$$

The expressions for nuclear reaction rates R_r in the above equations are:

$$R = \frac{1}{2}Y_{12}^2 \rho \lambda_{12,12} f_{12,12} \tag{6.230}$$

$$\lambda_{12,12} = \frac{4.27 \times 10^{26} T_9^{-2/3}}{(1 + 0.0396T_9)^{5/6}} \exp\left[-84.165\left(\frac{1 + 0.0396T_9}{T_9}\right)^{1/3}\right.$$
$$\left. - 2.12 \times 10^{-3} T_9^3\right] \tag{6.231}$$

$$f_{12,12} = \begin{cases} \exp(36\psi), & 36\psi \leq 0.1 \\ \exp(45.664\Phi), & 0.1 < 36\psi \leq 2 \end{cases} \tag{6.232}$$

$$R_{35} = Y_4 Y_{20} \rho \lambda_{20,4} f_{20,4} \tag{6.233}$$

$$\lambda_{20,4} = \frac{1}{1 + 5\exp\left(-\frac{18.960}{T_9}\right)}\left\{4.11 \times 10^{11} T_9^{-2/3}\exp\left[-\frac{46.766}{T_9^{1/3}}\right.\right.$$
$$\left.- \left(\frac{T_9}{2.219}\right)^2\right](1 + 0.009T_9^{1/3} + 0.882T_9^{2/3} + 0.055T_9 + 0.749T_9^{4/3}$$
$$+ 0.119T_9^{5/3}) + 5.27 \times 10^3 T_9^{-3/2}\exp\left(-\frac{15.869}{T_9}\right) + 6.51$$

$$\times 10^3 T_9^{1/2} \exp\left(-\frac{16.233}{T_9}\right) + 4.21 T_9^{-3/2} \exp\left(\frac{-9.115}{T_9}\right) + 3.20$$

$$\times 10^1 T_9^{-2/3} \exp\left(\frac{-9.383}{T_9}\right) \Bigg\}$$
(6.234)

$$f_{20,4} = \begin{cases} \exp(20\psi), & 20\psi \leq 0.1 \\ \exp(25.616\Phi), & 0.1 < 20\psi \leq 2 \end{cases}$$
(6.235)

$$R_{36} = Y_1 Y_{24} \rho \lambda_{24,1} f_{24,1}$$
(6.236)

$$\lambda_{24,1} = \frac{1}{1 + 5\exp\left(-\frac{15.882}{T_9}\right)} \left\{ 5.6 \times 10^8 T_9^{-2/3} \exp\left(-\frac{22.019}{T_9^{1/3}}\right) \right.$$

$$(1 + 0.019 T_9^{1/3} - 0.173 T_9^{2/3} - 0.023 T_9) + 1.48 \times 10^3 T_9^{-3/2}$$

$$\left. \exp\left(\frac{-2.484}{T_9}\right) + 4 \times 10^3 \exp\left(-\frac{4.180}{T_9}\right) \right\}$$
(6.237)

$$f_{24,1} = \begin{cases} \exp(12\psi), & 12\psi \leq 0.1 \\ \exp(15.325\Phi), & 0.1 < 12\psi \leq 2 \end{cases}$$
(6.238)

$$R_{38} = Y_1 Y_{25} \rho \lambda_{25,1} f_{24,1}$$
(6.239)

$$\lambda_{25,1} = 3.57 \times 10^9 T_9^{-2/3} \exp\left[\frac{-22.031}{T_9^{1/3}} - \left(\frac{T_9}{0.06}\right)^2\right](1 + 0.019 T_9^{1/3}$$

$$+ 7.669 T_9^{2/3} + 1.015 T_9 + 167.4 T_9^{4/3} + 56.35 T_9^{5/3}) + 3.07$$

$$\times 10^{-13} T_9^{-3/2} \exp\left(-\frac{0.435}{T_9}\right) + 1.94 \times 10^{-7} T_9^{-3/2} \exp$$

$$\left(-\frac{0.673}{T_9}\right) + 3.15 \times 10^{-5} T_9^{-3.4} \exp\left[-\frac{1.342}{T_9} - \left(\frac{T_9}{13}\right)^2\right] + 1.77$$

$$\times 10^4 T_9^{5/8} \exp\left[-\frac{3.049}{T_9} - \left(\frac{T_9}{13}\right)^2\right]$$
(6.240)

$$R_{40} = Y_1 Y_{22} \rho \lambda_{22,1} f_{22,1}$$
(6.241)

$$\lambda_{22,1} = 1.15 \times 10^9 T_9^{-2/3} \exp\left(\frac{-19.475}{T_9^{1/3}}\right) + 9.77 \times 10^{-12} T_9^{-3/2}$$

$$\exp\left(-\frac{0.348}{T_9}\right) + 8.96 \times 10^3 T_9^{-3/2} \exp\left(-\frac{4.840}{T_9}\right) + 6.52$$

$$\times 10^4 T_9^{-3/2} \exp\left(-\frac{5.319}{T_9}\right) + 7.97 \times 10^5 T_9^{-1/2} \exp\left(-\frac{7.418}{T_9}\right)$$

$$+ 1.63 \times 10^{-2} T_9^{-3/2} \exp\left(-\frac{1.775}{T_9}\right)$$
(6.242)

$$f_{22,1} = \begin{cases} \exp(10\psi), & 10\psi \leq 0.1 \\ \exp(13.051\Phi), & 0.1 < 10\psi \leq 2 \end{cases} \tag{6.243}$$

$$R_{41} = Y_1 Y_{23} \rho \lambda_{23,1} f_{23,1} \tag{6.244}$$

$$\begin{aligned} \lambda_{23,1} = &\left\{ 2.93 \times 10^8 T_9^{-2/3} \exp\left[\frac{-20.766}{T_9^{1/3}} - \left(\frac{T_9}{0.297}\right)^2 \right] (1 + 0.020 T_9^{1/3} \right. \\ &+ 1.61 T_9^{2/3} + 0.226 T_9 + 4.94 T_9^{4/3} + 1.76 T_9^{5/3}) + 9.34 \\ &\times 10^1 T_9^{-3/2} \exp\left(-\frac{2.789}{T_9}\right) + 1.89 \times 10^4 T_9^{-3/2} \exp\left(-\frac{3.434}{T_9}\right) \\ &\left. + 5.10 \times 10^4 T_9^{1/5} \exp\left(-\frac{5.510}{T_9}\right) \right\} \cdot \left[1 + 1.5 \exp\left(-\frac{5.105}{T_9}\right) \right]^{-1} \end{aligned} \tag{6.245}$$

$$f_{23,1} = \begin{cases} \exp(11\psi), & 11\psi \leq 0.1 \\ \exp(14.195\Phi), & 0.1 < 11\psi \leq 2 \end{cases} \tag{6.246}$$

$$R_{42} = Y_1 Y_{23} \rho \lambda'_{23,1} f_{23,1} \tag{6.247}$$

$$\begin{aligned} \lambda'_{23,1} = &\, 8.56 \times 10^9 T_9^{-2/3} \exp\left[\frac{-20.766}{T_9^{1/3}} - \left(\frac{T_9}{0.131}\right)^2 \right] (1 + 0.02 T_9^{1/3} \\ &+ 8.21 T_9^{2/3} + 1.15 T_9 + 44.36 T_9^{4/3} + 15.84 T_9^{5/3}) + 4.02 T_9^{-3/2} \\ &\exp\left(-\frac{1.99}{T_9}\right) + 1.18 \times 10^4 T_9^{-5/4} \exp\left(-\frac{3.148}{T_9}\right) + 8.59 \\ &\times 10^5 T_9^{4/3} \exp\left(-\frac{4.375}{T_9}\right) + 3.06 \times 10^{-13} T_9^{-3/2} \exp\left(-\frac{0.447}{T_9}\right) \end{aligned} \tag{6.248}$$

$$R_{43} = Y_4 Y_{23} \rho \lambda_{23,4} f_{23,4} \tag{6.249}$$

$$\lambda_{23,4} = 9.44 \times 10^6 (1 - 1.66 T_9^{1/2} + 1.43 T_9) \exp\left(-\frac{34.449}{T_9}\right) \tag{6.250}$$

$$f_{23,4} = \begin{cases} \exp(22\psi), & 22\psi \leq 0.1 \\ \exp(27.889\Phi), & 0.1 < 22\psi \leq 2 \end{cases} \tag{6.251}$$

$$R_{44} = Y_1 Y_{26} \rho \lambda_{26,1} f_{24,1} \tag{6.252}$$

$$\begin{aligned} \lambda_{26,1} = &\left\{ 7.39 \times 10^8 T_9^{-2/3} \exp\left[-\frac{22.042}{T_9^{1/3}} - \left(\frac{T_9}{0.299}\right)^2 \right] (1 + 0.019 T_9^{1/3} \right. \\ &+ 3.61 T_9^{2/3} + 0.478 T_9 + 9.78 T_9^{4/3} + 3.29 T_9^{5/3}) + 1.32 \times 10^{-10} T_9^{-3/2} \\ &\exp\left(-\frac{0.603}{T_9}\right) + 2.90 \times 10^{-5} T_9^{-3/2} \exp\left(-\frac{1.056}{T_9}\right) \end{aligned}$$

$$+ 6.45 \times 10^{-5} T_9^{-3/2} \exp\left(-\frac{1.230}{T_9}\right) + 5.64 \times 10^{-2} T_9^{-3/2} \exp\left(-\frac{1.694}{T_9}\right)$$

$$+ 2.86 \times 10^3 T_9^{-3/2} \exp\left(-\frac{3.265}{T_9}\right) + 7.99 \times 10^4 T_9^{-3/2} \exp\left(-\frac{3.781}{T_9}\right)$$

$$+ 4.23 \times 10^4 T_9^{1/2} \exp\left(-\frac{3.661}{T_9}\right) \Bigg\} \left[1 + 5 \exp\left(-\frac{20.99}{T_9}\right)\right]^{-1}$$

$$(6.253)$$

6.12 NEON BURNING

When the temperature in the star becomes $T \geq 1.5 \times 10^9$ K, neon burning can occur. The reaction equations and energies released in the neon burning process are listed in Table 6.5.

Table 6.5 The detailed equation, energy Q_r released and neutrino energy loss $(q_v)_r$ for nuclear reactions during neon burning (according to Georgeanne and Fowler 1988; Harris et al. 1983; Fowler et al. 1975).

r	Reaction	Q_r (MeV)	$(q_v)_r$ (MeV)
45	$^{20}Ne(\gamma, \alpha)^{16}O$	-4.734	
26	$^{16}O(\alpha, \gamma)^{20}Ne$	4.734	
35	$^{20}Ne(\alpha, \gamma)^{24}Mg$	9.312	
46	$^{24}Mg(\alpha, \gamma)^{28}Si$	9.984	
42	$^{23}Na(p, \alpha)^{20}Ne$	2.379	
43	$^{23}Na(\alpha, p)^{26}Mg$	1.822	
31	$^{22}Ne(\alpha, n)^{25}Mg$	-0.481	
47	$^{24}Mg(\alpha, p)^{27}Al$	-1.601	
48	$^{25}Mg(\alpha, n)^{28}Si$	2.653	
49	$^{26}Mg(\alpha, n)^{29}Si$	0.035	
50	$^{26}Mg(\alpha, \gamma)^{30}Si$	10.644	
44	$^{26}Mg(p, \gamma)^{27}Al$	8.272	
51	$^{27}Al(p, \gamma)^{28}Si$	11.585	
52	$^{30}Si(p, \gamma)^{31}P$	7.297	

The set of differential equations for changes in chemical compositions for the neon burning process are:

$$^{16}O: \frac{dY_{16}}{dt} = -R_{26} + R_{45} \tag{6.254}$$

$$^{20}Ne: \frac{dY_{20}}{dt} = R_{26} - R_{35} + R_{42} - R_{45} \tag{6.255}$$

$$^{22}Ne: \frac{dY_{22}}{dt} = -R_{31} \tag{6.256}$$

$$^{23}Na: \frac{dY_{23}}{dt} = -R_{42} - R_{43} \tag{6.257}$$

$$^{24}Mg: \frac{dY_{24}}{dt} = R_{35} - R_{46} - R_{47} \tag{6.258}$$

$$^{25}Mg: \frac{dY_{25}}{dt} = R_{31} - R_{48} \tag{6.259}$$

$$^{26}Mg: \frac{dY_{26}}{dt} = R_{43} - R_{44} - R_{49} - R_{50} \tag{6.260}$$

$$^{27}Al: \frac{dY_{27}}{dt} = R_{44} + R_{47} - R_{51} \tag{6.261}$$

$$^{28}Si: \frac{dY_{28}}{dt} = R_{46} + R_{48} + R_{51} \tag{6.262}$$

$$^{29}Si: \frac{dY_{29}}{dt} = R_{49} \tag{6.263}$$

$$^{30}Si: \frac{dY_{30}}{dt} = R_{50} - R_{52} \tag{6.264}$$

$$^{31}P: \frac{dY_{31}}{dt} = R_{52} \tag{6.265}$$

$$n: \frac{dY_{n}}{dt} = R_{31} + R_{48} + R_{49} \tag{6.266}$$

$$p: \frac{dY_{1}}{dt} = -R_{42} + R_{43} - R_{44} + R_{47} - R_{51} - R_{52} \tag{6.267}$$

$$\alpha : \frac{dY_4}{dt} = -R_{26} - R_{31} - R_{35} + R_{42} - R_{43} + R_{45} - R_{46} - R_{47} - R_{48}$$

$$- R_{49} - R_{50} \tag{6.268}$$

The energy generation rate during neon burning is

$$\varepsilon = 9.649 \times 10^{17}(Q_{26}R_{26} + Q_{31}R_{31} + Q_{35}R_{35} + Q_{42}R_{42} + Q_{43}R_{43} + Q_{44}R_{44}$$

$$Q_{45}R_{45} + Q_{46}R_{46} + Q_{47}R_{47} + Q_{48}R_{48} + Q_{49}R_{49} + Q_{50}R_{50} + Q_{51}R_{51}$$

$$+ Q_{52}R_{52}) \tag{6.269}$$

Substituting Q_r into the above equation, we have

$$\varepsilon = 9.649 \times 10^{17}(4.730R_{26} - 0.48R_{31} + 9.315R_{35} + 2.377R_{42} + 1.822R_{43}$$

$$+ 8.271R_{44} - 4.73R_{45} + 9.986R_{46} - 1.601R_{47} + 2.653R_{48} + 0.033R_{49}$$

$$+ 10.644R_{50} + 11.586R_{51} + 7.297R_{52}) \tag{6.270}$$

The expressions for nuclear reaction rates R_r in the above equations are:

$$R_{45} = Y_{20}\lambda_{20,\gamma} \tag{6.271}$$

$$\lambda_{20,\gamma} = 5.65 \times 10^{10} T_9^{3/2} \exp\left(-\frac{54.937}{T_9}\right)\lambda_{16,4} \tag{6.272}$$

$$R_{46} = Y_4 Y_{24}\rho\lambda_{24,4} f_{24,4} \tag{6.273}$$

$$\lambda_{24,4} = \left[1 + 5\exp\left(-\frac{15.882}{T_9}\right)\right]^{-1}\left\{4.78 \times 10^1 T_9^{-3/2}\exp\left(-\frac{13.506}{T_9}\right)\right.$$

$$+ 2.38 \times 10^3 T_9^{-3/2}\exp\left(-\frac{15.218}{T_9}\right) + 2.47 \times 10^2 T_9^{3/2}\exp$$

$$\left(-\frac{15.147}{T_9}\right) + 1.72 \times 10^{-10} T_9^{-3/2}\exp\left(-\frac{5.028}{T_9}\right) + 1.25$$

$$\left. \times 10^{-3} T_9^{-3/2}\exp\left(-\frac{7.929}{T_9}\right) + 2.43 \times 10^1 T_9^{-1}\exp\left(-\frac{11.52}{T_9}\right)\right\} \tag{6.274}$$

$$f_{24,4} = \begin{cases} \exp(24\psi), & 24\psi \leq 0.1 \\ \exp(30.136\Phi), & 0.1 < 24\psi \leq 2 \end{cases} \tag{6.275}$$

$$R_{47} = Y_4 Y_{24}\rho\lambda'_{24,4} f_{24,4} \tag{6.276}$$

$$\lambda'_{24,4} = 1.99 \times 10^8 T_9^{-2/3} \exp\left[-\frac{23.261}{T_9^{1/3}} - \left(\frac{T_9}{0.157}\right)^2 - \frac{18.575}{T_9} \right]$$

$$(1 + 0.018 T_9^{1/3} + 12.85 T_9^{2/3} + 1.61 T_9 + 89.87 T_9^{4/3} + 28.66 T_9^{5/3})$$

$$+ 2.33 \times 10^2 T_9^{-3/2} \exp\left(-\frac{21.092}{T_9}\right) + 1.02 \times 10^4 T_9^{7/2}$$

$$\exp\left(-\frac{21.996}{T_9}\right) + 3.62 \times 10^{-6} T_9^{-3/2} \exp\left(-\frac{19.433}{T_9}\right)$$

$$+ 5.09 \times 10^{-5} T_9^{-3/2} \exp\left(-\frac{19.543}{T_9}\right) \tag{6.277}$$

$$R_{48} = Y_4 Y_{25} \rho \lambda_{25,4} f_{24,4} \tag{6.278}$$

$$\lambda_{25,4} = \left\{ 3.59 \times 10^{20} (1 + 0.063 T_9)^{-5/6} T_9^{-2/3} \right.$$

$$\left. \exp\left[-53.410 \left(\frac{1 + 0.063 T_9}{T_9}\right)^{1/3} \right] \right\} \cdot \left[1 + \frac{10}{3} \exp\left(-\frac{13.180}{T_9}\right) \right]^{-1} \tag{6.279}$$

$$R_{49} = Y_4 Y_{26} \rho \lambda_{26,4} f_{24,4} \tag{6.280}$$

$$\lambda_{26,4} = \left\{ 2.93 \times 10^{20} (1 + 0.0628 T_9)^{-5/6} T_9^{-2/3} \exp \right.$$

$$\left. \left[-53.505 \left(\frac{1 + 0.0628 T_9}{T_9}\right)^{1/3} \right] \right\} \cdot \left[1 + 5 \exp\left(-\frac{20.99}{T_9}\right) \right]^{-1} \tag{6.281}$$

$$R_{50} = Y_4 Y_{26} \rho \lambda'_{26,4} f_{24,4} \tag{6.282}$$

$$\lambda'_{26,4} = \left\{ 4.55 \times 10^{-2} \exp\left[-\frac{53.51}{T_9^{1/3}} (0.0751 T_9 - 0.0105 T_9^2 + 5.57 \right. \right.$$

$$\left. \left. \times 10^{-4} T_9^3) \right] \lambda_{26,4} \right\} \cdot \left[1 + 5 \exp\left(-\frac{20.99}{T_9}\right) \right]^{-1} \tag{6.283}$$

$$R_{51} = Y_1 Y_{27} \rho \lambda_{27,1} f_{27,1} \tag{6.284}$$

$$\lambda_{27,1} = \left\{ 1.67 \times 10^8 T_9^{-2/3} \exp\left[-\frac{23.261}{T_9^{1/3}} - \left(\frac{T_9}{0.155}\right)^2 \right] (1 + 0.018 T_9^{1/3} \right.$$

$$+ 5.81 T_9^{2/3} + 0.728 T_9 + 27.31 T_9^{4/3} + 8.71 T_9^{5/3}) + 2.20 T_9^{-3/2}$$

$$\exp\left(-\frac{2.269}{T_9}\right) + 1.22 \times 10^1 T_9^{-3/2} \exp\left(-\frac{2.491}{T_9}\right) + 1.50 \times 10^4 T_9$$

$$\exp\left(-\frac{4.112}{T_9}\right) + 6.5 \times 10^{-11} T_9^{-3/2} \exp\left(-\frac{0.853}{T_9}\right)$$

$$+ 1.63 \times 10^{-10} T_9^{-3/2} \exp\left(-\frac{1.001}{T_9}\right)\Bigg\}\Bigg[1 + \frac{1}{3}\exp\left(-\frac{9.792}{T_9}\right)$$

$$\left. + \frac{2}{3}\exp\left(-\frac{11.773}{T_9}\right)\right]^{-1} \tag{6.285}$$

$$f_{27,1} = \begin{cases} \exp(13\psi), & 13\psi \leq 0.1 \\ \exp(16.442\Phi), & 0.1 < 13\psi \leq 2 \end{cases} \tag{6.286}$$

$$R_{52} = Y_1 Y_{30} \rho \lambda_{30,1} f_{30,1} \tag{6.287}$$

$$\lambda_{30,1} = 4.25 \times 10^8 T_9^{-2/3} \exp\left[-\frac{24.468}{T_9^{1/3}} - \left(\frac{T_9}{0.67}\right)^2\right](1 + 0.017 T_9^{1/3}$$

$$+ 0.15 T_9^{2/3} + 0.018 T_9 + 5.53 T_9^{4/3} + 1.68 T_9^{5/3}) + 1.86 \times 10^4 T_9^{-3/2}$$

$$\exp\left(-\frac{5.601}{T_9}\right) + 3.15 \times 10^5 T_9^{-3/2} \exp\left(-\frac{6.961}{T_9}\right)$$

$$+ 2.75 \times 10^5 T_9^{-1/2} \exp\left(-\frac{10.062}{T_9}\right) \tag{6.288}$$

$$f_{30,1} = \begin{cases} \exp(14\psi), & 14\psi \leq 0.1 \\ \exp(17.547\Phi), & 0.1 < 14\psi \leq 2 \end{cases} \tag{6.289}$$

6.13 OXYGEN BURNING

Oxygen undergoes thermonuclear reactions when $T \geq 2 \times 10^9$ K. The nuclear reaction chains for oxygen burning are as follows:

$$^{16}\text{O} \Big< \begin{matrix} (^{16}\text{O}, p)^{31}\text{P} \\ (^{16}\text{O}, \alpha)^{28}\text{Si} \\ (^{16}\text{O}, n)^{31}\text{S}(e^+\nu)^{31}\text{P} \end{matrix}$$

$$^{31}\text{P}(p, \alpha)^{28}\text{Si}(\alpha, \gamma)^{32}\text{S}$$

$$^{28}\text{Si}(\gamma, \alpha)^{24}\text{Mg}(\alpha, p)^{27}\text{Al}(\alpha, p)^{30}\text{Si}$$

$$^{32}\text{S}(n, \gamma)^{33}\text{S}(n, \alpha)^{32}\text{Si}(\alpha, \gamma)^{34}\text{S}$$

$$^{28}\mathrm{Si}(n,\gamma)^{29}\mathrm{Si}\begin{cases}(\alpha,n)^{32}\mathrm{S}(\alpha,p)^{35}\mathrm{Cl}\\(p,\gamma)^{30}\mathrm{P}(e^+\nu)^{30}\mathrm{Si}\end{cases}$$

$$^{33}\mathrm{S}(e^-,\nu)^{33}\mathrm{P}(p,n)^{33}\mathrm{S}$$

$$^{35}\mathrm{Cl}(e^-,\nu)^{35}\mathrm{S}(p,n)^{35}\mathrm{Cl}$$

$$^{32}\mathrm{S}(\alpha,\gamma)^{36}\mathrm{Ar}\begin{cases}(\alpha,p)^{39}\mathrm{K}\\(n,\gamma)^{37}\mathrm{Ar}(e^+\nu)^{37}\mathrm{Cl}\end{cases}$$

$$^{35}\mathrm{Cl}\begin{cases}(\gamma,p)^{34}\mathrm{S}(\alpha,\gamma)^{38}\mathrm{Ar}\begin{cases}(P,\gamma)^{39}\mathrm{K}(p,\gamma)^{40}\mathrm{Ca}\\(\alpha,\gamma)^{42}\mathrm{Ca}\begin{cases}(\alpha,\gamma)^{46}\mathrm{Ti}\\(\alpha,p)^{45}\mathrm{Sc}(p,\gamma)^{46}\mathrm{Ti}\end{cases}\end{cases}\\(e^-,\nu)^{35}\mathrm{S}(\gamma,p)^{34}\mathrm{S}\end{cases}$$

$$^{31}\mathrm{P}(e^-,\nu)^{31}\mathrm{S}$$

$$^{31}\mathrm{P}(n,\gamma)^{32}\mathrm{P}$$

$$^{32}\mathrm{S}(e^-,\nu)^{32}\mathrm{P}(p,n)^{32}\mathrm{S}$$

$$^{33}\mathrm{P}(p,\alpha)^{30}\mathrm{Si}$$

The major nuclear reactions of oxygen burning are the fusion of oxygen atoms into the elements $^{32}\mathrm{S}, ^{31}\mathrm{P}, ^{28}\mathrm{Si}, ^{31}\mathrm{S}$ etc. A special feature of the oxygen burning process is that nuclei heavier than iron formed in several S-processes (or slow neutron capture processes) during the previous helium, carbon and neon burning have undergone photo-disintegration under extremely high temperatures to form iron group elements. Besides, the neutron excess value η increases rapidly during oxygen burning. The definition of η is

$$\eta = \sum_r (N_r - Z_r)Y_r \tag{6.290}$$

where N_r, Z_r and Y_r are the neutron number, the proton number, and the number abundance of the element r, respectively.

When the nuclear reaction rates for the various elements are equal to their inverse reaction rates, the abundance of elements are under equilibrium and can be determined using the statistical equilibrium equation. In such a case, the abundance of elements are determined by three parameters: gas temperature T, density ρ and neutron excess value η (see §10.3).

At a later stage of oxygen burning, there are also isolated "quasi-equilibrium groups" (see §6.14 below). For example, the reactions such as $^{29}Si(\gamma, n)^{28}Si$, $^{28}Si(n, \gamma)^{27}Si$, $^{30}P(\gamma, p)^{29}Si$ and $^{29}Si(p, \gamma)^{30}P$ have reached equilibrium and are transitting to silicon burning.

6.14 SILICON BURNING

The thermonuclear reactions with the direct fusion of two silicon nuclei (^{28}Si) into other nuclei will not occur inside the stars, since the high temperature required to overcome the Coulomb barrier of two ^{28}Si nuclei can already cause photodisintegration of all elements. However, when the temperature inside the star reaches 3×10^9 K, part of the ^{28}Si will undergo a series of photodisintegrations and "melt" to form neutrons, protons and α particles. The major photodisintegration reaction series for zero neutron excess value η is

$$^{28}Si(\gamma, \alpha)^{24}Mg(\gamma, \alpha)^{20}Ne(\gamma, \alpha)^{16}O(\gamma, \alpha)^{12}C(\gamma, 2\alpha)\alpha$$

and that for a larger neutron excess value η is

$$^{25}Mg(p, \alpha)^{22}Na, \ ^{26}Mg(p, \alpha)^{23}Na, \ ^{25}Mg(n, \alpha)^{22}Ne$$

These released neutrons, protons and α particles will add to the quasi-equilibrium group with elements heavier than ^{28}Si and gradually increase the mean atomic weight for the nuclides to finally reach those of the iron group. At the same time, the abundance of the element ^{28}Si has become very small. This is the silicon burning process. The elements heavier than ^{24}Mg which are under equilibrium are called the quasi-equilibrium group. However, they are not in equilibrium with ^{28}Si. The abundance of the nuclides in a quasi-equilibrium group are determined by four parameters: temperature T, density ρ, neutron excess value η and the silicon abundance Y_{28}.

During the silicon burning process, the quasi-equilibrium group is separated into two groups at $Z = 20$: the first (group I) is $12 \leq Z < 20$ and the second (group II) is $22 \leq Z < 28$. The nuclear reaction chains linking groups I and II are

$$\left.\begin{array}{l} ^{38}Ar(\alpha, \gamma) \\ ^{39}K(\alpha, p) \end{array}\right\rangle {}^{42}Ca(\alpha, p)^{45}Sc$$

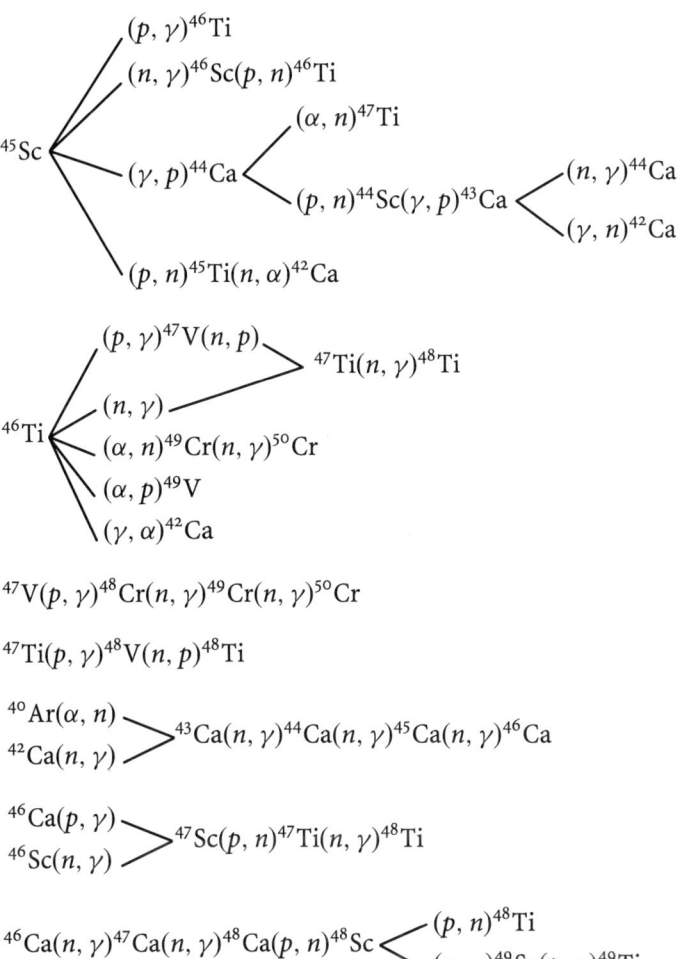

^{45}Sc
- $(p, \gamma)^{46}$Ti
- $(n, \gamma)^{46}$Sc$(p, n)^{46}$Ti
- $(\gamma, p)^{44}$Ca
  - $(\alpha, n)^{47}$Ti
  - $(p, n)^{44}$Sc$(\gamma, p)^{43}$Ca
    - $(n, \gamma)^{44}$Ca
    - $(\gamma, n)^{42}$Ca
- $(p, n)^{45}$Ti$(n, \alpha)^{42}$Ca

^{46}Ti
- $(p, \gamma)^{47}$V(n, p) ^{47}Ti$(n, \gamma)^{48}$Ti
- (n, γ)
- $(\alpha, n)^{49}$Cr$(n, \gamma)^{50}$Cr
- $(\alpha, p)^{49}$V
- $(\gamma, \alpha)^{42}$Ca

^{47}V$(p, \gamma)^{48}$Cr$(n, \gamma)^{49}$Cr$(n, \gamma)^{50}$Cr

^{47}Ti$(p, \gamma)^{48}$V$(n, p)^{48}$Ti

^{40}Ar(α, n)
^{42}Ca(n, γ) $\Big\rangle$ ^{43}Ca$(n, \gamma)^{44}$Ca$(n, \gamma)^{45}$Ca$(n, \gamma)^{46}$Ca

^{46}Ca(p, γ)
^{46}Sc(n, γ) $\Big\rangle$ ^{47}Sc$(p, n)^{47}$Ti$(n, \gamma)^{48}$Ti

^{46}Ca$(n, \gamma)^{47}$Ca$(n, \gamma)^{48}$Ca$(p, n)^{48}$Sc
- $(p, n)^{48}$Ti
- $(n, \gamma)^{49}$Sc$(p, n)^{49}$Ti

In the above quasi-equilibrium reactions chains, argon (^{38}Ar, ^{40}Ar) and potassium (^{39}K) are quasi-equilibrium group I elements, while titanium (^{48}Ti,^{49}Ti), vanadium (^{49}V) and chromium (^{50}Cr) are quasi-equilibrium group II elements.

When silicon burning is completed, the temperature $T \geq 5 \times 10^9$ K and the density $\rho > 3 \times 10^6$g/cm^3, and there is statistical equilibrium among the atomic nuclei. A large amount of collisions occur between high energy photons and atomic nuclei. On one hand, the nuclei are shattered and on the other hand, these shattered nuclei combine with other particles rapidly. Finally, there is statistical equilibrium between shattering and formation of nuclei. When the nuclear matter is under statistical equilibrium, the abundance of each nuclide can be evaluated using statistical equilibrium methods, and are determined by three parameters: temperature T, density ρ and neutron excess value η (see §10.3).

6.15 NEUTRINOS

The interaction cross section σ_ν between neutrino and matter is extremely small. That for neutrinos with energy E_ν can be approximated as $\sigma_\nu \approx \left(\frac{E_\nu}{m_e c^2}\right)^2 \times 10^{-44}$ (cm²) (see Clayton 1968). Therefore, the cross section for neutrinos with MeV energies is $\sigma_\nu \approx 10^{-44}$ cm², which is 10^{-18} times the interaction cross section for photons. In a medium with density ρ, the mean free path of the neutrino is

$$l_\nu = \frac{1}{n\sigma_\nu} = \frac{\mu m_\nu}{\rho\sigma_\nu} \cong \frac{2 \times 10^{20}}{\rho} \text{ (cm)} \qquad (6.291)$$

If we take $\mu \approx 1$ and $\rho \approx 1$ g/cm³, $l_\nu \approx 100$ pc from Eq. (6.291) which is greater than the stellar radius by a few orders of magnitude. Therefore, it is certain that the neutrino can penetrate through a star without making any collisions except for the collapse stage before supernova explosion where $\rho = 10^{14}$ g/cm³.

In §6.9 to §6.14, we have shown that some nuclear reactions produce neutrinos. For example, the $^1\mathrm{H}(\mathrm{p}, e^+\nu)^2\mathrm{H}$, $^7\mathrm{Be}(e^-, \nu)^7\mathrm{Li}$, $^{13}\mathrm{N}(e^+\nu)^{13}\mathrm{C}$, $^{15}\mathrm{O}(e^+\nu)^{15}\mathrm{N}$, $^{18}\mathrm{F}(e^+\nu)^{18}\mathrm{O}$, $^{13}\mathrm{N}(e^+\nu)^{13}\mathrm{C}$, $^{25}\mathrm{Al}(e^+\nu)^{25}\mathrm{Mg}$ and the $^{26}\mathrm{Al}(e^+\nu)^{26}\mathrm{Mg}$ reactions during the hydrogen, helium and carbon burning processes will produce neutrinos. We have also accounted for the energy loss due to neutrinos in calculations of energy generation rates.

Besides neutrino production by nuclear reactions, there are also some weak interactions in the star which can produce a large amount of neutrinos when the temperature and density reach certain ranges, e.g., $10^2 \leq \rho$ (gcm⁻³) $\leq 10^9$ and $10^8 \leq T(\mathrm{K}) \leq 5 \times 10^9$. These weak interactions include the electron pair annihilation neutrino process, the photoneutrino process, the plasmaneutrino process and the bremsstrahlung neutrino process. Figure 6.7 shows regions in the ρ-T plane for different neutrino-producing weak interaction processes.

6.15.1 Pair Annihilation Neutrinos

When the temperature is greater than 10^9 K, there are a lot of high energy photons inside the star, which can create positive and negative electron

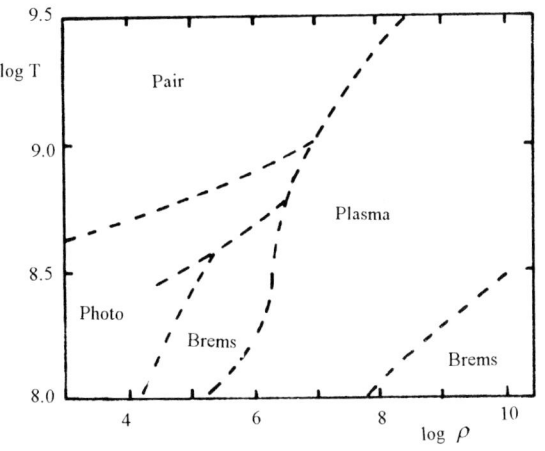

Fig. 6.7 Regions in the ρ-T plane for different neutrino-producing weak interaction processes.

pairs (e^+ e^-). The electron pairs will annihilate quickly and each electron pair will produce a pair of photons on annihilation. The timescales for formation and annihilation of electron pairs are very short compared to the astrophysical timescale, so the formation and the annihilation are at equilibrium, and there is always a considerable amount of electron pairs inside the star. On annihilation of an electron pair, besides the production of a photon pair, there is also an extremely small probability (10^{-19}) of producing a neutrino pair, i.e.,

$$\gamma + \gamma \rightleftarrows e^+ + e^- \longrightarrow \nu + \bar{\nu} \tag{6.292}$$

This happens only when the temperature is extremely high while the density is not so high (see Fig. 6.7). For the non-degenerate condition, the energy loss rate due to production of neutrinos through electron pair annihilation is (Kippenhahn and Weigert 1991)

$$\varepsilon_\nu^{(pair)} = \begin{cases} \frac{4.9\times10^8}{\rho} T_9^3 e^{-11.86/T_9}, & T_9 < 1 \\ \frac{4.45\times10^{15}}{\rho} T_9^9, & T_9 > 3 \end{cases} \tag{6.293}$$

where $T_9 = T/10^9$. When the electron gas is degenerate (relativistic degenerate), we have

$$\varepsilon_\nu^{(pair)} = 1.43 \times 10^{15} \frac{1}{\rho}(1 + 5kT/E_F)T_{10}^4 e^{-E_F/kT}$$

where

$$E_F = c\sqrt{(3\pi^2 n_e)^{2/3} + m_e^2 c^2}$$

is the Fermi energy of the relativistic electron.

6.15.2 Photoneutrinos

The reaction equation for this process can be written as

$$\gamma + e^- \left\langle \begin{array}{l} e^- + \gamma \\ e^- + \nu + \bar{\nu} \end{array} \right. \tag{6.294}$$

In fact, this is the analogue of the normal Thompson scattering of a photon by a free electron. In a very few cases, the photon is replaced by a neutrino pair ($\nu\bar{\nu}$) after scattering. For the photoneutrino process, the energy loss rate greatly depends on whether the electrons are degenerate or relativistic. Crude interpolation equations are (see Petrosian *et al.* 1967)

$$\varepsilon_\nu^{(phot)} = \varepsilon_1 + \varepsilon_2(\mu_e + \bar{\rho})^{-1} \tag{6.295}$$

$$\varepsilon_1 = 1.103 \times 10^{13} \rho^{-1} T_9^9 e^{-5.93/T_9}$$

$$\varepsilon_2 = 0.976 \times 10^8 T_9^8 (1 + 4.2T_9)^{-1}$$

$$\bar{\rho} = 6.446 \times 10^{-6} \rho T_9^{-1}(1 + 4.2T_9)^{-1}$$

where ε and ρ are in cgs units.

6.15.3 Plasmaneutrinos

This process is very important for lower temperatures and higher densities (see Fig. 6.7). It involves the interaction between radiation and a plasma with the formation of neutrinos. When a beam of electromagnetic wave with frequency ω passes through a plasma with frequency ω_p, from the dispersion relation

$$\omega^2 = K^2 c^2 + \omega_p^2 \tag{6.296}$$

we see that the wave can propagate only when $\omega > \omega_p$. Multiplying the above equation by h^2, we obtain the square of the energy of a quantum which resembles a relativistic particle with a rest mass equivalent to an energy $h\omega_p$. We call this quantum a "plasmon", which can be spontaneously converted into a neutrino pair through

$$\gamma_{plasma} \longrightarrow \nu\bar{\nu} \tag{6.297}$$

The "plasmon" can also be classified as latitudinal and longitudinal, so the neutrino energy loss is the sum of the latitudinal and longitudinal components (Kippenhahn and Weigert 1991, Zaidi 1966), i.e.,

$$\varepsilon_\nu^{(plasma)} = \varepsilon_\nu^{(latitudinal)} + \varepsilon_\nu^{(longitudinal)} \qquad (6.298)$$

where

$$\varepsilon_\nu^{(latitudinal)} = \begin{cases} 2.952 \times 10^{22} \left(\dfrac{\hbar\omega_p}{m_e c^2}\right)^6 \rho^{-1} \left(\dfrac{T_9}{5.93}\right)^3, & \dfrac{\hbar\omega_p}{kT} \ll 1 \\[3mm] 1.539 \times 10^{22} \left(\dfrac{\hbar\omega_p}{m_e c^2}\right)^{7.5} \rho^{-1} \left(\dfrac{T_9}{5.93}\right)^{1.5} e^{-\frac{\hbar\omega_p}{kT}}, & \dfrac{\hbar\omega_p}{kT} \gg 1 \end{cases} \qquad (6.299)$$

$$\varepsilon_\nu^{(longitudinal)} = 3.15 \times 10^{20} \left(\frac{\hbar\omega_p}{m_e c^2}\right)^4 \rho^{-1} \frac{1}{e^{\hbar\omega_p/kT} - 1} \quad (\text{erg g}^{-1}\text{s}^{-1}) \quad (6.300)$$

where

$$\left.\begin{aligned} \frac{\hbar\omega_p}{kT} &= \left(\frac{\hbar\omega_p}{m_e c^2}\right)\frac{m_e c^2}{kT} = \left(\frac{\hbar\omega_p}{m_e c^2}\right)\frac{5.930}{T_9} \\[2mm] \frac{\hbar\omega_p}{m_e c^2} &= 3.97 \times 10^{-5} \frac{(1 + X_H)^{1/2}\rho^{1/2}}{[1 + 6.40 \times 10^{-5}(1 + X_H)^{2/3}\rho^{2/3}]^{1/4}} \\[2mm] \omega_p &= \begin{cases} \dfrac{4\pi e^2 n_e}{m_e}, & \text{non-degenerate} \\[3mm] \dfrac{4\pi e^2 n_e}{m_e}\left[1 + \left(\dfrac{\hbar}{m_e c}\right)^2 (3\pi^2 n_e)^{2/3}\right]^{-1/2}, & \text{degenerate} \end{cases} \end{aligned}\right\} \qquad (6.301)$$

6.15.4 Bremsstrahlung Neutrinos

This process is very important when the temperature is relatively lower and the density is very high (Fig. 6.7). When an electron has an inelastic scattering (deceleration) in a Coulomb field, bremsstrahlung radiation photons (free-free emission) will be produced in general. In extremely rare cases, however, a pair of neutrinos is produced instead of the bremsstrahlung radiation. At a density $\rho \geq 10^8$ g/cm³, the energy loss rate by bremsstrahlung neutrinos is

$$\varepsilon_\nu^{(brems)} = 0.76 Z^2 A^{-1} T_8^6 \quad (\text{erg g}^{-1}\text{s}^{-1}) \qquad (6.302)$$

where Z and A are the electric charge number and the mass number of the nucleus. For a low density $\rho = 10^4$ g/cm^3, the energy loss rate by bremsstrahlung neutrinos is smaller than the results of Eq. (6.302) by a factor of about 10.

CHAPTER SEVEN

MODELS OF STELLAR STRUCTURE AND EVOLUTION

7.1 BASIC TASKS AND ASSUMPTIONS

The theory of stellar structure and evolution studies physical processes in the stars, and the physical and chemical parameters determined by these physical processes such as density, pressure, temperature, radiation and chemical composition as well as their distribution and their temporal variations. The interior of a star cannot be observed so the internal structure can only be studied theoretically. In practice, for a complicated star, some reasonable simplified assumptions are adopted. The physical state and chemical composition are then analyzed for the interior of the star for establishment of a set of basic equations to reflect the physical and chemical structure and their variations in the interior of the star, and the corresponding boundary conditions and material function related to the stellar matter. In other words, a model of stellar structure and evolution of the star is built. On solving the model, the physical and chemical processes taking place at different time in the interior of the star as well as the distribution of the various physical and chemical parameters can be obtained. The theoretically obtained stellar structure and evolution are then compared with the observational results. If the results agree well, we can say that the theoretically obtained stellar structure and evolution have truly reflected the structure and evolution of the interior of the star.

The physical state of the interior of a star is very complicated, and model study is very difficult without some simplified assumptions. In practice, the following simplified assumptions will be taken for the physical conditions and geometrical structure for the interior of stars.

(A) Spherical symmetry is assumed for the star

The interior of a star can be viewed as consisting of a number of concentric spherical shells, and the physical and chemical properties on each spherical shell can be assumed to be uniform. In other words, the temperature, pressure, density, luminosity, opacity and chemical composition on the same shell are identical. This assumption transforms the stellar structure problem into a one-dimensional problem for which all physical quantities and chemical compositions are only functions of the radius.

(B) The star is assumed to be stable

Therefore, the interior of the star fulfills hydrostatic equilibrium. In practice, the magnetic force, tidal force and rotational effects are not considered, so the hydrostatic equilibrium is in fact the equilibrium between the gravitational force and the pressure.

7.2 BASIC EQUATIONS

7.2.1 Equation of Mass Distribution

Under spherical symmetry for the star, the mass M_r of the sphere with a radius r can be written as

$$M_r = \int_0^r 4\pi r^2 \rho dr \tag{7.1}$$

where ρ is the density. Expressing in the differential form, the mass for the spherical shell with radius between r and $r + dr$ is

$$dM_r = 4\pi r^2 \rho dr \tag{7.2}$$

or

$$\frac{dM_r}{dr} = 4\pi r^2 \rho \tag{7.3}$$

which is the equation of mass distribution. In realistic computations of the stellar model, it is common not to choose r as the independent variable, but to choose the mass M_r of the sphere with radius r as the independent variable. When M_r replaces r as the independent variable, we have

$$\frac{dr}{dM_r} = \frac{1}{4\pi r^2 \rho} \tag{7.4}$$

7.2.2 Equation of Hydrostatic Equilibrium

Consider a volume element at a distance r from the center of the star, with a surface area of df (see Fig. 7.1). When the magnetic force, tidal force and rotational effects are excluded, the forces acting on the volume element are only the gravitational force $g\rho df dr$ pointing towards the center of the star and the pressure $-dP df$ in the opposite direction (P is the total pressure including the gas pressure and the radiation pressure). Only when these two forces are in equilibrium, the volume element can be in stable equilibrium. Therefore, under hydrostatic equilibrium, we have

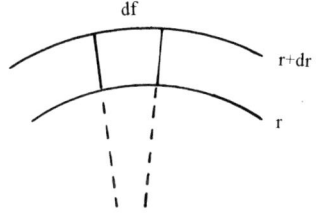

$$dP = -g\rho dr \tag{7.5}$$

Fig. 7.1 The geometry for establishment of the equation of hydrostatic equilibrium.

The gravitational acceleration g in the equation is

$$g = \frac{GM_r}{r^2} \tag{7.6}$$

From Eqs. (7.5) and (7.6), we can obtain the equation of hydrostatic equilibrium as

$$\frac{dP}{dr} = -\frac{GM_r}{r^2} \rho \tag{7.7}$$

If M_r replaces r as the independent variable, the above equation becomes

$$\frac{dP}{dM_r} = -\frac{GM_r}{4\pi r^4} \tag{7.8}$$

7.2.3 Energy Equation

For the star to maintain a stable state, the total energy emitted per second from the stellar surface should be compensated by the total energy generated per second in the interior of the star. If the luminosity L represents the total energy emitted per second from the stellar surface, and the energy generation rate ε represents the total energy produced per second per gram of the stellar matter, we have

$$L = \int_0^M \varepsilon \, dM_r \qquad (7.9)$$

The conservation of energy should be valid within each spherical shell in the interior of the star. For a spherical shell with radius r to $r + dr$, the energy L_r flowing out from the surface of the shell per second should be compensated by the energy produced per second inside the shell, so we have the differential form of the energy equation as

$$\frac{dL_r}{dM_r} = \varepsilon \qquad (7.10)$$

The mechanisms of energy generation in the interior of a star include the following:

(A) energy generation from thermonuclear reactions. The energy generation rate is represented by ε_n.
(B) energy released from the change in the thermal state of matter, for example, when internal energy changes to work done in expansion. The energy generation rate is represented by ε_g.
(C) energy loss by neutrino emissions. The energy generation rate is represented by $-\varepsilon_\nu$.
(D) gravitational energy released during collapse of the star.

Under stable conditions for the star, only the first three energy generation mechanisms exist. Therefore, Eq. (7.10) becomes

$$\frac{dL_r}{dM_r} = \varepsilon_n + \varepsilon_g - \varepsilon_\nu \qquad (7.11)$$

where the nuclear energy generation rate ε_n and the neutrino energy loss rate $-\varepsilon_\nu$ have been discussed in detail in Chapter 6. The expression for ε_g can be obtained from Eq. (3.32) as

$$\varepsilon_g = -\frac{dQ}{dt} = -C_P \dot{T} + \frac{\delta}{\rho}\dot{P} \tag{7.12}$$

so the energy equation (7.11) can be written as

$$\frac{dL_r}{dM_r} = \varepsilon_n - \varepsilon_\nu - C_P \dot{T} + \frac{\delta}{\rho}\dot{P} \tag{7.13}$$

From discussions in Chapter 6, we know that neutrinos can be produced from some thermonuclear reactions, or from some high energy weak interactions when the temperature and density reach certain levels (such as the pair annihilation neutrino process, photoneutrino process and the plasmaneutrino process etc). Usually, the energy loss by the neutrinos formed in nuclear reactions has been included in the energy generation rate ε_n for nuclear reactions. In this way, the $-\varepsilon_\nu$ term in Eq. (7.13) only refers to the energy loss by neutrinos formed in high energy weak interactions.

In practice, the density of stellar matter is a function of pressure, temperature and chemical composition, i.e., $\rho = \rho(P, T, \mu)$ where μ is the mean molecular weight. Therefore, we have

$$d\rho = \alpha dP - \delta dT + \psi d\mu \tag{7.14}$$

where

$$\alpha = \left(\frac{\partial \ln \rho}{\partial \ln P}\right)_{T,\mu} \tag{7.15}$$

$$\delta = \left(\frac{\partial \ln \rho}{\partial \ln T}\right)_{P,\mu} \tag{7.16}$$

$$\psi = \left(\frac{\partial \ln \rho}{\partial \ln \mu}\right)_{P,T} \tag{7.17}$$

In the derivation for Eq. (7.13), we have treated the density ρ as a function of pressure P and temperature T, which is in fact only valid for regions where the variation of chemical composition is very small, i.e., $d\mu = 0$. Sometimes, there are regions with large variations of μ. In these cases, Eq. (7.13) should also take into account the term for variation of μ, so Eq. (7.13) becomes

$$\frac{dL_r}{dM_r} = \varepsilon_n - \varepsilon_\nu - C_P\dot{T} + \frac{\chi_T\alpha}{\mu}\dot{P} + \frac{P}{\mu\rho}\chi_T\psi\dot{\mu} \tag{7.18}$$

where

$$\chi_T \equiv \left(\frac{\partial\ln P}{\partial\ln T}\right)_{\rho,\mu} \tag{7.19}$$

7.2.4 Equation of Energy Transport

The temperature gradient is commonly expressed as

$$\nabla \equiv \frac{d\ln T}{d\ln P} = \frac{P}{T}\frac{dT}{dP} \tag{7.20}$$

Substituting the hydrostatic equilibrium equation (7.8) into Eq. (7.20), we have

$$\frac{dT}{dM_r} = -\frac{GM_r}{4\pi r^4}\frac{T}{P}\cdot\nabla, \quad \nabla = \begin{cases} \nabla_R, & \text{radiative equilibrium regions} \\ \nabla_{con}, & \text{convective regions} \end{cases} \tag{7.21}$$

Equation (7.21) is the equation of energy transport. For radiative equilibrium regions, the temperature gradient ∇_R is given by Eq. (3.47) as

$$\nabla_R = \frac{3}{16\pi acG}\frac{\bar{\kappa}L_r P}{M_r T^4} \tag{7.22}$$

For convective regions, the temperature gradient ∇_{con} has been discussed in §3.3. ∇_{con} varies with the position of the convective region and can be obtained from Eq. (3.74), thus

$$\nabla_{con} = \begin{cases} \nabla_{ad}, \text{ interior of the star} \\ \nabla_{ad} + \frac{1}{W^2}\left(W^2 + \frac{19}{27}WU - \frac{E}{3}\right)^2 - U^2, \text{ envelope of the star} \end{cases} \tag{7.23}$$

where the expressions of U, W and E can be obtained from Eqs. (3.72) and (3.75).

7.2.5 Equation of Change in Chemical Composition

The change in chemical composition due to thermonuclear reactions can be obtained from Eq. (6.29) as

$$\frac{dY_i}{dt} = -\rho N_A \sum_j \frac{a_i}{1+\delta_{ij}} Y_i Y_j \langle \sigma v \rangle_{ij} + \rho N_A \sum_{k,l} \frac{b_i}{1+\delta_{kl}} Y_k Y_l \langle \sigma v \rangle_{kl} \quad (7.24)$$

where Y_i is the number abundance of element i, which is related to the mass abundance X_i for the same element by

$$Y_i = X_i/A_i \quad (7.25)$$

where A_i is the atomic number of element i. The detailed expression for Eq. (7.24) for different nuclear burning processes can be found from §6.9 to §6.12.

If thermonuclear reactions take place in a convective region, the chemical composition of the region will be made uniform by convective mixing. Therefore, the chemical composition for this convective region in which thermonuclear reactions are taking place can be calculated by

$$\frac{dY_i}{dt} = \frac{\int \left(\frac{dY_i}{dt}\right)_{M_r} dM_r}{\int dM_r} \quad (7.26)$$

where $(dY_i/dt)M_r$ in the integration represents the change in chemical composition in the mass shell M_r due to nuclear reactions, and can be computed by Eq. (7.24). The integration is over the entire convective region.

If thermonuclear reactions take place in a semi-convective region (see §3.5), the chemical composition in this region will be gradually mixed up. If the diffusion method is employed to compute the mixing by semi-convection, the chemical composition can be calculated by

$$\frac{dY_i}{dt} = \left(\frac{dY_i}{dt}\right)_{sc} + \frac{d}{dM_r}\left(\alpha \frac{dX_i}{dM_r}\right) \quad (7.27)$$

where $(dY_i/dt)_{sc}$ represents the change in the chemical composition due to thermonuclear reactions and can be computed by Eq. (7.24), $\frac{d}{dM_r}\left(\alpha \frac{dX_i}{dM_r}\right)$ is the diffusion term in which α is the diffusion coefficient.

The diffusion term signifies the use of the diffusion method in dealing with the change in chemical composition, and the value of α will determine the mixing rate of the chemical composition.

7.3 MATERIAL FUNCTIONS AND BOUNDARY CONDITIONS

For derivations of the basic equations, we have treated physical quantities in the interior of the star as functions of M_r only. When we perform model evolution calculations, all physical quantities and abundance of chemical elements should be treated as functions of M_r and t, so the basic equations should be written as partial derivative forms. Therefore the basic equations for the stellar evolution model should be

$$\frac{\partial r}{\partial M_r} = \frac{1}{4\pi r^2 \rho} \tag{7.28}$$

$$\frac{\partial P}{\partial M_r} = -\frac{GM_r}{4\pi r^4} \tag{7.29}$$

$$\frac{\partial L_r}{\partial M_r} = \varepsilon_n - \varepsilon_\nu - C_P \dot{T} + \frac{\delta}{\rho} \dot{P} \tag{7.30}$$

$$\frac{\partial T}{\partial M_r} = -\frac{GM_r T}{4\pi r^4 P} \cdot \nabla \tag{7.31}$$

$$\nabla = \begin{cases} \nabla_R, \text{ when } \nabla_R < \nabla_{ad} \text{ (radiative equilibrium regions)} \\ \nabla_{con}, \text{ when } \nabla_R > \nabla_{ad} \text{ (convective regions)} \end{cases} \tag{7.32}$$

$$\nabla_R = \frac{3}{16\pi ac} \frac{\bar{\kappa} L_r P}{M_r T^4} \tag{7.33}$$

$$\nabla_{ad} = \frac{\delta P}{C_P T \rho} \tag{7.34}$$

$$\nabla_{con} = \begin{cases} \nabla_{ad}, \text{ interior of the star} \\ \nabla_{ad} + \frac{1}{W^2}\left(W^2 + \frac{19}{27}WU - \frac{E}{3}\right)^2 - U^2, \text{ envelope of the star} \end{cases} \tag{7.35}$$

$$\frac{\partial Y_i}{\partial t} = \begin{cases} -\rho N_A \sum_j \frac{a_i}{1+\delta_{ij}} Y_i Y_j \langle \sigma v \rangle_{ij} + \rho N_A \sum_{k,l} \frac{b_i}{1+\delta_{kl}} Y_k Y_l \langle \sigma v \rangle_{kl} \\ \qquad\qquad\qquad\qquad \text{(radiative equilibrium regions)} \\ \int \left(\frac{\partial Y_i}{\partial t}\right)_{M_r} dM_r / \int dM_r \qquad \text{(convective regions)} \end{cases} \tag{7.36}$$

In the above equations, some functions are related to properties of materials of the star, which are called material functions:

$$\rho = \rho(P, T, X_i)$$
$$\varepsilon_n = \varepsilon_n(P, T, X_i)$$
$$\varepsilon_\nu = \varepsilon_\nu(P, T, X_i) \qquad (7.37)$$
$$\kappa = \kappa(P, T, X_i)$$

Furthermore, there are also some thermodynamic quantities in the equations such as C_P, δ, ∇_{ad} etc. All material functions and thermodynamic quantities are functions of P, T and X_i, and have been discussed in detail in Chapters 4, 5 and 6 which are assumed to be known. Substituting the material functions and thermodynamic quantities into the basic equations, five partial differential equations are obtained for five unknowns which are $T(M_r, t)$, $P(M_r, t)$, $r(M_r, t)$, $L_r(M_r, t)$ and $Y_i(M_r, t)$. For simplicity, we treat Eq. (7.36) as a single equation, and treat X_i as the only unknown. In fact, there are different compositions inside a star, so Eq. (7.36) should be a set of equations and Y_i are the abundances for many chemical elements.

In practical calculations for stellar evolution, the calculations for the stellar structure and for the change in the abundance of chemical elements can be separated. For example, when the total mass of the star and the chemical composition are given at a certain time t_0, we can first integrate Eqs. (7.28) to (7.31) to obtain the stellar structure at t_0, i.e., the quantities $T(M_r, t_0)$, $P(M_r, t_0)$, $r(M_r, t_0)$, and $L_r(M_r, t_0)$. We then calculate the change in chemical composition in the interior of the star after a time interval Δt, i.e., at $t_1 = t_0 + \Delta t$, by integrating Eq. (7.36) to obtain $X_i(M_r, t_1)$. According to the total mass of the star and $X_i(M_r, t_1)$ at t_1, the stellar structure of the star at t_1 can be computed. The properties of stellar evolution at different times can be obtained by such computations.

The computation of stellar structure requires integration of the four basic equations (7.28) to (7.31), which involve four unknown parameters $T(M_r, t_0)$, $P(M_r, t_0)$, $r(M_r, t_0)$, and $L_r(M_r, t_0)$ so four boundary conditions should be present. However, these four boundary conditions are not given at a single boundary in the star (such as the surface or the center) but are distributed over two boundaries. Two boundary conditions are specified at the center of the star where $M_r = 0$ are $r = 0$ and $L_r = 0$; while two at the surface of the star where $M_r = M$ are $P = 0$ and $T = 0$ (the latter being called the "zero boundary conditions").

The center ($M_r = 0$) and the surface ($M_r = M$) of the star are singularities for Eqs. (7.28) to (7.31), so the following have been considered for the choice of boundary conditions.

(A) Internal boundary condition

In Eqs. (7.28), (7.29) and (7.31), r appears in the denominator. If the center ($M_r = 0$ and $r = 0$) is chosen as a boundary, $(\partial r/\partial M_r)$, $(\partial P/\partial M_r)$ and $(\partial T/\partial M_r)$ will become infinite and the basic equations cannot be integrated. To avoid such a singularity, we choose a point near the center as the internal boundary point. The values for a point near the center can be obtained through linear expansions, which are

$$r = \left(\frac{3}{4\pi\rho}\right)_c^{1/3} M_r^{1/3} \tag{7.38}$$

$$P = P_c - \frac{1}{2}\left(\frac{4\pi}{3}\right)^{1/3} G\rho_c^{4/3} M_r^{2/3} \tag{7.39}$$

$$L_r = M_r\left(\varepsilon_n - \varepsilon_v - C_P \dot{T} + \frac{\delta}{\rho}\dot{P}\right) \tag{7.40}$$

$$T = T_c - \frac{T_c}{P_c}\left(\frac{d\ln T}{d\ln P}\right)_c (P_c - P) \tag{7.41}$$

$$\left(\frac{d\ln T}{d\ln P}\right)_c = \begin{cases} \left(\frac{3}{16\pi acG}\frac{\bar{\kappa}L_r P}{M_r T^4}\right)_c, & \text{when } \nabla_R < \nabla_{ad} \\ \nabla_{ad}, & \text{when } \nabla_R > \nabla_{ad} \end{cases}$$

Equations (7.38) to (7.41) are the internal boundary conditions and the subscript "c" signifies values at the center.

(B) Surface boundary condition

Since P appears in the denominator of Eq. (7.31), if we take $P = 0$ and $T = 0$ at $M_r = M$ as the surface boundary conditions, $(\partial T/\partial M_r)$ will become infinite. In realistic cases, P and T will not be zero. Therefore, we only need to take realistic values of P and T at $M_r = M$. Using the approximate expression for the gray atmospheric temperature from the theory of stellar atmosphere, and using the equation for radiation pressure, we have

$$T^4 = \frac{3}{4}T_{eff}^4\left(\tau + \frac{2}{3}\right), \quad (\tau \text{ is the optical depth})$$

$$P_R = \frac{a}{3}T^4,$$

We can obtain the surface boundary conditions as

$$M_r = M: \quad T_{(\tau=0)} = \left(\frac{1}{2}\right)^{1/4} T_{eff}, P = \frac{a}{3}T_{(\tau=0)}^4 \tag{7.42}$$

7.4 NUMERICAL METHODS FOR THE STELLAR STRUCTURE MODEL

The stellar structure is composed of the four basic Eqs. (7.28) to (7.31), together with the four internal and surface boundary conditions. In practice, for this special case in which the boundary conditions are separately defined at two boundaries, the use of the Henyey method for numerical solutions has been found to be the most convenient. The special feature of the Henyey method is that a set of approximate solutions is first given, and the final correct solution is obtained through many iterative revisions. In the following, we will describe in detail the calculation methods for the stellar structure model developed by Kippenhahn, Weigert and Hofmeister (1967).

7.4.1 Building the Difference Equations

For numerical integration of the stellar structure model, it is necessary to divide the star into many concentric spherical shells, and to take the boundaries of these spherical shells as integral grids. In each spherical shell, the basic equations are written as difference equations to be solved. Therefore, when dividing the star into concentric spherical shells, the change in physical quantities between two spherical shells, such as P, T, r and L_r should be kept small in order to minimize the error introduced by writing the basic equations as difference equations. The integral grids are numbered from the outside to the inside, i.e., the number of the outermost boundary is 1 while that for the center of the star is m.

We now discuss the rewriting of the differential Eqs. (7.28) to (7.31) as difference equations for the spherical shell defined by the integral grid with numbers j and $j+1$. The variation in physical quantities in a star can be tremendous and reach a few orders of magnitude. For convenience in calculations, we introduce some new variables to replace M_r, P, T, L_r and r, which are

$$\left.\begin{aligned}
\xi &= \ln\left(1 - \frac{M_r}{M}\right) \\
p &= \ln P \\
t &= \ln T \\
l &= \ln\left(1 + \frac{L_r}{\alpha L}\right) \\
x &= \ln r
\end{aligned}\right\} \qquad (7.43)$$

where α is a constant with a small value, so Eqs. (7.28) to (7.31) can be rewritten as the difference equations

$$\frac{x_{j+1} - x_j}{\xi_{j+1} - \xi_j} = -\frac{M}{4\pi \rho_{j+\frac{1}{2}}} \exp\left(\xi_{j+\frac{1}{2}} - 3x_{j+\frac{1}{2}}\right) \tag{7.44}$$

$$\frac{p_{j+1} - p_j}{\xi_{j+1} - \xi_j} = \frac{GM^2}{4\pi} \exp\left(\xi_{j+\frac{1}{2}}\right)\left[1 - \exp\left(\xi_{j+\frac{1}{2}}\right)\right]\exp\left(-4x_{j+\frac{1}{2}} - p_{j+\frac{1}{2}}\right) \tag{7.45}$$

$$\frac{l_{j+1} - l_j}{\xi_{j+1} - \xi_j} = -\left[\varepsilon_{n_{j+\frac{1}{2}}} - \varepsilon_{v_{j+\frac{1}{2}}} - C_{P_{j+\frac{1}{2}}}\left(\frac{\exp\left(t_{j+\frac{1}{2}}\right) - \exp\left(t_{j+\frac{1}{2}}^{n-1}\right)}{\tau}\right)\right.$$

$$\left. + \frac{\delta_{j+\frac{1}{2}}}{\rho_{j+\frac{1}{2}}}\left(\frac{\exp\left(p_{j+\frac{1}{2}}\right) - \exp\left(p_{j+\frac{1}{2}}^{n-1}\right)}{\tau}\right)\right] \cdot M \exp\left(\xi_{j+\frac{1}{2}} - l_{j+\frac{1}{2}}\right)\frac{1}{\alpha L} \tag{7.46}$$

$$\frac{t_{j+1} - t_j}{\xi_{j+1} - \xi_j} = \frac{3M\alpha L}{64\pi^2 ac}\bar{\kappa}_{j+\frac{1}{2}} \exp\left(\xi_{j+\frac{1}{2}} - 4t_{j+\frac{1}{2}} - 4x_{j+\frac{1}{2}}\right)\left[\exp\left(l_{j+\frac{1}{2}}\right) - 1\right] \tag{7.47a}$$

$$\frac{t_{j+1} - t_j}{\xi_{j+1} - \xi_j} = \frac{GM^2}{4\pi}(\nabla_{ad})_{j+\frac{1}{2}}\left[1 - \exp\left(\xi_{j+\frac{1}{2}}\right)\right]\exp\left(\xi_{j+\frac{1}{2}} - 4x_{j+\frac{1}{2}} - p_{j+\frac{1}{2}}\right) \tag{7.47b}$$

where τ is the evolution time interval, $t_{j+1/2}^{n-1}$ and $p_{j+1/2}^{n-1}$ respectively, represent $t_{j+1/2}$ and $p_{j+1/2}$ in the last evolution model. Furthermore,

$$f_{j+\frac{1}{2}} = \frac{1}{2}(f_{j+1} + f_j) \tag{7.48}$$

but the energy generation rate is

$$\varepsilon_{j+\frac{1}{2}} = \sqrt{\varepsilon_{j+1} \cdot \varepsilon_j} \tag{7.49}$$

Similarly, the internal boundary conditions in Eqs. (7.38) to (7.41) can be written as difference equations. The label for the center of the star is m and that for the internal boundary is $m - 1$, so the difference equation for the internal boundary condition is

$$l_{m-1} = \ln\left\{1 + \frac{M}{\alpha L}[1 - \exp(\xi_{m-1})]\left[\varepsilon_{n_{m-1/2}} - \varepsilon_{v_{m-1/2}} - C_{P_{m-1/2}}\right.\right.$$

$$\left.\left. \cdot \frac{1}{\tau}(e^{t_{m-1/2}} - e^{t_{m-1/2}^{n-1}}) + \frac{\delta_{m-1/2}}{\rho_{m-1/2}}\frac{1}{\tau}(e^{p_{m-1/2}} - e^{p_{m-1/2}^{n-1}})\right]\right\} \tag{7.50}$$

$$\exp(p_{m-1}) = \exp(p_m) - \frac{1}{2}\left(\frac{4\pi}{3}\right)^{1/3} G\rho_m^{4/3} M^{2/3}[1 - \exp(\xi_{m-1})]^{2/3} \qquad (7.51)$$

$$3x_{m-1} = \ln \frac{3M}{4\pi} + \ln[1 - \exp(\xi_{m-1})] - \ln\rho_m \qquad (7.52)$$

$$\frac{t_m - t_{m-1}}{p_m - p_{m-1}} = (\nabla_R)_{m-1/2} \qquad (7.53a)$$

$$\frac{t_m - t_{m-1}}{p_m - p_{m-1}} = (\nabla_{ad})_{m-1/2} \qquad (7.53b)$$

7.4.2 Fitting Point and Boundary Conditions

Regarding integrations of the four basic equations, we do not integrate from the center out to the stellar surface and not from the surface to the center, since it will be difficult for such integrations to converge because there are only two boundary conditions for each boundary. In practice, a "reference point" is chosen at a certain position of the star. The region interior to this point is called "internal" while the region exterior to this point is called "outside". The integrations for the four basic equations will be performed separately in the internal and outside regions, but the results should be fitting at the reference point, and the reference point is therefore referred to as the fitting point. The fitting point is usually in the envelope of the star. We require the regions exterior to the fitting point not to have thermonuclear reactions and the fitting point should be below the photosphere. For example, one can choose $M_r = 0.97M$ as the fitting point. In order to fit the results from "outside" integrations and "internal" integrations, the best way is to use the results from "outside" integrations as boundary conditions for "internal" integrations.

At the surface of the star where $M_r = M$, there are only two boundary conditions and only the values for pressure P and temperature T are given. The other two parameters R and L are unknown. For the integration, a set of R and L values should be assumed. Different values of P_F, T_F, r_F and l_F are obtained at the fitting point from outside integrations for different values of R and L. If r_F and l_F are plotted against P_F and T_F, we can obtain two different surfaces as shown in Fig. 7.2.

When a surface is continuous, a small region around a point on the surface can be approximated by a plane for which the following

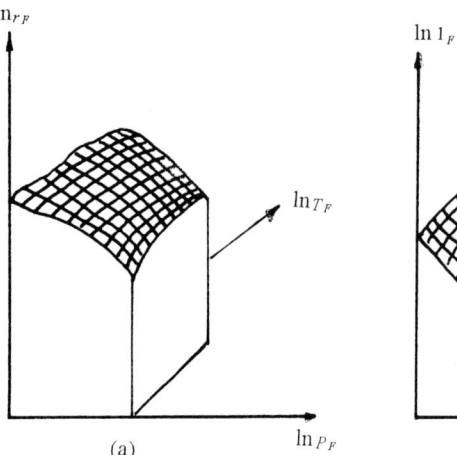
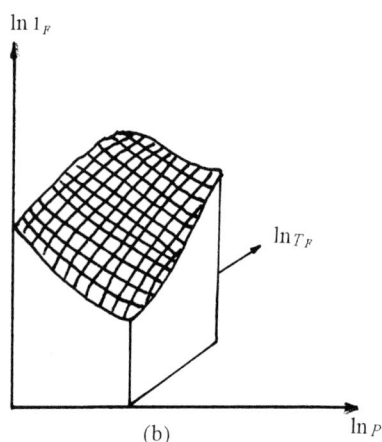

Fig. 7.2 The display of r_F and l_F as functions of P_F and T_F.

(a)   (b)

relationships are valid:

$$\ln r_F = a_1 \ln P_F + b_1 \ln T_F + c_1 \tag{7.54}$$

$$\ln l_F = a_2 \ln P_F + b_2 \ln T_F + c_2 \tag{7.55}$$

Equations (7.54) and (7.55) are the relationships between the physical quantities at the fitting point obtained from outside integrations. To determine the coefficients a_1, b_1, c_1, a_2, b_2 and c_2, we have to choose three different sets of R and L values and perform outside integrations to obtain three different sets of P_F, T_F, r_F and l_F. These three sets of R and L values should be very close so that the three sets of P_F, T_F, r_F and l_F are also very close.

In order to fit the results from outside integrations and internal integrations, Eqs. (7.54) and (7.55) can be used as boundary conditions for internal integrations.

7.4.3 Internal Integration

For "internal" regions interior to the fitting point, the Henyey method is used to integrate the basic equations. Approximate solutions to the basic equations are first given to which repeated modifications are applied to obtain the final correct solutions. The initial approximate solutions are called the "initial model", which include values of p, t, r and l at positions from the center of the star to the fitting point, and the corresponding material function and thermodynamic quantities.

From the difference Eqs. (7.44) to (7.47), we can define the following null functions:

$$
G_{1j} \equiv \frac{l_{j+1} - l_j}{\xi_{j+1} - \xi_j} + \left[\varepsilon_{n_{j+\frac{1}{2}}} - \varepsilon_{v_{j+\frac{1}{2}}} - C_{P_{j+\frac{1}{2}}} \left(\frac{\exp\left(t_{j+\frac{1}{2}}\right) - \exp\left(t_{j+\frac{1}{2}}^{n-1}\right)}{\tau} \right) \right.
$$
$$
\left. + \frac{\delta_{j+\frac{1}{2}}}{\rho_{j+\frac{1}{2}}} \left(\frac{\exp\left(p_{j+\frac{1}{2}}\right) - \exp\left(p_{j+\frac{1}{2}}^{n-1}\right)}{\tau} \right) \right] \cdot M \exp\left(\xi_{j+\frac{1}{2}} - l_{j+\frac{1}{2}}\right) \frac{1}{\alpha L}
$$
$$
(7.56)
$$

$$
G_{2j} \equiv \frac{p_{j+1} - p_j}{\xi_{j+1} - \xi_j} - \frac{GM^2}{4\pi} \exp\left(\xi_{j+\frac{1}{2}}\right) \left[1 - \exp\left(\xi_{j+\frac{1}{2}}\right) \right] \exp\left(-4x_{j+\frac{1}{2}} - p_{j+\frac{1}{2}}\right)
$$
$$
(7.57)
$$

$$
G_{3j} \equiv \frac{x_{j+1} - x_j}{\xi_{j+1} - \xi_j} + \frac{M}{4\pi \rho_{j+\frac{1}{2}}} \exp\left(\xi_{j+\frac{1}{2}} - 3x_{j+\frac{1}{2}}\right) \tag{7.58}
$$

$$
G_{4j} \equiv \frac{t_{j+1} - t_j}{\xi_{j+1} - \xi_j} - \frac{3M\alpha L}{64\pi^2 ac} \bar{\kappa}_{j+\frac{1}{2}} \exp\left(\xi_{j+\frac{1}{2}} - 4t_{j+\frac{1}{2}} - 4x_{j+\frac{1}{2}}\right) \left[\exp\left(l_{j+\frac{1}{2}}\right) - 1 \right],
$$
$$
\left(\frac{d \ln T}{d \ln P} < \nabla_{ad} \right) \tag{7.59a}
$$

$$
G_{4j} \equiv \frac{t_{j+1} - t_j}{\xi_{j+1} - \xi_j} - \frac{GM^2}{4\pi} (\nabla_{ad})_{j+\frac{1}{2}} \left[1 - \exp\left(\xi_{j+\frac{1}{2}}\right) \right] \exp\left(\xi_{j+\frac{1}{2}} - 4x_{j+\frac{1}{2}} - p_{j+\frac{1}{2}}\right)
$$
$$
\left(\frac{d \ln T}{d \ln P} > \nabla_{ad} \right) \tag{7.59b}
$$

From the internal boundary conditions in Eqs. (7.50) to (7.53), we can define the following null functions:

$$
Z_1 \equiv l_{m-1} - \ln\left\{ \exp(l_m) + \frac{M}{\alpha L} \left(e^{\xi_m} - e^{\xi_{m-1}} \right) \left[\varepsilon_{n_{m-1/2}} - \varepsilon_{v_{m-1/2}} - C_{P_{m-1/2}} \right. \right.
$$
$$
\left. \left. \cdot \frac{1}{\tau} \left(e^{t_{m-1/2}} - e^{t_{m-1/2}^{n-1}} \right) + \frac{\delta_{m-1/2}}{\rho_{m-1/2}} \frac{1}{\tau} \left(e^{P_{m-1/2}} - e^{P_{m-1/2}^{n-1}} \right) \right] \right\} \tag{7.60}
$$

$$
Z_2 \equiv \exp(p_{m-1}) - \exp(p_m) + \frac{1}{2} \left(\frac{4\pi}{3} \right)^{1/3} G \rho_m^{4/3} M^{2/3} [1 - \exp(\xi_{m-1})]^{2/3} \tag{7.61}
$$

$$
Z_3 \equiv \ln \frac{3M}{4\pi} + \ln[1 - \exp(\xi_{m-1})] - \ln \rho_m - 3x_{m-1} \tag{7.62}
$$

$$
Z_4 = \frac{t_m - t_{m-1}}{p_m - p_{m-1}} - (\nabla_R)_{m-1/2}, \quad \left(\frac{d \ln T}{d \ln P} < \nabla_{ad} \right) \tag{7.63a}
$$

$$Z_4 = \frac{t_m - t_{m-1}}{p_m - p_{m-1}} - (\nabla_{ad})_{m-1/2}, \quad \left(\frac{d \ln T}{d \ln P} > \nabla_{ad} \right) \qquad (7.63b)$$

From the boundary conditions at the fitting point in Eqs. (7.54) and (7.55), we can define the following two null functions:

$$R_1 \equiv a_1 p_1 + b_1 t_1 + c_1 - x_1 \qquad (7.64)$$

$$R_2 \equiv a_2 p_1 + b_2 t_1 + c_2 - l_1 \qquad (7.65)$$

If the initial model is the correct solution to the basic equations and the p, t, x and l values given by the initial model are substituted into the G, Z and R functions for each shell defined above, all values for these functions should be zero. However, since the initial model is only an approximate solution, the G, Z and R functions are non-zero. Our task is to look for corrected values for p, t, x and l values for each shell so that the G, Z and R functions become zero. At this time, we obtain the correct solutions after corrections.

To obtain the corrected quantities, the G, Z and R functions are expanded by Taylor expansion and only the linear terms are retained, i.e.,

$$\begin{cases} G_{ij} + \sum_k \left(\frac{\partial G_{ij}}{\partial q_{kj}} \right) \delta q_{kj} + \sum_k \left(\frac{\partial G_{ij}}{\partial q_{kj+1}} \right) \delta q_{kj+1} = 0, \ (i=1, ..., 4; j = 1, ..., m) \\[2mm] Z_i + \sum_k \left(\frac{\partial Z_i}{\partial q_{km}} \right) \delta q_{km} + \sum_k \left(\frac{\partial Z_i}{\partial q_{km-1}} \right) \delta q_{km-1} = 0, \ (i = 1, ..., 4) \\[2mm] R_i + \sum_k \left(\frac{\partial R_i}{\partial q_{k1}} \right) \delta q_{k1} = 0, \ (i=1, 2) \end{cases}$$

$$(7.66)$$

where q_{kj} represents p_j, t_j, x_j and l_j. When the initial model is given, the p, t, x and l values for the initial model are substituted into the G, Z and R functions and their partial differential forms, so in Eq. (7.66), G_{ij}, Z_i, R_i and $\frac{\partial G_{ij}}{\partial q_{kj}}$, $\frac{\partial G_{ij}}{\partial q_{kj+1}}$, $\frac{\partial Z_i}{\partial q_{km}}$, $\frac{\partial Z_i}{\partial q_{km-1}}$, $\frac{\partial R_i}{\partial q_{k1}}$ are all known, and Eq. (7.66) are only equations in the correction terms $\delta q_{kj}(k = 1, ..., 4; j = 1, ..., m)$. On solving the equations (7.66), the correction terms δp_j, δt_j, δx_j and δl_j can be obtained.

The procedures for solving the equations (7.66) are as follows. In the first shell (at the fitting point), there are six equations, namely

$$
\left.
\begin{aligned}
R_i + \sum_k \left(\frac{\partial R_i}{\partial q_{k1}}\right)\delta q_{k1} &= 0, \ (i=1,\, 2;\ k=1,\, \ldots,\, 4) \\
G_{i1} + \sum_k \left(\frac{\partial G_{i1}}{\partial q_{k1}}\right)\delta q_{k1} + \sum_k \left(\frac{\partial G_{i1}}{\partial q_{k2}}\right)\delta q_{k2} &= 0, \ (i=1,\, \ldots,\, 4;\ k=1,\, \ldots,\, 4)
\end{aligned}
\right\}
$$

$$(7.67)$$

which have eight unknowns, δp_1, δt_1, δx_1, δl_1, δp_2, δt_2, δx_2 and δl_2. In solving the equations, two unknowns (such as δp_2 and δt_2) can be taken as references, and the other six unknowns can be expressed as their functions, so we have

$$
\left.
\begin{aligned}
\delta p_1 &= \alpha_{11}\delta p_2 + \beta_{11}\delta t_2 + \gamma_{11} \\
\delta t_1 &= \alpha_{21}\delta p_2 + \beta_{21}\delta t_2 + \gamma_{21} \\
\delta x_1 &= \alpha_{31}\delta p_2 + \beta_{31}\delta t_2 + \gamma_{31} \\
\delta l_1 &= \alpha_{41}\delta p_2 + \beta_{41}\delta t_2 + \gamma_{41} \\
\delta x_2 &= \alpha_{51}\delta p_2 + \beta_{51}\delta t_2 + \gamma_{51} \\
\delta l_2 &= \alpha_{61}\delta p_2 + \beta_{61}\delta t_2 + \gamma_{61}
\end{aligned}
\right\}
$$

$$(7.68)$$

In the second shell, there are four equations, namely

$$
G_{i2} + \sum_k \left(\frac{\partial G_{i2}}{\partial q_{k2}}\right)\delta q_{k2} + \sum_k \left(\frac{\partial G_{i2}}{\partial q_{k3}}\right)\delta q_{k3} = 0, \ (i=1,\, \ldots,\, 4;\ k=1,\, \ldots,\, 4)
$$

$$(7.69)$$

which have eight unknowns, δp_2, δt_2, δx_2, δl_2, δp_3, δt_3, δx_3 and δl_3. According to eq. (7.68), δx_2 and δl_2 are also functions of δp_2 and δt_2 so Eq. (7.69) in fact represents four equations with six unknowns. In solving the equations, two unknowns (such as δp_3 and δt_3) can be taken as references, and the other four unknowns can be expressed as their functions, so we have

$$
\left.
\begin{aligned}
\delta p_2 &= \alpha_{12}\delta p_3 + \beta_{12}\delta t_3 + \gamma_{12} \\
\delta t_2 &= \alpha_{22}\delta p_3 + \beta_{22}\delta t_3 + \gamma_{22} \\
\delta x_3 &= \alpha_{32}\delta p_3 + \beta_{32}\delta t_3 + \gamma_{32} \\
\delta l_3 &= \alpha_{42}\delta p_3 + \beta_{42}\delta t_3 + \gamma_{42}
\end{aligned}
\right\}
$$

$$(7.70)$$

Starting from the second shell, every shell has four equations so entirely similar methods can be used. For the shell sandwiched between the grids $n+1$ and n, we have the following relationships:

$$\left.\begin{aligned}
\delta p_n &= \alpha_{1n}\delta p_{n+1} + \beta_{1n}\delta t_{n+1} + \gamma_{1n}\\
\delta t_n &= \alpha_{2n}\delta p_{n+1} + \beta_{2n}\delta t_{n+1} + \gamma_{2n}\\
\delta x_{n+1} &= \alpha_{3n}\delta p_{n+1} + \beta_{3n}\delta t_{n+1} + \gamma_{3n}\\
\delta l_{n+1} &= \alpha_{4n}\delta p_{n+1} + \beta_{4n}\delta t_{n+1} + \gamma_{4n}
\end{aligned}\right\} \qquad (7.71)$$

For the innermost shell sandwiched between the grids $m-1$ and m, we have the four equations

$$Z_i + \sum_k \left(\frac{\partial Z_i}{\partial q_{km-1}}\right)\delta q_{km-1} + \sum_k \left(\frac{\partial Z_i}{\partial q_{km}}\right)\delta q_{km} = 0,$$

$$(i = 1, ..., 4; k = 1, ..., 4) \qquad (7.72)$$

which have six unknowns $\delta p_{m-1}, \delta t_{m-1}, \delta x_{m-1}, \delta l_{m-1}, \delta p_m$ and δt_m. However, from Eq. (7.71), δx_{m-1} and δl_{m-1} are functions of δp_m and δt_m so there are in fact four unknowns. In this way, the four equations can uniquely determine the four unknowns.

The realistic computational procedures are as follows. The four correction terms $\delta p_{m-1}, \delta t_{m-1}, \delta p_m$ and δt_m are first computed from Eq. (7.72). Then δp_{m-1} and δt_{m-1} are substituted to the right of Eq. (7.71) and $\delta p_{m-2}, \delta t_{m-2}, \delta x_{m-1}$ and δl_{m-1} are determined. After that, δp_{m-1} and δt_{m-1} are substituted into Eq. (7.71) and $\delta p_{m-3}, \delta t_{m-3}, \delta x_{m-2}$ and δl_{m-2} are determined. The correction terms are determined outwards shell by shell, until the fitting point is reached. In this way, the correction terms $\delta p_j, \delta t_j, \delta x_j$ and δl_j ($j = 1, ..., m$) for all shells have been determined.

After corrections, the values $p_j + \delta p_j, t_j + \delta t_j, x_j + \delta x_j$ and $l_j + \delta l_j$ ($j = 1, ..., m$) are treated as a new approximate solution (initial model), and the above procedures are repeated for the second correctional computations. The maximum value for correction terms $\delta p'_j, \delta t'_j, \delta x'_j$ and $\delta l'_j$ from the second correctional computations should be smaller than the maximum value for correction terms $\delta p_j, \delta t_j, \delta x_j$ and δl_j obtained from the first correctional computations. If this is true, the maximum value for the correction terms after a few correctional computations should be within acceptable ranges. At this time, we have obtained the internal solution. If the maximum value for correction terms in a correctional computation is larger than the maximum value for correction terms in a previous correctional computation, i.e., the computation does not converge, we might have the following problems: (1) the earliest initial model deviates too far from the correct solution, and the initial model should be modified; (2) the evolutionary time interval is too large, or the choice of integral grids for the stellar shells are not correct.

When performing internal integrations, we treat the results of outside integrations as boundary conditions at the fitting point. However, the values of p'_F, t'_F, x'_F and l'_F at the fitting point obtained in the results of internal integrations can be different from the values of p_F, t_F, x_F and l_F obtained from outside intergrations and used as boundary conditions. In this case, the results from internal integrations and outside intergrations cannot fit satisfactorily. The method to fit these results will be discussed in §7.4.5.

7.4.4 Outside Integration

The outside integration refers to the inward integration from the stellar surface to the fitting point, which is divided into two parts: integration from the stellar surface (where the optical depth $\tau = 0$) to a place where $\tau = 2/3$ (photosphere) which is integration in the stellar atmosphere; and integration from the place where $\tau = 2/3$ to the fitting point.

(A) Integration from Stellar Surface to the place where $\tau = 2/3$

When we choose the fitting point, we have made sure that there are no thermonuclear reactions exterior to this point, so we can suppose that the luminosity L_F at the fitting point is equal to the luminosity L at the stellar surface.

The differential equations for this region are

$$\frac{\partial r}{\partial M_r} = \frac{1}{4\pi R^2 \rho} \tag{7.73}$$

$$\frac{\partial P}{\partial M_r} = -\frac{GM}{4\pi R^4} \tag{7.74}$$

$$d\tau = -\bar{\kappa}\rho dr \tag{7.75}$$

Furthermore, we employ the temperature distribution for gray atmosphere in Eq. (13.49) in the theory for stellar atmosphere

$$T^4 = \frac{3}{4} T^4_{eff}\left(\tau + \frac{2}{3}\right) \tag{7.76}$$

where

$$T^4_{eff} = \frac{L}{4\pi\sigma R^2} \tag{7.77}$$

The mass of stellar atmosphere is extremely small when compared to the total stellar mass, so we can assume that M_r can be replaced by the total mass at all positions in the stellar atmosphere. Similarly, r can be replaced by the total radius R at all positions in the stellar atmosphere. Besides, $p = \ln P$ is chosen as the independent variable. The procedures for integration are as follows.

From Eqs. (7.73) to (7.75), the gradient of τ is

$$\frac{d\tau}{d\ln P} = \frac{\bar{\kappa}PR^2}{MG} \tag{7.78}$$

Then, from the relationship

$$\tau = \tau(0) + d\ln P\frac{d\tau}{d\ln P} \tag{7.79}$$

and from $\tau(0)$, we can obtain the τ value for the next point.

The integration starts from the surface where $\tau = 0$. A set of L and R values should first be chosen, and the effective temperature T_{eff} can be obtained from Eq. (7.77). Using the surface boundary conditions in Eq. (7.42),

At $\tau = 0$:

$$T(\tau = 0) = \left(\frac{1}{2}\right)^{1/4} T_{eff},$$

$$P(\tau = 0) = \frac{a}{3}T^4(\tau = 0)$$

the values of T and P can be obtained for the stellar surface. From the values of T and P, the density ρ and the opacity $\bar{\kappa}$ at the stellar surface can also be obtained. Substituting these quantities into Eq. (7.78), we get the value of $(d\tau/d\ln P)$ at the stellar surface. Thus, from Eq. (7.79), we can get the τ value for the next point. After that, the temperature for that point can be given by Eq. (7.76). Since $\ln P$ is the independent variable, its value at that point is $\ln P(\tau = 0) + d\ln P$. When the values for T and P for that point are known, the above computations can be repeated to give the $(d\tau/d\ln P)$ value at that point, and then the τ value for the next point. The integrations are performed step by step until the place where $\tau = 2/3$ is reached.

It should be noted that the integrations are in general performed using the prediction-correction method. When the integration comes to the proximity of $\tau = 2/3$, it is possible to have the next step exceeding $\tau = 2/3$ due to choice of the integration step length. In this case, we should interpolate the values to the place exactly at $\tau = 2/3$.

(B) Integration from the Place where $\tau = 2/3$ to the Fitting Point

The equations in this region are

$$\frac{\partial \ln r}{\partial \ln P} = -\frac{rP}{G\rho M_r} \tag{7.80}$$

$$\frac{\partial \ln M_r}{\partial \ln P} = -\frac{4\pi r^4 P}{GM_r^2} \tag{7.81}$$

$$\frac{\partial \ln T}{\partial \ln P} = \begin{cases} \nabla_R = \dfrac{3}{16\pi acG}\dfrac{\bar{\kappa}LP}{M_r T^4}, \ \nabla_R < \nabla_{ad} \\ \nabla_{con}, \ \nabla_R > \nabla_{ad} \end{cases} \tag{7.82}$$

which are all differential equations with $p = \ln P$ as the independent variable. The integrations are also performed using the prediction-correction method. When the values of T, P, ρ and $\bar{\kappa}$ are known at $\tau = 2/3$, the values of $(\partial \ln r/\partial \ln P)$, $(\partial \ln M_r/\partial \ln P)$ and $(\partial \ln T/\partial \ln P)$ can be obtained through Eqs. (7.80) to (7.82). The values of $\ln r$, $\ln M_r$ and $\ln T$ for the next point are then obtained by

$$\left.\begin{array}{l} \ln r = \ln r(0) + d \ln P \dfrac{\partial \ln r}{\partial \ln P} \\[2mm] \ln M_r = \ln M_r(0) + d \ln P \dfrac{\partial \ln M_r}{\partial \ln P} \\[2mm] \ln T = \ln T(0) + d \ln P \dfrac{\partial \ln T}{\partial \ln P} \end{array}\right\} \tag{7.83}$$

The integrations are then performed step by step until the fitting point is reached.

7.4.5 Criterion on Goodness of Fit

From the above introduction of outside integrations, we know that a set of L and R values should be chosen and the surface boundary conditions should be used for the integrations. Since $T_{eff}^4 = L/(4\pi\sigma R^2)$, a choice in L and T_{eff} is equivalent to a choice in L and R. In practice, the former choice is more convenient. The choice of certain values of L and T_{eff} is equivalent to a choice of the position in the HR diagram. From §7.4.2, to obtain the coefficients in the boundary conditions in Eqs. (7.54) and (7.55) for internal integrations, we need three different sets of P_F, T_F, r_F and l_F values obtained from outside integrations

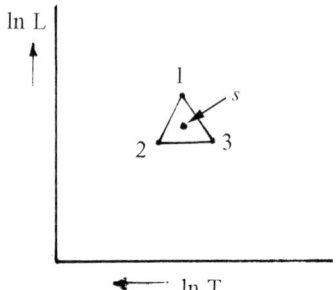

Fig. 7.3 The geometry for the choice of three close sets of values of L and T_{eff} in the HR diagram.

which should be very close to each other. In other words, we have to choose three close sets of values of L and T_{eff} and perform outside integrations to obtain three close sets of p_F, t_F, r_F and l_F at the fitting point. The method to choose three close sets of L and T_{eff} in the HR diagram is described as follows. For a given total stellar mass, a set of L and T_{eff} can be estimated, which corresponds to a point s in the HR diagram (Fig. 7.3). An equilateral triangle can be constructed with point s as the centre of gravity. The vertices of this triangle will then correspond to three sets of L and T_{eff}. Since the three sets of values should be close to each other, the triangle should be very small and there should be some restriction to the characteristic lengths of the triangle $\Delta \ln T_{eff}$ and $\Delta \ln L$.

By using methods similar to those for determining the boundary conditions in Eqs. (7.54) and (7.55) at the fitting point, we can establish relations between values on the surface and at the fitting point. Since the L_F value at the fitting point equals the L value at the surface, we only need to establish relations between the surface T_{eff} value and the p_F and t_F values at the fitting point. We suppose

$$\ln T_{eff} = a_3 p_F + b_3 t_F + c_3 \tag{7.84}$$

We take the $\ln L$ and $\ln T_{eff}$ values for the three vertices of the triangle in the HR diagram and perform outside integrations to obtain the three sets of p_F and t_F at the fitting point. Substituting these three sets of p_F and t_F values and $\ln T_{eff}$ into Eq. (7.84), the coefficients a_3, b_3 and c_3 can be obtained and relations between the values on the surface and at the fitting point can be established.

We now discuss the criterion for a good fit between results of outside and internal integrations at the fitting point. We choose a point s in the HR diagram, construct a small equilateral triangle and use the $\ln L$ and $\ln T_{eff}$ values at the vertices of the triangle for outside integrations to obtain the two boundary conditions at the fitting point as

$$\left. \begin{array}{l} a_1 p_F + b_1 t_F + c_1 - x_F = 0 \\ a_2 p_F + b_2 t_F + c_2 - l_F = 0 \end{array} \right\} \tag{7.85}$$

and one relation between values on the surface and at the fitting point as

$$a_3 p_F + b_3 t_F + c_3 - \ln T_{eff} = 0 \tag{7.86}$$

Using the boundary condition in Eq. (7.85), the internal integrations can be performed. After corrections to the initial model, the quantities at the fitting point are $p_F' = p_F + \delta p_F$ and $t_F' = t_F + \delta t_F$. The obtained p_F' and t_F' values are then substituted back into Eq. (7.86) to compute the corresponding value of $\ln T_{eff}'$ and to locate a corresponding point s' on the HR diagram. If s' is very near s and is within the original small triangle, we can say that the internal solution fits the outside solution at the fitting point. If s' is far away from s and is outside the original small triangle, we conclude that the internal solution does not fit the outside solution at the fitting point. In this situation, we construct another equilateral triangle around s' and repeat the above computational procedures until the internal solution finally fits the outside solution at the fitting point, i.e., s' falls into the triangle constructed around the previous point s in the HR diagram.

7.5 ON THE UNIQUENESS OF SOLUTIONS

The stellar structure model is composed of the four basic equations (7.28) to (7.31) and the four boundary conditions in Eqs. (7.38) to (7.42), which have four unknowns P, T, r and L_r. When the total mass and the chemical composition for the star are given, the solution can be obtained by using the method outlined in §7.4. We now discuss the uniqueness of such a solution.

If the four boundary conditions are given on the same boundary, the four unknowns determined from the four differential equations will satisfy the Lipschitz uniqueness condition and the obtained solution will be unique. However, the four boundary conditions for the stellar structure model are given separately on surface and internal boundaries, so the Lipschitz uniqueness condition is not satisfied and the obtained solution is not unique, which has been verified in many realistic computations of the stellar structure (Biermann and Kippenhahn 1971; Gabriel and Ledoux 1967; Paczynski 1972; Roth and Wiegert 1972). In practice, however, the multiple solutions obtained have a special feature: they have large discrepancies and no two solutions can be very close to each other. Kähler (1972) proved through rigorous mathematical methods that no two infinitesimally close equilibrium solutions can be obtained for the stellar structure for a given total mass and chemical composition of a star. In other words, in the small region surrounding a point (L, T_{eff}) in the HR diagram, there is only one equilibrium solution for a given total mass and chemical composition of a star. This shows that a generally unique theory for the solution for stellar structure does not exist, but a locally unique theory does exist. Thanks to this, the solution to the stellar structure can be meaningful.

7.6 VIRIAL THEOREM

The major energy sources in a star are thermonuclear energy, gravitational energy and thermal energy. When thermonuclear reactions cease, the star will collapse and release gravitational energy to maintain the external radiation. The released gravitational energy will be partially used to maintain the external radiation, with the rest converted into thermal energy which increases the temperature of the interior of the star. There is a certain relation between the gravitational energy E_G and the thermal energy E_T, which will be studied in the following for various conditions.

7.6.1 When the Surface Pressure is Zero

Multiplying the hydrostatic equilibrium equation (7.8) by $4\pi r^3$, and integrating over the entire star, the left of the equation becomes

$$\int_0^M 4\pi r^3 \frac{dP}{dM_r} dM_r = \left[4\pi r^3 P\right]_0^M - \int_0^M 12\pi r^2 \frac{dr}{dM_r} P dM_r \qquad (7.87)$$

The bracketed term in the above expression is zero because $r = 0$ at the centre of the star and $P = 0$ at the stellar surface. Substituting Eq. (7.4) into the integration term on the right of the above equation, we have

$$\int_0^M 4\pi r^3 \frac{dP}{dM_r} dM_r = -\int_0^M 3\frac{P}{\rho} dM_r \qquad (7.88)$$

After multiplying the hydrostatic equilibrium equation (7.8) by $4\pi r^3$, and integrating over the entire star, the right-hand side of the equation becomes

$$-\int_0^M \frac{GM_r}{4\pi r^4} 4\pi r^3 dM_r = -\int_0^M \frac{GM_r}{r} dM_r = E_G \qquad (7.89)$$

From Eqs. (7.88), (7.89) and (7.8), we get the relation

$$3\int_0^M \frac{P}{\rho} dM_r = \int_0^M \frac{GM_r}{r} dM_r = -E_G \qquad (7.90)$$

If the star is entirely composed of an ideal monatomic gas, the thermal energy of one gram of material is

$$\varepsilon_{kin} = \frac{3}{2}\frac{\Re}{\mu} T = \frac{3}{2}\frac{P}{\rho} \qquad (7.91)$$

so

$$E_T = \int_0^M \varepsilon_{kin} dM_r = \frac{3}{2}\int_0^M \frac{P}{\rho} dM_r \qquad (7.92)$$

Substituting Eq. (7.92) into Eq. (7.90), we have

$$E_G = -2E_T \qquad (7.93)$$

which is the Virial theorem. From Eq. (7.93), we see that when the star collapses and releases gravitational energy, its thermal energy E_T will change at the same time. Half of the released gravitational energy will be converted into thermal energy.

The above derivations for the Virial theorem are for a monatomic gas. If the gas is polyatomic, e.g., during the collapse of an interstellar cloud to form a star, the thermal energy of one gram of material is

$$\varepsilon_{kin} = \frac{f}{2\rho} nkT \qquad (7.94)$$

where f is the degree of freedom. From $P = nkT$ and $\gamma - 1 = 2/f (\gamma = C_P/C_V)$, we obtain

$$\frac{P}{\rho} = (\gamma - 1)\varepsilon_{kin} \qquad (7.95)$$

Substituting Eq. (7.95) into Eq. (7.90), we get

$$3(\gamma - 1)E_T + E_G = 0 \tag{7.96}$$

which is the Virial theorem for a polyatomic gas.

7.6.2 When the Surface Pressure is Non-Zero

We now consider a special case in which the internal material of a sphere is isothermal and the surface pressure of the sphere has a non-zero value P_f. This special case can occur when an interstellar cloud collapses to form a star. Besides, during stellar evolution, the core can also fulfil the above conditions. For example, when hydrogen in the core is burnt up and before helium burning starts, the material inside the core can be treated as isothermal and the surface pressure for the core is non-zero. We use methods similar to those in the above derivations for these conditions. From Eqs. (7.87) to (7.89),

$$\left[4\pi r^3 P\right]_0^M - \int_0^M 3\frac{P}{\rho}\,dM_r = -\int_0^M \frac{GM_r}{r}\,dM_r \tag{7.97}$$

Introducing two mean quantities

$$C_1 \equiv \int_0^M \frac{M_r\,dM_r}{r} \Big/ \frac{M^2}{R} \tag{7.98}$$

$$\bar{T} \equiv \int_0^M T\,dM_r / M \tag{7.99}$$

and using the equation of state for ideal gas, from Eq. (7.97), we have

$$4\pi R^3 P_f - 3\frac{M\mathfrak{R}}{\mu}\bar{T} = -C_1\frac{GM^2}{R} \tag{7.100}$$

from which we get

$$P_f = \frac{3M\mathfrak{R}\bar{T}}{4\pi\mu R^3} - \frac{C_1 G}{4\pi}\frac{M^2}{R^4} \tag{7.101}$$

which is very useful in determining whether a collapse of the central core of a star will occur.

7.7 TIMESCALES

The various physical processes in a star will take place at different rates, which are referred to as having different timescales. For example, thermonuclear reactions in the core of stars are in general slow, so the changes of chemical composition in central regions of stars take place with a long timescale [see dY_i/dt in Eq. (7.24)]. On the other hand, the changes in pressure and temperature due to expansion and contraction during stellar evolution are fast relative to thermonuclear reactions, i.e., they take place with a shorter timescale [see dP/dt and dT/dt in Eq. (7.12)]. The concept of timescales is very important for the understanding of intrinsic properties of some physical phenomena in stars. Three basic and common timescales will be discussed in the following.

7.7.1 Hydrodynamic Timescale

When a star is under hydrostatic equilibrium, the gravitational force pointing towards the centre and the pressure pointing in the opposite direction are balanced for each fluid element inside the star (when the other forces such as the magnetic force and centrifugal force are neglected). If some perturbations lead to an upset of the hydrostatic equilibrium, the fluid element will move until a new balance is attained between the gravitational force and the pressure. The time taken to set up this new balance is called the hydrodynamic timescale. The following physical processes take place on the hydrodynamic timescale.

(A) Free falling collapse

If a fluid element does not fulfil hydrostatic equilibrium and has motion, the equation of motion according to Newton's law is

$$\ddot{r} = -\frac{GM_r}{r^2} - \frac{1}{\rho}\frac{dP}{dr} \qquad (7.102)$$

If this fluid element is near the stellar surface, $r = R$ and $M_r = M$, so Eq. (7.102) becomes

$$\ddot{R} = -\frac{GM}{R^2} - \frac{1}{\rho}\frac{dP}{dR} \tag{7.103}$$

If the perturbation emerges because the outward pressure is far less than the gravitational force, i.e., $\frac{1}{\rho}\frac{dP}{dR} << \frac{GM}{R^2}$, free falling collapse towards the centre of the star will take place, and Eq. (7.103) can be approximated by

$$-\ddot{R} \approx -\frac{GM}{R^2} \tag{7.104}$$

Assuming that the distance travelled by the fluid element is R and that the acceleration is constant over this distance, we have

$$R = -\left(\frac{\ddot{R}}{2}\right)t_{ff}^2 \tag{7.105}$$

Substituting Eq. (7.104), we get

$$t_{ff} = \left(\frac{2R^3}{GM}\right)^{1/2} = \frac{1}{\left(\frac{2}{3}\pi G\right)^{1/2}}(\bar{\rho})^{-1/2} \approx 0.^d04\left(\frac{\bar{\rho}_\odot}{\bar{\rho}}\right)^{1/2} \tag{7.106}$$

For the Sun, the free falling timescale t_{ff} is about one hour.

(B) Stellar oscillation

We consider the case when the motion of the fluid element after the upset of hydrostatic equilibrium is oscillation around its equilibrium position. Apparently, the oscillation period T is of the same order of magnitude as the hydrodynamic timescale. From Eq. (7.106), we obtain an approximate rule as

$$T \cdot \sqrt{\bar{\rho}} = const \tag{7.107}$$

(3) Propagation of acoustic waves

If the perturbation emerges because the gravitational force is far less than the pressure, i.e., $\frac{GM}{R^2} << \frac{1}{\rho}\frac{dP}{dR}$, Eq. (7.103) can be approximated by

$$\frac{R}{t_s^2} \cong \frac{1}{\rho}\frac{dP}{dR} \cong \frac{1}{\rho}\frac{P}{R} \tag{7.108}$$

or

$$\frac{R}{t_s} \cong \left(\frac{P}{\rho}\right)^{1/2} \tag{7.109}$$

The left-hand side of the above equation has the dimension of velocity and the expression on the right-hand side differs from the speed of sound C_s by a coefficient, so

$$t_s \cong \frac{R}{C_s} \tag{7.110}$$

From this we know that t_s is the time needed for the acoustic wave (pressure wave) to traverse the radius of the star. For an ideal gas,

$$C_s \cong \sqrt{\frac{P}{\rho}} \approx \sqrt{T} \tag{7.111}$$

In the interior of the star, the temperature is very high, so C_s is very large; in the envelope of the star, the temperature is relatively lower so C_s is relatively smaller. Therefore, t_s is in fact the time needed for acoustic wave to pass through the envelope of the star.

7.7.2 Thermal Timescale (Kelvin–Helmholtz timescale)

During stellar evolution, there may or may not be thermonuclear reactions. If there are no thermonuclear reactions in the star, the external radiation will rely on the dissipation of the gravitational energy and the internal energy. The time required to dissipate all the gravitational energy and the internal energy in a star is defined as the thermal timescale or the Kelvin–Helmholtz timescale.

When there are no thermonuclear reactions in the star, the total energy (when the magnetic energy is neglected) is

$$E = E_G + E_T \tag{7.112}$$

where E_G and E_T are, respectively, the gravitational energy and the thermal energy. Substituting Eq. (7.93) into Eq. (7.112), we have

$$E = \frac{1}{2} E_G \tag{7.113}$$

where E_G is given by

$$E_G = -G \int_0^M \frac{M_r dM_r}{r} = -q \frac{GM^2}{R} \tag{7.114}$$

where

$$q \equiv \int_0^1 \frac{M\left(\frac{r}{R}\right)/M}{\frac{r}{R}} \frac{dM\left(\frac{r}{R}\right)}{M} \tag{7.115}$$

is a dimensionless quantity dependent only on the concentration of mass in the star. Substituting Eq. (7.114) into Eq. (7.113), we get

$$E = -\frac{1}{2} q \frac{GM^2}{R} \tag{7.116}$$

The total energy radiated by a star per second is L, so the time required to dissipate all the energy E of the star, or the thermal timescale, is

$$t_K = \frac{E}{L} = \frac{1}{2} q \frac{GM^2}{RL} \tag{7.117}$$

In general, we take $q = 1.5$ for a star. We further take $L_\odot$, $M_\odot$ and $R_\odot$ as the units for L, M and R, so

$$t_K = 2 \times 10^7 \frac{M^2}{LR} \quad (\text{yr}) \tag{7.118}$$

For the Sun, $t_K = 2 \times 10^7$ yr.

7.7.3 Nuclear Timescale

If the external radiation L is mainly supported by energy generation from nuclear reactions, the time needed to dissipate all the nuclear energy E_N in the star is

$$t_N = \frac{E_N}{L} \tag{7.119}$$

which is called the nuclear timescale. Since hydrogen is the major constituent of stellar material and hydrogen burning releases the largest amount of energy, we assume the whole star is composed by hydrogen. The energy released by the nuclear reaction $^1H + {}^1H \rightarrow {}^4He$ is $Q = 6.3 \times 10^{18}$erg/g. Thus, the total nuclear energy for a star with a mass of M (g) and a luminosity of L (erg/g) is $E_N = 6.3 \times 10^{18} M$ (erg). From Eq. (7.119), the nuclear timescale can be obtained as

$$t_N = 6.3 \times 10^{18} \frac{M}{L} \approx 10^{10} \left(\frac{M}{M_\odot}\right)\left(\frac{L_\odot}{L}\right) \quad (\text{yr}) \qquad (7.120)$$

For the Sun, $t_N \cong 10^{10}$ yr.

Among the above three timescales, the nuclear timescale is the longest, the thermal timescale is the second longest and the hydrodynamic timescale is the shortest, i.e.,

$$t_N >> t_K >> t_{ff}$$

7.8 THE SIMPLEST STELLAR MODEL–POLYTROPIC MODEL

The polytropic model is the simplest and the earliest method in computing the internal structure of the star. The method only requires two out of the five basic equations, i.e., the equation of mass distribution of Eq. (7.4) and the hydrostatic equilibrium equation (7.8), and assumes that the pressure and the density for the entire star fulfil the polytropic relation

$$P = K \cdot \rho^{1+\frac{1}{n}} \qquad (7.121)$$

where K and n are constants for the entire star and n is the polytropic index. From Eqs. (7.4), (7.8) and (7.121), one can establish a differential equation called the Emden differential equation. On solving the Emden differential equation, the distribution of internal prarameters of the star such as pressure, density, temperature, i.e., the internal structure of the star can be obtained. However, the polytropic model does not make use of the other basic equations related to energy generation and transfer in the interior of the star. Therefore, the polytropic model cannot give complicated information on energy generation and transfer, convection, ionization and degeneracy.

Nevertheless, the polytropic model has still got applications until today. For some special stars, or for some stars reaching a certain evolutionary stage, the internal pressure and the density can fulfil the polytropic relation in Eq. (7.118), so that the simple internal structure can be studied with the assistance of the polytropic model. The following are some examples in which we can use the polytropic model.

For compact stellar objects (white dwarfs and neutron stars) formed in the late stage of stellar evolution, the electrons inside have become degenerate. From §4.3, when degenerate electrons are non-relativistic, the relation between pressure and density fulfils $P = K\rho^{5/3}$, or $n = 3/2$. When electrons are relativistic degenerate, the relation between pressure and density fulfils $P = K\rho^{4/3}$, or $n = 3$. The relation in Eq. (7.121) is satisfied in both of these two cases.

Supermassive stars have very strong radiation pressure, and the ratio between the gas pressure and the total pressure can be taken as a constant ($\beta = P_g/P = $ constant), so the equation of state can be written as

$$P = \frac{\Re}{\mu}\rho T + \frac{a}{3}T^4 = \frac{\Re}{\mu\beta}\rho T \qquad (7.122)$$

In the derivation of the above equation, we have used the relation

$$(1 - \beta)P = P_R = \frac{a}{3}T^4 \qquad (7.123)$$

Since $\beta = $ constant, we can see from Eq. (7.123) that $T^4 \sim P$ for the entire star. Substituting this into Eq. (7.122), we have

$$P = \left(\frac{3\Re^4}{a\mu^4}\right)^{1/3}\left(\frac{1 - \beta}{\beta^4}\right)^{1/3}\rho^{4/3} \qquad (7.124)$$

This is a polytropic relation in which $K = \left(\frac{3\Re^4}{a\mu^4}\right)^{1/3}\left(\frac{1-\beta}{\beta}\right)^{\frac{1}{3}}$ and $n = 3$. For supermassive stars with different masses, K can have different values; for the same supermassive star, K is constant. Therefore, the internal structure of a supermassive star can be studied using the polytropic model.

Red giants have very low radiation pressures which can be neglected, and their interior are almost completely convective. Thus, except for the surface, their internal temperature gradient can be written as $\nabla = (d\ln T/d\ln P) = \nabla_{ad}$. When the gas is completely ionised, $\nabla_{ad} = 0.4$ so $T \sim P^{2/5}$ for the entire star. Supposing the stellar materials satisfy the equation of state for an ideal gas and $\mu = $ constant, $T \sim P/\rho$, so $P \sim \rho^{5/3}$ which fulfils the polytropic equation (7.121) with $n = 3/2$. Thus, the internal structure of red giants can be studied using the polytropic model.

When the star has evolved to a certain stage, the central core is isothermal $(T = T_0)$. This occurs when hydrogen burning has completed but helium burning has not started in the core of the star. The equation of state in the central core $P = \Re T_0 \rho / \mu$ is itself a polytropic relation, where $K = \Re T_0 / \mu$ and $n = \infty$. Thus, some properties of the isothermal central core can be studied by the polytropic model.

7.8.1 Emden Differential Equation

If r is chosen as the independent variable, the hydrostatic equilibrium equation (7.5) can be written as

$$dP + g_r \rho dr = 0 \tag{7.125}$$

where g_r can be rewritten through the mass distribution equation (7.4) as

$$g_r = \frac{G}{r^2} \int_0^r 4\pi r^2 \rho dr \tag{7.126}$$

From Eqs. (7.125) and (7.126),

$$\frac{r^2}{\rho} \frac{dP}{dr} + G \int_0^r 4\pi r^2 \rho dr = 0 \tag{7.127}$$

Integrating the above equation with respect to r,

$$\frac{1}{r^2} \frac{d}{dr} \left(\frac{r^2}{\rho} \frac{dP}{dr} \right) + 4\pi G \rho = 0 \tag{7.128}$$

Introducing the gravitational potential Ψ, so that

$$g_r = -\frac{d\psi}{dr} \tag{7.129}$$

From Eq. (7.126),

$$\frac{d^2\psi}{dr^2} = -\frac{dg_r}{dr} = -4\pi G\left[\rho - \frac{2}{r^3}\int_0^r r^2\rho\, dr\right]$$

$$= -4\pi G\rho + \frac{2}{r}g_r \tag{7.130}$$

Substituting Eq. (7.129), we get

$$\frac{d^2\psi}{dr^2} + \frac{2}{r}\frac{d\psi}{dr} + 4\pi G\rho = 0 \tag{7.131}$$

This equation only differs from Eq. (7.127) by changing P to ψ. It is the combined result of the hydrostatic equilibrium equation and the mass distribution equation. There are still two unknowns ψ and ρ in the equation. To eliminate one of these two unknowns, we need a relation between ψ and ρ (i.e., a relation between P and ρ). Assuming the polytropic relation (7.121) is satisfied in the interior of the star, we have from Eq. (7.121)

$$dP = K\left(1 + \frac{1}{n}\right)\rho^{1/n}d\rho \tag{7.132}$$

From Eq. (7.125) and (7.129), we get

$$dP = \rho\, d\psi \tag{7.133}$$

Substituting this into Eq. (7.132), we obtain

$$d\psi = K\left(1 + \frac{1}{n}\right)\rho^{\left(\frac{1}{n}-1\right)}d\rho \tag{7.134}$$

Integrating, we have

$$\psi + const = K\left(1 + \frac{1}{n}\right)\rho^{1/n} \tag{7.135}$$

Making use of the conditions $\psi = 0$ and $\rho = 0$ at the stellar surface, we can evaluate the integration constant in Eq. (7.135) as zero, so

$$\rho = \left[\frac{\psi}{K(1+n)}\right]^n \tag{7.136}$$

Substituting this into Eq. (7.131), ρ can be eliminated and

$$\frac{d^2\psi}{dr^2} + \frac{2}{r}\frac{d\psi}{dr} + \frac{4\pi G}{[K(1+n)]^n}\psi^n = 0 \tag{7.137}$$

We introduce two variables Z and u to replace r and ψ:

$$\left.\begin{aligned} Z &\equiv Ar,\ A^2 \equiv \frac{4\pi G}{[K(1+n)]^n}\psi_c^{n-1} = \frac{4\pi G}{K(1+n)}\rho_c^{\frac{n-1}{n}} \\ u &\equiv \frac{\psi}{\psi_c} = \left(\frac{\rho}{\rho_c}\right)^{1/n} \end{aligned}\right\} \tag{7.138}$$

where ψ_c and ρ_c represent the values of ψ and ρ at the centre of the star. On introduction of these two variables, Eq. (7.137) can be written as

$$\frac{d^2u}{dZ^2} + \frac{2}{Z}\frac{du}{dZ} + u^n = 0$$

or

$$\frac{1}{Z^2}\frac{d}{dZ}\left(Z^2\frac{du}{dZ}\right) + u^n = 0 \tag{7.139}$$

which is called the Emden differential equation with index n and describes the internal structure of a star satisfying the hydrostatic equilibrium condition and the polytropic relation.

At the centre of the star, i.e., $Z = 0$, we are only interested in the solutions with finite values for u. From Eq. (7.139), it is not difficult to see that when $Z = 0$, finite solutions can only be obtained when $du/dZ = 0$. Thus, the boundary conditions for the Emden differential equation are

$$\left.\begin{aligned} &\text{At the centre of the star } Z = 0: u = 1,\ u' = 0 \\ &\text{At the stellar surface } Z = Z_n = RA: u = 0 \end{aligned}\right\} \tag{7.140}$$

7.8.2 Properties of the Solution

The Emden differential equation has a singularity at $Z = 0$. To understand properties of the solution near the singularity, we expand the solution $u(Z)$ in the polynomial form

$$u(Z) = 1 + a_1 Z + a_2 Z^2 + a_3 Z^3 + \ldots \qquad (7.141)$$

where $a_1 = u'(0)$, $a_2 = (1/2)u''(0)$, ... At the centre of the star, $-g_r = (d\psi/dr) \sim (du/dZ)$ should be zero, so $a_1 = 0$. Substituting Eq. (7.141) into Eq. (7.139) and comparing the coefficients, we have

$$u(Z) = 1 - \frac{1}{6} Z^2 + \frac{n}{120} Z^4 + \ldots \qquad (7.142)$$

Here, we have not considered the case for $n = \infty$. From Eq. (7.142), we know $u(Z)$ has a maximum value at $Z = 0$. Using analytical methods, we can only get the solutions for three values of n, i.e.,

$$n = 0 : u(Z) = 1 - \frac{1}{6} Z^2 \qquad (7.143)$$

$$n = 1 : u(Z) = \frac{\sin Z}{Z} \qquad (7.144)$$

$$n = 5 : u(Z) = \frac{1}{\left(1 + \frac{Z^2}{3}\right)^{1/2}} \qquad (7.145)$$

From the surface boundary condition in Eq. (7.140), for $n = 0$ and $n = 1$, we can have finite values of $Z = Z_n$ which fulfil the surface boundary condition $u(Z_n) = 0$. However, if $n = 5$, only when $Z = Z_n = \infty$ will the surface boundary condition $u(Z_n) = 0$ be satisfied. In other words, if $n = 5$, only when the radius of the polytropic sphere becomes infinite will the surface boundary condition be satisfied. A more general condition can be proved: if $n < 5$, the radius of the polytropic sphere is finite; if $n \geq 5$, the radius of the polytropic sphere is infinite and the extreme condition for $n = \infty$ is also valid.

Besides using the analytical method to solve for three values of n, the Emden differential equation can also be solved using numerical methods. Figure 7.4 shows the solutions for $n = 3$ and $n = 3/2$. It can be seen that the curves have maxima near $Z = 0$ where $u = 1$ and $u' = 0$. When Z increases, u decreases. At $Z = Z_n$ (i.e., at the surface), $u = 0$. From Eqs. (7.143) to (7.145), it is not difficult in getting the values of Z_n as

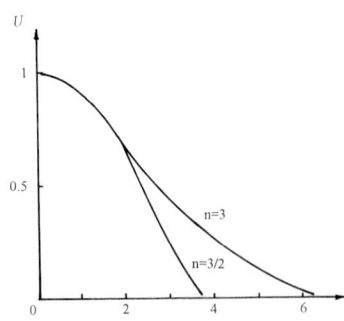

Fig. 7.4 The display of functions $u(Z)$ for $n = 3$ and $n = 3/2$.

$$Z_0 = \sqrt{6},\ Z_1 = \pi,\ Z_5 = \infty$$

In general, for $n < 5$, Z_n corresponds to finite values; for $n \geq 5$, $Z_n = \infty$. Table 7.1 gives some Z_n values and values for some functions, which are very helpful for using the polytropic model to compute the stellar model.

7.8.3 Determination of the Stellar Structure using solutions of the Emden Differential Equation

We now discuss how to determine the distribution of density, mass, pressure and temperature in the interior of a star from the solutions of the Emden differential equation for $n < 5$ and for the given total mass M, radius R and chemical composition of the star. For a given n, the Emden differential equation can be solved to give the distribution of u and (du/dZ) in the star, and the values of Z_n and $\left(-\frac{Z_n}{3}\frac{du}{dZ}\right)_{Z=Z_n}$ at the stellar surface. From Eq. (7.138), we get

$$\rho = \rho_c u^n \tag{7.146}$$

Table 7.1 The Z_n values for some polytropic index n and values for some functions (from Chandrasekhar 1939).

n	Z_n	$\left(-Z^2\dfrac{du}{dZ}\right)_{Z=Z_n}$	$\rho_c/\bar{\rho}$
0	2.4494	4.8988	1.0000
1	3.14159	3.14159	3.28987
1.5	3.65375	2.71406	5.99071
2	4.35287	2.41105	11.40254
3	6.89685	2.01824	54.1825
4	14.97155	1.79723	622.408
4.5	31.8365	1.73780	6189.47
5	∞	1.73205	∞

According to the polytropic relation in Eq. (7.121), we have

$$P = K\rho^{1+\frac{1}{n}} = K\rho_c^{1+\frac{1}{n}} u^{1+n} \tag{7.147}$$

From Eqs. (7.146) and (7.147), we see that the distribution of density and pressure can be obtained when the distribution of u inside a star and the density ρ_c at the center of the star are known. The density ρ_c can be obtained as follows. From Eqs. (7.4) and (7.146), we have

$$M_r = \int_0^r 4\pi \rho r^2 dr = 4\pi \rho_c \int_0^r u^n r^2 dr$$

Using the relation between r and Z in Eq. (7.138), and considering that $(r^3/Z^3) = A^{-3}$ is a constant which can be taken out of the integration sign, the above equation becomes

$$M_r = 4\pi \rho_c \frac{r^3}{Z^3} \int_0^r u^n Z^2 dZ \tag{7.148}$$

From the Emden differential equation (7.139), we see that the term $u^n Z^2$ within the integration on the right-hand side of Eq. (7.148) is a differential which will become $-Z^2(du/dZ)$ after integration. Therefore, Eq. (7.148) can be written as

$$M_r = 4\pi \rho_c r^3 \left(-\frac{1}{Z} \frac{du}{dZ} \right) \tag{7.149}$$

For the stellar surface, we get from Eq. (7.149)

$$M = 4\pi \rho_c R^3 \left(-\frac{1}{Z} \frac{du}{dZ} \right)_{Z=Z_n} \tag{7.150}$$

We introduce the mean density $\bar{\rho}$:

$$\bar{\rho} = \frac{3M}{4\pi R^3} \tag{7.151}$$

From Eqs. (7.150) and (7.151), we have

$$\rho_c = \bar{\rho} \bigg/ \left(-\frac{3}{Z}\frac{du}{dZ} \right)_{Z=Z_n} \tag{7.152}$$

Therefore, for given M and R, we can obtain from Eq. (7.151) the mean density $\bar{\rho}$ and compute the density ρ_c at the center through Eq. (7.152). The term $\left(-\frac{3}{Z}\frac{du}{dZ} \right)_{Z=Z_n}$ in Eq. (7.152) can be obtained from the solution to the Emden differential equation. By further using the three equations (7.146), (7.147) and (7.149), we can compute the distribution of the density, pressure and mass in the star. The distribution of temperature is then obtained by the equation of state for an ideal gas and from the given chemical composition.

As an example, we compute the sun model with $n = 3$. It is known that $M_\odot = 1.989 \times 10^{33}$ g and $R_\odot = 6.96 \times 10^{10}$ cm. From Table 7.1, we see $Z_3 = 6.89685$ and $\rho_c/\bar{\rho} = 54.1825$ for $n = 3$. From the values of $M_\odot$ and $R_\odot$, we obtain $\bar{\rho} = 1.41$ gcm^{-3}, so $\rho_c = 76.39$ gcm^{-3} and $A = Z_3/R_\odot = 9.91 \times 10^{-11}$. For known values of A and ρ_c, we have $K = 3.85 \times 10^{14}$ from Eq. (7.135), and $P_c = 1.24 \times 10^{17}$ dyn/cm^2 from Eq. (7.144). The chemical composition of the sun is taken as $X_H = 0.7$ and $X_{He} = 0.3$, so $\mu = 0.62$. From the equation of state for an ideal gas, the temperature at the center of the sun is obtained as $T_c = 1.2 \times 10^7$ K. The temperature T_c obtained from the sun model using the whole set of stellar structure equations is $T_c = 1.4 \times 10^7$ K, which is very close to the results obtained from the polytropic model with $n = 3$.

CHAPTER EIGHT

EARLY STELLAR EVOLUTION

The early stage of stellar evolution refers to the evolutionary stage from the collapse of the interstellar gas cloud to form a protostar, until the occurrence of core hydrogen burning, i.e., before entering the main sequence stage.

8.1 STAR FORMATION

Some astronomical phenomena can give clues as to how stars are formed. For example, people know that more massive and more luminous stars are younger and have just been formed. These stars are concentrated in the spiral arms of the galactic plane and their surroundings are fully filled by interstellar gas and dust. These lead us to assert that stars are formed from interstellar gas and dust. Of the observed interstellar gas clouds, some have large mass, large radius, small density, and are gas clouds composed of hydrogen atoms; while others have large mass, small radius, high density and are molecular clouds composed of molecules. The latter always stay together in form of clusters. At the same time, stars are also observed to stay together in groups, clusters or associations. These lead us to depict star formation processes as those shown in Fig. 8.1. A massive gas

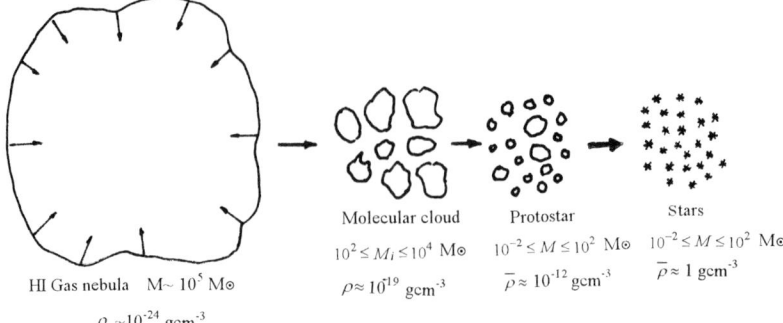

Fig. 8.1 Schematic diagram showing the collapse of the interstellar gas cloud, and its fragmentation to form stars.

cloud collapses due to gravitational instabilities and fragments during the collapse to form many smaller molecular clouds, which continue to collapse due to gravitational instabilities and fragment during the collapse to form protostars. The protostars continue to contract and form stars.

The above processes have observational supports and comply with theoretical predictions for formation of stars. In this section, we will further explain theoretically the emergence of gravitational instabilities under certain conditions which leads to the repeated collapse and fragmentation and the final formation of a group of protostars.

8.1.1 Gravitational Instability of Interstellar Gas Clouds

We consider an infinite uniform gas cloud under hydrostatic equilibrium. Obviously, the density and temperature are constants everywhere in the cloud, and the gas velocity is zero, i.e., $\rho = \rho_0 = $ const, $T = T_0 = $ const and $\vec{v}_0 = 0$.

If there is motion inside the gas cloud, the basic equations for hydrodynamics and the equation of state should be satisfied, i.e.,

Continuity equation: $$\frac{\partial \rho}{\partial t} + \nabla \cdot (\rho \vec{v}) = 0 \qquad (8.1)$$

Equation of motion: $$\frac{\partial \vec{v}}{\partial t} + (\vec{v} \cdot \nabla)\vec{v} = -\frac{1}{\rho}\nabla P - \nabla \Phi \qquad (8.2)$$

Poisson equation $$\nabla^2 \Phi = 4\pi G \rho \qquad (8.3)$$

Equation of state for an ideal gas $$P = \frac{\Re}{\mu}\rho T = C_s^2 \rho \qquad (8.4)$$

where C_s is the speed of sound.

We go on to discuss whether instabilities will arise from perturbations in a gas cloud. The state of the gas cloud under perturbation is called the perturbed state. All physical quantities such as density, pressure, gravitational potential and velocities etc., corresponding to the perturbed state can be viewed as the sum of the perturbation values and the equilibrium values, i.e.,

$$\left. \begin{array}{l} \rho = \rho_0 + \rho_1 \\ P = P_0 + P_1 \\ \Phi = \Phi_0 + \Phi_1 \\ \vec{v} - \vec{v}_0 + \vec{v}_1 \end{array} \right\} \tag{8.5}$$

where the subscript "0" refers to the equilibrium values, and the subscript "1" refers to the perturbation values. Here, we assume there is no perturbation to the speed of sound.

If the physical quantities for the perturbed state given in Eq. (8.5) are substituted into Eqs. (8.1) to (8.3), and considering that the equilibrium values will satisfy Eqs. (8.1) to (8.3), we can set up a set of linear differential equations for the perturbation values:

$$\frac{\partial \vec{v}_1}{\partial t} = -\nabla \left(\Phi_1 + \frac{C_s^2}{\rho_0} \rho_1 \right) \tag{8.6}$$

$$\frac{\partial \rho_1}{\partial t} + \rho_0 \nabla \cdot \vec{v}_1 = 0 \tag{8.7}$$

$$\nabla^2 \Phi_1 = 4\pi G \rho_1 \tag{8.8}$$

Taking divergence for Eq. (8.6), we have

$$\nabla \cdot \frac{\partial \vec{v}_1}{\partial t} = \frac{\partial}{\partial t} (\nabla \cdot \vec{v}_1) = -\nabla^2 \Phi_1 - \left(\frac{C_s^2}{\rho_0} \right) \nabla^2 \rho_1 \tag{8.9}$$

Partial differentiating Eq. (8.7) with respective to time, we have

$$\frac{\partial}{\partial t} (\nabla \cdot \vec{v}_1) = -\frac{1}{\rho_0} \frac{\partial^2 \rho_1}{\partial t^2} \tag{8.10}$$

Substituting Eq. (8.8) into Eq. (8.9) to eliminate $\nabla^2 \Phi_1$, and substituting Eq. (8.10) into Eq. (8.9) to eliminate $\partial(\nabla \cdot \vec{v}_1)/\partial t$, we can obtain from Eq. (8.9)

$$\left(\nabla^2 - \frac{1}{C_s^2}\frac{\partial^2}{\partial t^2} + \frac{4\pi G\rho_0}{C_s^2}\right)\rho_1 = 0 \tag{8.11}$$

which is a typical oscillation equation with the solution

$$\rho_1 = Ae^{i(\vec{k}\cdot\vec{r}-\omega t)} \tag{8.12}$$

which represents a planar density wave with wave vector $\vec{k}$ and angular velocity ω fulfilling the dispersion relation

$$\omega^2 = k^2 C_s^2 - 4\pi G\rho_0 \tag{8.13}$$

From Eq. (8.13), it is not difficult to observe that ω is real when k is larger than a critical value k_j given by

$$k_j = \frac{(4\pi G\rho_0)^{1/2}}{C_s} \tag{8.14}$$

or when the wavelength λ is smaller than a critical wavelength λ_j given by

$$\lambda_j = \frac{2\pi}{k_j} = \left(\frac{\pi}{G\rho_0}\right)^{1/2}C_s \tag{8.15}$$

This implies that the density perturbation is time dependent, but the amplitude will not grow. Therefore, the gas cloud is stable when k is very large ($k > k_j$) or the wavelength is very small, which is equivalent to a density perturbation in a small region in the gas. The perturbed gas will only oscillate periodically around the equilibrium state.

If the wave number k is smaller than k_j, i.e.,

$$k < k_j = \frac{(4\pi G\rho_0)^{1/2}}{C_s} \tag{8.16}$$

or if the wavelength is larger than the critical wavelength λ_j, i.e.,

$$\lambda > \lambda_j = \left(\frac{\pi}{G\rho_0}\right)^{1/2}C_s \tag{8.17}$$

ω is imaginary ($\omega = \pm i\eta$), so the density perturbation ρ_1 will grow exponentially with time. Therefore, when k is small ($k < k_j$) or the wavelength is very large, or equivalently the region of density perturbation is large in the gas, the gas cloud is unstable. Equation (8.16) was first given by Jeans (1928) and is thus known as the Jeans criterion.

It should be noted that there are difficulties in solving the above equation set, which can be observed from the non-perturbed equilibrium equation set. Since ρ_0 and P_0 are constants, $\nabla P_0 = 0$, so $\rho_0 = 0$ or $\nabla \Phi_0 = 0$ according to the hydrostatic equilibrium equation $\nabla P_0 + \rho_0 \nabla \Phi_0 = 0$. Assuming $\rho_0 \neq 0$, we have $\nabla \Phi_0 = 0$ and $\nabla^2 \Phi_0 = 0$. However, according to the Poisson equation, we should have $\rho_0 = 0$. Therefore, strictly speaking, Eqs. (8.6) to (8.8) can only have a special solution, i.e., $\rho_0 = 0$ and $k_j = 0$, which is nevertheless a solution without physical meanings. Notwithstanding, it should be pointed out that certain gas clouds with special structures do exist with $\rho_0 =$ constant and a solution similar to Eq. (8.14) carrying different coefficients. As an example, we consider a disc shaped gas cloud at equilibrium having an infinite plane parallel structure with uniform physical quantities in each layer. In this way, all physical quantities are functions of the height Z, i.e.,

$$\rho_0 = \rho_0(Z), P_0 = P_0(Z), \Phi_0 = \Phi_0(Z), \vec{v}_0 = 0 \qquad (8.18)$$

and the equilibrium hydrostatic equation, equation of state and Poisson equation can be respectively written as

$$\frac{dP_0}{dZ} = -\rho_0 \frac{d\Phi_0}{dZ} \qquad (8.19)$$

$$P_0 = C_s^2 \rho_0 \qquad (8.20)$$

$$\frac{d^2 \Phi_0}{dZ^2} = 4\pi G \rho_0 \qquad (8.21)$$

From Eqs. (8.19) and (8.20), we get

$$C_s^2 \frac{d \ln \rho_0}{dZ} = -\frac{d\Phi_0}{dZ} \qquad (8.22)$$

Differentiating Eq. (8.22) and putting into Eq. (8.21), we obtain

$$\frac{d^2 \ln \rho_0}{dZ^2} = -\frac{4\pi G}{C_s^2} \rho_0 \qquad (8.23)$$

Employing the boundary condition $\rho_0 = 0$ at $Z = \pm\infty$, the solution of Eq. (8.23) becomes

$$\rho_0(Z) = \rho_0(0)\,\text{sech}^2(Z/H) \tag{8.24}$$

where

$$H = \left(\frac{\Re T}{2\pi\mu G\rho_0(0)}\right)^{1/2} = \frac{C_s}{[2\pi G\rho_0(0)]^{1/2}} \tag{8.25}$$

Substituting Eqs. (8.24) and (8.25) into Eq. (8.23), we see that $\rho_0(Z) = 0$ is not a necessary result. We further study the stability of the gas disc. We adopt the plane rectangular frame (x, y, z) in which the x and y axes are contained by the disc plane. Supposing the density perturbation is time dependent and propagates in the x direction, $\rho_1(Z)$ can be written in the form

$$\rho_1(Z) = \rho_0(Z)e^{i[kx+\omega t]} \tag{8.26}$$

where $\rho_0(Z)$ fulfils Eq. (8.24). Introducing a new variable ξ

$$\xi = \tanh\left(\frac{Z}{H}\right) \tag{8.27}$$

Eq. (8.26) becomes

$$\rho_1(\xi) = \rho_0(\xi)\theta(\xi)e^{i[kx+\omega t]} \tag{8.28}$$

We study the situation for $\omega = 0$. Since $\omega = 0$ represents the critical stability, it can give the critical wave number k_j for instabilities. When $\omega = 0$, Eq. (8.28) becomes

$$\rho_1(\xi) = \rho_0(\xi)\theta(\xi)e^{ik_j x} \tag{8.29}$$

Substituting Eq. (8.29) into the perturbation Eqs. (8.6) and (8.8), the differential equation for the function $\theta(\xi)$ can be obtained as

$$\frac{d^2\theta}{d\xi^2} - \frac{2\xi}{1-\xi^2}\frac{d\theta}{d\xi} + \left[\frac{2}{1-\xi^2} - \frac{(k_j H)^2}{(1-\xi^2)^2}\right]\theta = 0 \tag{8.30}$$

with a general solution

$$\theta(\xi) = A\left(\frac{1+\xi}{1-\xi}\right)^{k_j H/2}(k_j H - \xi) + B\left(\frac{1-\xi}{1+\xi}\right)^{k_j H/2}(k_j H + \xi) \tag{8.31}$$

Employing the boundary condition $\rho_0 = 0$ at $Z = \pm\infty$, we can obtain for Eq. (8.31) $k_j H = 1$ (see Spitzer 1968), so that

$$k_j = \frac{1}{H} = \frac{(2\pi G\rho_0)^{1/2}}{C_s} \tag{8.32}$$

Comparing Eqs. (8.32) and (8.14), we notice that the only difference in the two equations is in the coefficients, which arise from the geometry of the gas cloud. Therefore, the infinite gas disc is unstable when $k < k_j$ and the density perturbation will grow exponentially with time. Conversely, the disc is stable when $k > k_j$ and the density perturbation is time dependent but its amplitude will not grow.

We have discussed the stability of the gas cloud in the above discussions. When the gas cloud is in a state of equilibrium, the gravitational force and the gaseous pressure are at equilibrium. If there is a perturbation, for example, when the gravitational force increases to compress part of the gas, the gas might develop gravitational instabilities and collapse depending on the value of k. If k is very large ($k > k_j$) or the wavelength is very small, or equivalently the region of compression is small, the gas cloud is stable, and will only oscillate periodically around the equilibrium values but will not collapse. Conversely, if k is very small ($k < k_j$) or the wavelength is very large, or equivalently the region of compression is large, the gas cloud is unstable and will collapse.

8.1.2 Gravitational Instability for a Sphere in the Interstellar Gas Cloud

In the earlier section, we have discussed the gravitational instability of the gas cloud with a special structure. To understand the possibility of gravitational instabilities in a gas cloud in a more general way, we consider gravitational instabilities for a gas sphere inside the cloud (Fig. 8.2). Suppose the sphere is composed of an ideal gas, which has a mass M_c, radius R, a uniform temperature, and a pressure P^* from the gas external to the sphere. For an isothermal sphere with non-zero surface pressure, according to the Virial theorem in Eq. (7.101), the surface pressure P_0 at the inner surface of the sphere is

$$P_0 = A\frac{M_c T}{R^3} - B\frac{M_c^2}{R^4} \tag{8.33}$$

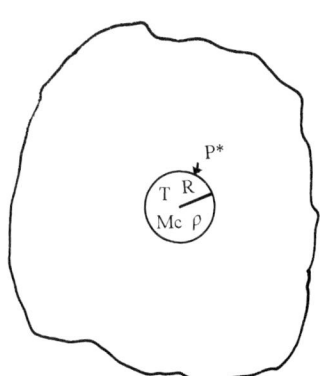

Fig. 8.2 Schematic diagram showing the gravitational instability for a sphere in an interstellar gas cloud.

where

$$A = \frac{3\Re}{4\pi\mu} > 0, B = \frac{C_1 G}{4\pi} > 0 \qquad (8.34)$$

The first term on the right-hand side of Eq. (8.33) is the gas pressure which leads to the outward expansion of the sphere. The second term is the contribution from self-gravitation, which leads to the collapse of the sphere towards the center. When M_c and T are fixed, P_0 is only a function of R and has a maximum value P_{max} at a certain value of R_j (Fig. 8.3).

From the condition for maximum P_0, i.e., $\partial P_0/\partial R = 0$, we obtain

$$R_j = \frac{4B}{3A}\frac{M_c}{T} = \frac{4C_1}{9}\frac{G\mu M_c}{\Re T} \qquad (8.35)$$

Suppose the gas cloud is at equilibrium, so the sphere is also at equilibrium with its surroundings, i.e., $P_0 = P^*$. For $R < R_j$, P_0 decreases when R decreases. This means if there is a small compression in the sphere so that $P_0 < P^*$, P_0 will keep decreasing and the gas keeps being compressed. Thus, the gas is unstable. Conversely, for $R > R_j$, P_0 increases when R decreases. Therefore, if there is a small compression in the sphere so that $P_0 < P^*$, P_0 will increase and the gas expands. Thus, the gas is stable. Replacing M_c in the expression of R_j in Eq. (8.35) by $(4/3)\pi R_j^3 \bar{\rho}$ ($\bar{\rho}$ is the average density), we have

$$R_j = \left(\frac{27}{16\pi C_1 G \bar{\rho}}\right)^{1/2} \left(\frac{\Re T}{\mu}\right)^{1/2} \qquad (8.36)$$

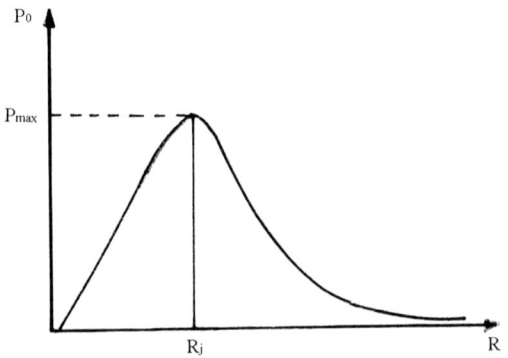

Fig. 8.3 Schematic diagram showing that P_0 is only a function of R and has a maximum value P_{max} at a certain value of R_j.

We define a mass M_j known as the Jeans mass:

$$M_j \equiv \frac{4}{3}\pi R_j^3 \bar{\rho} = \frac{27}{16}\left(\frac{3}{\pi}\right)^{1/2}\left(\frac{\Re}{C_1 G}\right)^{3/2}\left(\frac{T}{\mu}\right)^{3/2}\left(\frac{1}{\bar{\rho}}\right)^{1/2} \tag{8.37}$$

From Eqs. (8.36) and (8.37), we see that R_j and M_j are only determined by $\bar{\rho}$ and T.

We now study a gas cloud at equilibrium with a mass M, temperature T and average density $\bar{\rho}$. An M_j value can be obtained from Eq. (8.37). If $M > M_j$, the situation shown in Fig. 8.2 will happen, that is, there is matter exterior to the sphere with radius R_j, so the external surface of the sphere has a pressure $P^* > 0$. At this time, the regions with $R < R_j$ are unstable and will collapse inwards. Therefore, when $M > M_j$, the gas cloud is unstable as a whole. If $M < M_j$, there is no matter exterior to the sphere with radius R_j so $P^* = 0$. At this time, the gas cloud will not collapse inwards and is stable as a whole. These results are derived from the Virial theorem under hydrostatic equilibrium, which are consistent with those derived from hydrodynamic equations and linear perturbation theory. It is interesting to note that, according to $C_s^2 = \Re T/\mu$, Eq. (8.36) can be written as

$$R_j = \left(\frac{27}{16\pi C_1}\frac{1}{G\bar{\rho}}\right)^{1/2}C_s \tag{8.38}$$

Comparing Eqs. (8.38) and (8.15), we see that the expressions for R_j and λ_j are very similar with the same order of magnitude only differing in their coefficients.

If the Jeans mass M_j is defined by the results from the hydrodynamic instability theory, i.e., by using $M_j \equiv (4/3)\pi\lambda_j^3\rho$, we can obtain a result which is similar to Eq. (8.37) with the same order of magnitude only differing in their coefficients. In general, the expression adopted for the Jeans mass obtained from the hydrodynamic instability theory is

$$M_j = \left(\frac{\pi\Re}{G\mu}\right)^{3/2}T^{3/2}\rho^{-1/2}$$

$$= 1.2 \times 10^5 M_\odot \left(\frac{T}{100K}\right)^{3/2}\left(\frac{\rho}{10^{-24}\ gcm^{-3}}\right)^{-1/2}\mu^{-3/2} \tag{8.39}$$

Supposing $\rho = 10^{-24}$ gcm^{-3}, $T = 100$ K and $\mu = 1$ (typical parameters for HI clouds), we have $M_j \approx 10^5 M_\odot$, a value much greater than the mass for a common star, which is 0.1 to 120$M_\odot$.

8.1.3 Fragmentation of Collapsing Clouds

For a HI interstellar gas cloud with density $\rho = 10^{-24}$ gcm^{-3}, temperature $T = 100\ K$ and $\mu = 1$, we know from the above discussions that gravitational instabilities leading to the collapse will only occur when the total mass $M > 10^5 M_\odot$. From observations, the mass of a star ranges from $0.1 \sim 120 M_\odot$ which is much smaller than $10^5 M_\odot$. At the same time, observations show that stars form groups, clusters and associations. These have led people to assert that the fragmentation of interstellar clouds during their collapse forms small cloud clusters which will individually continue their collapse and fragmentation, and form cloud clusters with masses of stars which finally collapse to form stars.

For the above scenario to be true, it is most important that M_j should not be a constant during the collapse of the interstellar cloud. If M_j decreases when the density increases during the collapse, the mass of local regions in the cloud will be greater than M_j, which leads to local instabilities and separated collapse of individual local regions, and to the fragmentation of the cloud into many small cloud clusters. From Eqs. (8.37) and (8.39), we know that $M_j \sim \rho^{-1/2}$ for an isothermal collapse, i.e., M_j really decreases when the density increases. The situation is opposite for an adiabatic collapse where $\nabla_{ad} = \left(\frac{d \ln T}{d \ln P}\right)_{ad} = \frac{2}{5}$ for an ideal gas or $T \sim P^{2/5}$, so from the equation of state $P \sim \rho T$ or $T \sim \rho^{2/3}$, we have $M_j \sim T^{2/3}\rho^{-1/2} \sim \rho^{1/2}$; in other words, M_j increases with the density for adiabatic collapse.

To ensure the collapse is isothermal, the gravitational energy released during the collapse of the interstellar cloud should be entirely radiated out in order not to increase the internal energy of the gas. If τ represents the collapse timescale of the cloud, and τ_{adj} represents the thermal adjustment timescale (which is the time required to radiate out all the internal energy inside the interstellar cloud), the collapse will be isothermal only when $\tau >> \tau_{adj}$. The interstellar cloud will have free-fall type collapse when the gas pressure is far less than the gravitation. Therefore, from the free-fall timescale in Eq. (7.106) or $\tau \approx (G\rho)^{-1/2}$, or from Eq. (8.13) assuming $k^2 C_s^2 << 4\pi G\rho_0$, we have $i\omega \approx (G\rho_0)^{-1/2}$ so $\tau \approx (G\rho_0)^{-1/2}$. For an interstellar cloud with density $\rho = 10^{-24}$ gcm^{-3}, we obtain $\tau \approx 10^8 y$. Spitzer (1968) and Low and Lynden–Bell (1976) calculated the thermal adjustment timescale for the interstellar cloud to be $\tau_{adj} \approx 100y$, which obviously fulfils $\tau >> \tau_{adj}$, showing that the collapse of an interstellar cloud is indeed isothermal and that $M_j \sim \rho^{-1/2}$.

Another question is whether the fragmentation of the interstellar cloud will continue to give cloud clusters with mass smaller than the stellar mass. To answer this question, we have to know whether the collapse of the interstellar cloud will change from isothermal to adiabatic.

We first study the energy radiated per second during the isothermal collapse of the interstellar cloud. It is known that the collapse timescale is $\tau \approx (G\rho)^{-1/2}$. The total gravitational energy released in the collapse is $E_g = GM^2/R$, where M and R are the total mass and the radius of the interstellar cloud. Thus, the total energy radiated per second during the isothermal collapse of the interstellar cloud is

$$L_G \cong \frac{GM^2}{R} / (G\rho)^{-1/2} = \left(\frac{3}{4\pi}\right)^{1/2} \frac{G^{3/2} M^{5/2}}{R^{5/2}} \tag{8.40}$$

The energy radiated per second in the collapse of the cloud should be smaller than that of a black body with the same temperature and volume, and can be written as

$$L_c = f \cdot 4\pi R^2 \sigma T^4 \tag{8.41}$$

where $f < 1$.

Apparently, $L_c \gg L_G$ for an isothermal collapse while $L_c \approx L_G$ for an adiabatic collapse. From Eqs. (8.40) and (8.41), when $L_c = L_G$, we have

$$M = \left(\frac{64\pi^3 \sigma^2}{3G^3}\right)^{1/5} f^{2/5} T^{8/5} R^{9/5} \tag{8.42}$$

We take M in Eq. (8.42) to be the mass limit M_j for the collapse, since M_j will increase with density after the collapse turns adiabatic. Using $R = \left(\frac{3}{4\pi}\right)^{1/3} \frac{M_j^{1/3}}{\rho^{1/3}}$, from Eq. (8.42), we obtain

$$M_j = \left(\frac{\pi^9}{9}\right)^{1/4} (\sigma G^3)^{-1/2} \left(\frac{\Re}{\mu}\right)^{9/4} f^{-1/2} T^{1/4}$$
$$= 0.02 f^{-1/2} T^{1/4} \quad (M_\odot) \tag{8.43}$$

where we have taken $\mu = 1$ and the dimension of T is K.

From Eq. (8.43), the mass limit M_j is related to the temperature T and the coefficient f. If the smallest cloud cluster after fragmentation is supposed to have a temperature of 1000 K and $f = 0.1$, we have $M_j = (1/3)M_\odot$. Thus, the mass of the smallest cloud cluster is comparable to the stellar mass.

8.1.4 Model Calculation for Formation of Protostars

In the above, we have employed a linear theory to study the condition for occurrence of gravitational instabilities leading to the collapse. However, the linear theory has a limitation of not being able to tell the detailed process and the final products of the collapse of the cloud. A non-linear theory is needed for these. In this section, we will describe the model calculations for the collapse of the interstellar cloud with a mass of $1M_\odot$ and chemical compositions $X_H = 0.651$, $X_{He} = 0.324$ and $X_Z = 0.025$ first carried out by Larson (1969).

We consider a spherical interstellar cloud which collapses towards the center. Here, the motion of a mass layer should satisfy the following equations:

$$\text{Continuity equation:} \quad \frac{\partial m}{\partial t} + 4\pi r^2 v \rho = 0 \tag{8.44}$$

where v is the radial velocity with positive values defined to be outward velocities, m is the mass of the sphere with a radius r, and ρ is the density.

$$\text{Equation of motion:} \quad \frac{\partial v}{\partial t} + v \frac{\partial v}{\partial r} + \frac{Gm}{r^2} + \frac{1}{\rho} \frac{\partial P}{\partial r} = 0 \tag{8.45}$$

where the derivatives of the velocity v have made use of the relationship

$$\frac{d}{dt} = \frac{\partial}{\partial t} + v \frac{\partial}{\partial r}$$

$$\text{Energy equation:} \quad \frac{\partial u}{\partial t} + v \frac{\partial u}{\partial r} + P \frac{\partial}{\partial t}\left(\frac{1}{\rho}\right) + Pv \frac{\partial}{\partial r}\left(\frac{1}{\rho}\right) + \frac{1}{4\pi \rho r^2} \frac{\partial l}{\partial r} = 0 \tag{8.46}$$

where u is the internal energy per unit mass, l is the luminosity, and the first four terms on the left-hand side are in fact $\frac{du}{dt} + P\frac{d}{dt}\left(\frac{1}{\rho}\right)$.

$$\text{Mass distribution equation:} \quad \frac{\partial m}{\partial r} = 4\pi r^2 \rho \tag{8.47}$$

$$\text{Radiative transfer equation:} \quad l = -\frac{16\pi a c r^2}{3\kappa\rho} T^3 \frac{\partial T}{\partial r} \tag{8.48}$$

Here, the radiative transfer equation is taken to be the one for a radiative equilibrium region in the interior of the star. For an optically thin interstellar cloud, this equation is not very appropriate, but will only lead to quantitative and not qualitative differences.

The five equations (8.44) to (8.48) involve five unknowns: $m(r, t)$, $v(r, t)$, $P(r, t)$, $T(r, t)$ and $l(r, t)$. The equations also involve some functions related to properties of matter: density ρ, opacity κ and internal energy u which are functions of P, T and chemical composition. The computation for κ is a bit different from that in the case for stars. According to Gaustadt (1963), absorption by dust particles is the most important when the cloud temperature is less than 1000 K, and absorption by molecules is the most important when the temperature is higher than 1000 K. The computation of the density ρ can make use of the equation of state for an ideal gas, taking into consideration effects of dissociation and ionization. The boundary conditions for the equations (8.44) and (8.48) are

outside boundary condition: at $r = R$, $v(R, t) = 0$ (8.49)

internal boundary condition: at $r = 0$, a real solution required.

(8.50)

To integrate the equations (8.44) to (8.48), we must also give the initial model values at $t = 0$. To ensure that the interstellar cloud is gravitationally unstable initially, we require the density of the interstellar cloud to be sufficiently large and the radius not to be very large. Using Eq. (8.35), an interstellar cloud with a mass of $1M_\odot$ is gravitationally unstable when $R = 1.63 \times 10^{17}$ cm and $\rho = 10^{-19}$ gcm^{-3}, which can be taken to be the initial values of radius and density. Initially, $v(r, 0) = 0$, $P(r, 0)$ and $T(r, 0)$ are constants, so $l(r, 0) = 0$. $T(r, 0)$ can be taken as 10 K.

The numerical methods for the models will not be discussed in detail here. In the following, only the results from model calculations, that is, properties at different stages in the collapse, will be described.

(A) Optically thin stage and formation of central core

At the early stage of the collapse, the entire interstellar cloud is optically thin and is isothermal ($T = 10$ K). The collapse starts from the inside of the interstellar cloud. The internal matter first collapses inwards to the center, which is followed by the collapse of the outer matter. Therefore, after the commencement of the collapse, it can be found that the density increases towards the center. The free-falling timescale at a distance r from the center is $\tau \sim [G\bar{\rho}(r)]^{-1/2}$. Thus, after the commencement of the collapse, the density $\bar{\rho}$ increases towards the

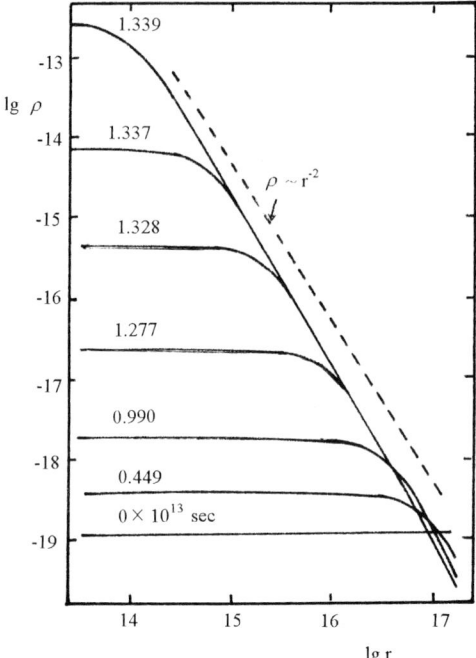

Fig. 8.4 The relationship between the density (gcm^{-3}) and the distance from the center r (cm) within the interstellar cloud with a mass of 1M$_\odot$ after the commencement of the collapse. The solid lines refer to the density distributions corresponding to different times with the attached numbers referring to the time after commencement of the collapse (in units of 10^{13} s). In a completely free-falling region, the relationship between density and distance is close to $\rho \sim r^{-2}$.

center and the local free-falling timescale will decrease. In other words, the fall of interior regions are faster than those of the exterior regions. This further increases the density $\bar{\rho}$ towards the center. Figure 8.4 shows the relationship between density and r within the interstellar cloud after commencement of the collapse.

However, when the density at the center increases to 10^{-13} gcm^{-3}, it becomes optically opaque. At this time, the continuous increase of the density will lead to an adiabatic increase of temperature at the center, and to the continuous increase of the gas pressure. Ultimately, the free-falling processes at the center will cease. In this way, a core under hydrostatic equilibrium is formed at the center region with an initial radius of 6×10^{13} cm, a mass of 10^{31} g, a core density of $\rho_c = 2 \times 10^{-10}$ gcm^{-3} and a core temperature of $T_c = 170$ K. In the outside region of the core, the matter continues to fall with a velocity of 75 km/s, which exceeds the local speed of sound, so a spherical shock front is formed at the surface of the central core which separates the core inside under hydrostatic equilibrium and the outside falling matter.

(B) Re-collapse of the central core

Due to the continuous in-fall of matter exterior to the core, the surface pressure of the core increases continuously, so the core keeps contracting and the internal temperature keeps rising. For an isothermal sphere constituted by a polyatomic gas under hydrostatic equilibrium, from the Virial theorem Eq. (7.96), the total energy can be written as

$$E = E_T + E_G = (4 - 3\gamma)E_T \qquad (8.51)$$

where $\gamma = C_P/C_V$. From Eq. (8.51), if $\gamma > 4/3$, the core has a negative total energy so it is stable. If $\gamma < 4/3$, the core has a positive total energy so it is unstable. Therefore, $\gamma = 4/3$ is the criterion to have dynamic instabilities in the core.

When the central core contracts, the internal temperature keeps increasing. When the temperature is lower than 2000 K, the hydrogen in the core is molecular. When the temperature reaches 2000 K, part of the molecules will be dissociated into atoms. When the gas inside the core is under a molecular state, $\gamma = 7/5 = 1.40$, which is already close to 4/3 or the critical value for dynamic instabilities in the core. If part of the hydrogen becomes atoms, the γ value will become smaller than 4/3 which leads to dynamic instabilities in the core, and the central core will once again collapse.

The calculation results of Larson show that dynamic instabilities and re-collapse will occur when the core mass is twice the initial mass and the radius is half the initial radius. The outcome of the collapse is the emergence of a new core in the central region with a larger density, which is constituted by atomic hydrogen. The initial parameters for this new core are: the mass is $1.5 \times 10^{-3} M_\odot$, the radius is $1.3\ R_\odot$, the central density is $2 \times 10^{-2}\ \mathrm{gcm}^{-3}$ and the central temperature is 2×10^4 K. Similarly, there will be a new shock front at the surface of this new core which separates the core from the external continuously falling matter.

The new core formed during the second collapse is called the protostar. Although the masses of a protostar and a star are of the same magnitude, the central density and temperature of the protostar are far less than those of the star with a similar mass, and there are no thermonuclear reactions at the center of the protostar. The energy radiated out from the protostar is compensated by the energy released when the in-falling matter reaches the shock front. Furthermore, hydrostatic equilibrium is not attained at every corner inside the protostar. Its surface is still subjected to the pressure caused by the continuously falling exterior matter.

Due to the continuously falling exterior matter, the mass of the protostar keeps growing and the radius keeps diminishing. When the accretion timescale τ_{acc} is smaller than the thermal timescale (Helmholtz timescale), the compression is adiabatic. However, with the gradual exhaust of the exterior matter, the accretion rate is lowered and the accretion timescale τ_{acc} will gradually increase. When τ_{acc} is comparable to the thermal timescale, the adiabatic process ceases. At this time, since the accretion rate is very small, the core mass is effectively unchanged, the surface shock front disappears and hydrostatic equilibrium is attained inside the protostar. Here, the protostar has turned into a pre-main-sequence star.

It should be noted that the stated model calculations for the collapse of the interstellar cloud are in fact extremely simplified. The initial velocities of the particles in the interstellar cloud (e.g., the velocity caused by the rotation of the interstellar cloud about the center of the galactic center) and the initial magnetic field of the gas have not been considered. The calculations for opacity in the study of radiative transfer are also very crude, and the entire model employs one-dimensional calculations. Therefore, the obtained results can only be treated as qualitative only. The currently observed stellar phenomena indicate more complicated conditions in the formation of stars. For example, the presence of binaries and planetary systems show that they can be split from a star during its formation. Another example is found in the discovery of jet phenomena in some young stars, which indicates the presence of the disc structure.

8.2 PRE MAIN SEQUENCE EVOLUTION

After a certain time from the collapse of the interstellar cloud to form the protostar, the exterior infalling matter is exhausted and the surface shock front disappears, the total mass does not increase any more, and the interior gradually reaches hydrostatic equilibrium and is at a convective state. At this time, the stellar object is called a pre main sequence star. In order to explain the position of a pre main sequence star in the HR diagram, its evolution, and the characteristic changes in its internal structure during the evolution, we first introduce the features of the Hayashi line in the HR diagram and describe features of the internal structure of a pre main sequence star during evolution.

8.2.1. Hayashi Line

In the effective temperature range in the HR diagram which is about 3000 ~ 5000 K, there is a line known as the Hayashi line which is almost perpendicular to the main sequence (see Fig. 8.5). The position of the Hayashi line is dependent on the stellar mass; the Hayashi line will move towards higher temperature regions for more massive stars.

The Hayashi line has two important features. First, when an interstellar cloud collapses to form a protostar, and hydrostatic equilibrium is attained in the interior region which is fully convective, it can be called a pre main sequence star and its position in the HR diagram must fall on the Hayashi line. Second, for a given mass and chemical composition of the star, the Hayashi line is a boundary on the HR diagram. According to whether the stellar object fulfils hydrostatic equilibrium, the HR diagram is divided into "forbidden" and "allowed" regions. The forbidden region is to the right of the Hayashi line in which there is no star under hydrostatic equilibrium.

The following discussions showing the above two features are based on Kippenhahn and Weigert (1991).

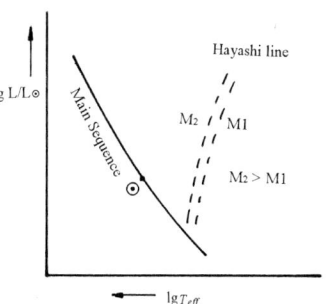

Fig. 8.5 Schematic diagram for the position of the Hayashi line in the HR diagram.

(1) Completely convective and hydrostatic equilibrium

For a star with fixed mass and composition, if its interior is completely convective and under hydrostatic equilibrium, it will be on the Hayashi line in the HR diagram.

If the entire interior of the star is convective, the difference between the temperature gradient ∇ and the adiabatic gradient ∇_{ad} is very small everywhere inside the entire star, i.e., $|\nabla - \nabla_{ad}| < 10^{-7}$ (see §3.3). Therefore, for the entire star, we have

$$\nabla \equiv \frac{d \ln T}{d \ln P} = \nabla_{ad} \tag{8.52}$$

For a completely ionized gas, $\nabla_{ad} = 0.4$. Integrating Eq. (8.52), we have

$$P = C \cdot T^{1/\nabla_{ad}} \tag{8.53}$$

Replacing T in the above equation by the equation of state for an ideal gas, $T = (\mu/\Re)(P/\rho)$, we obtain

$$P = K \cdot \rho^{1+\frac{1}{n}} \tag{8.54}$$

where

$$\left.\begin{array}{l} n = \dfrac{1}{\nabla_{ad}} - 1 \\[2mm] K^n = C^{-1}\left(\dfrac{\Re}{\mu}\right)^{n+1} \end{array}\right\} \tag{8.55}$$

From Eq. (8.54), we know that a star of which the interior is completely convective will fulfil the polytropic equation (7.121) with the polytropic index $n = 3/2$ (when $\nabla_{ad} = 0.4$ is adopted). Therefore, we can directly import some results from the polytropic model. From Eqs. (7.139), (7.152) and (7.153), we have

$$
\left.
\begin{aligned}
A^2 &= \frac{4\pi G}{K(1+n)}\rho_c^{\frac{n-1}{n}} \\
\rho_c &= \bar{\rho}\Big/\left(-\frac{3}{Z}\frac{du}{dZ}\right)_{Z=Z_n} \\
\bar{\rho} &= \frac{3M}{4\pi R^3}
\end{aligned}
\right\}
\tag{8.56}
$$

where $\left(-\frac{3}{Z}\frac{du}{dZ}\right)_{Z=Z_n}$ is a constant for a stellar model. From Eq. (8.56), we get

$$
K = K'(n)\left(\frac{R^{3-3n}}{M^{1-n}}\right)^{1/n}
\tag{8.57}
$$

Combining Eqs. (8.53),(8.55) and (8.57), we obtain

$$
P = C'(n)\mu^{-1-n}M^{1-n}R^{3n-3}T^{1+n}
\tag{8.58}
$$

Using $R_\odot$ and $M_\odot$ as units, the above equation can be written as

$$
C'(n) = \frac{\Re^{n+1}}{4\pi G^n}(1+n)^n\left[-\left(\frac{du}{dZ}\right)_{Z=Z_n}\right]^{n-1}Z_n^{n+1}R_\odot^{3n-3}M_\odot^{1-n}
\tag{8.59}
$$

Since the entire star is convective except in the atmosphere, Eq. (8.58) is applicable from the star center to the photosphere. Substituting the values $P_{\tau=2/3}$ and T_{eff} in the photosphere ($\tau = 2/3$) into Eq. (8.58), we should have

$$
P_{\tau=\frac{2}{3}} = C'(n)\mu^{-1-n}M^{1-n}R^{3n-3}T_{eff}^{1+n}
\tag{8.60}
$$

On the other hand, the following relationships should also be valid in the stellar atmosphere from the stellar surface to the photosphere ($\tau = 2/3$) [see Eqs. (7.73) to (7.76)]:

$$
T^4 = \frac{3}{4}T_{eff}^4\left(\tau + \frac{2}{3}\right)
\tag{8.61}
$$

$$d\tau = -\bar{\kappa}\rho dr \qquad (8.62)$$

$$\frac{dr}{dM_r} = \frac{1}{4\pi r^2 \rho} \qquad (8.63)$$

$$\frac{dP}{dM_r} = -\frac{GM_r}{4\pi r^4} \qquad (8.64)$$

We also assume

$$\bar{\kappa} = \kappa_0 P^a T^b \qquad (8.65)$$

In the above equations, if we replace r by R, M_r by M, we get from Eqs. (8.62) to (8.65)

$$\frac{d\tau}{dP} = \frac{\bar{\kappa}R^2}{GM} = \frac{\kappa_0 P^a T^b R^2}{GM} \qquad (8.66)$$

Differentiating Eq. (8.61), we have

$$4T^3 dT = \frac{3}{4}T_{eff}^4 d\tau \qquad (8.67)$$

Substituting Eq. (8.66) into (8.67) and integrating, and further inserting the values $P_{\tau=2/3}$ and T_{eff} in the photosphere, we have

$$P_{\tau=\frac{2}{3}} = const\left(\frac{M}{R^2}T_{eff}^{-b}\right)^{\frac{1}{a+1}} \qquad (8.68)$$

Combining Eqs. (8.60) and (8.68) to eliminate $P_{\tau=2/3}$, and employing the relationship among effective temperature, luminosity and radius ($R^2 \sim L/T_{eff}^4$), and taking $n = 3/2$, we have

$$\lg T_{eff} = \frac{0.75a - 0.25}{b + 5.5a + 1.5}\lg L + \frac{0.5a + 1.5}{b + 5.5a + 1.5}\lg M + const \qquad (8.69)$$

Equation (8.69) gives the positions in the HR diagram for internally convective stars with given masses and chemical compositions, which is the Hayashi line, but the coefficients a and b are yet to be determined. It is already known that the atmosphere for a completely convective star is cold, so the Hayashi line should be in a region with lower values of T_{eff}. For an atmosphere with a relatively lower temperature, the opacity is mainly contributed by negative hydrogen ions (H^-), so $\bar{\kappa}$ is dependent on n_H and n_e (where n_H is the number of H^- ions per unit volume and n_e is the number of electrons per unit volume), that is, κ is dependent on the chemical composition and the temperature. In fact, for $T < 5000$ K, hydrogen cannot be ionized. Here, the electron number is mainly contributed by metal ionisation. Roughly speaking, $a \approx 1$ and $b \approx 3$. Using these values in Eq. (8.69), we have

$$\lg T_{eff} = 0.05\lg L + 0.2\lg M + const \qquad (8.70)$$

From Eq. (8.70), we notice that $\lg T_{eff}$ is only slightly influenced by $\lg L$ due to the small coefficient. This implies an almost vertical Hayashi line for very small values of T_{eff}. Moreover, the position of the Hayashi line is dependent on the stellar mass.

In the above derivations, we have assumed that the condition $(d\ln T / d\ln P) = \nabla_{ad}$ is fulfilled throughout the interior of the star. In fact, there are partially ionized H and He regions in the envelope of the star close to the photosphere for which there are discrepancies between the ∇_{con} and ∇_{ad} values. Because of this, the Hayashi line is a curve instead of a vertical line. Furthermore, the position of the Hayashi line is also dependent on the mixing length l in the convection theory and the pressure scale height H_P. Figures 8.6 and 8.7 show the

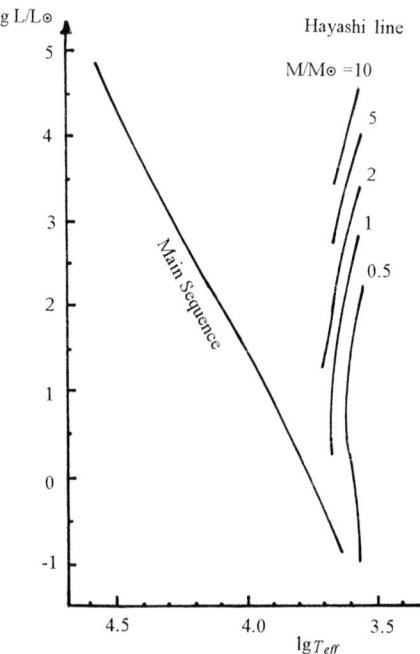

Fig. 8.6 The Hayashi line in the HR diagram for stars with $M = 0.5, ..., 10M_\odot$, $X_H = 0.739$, $X_2 = 0.021$ (according to Ezer and Cameron 1967).

relationships between the position of the Hayashi line and stellar mass, and the ratio between the mixing length and the pressure scale height, respectively.

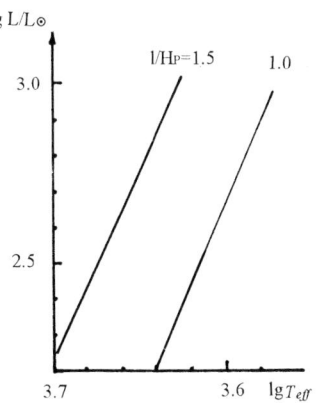

lg L/L⊙

Fig. 8.7 The Hayashi line in the HR diagram for stars with a mass of 5M⊙ with different l/H_p values (according to Henyey *et al.* 1965).

(2) Separating the HR diagram into "allowed" and "forbidden" regions
The HR diagram can be separated into "allowed" and "forbidden" regions with the Hayashi line as the boundary. For a star with a fixed mass and chemical composition, if it attains hydrostatic equilibrium, it has to be situated in the "allowed" region to the left of the Hayashi line, and will not stay in the "forbidden" region to the right of the Hayashi line.

In the discussions above, it has been explained that a star with fixed mass and chemical composition will fall on the Hayashi line if everywhere from its center to the photosphere is convective. If some regions are not entirely convective, its position on the HR diagram might be near the Hayashi line. For such a star, an average temperature gradient $\bar{\nabla}$ can be given. Besides, we can suppose there are many possible states, each corresponding to one $\bar{\nabla}$ value. These different states can be described by a series of polytropic models each having a different value of $n(\bar{\nabla} = 1/(n+1))$ where the model with $n = 3/2$ corresponds to the star on the Hayashi line. We will discuss the position of these models in the HR diagram using methods in the above discussions.

Taking logarithm of Eq. (8.58), we have

$$\lg P = \lg C'(n) - (1+n)\lg\mu + (1-n)\lg M + (3n-3)\lg R + (1+n)\lg T$$
$$(8.71)$$

Substituting values of $P_{\tau=2/3}$ and T_{eff} in the photosphere, we obtain

$$\lg P_{\tau=\frac{2}{3}} = \lg C'(n) - (1+n)\lg\mu + (1-n)\lg M + (3n-3)\lg R + (1+n)\lg T_{eff}$$
$$(8.72)$$

On the other hand, from the logarithm of Eq. (8.68) for the atmosphere, we get

$$\lg P_{\tau=\frac{2}{3}} = \frac{1}{a+1}\lg M - \frac{2}{a+1}\lg R - \frac{b}{a+1}\lg T_{eff} + C_1 \qquad (8.73)$$

Eliminating $P_{\tau=2/3}$ from Eqs. (8.72) and (8.73), and using the relationship among effective temperature, luminosity and radius

$$\lg R = \frac{1}{2}\lg L - 2\lg T_{eff} + C_2 \qquad (8.74)$$

and further taking $a = 1$ and $b = 3$, we have

$$\lg T_{eff} = \alpha_1 \lg L + \alpha_2 \lg \mu + \alpha_3 \lg M + \alpha_4 \lg C'(n) + \alpha_5 C_2 + \alpha_6 C_1 \qquad (8.75)$$

where

$$\alpha_1 = \frac{2-n}{13-2n} \ , \quad \alpha_2 = \frac{2+2n}{13-2n} \ , \quad \alpha_3 = \frac{2n-1}{13-2n}$$
$$\alpha_4 = \frac{-2}{13-2n} \ , \quad \alpha_5 = \frac{4-2n}{13-2n} \ , \quad \alpha_6 = \frac{2}{13-2n} \qquad (8.76)$$

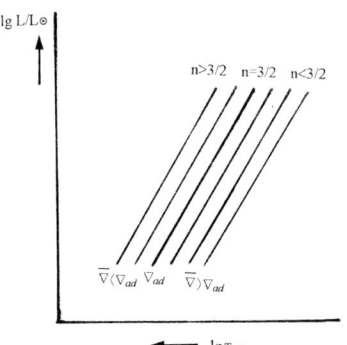

Fig. 8.8 Positions of star models with different values of n in the HR diagram. The model with $n = 3/2$ lies precisely on the Hayashi line.

Equation (8.75) gives the positions of star models with different values of n in the HR diagram. The model with $n = 3/2$ lies precisely on the Hayashi line (see Fig. 8.8).

In the following, we will discuss the positions of models with different n values in the HR diagram. We first note the change of α_1 with respect to n in Eq. (8.75) as

$$\left(\frac{d\alpha_1}{dn}\right)_{n=\frac{3}{2}} = -0.09 \qquad (8.77)$$

This shows that a small change in n will not affect α_1 near the Hayashi line ($n = 3/2$), i.e., the lines with different n will be parallel to the Hayashi line. We now study the relationship between T_{eff} and n. Without loss of generality, we take $L = \mu = M = 1$ in Eq. (8.75). In this way, the variation of $\lg T_{eff}$ with n mainly comes from the change of $\lg C'(n)$ with n. From Eq. (8.59), it is not difficult to arrive at

$$\left(\frac{d\lg C'(n)}{dn}\right)_{n=1,2} = -6.7 \qquad (8.78)$$

so

$$\frac{\Delta\lg T_{eff}}{\Delta n} \simeq \left(\alpha_4 \frac{d\lg C'(n)}{dn}\right)_{n=\frac{3}{2}} = 1.34 \qquad (8.79)$$

This relation shows that n decreases when T_{eff} decreases and vice versa. Therefore, on the right of the Hayashi line, n decreases when T_{eff}

decreases; and on the left of the Hayashi line, n increases when T_{eff} increases. Furthermore, since $\bar{\nabla} = 1/(1 + n)$, $\bar{\nabla}$ will increase to the right of the Hayashi line and decrease to the left.

Since $\bar{\nabla} < \nabla_{ad}$ to the left of the Hayashi line, the non-convective regions inside the star become radiative equilibrium, since the condition for radiative equilibrium is $\bar{\nabla} < \nabla_{ad}$. Only when this is true will the average temperature gradient for the star have $\bar{\nabla} < \nabla_{ad}$.

On the other hand, since $\bar{\nabla} > \nabla_{ad}$ to the right of the Hayashi line, some regions inside the star become super adiabatic convective regions. To transfer the energy required for the radiation of the star, we only need a temperature gradient slightly larger than the adiabatic temperature gradient ($\bar{\nabla} - \nabla_{ad} \approx 10^{-7}$). If there is a super adiabatic region inside the star, the $\bar{\nabla} - \nabla_{ad}$ value for this super adiabatic region is relatively large, so the velocity of a convective element is also relatively large [$v \sim (\bar{\nabla} - \bar{\nabla}_{ad})^{1/2}$] and the energy transferred by convection is also large. As a result, the interior of the convective region is cool while the exterior becomes hot, so the equilibrium inside the star is upset. Nonetheless, the star will start to regulate at once, until $\bar{\nabla} = \nabla_{ad}$ is restored inside the star, i.e., the star returns to the Hayashi line. The time needed for the regulation is very small.

From the above discussions, when hydrostatic equilibrium is fulfilled and $\bar{\nabla} < \nabla_{ad}$ inside a star with a fixed mass and chemical composition, a polytropic model can be employed for the description. The star will be inside the "allowed" region to the left of the Hayashi line. If $\bar{\nabla} > \nabla_{ad}$ for some reasons, the hydrostatic equilibrium will be upset inside the star, and the star will be in the region to the right of the Hayashi line. Prompt regulation will occur in the star to restore hydrostatic equilibrium and to bring the star back to the Hayashi line. Therefore, we see that the region to the right of the Hayashi line is in fact a "forbidden" region for stars which fulfil hydrostatic equilibrium.

From the above discussions, the protostar formed by the collapse of the interstellar cloud should be in the region to the right of the Hayashi line since hydrostatic equilibrium has not been completely established. After a short time, however, when hydrostatic equilibrium is set up, and when the entire interior region is under convection, it will get onto the Hayashi line. At this time, the stellar object is called the pre main sequence star.

8.2.2 Evolution from the Hayashi Line to the Main Sequence

The pre main sequence star cannot have thermonuclear reactions because the internal temperature is too low. To compensate the energy radiated per second from the surface, the star has to contract to release the gravitational energy. According to the Virial theorem, part of the energy is used to compensate for radiation and part of the energy will be used to increase the thermal energy to raise the internal temperature. Since the inside has reached hydrostatic equilibrium, the star contracts in a quasi-static and uniform way, which is the most basic feature in the evolution of the pre main sequence star. Therefore, the time scale for the pre main sequence evolution is the Kelvin–Helmholtz time scale. From Eq. (7.117),

$$\tau_{KH} = \frac{GM^2}{RL} \tag{8.80}$$

From Eq. (8.80), it is easy to note that the more massive stars will have larger products of the luminosity L and the radius R ($L \sim M^{2.7-4.0}$, $R \sim M^{0.6}$), and thus faster evolution and shorter life in the pre main sequence stage. Those with smaller mass will have slower evolution and thus longer life in the pre main sequence stage. However, the Kelvin–Helmholtz time scale is far smaller than the nuclear time scale, so the life in the pre main sequence stage for both massive and low mass stars are shorter than the life in the main sequence. From Eq. (8.80), it can also be seen that a star with a fixed mass will have larger values of RL and evolve faster near the Hayashi line, and have smaller values of RL and evolve slower near the main sequence.

Figure 8.9 gives the evolutionary track of a $1M_\odot$ pre main sequence star in the HR diagram calculated by Sengbusch (1968). The star first moves down along the Hayashi line. With the decrease in the luminosity L, the internal temperature gradient diminishes ($\nabla \sim L_r$). When the temperature gradient near the center decreases below the adiabatic temperature gradient ∇_{ad}, a radiative equilibrium region is formed in the central part. At this time, the average temperature gradient is $\bar\nabla < \nabla_{ad}$ for the entire star, so the star starts to leave the Hayashi line to the left side where T_{eff} is higher (see previous discussions on the features of the Hayashi line). Figure 8.10 gives the temporal change of the internal structure of this star. In the figure, regions with clouds have convection. It can be seen that the central radiative equilibrium region grows with time after the star leaves the Hayashi line. The broken lines refer to $\log T = $ constant, which give the internal temperature distribution during contraction of the star. Their slopes change abruptly at point E. At this point, central hydrogen burning ignites and stellar evolution slows down. When central hydrogen burning ignites, the star has evolved onto the main sequence in the HR diagram.

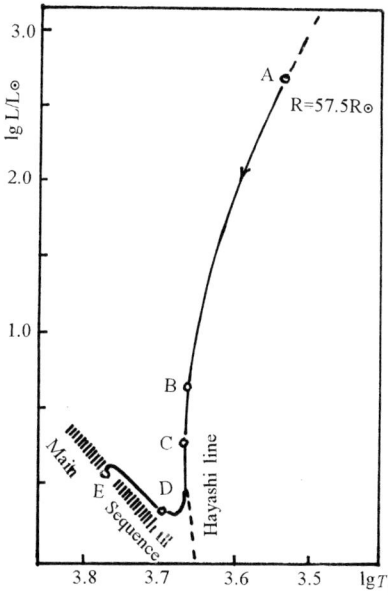

Fig. 8.9 The evolutionary track of a 1$M_\odot$ pre main sequence star in the HR diagram (according to Sengbusch 1968).

Figure 8.11 shows the evolutionary tracks in the HR diagram for stars with masses 0.5, ..., 15$M_\odot$ during the pre main sequence stage calculated by Iben (1965), and the points on the tracks give the age (see Table 8.1). It can be clearly seen that the time needed to evolve to the main sequence for massive stars is much shorter than those needed for low mass stars. When evolved to the main sequence, the massive stars will be in the upper left region of the main sequence, while the low mass

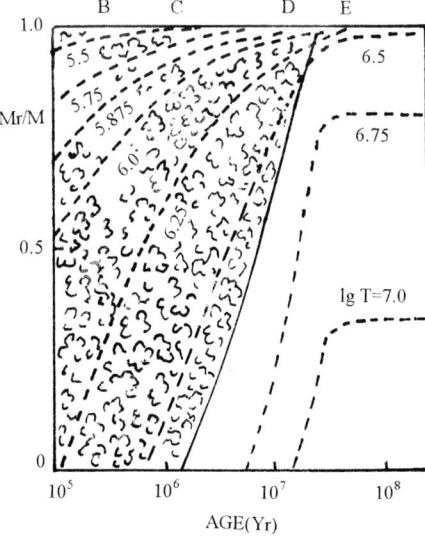

Fig. 8.10 The temporal change of the internal structure of a 1$M_\odot$ pre-main-sequence star in the HR diagram (according to Sengbusch 1968).

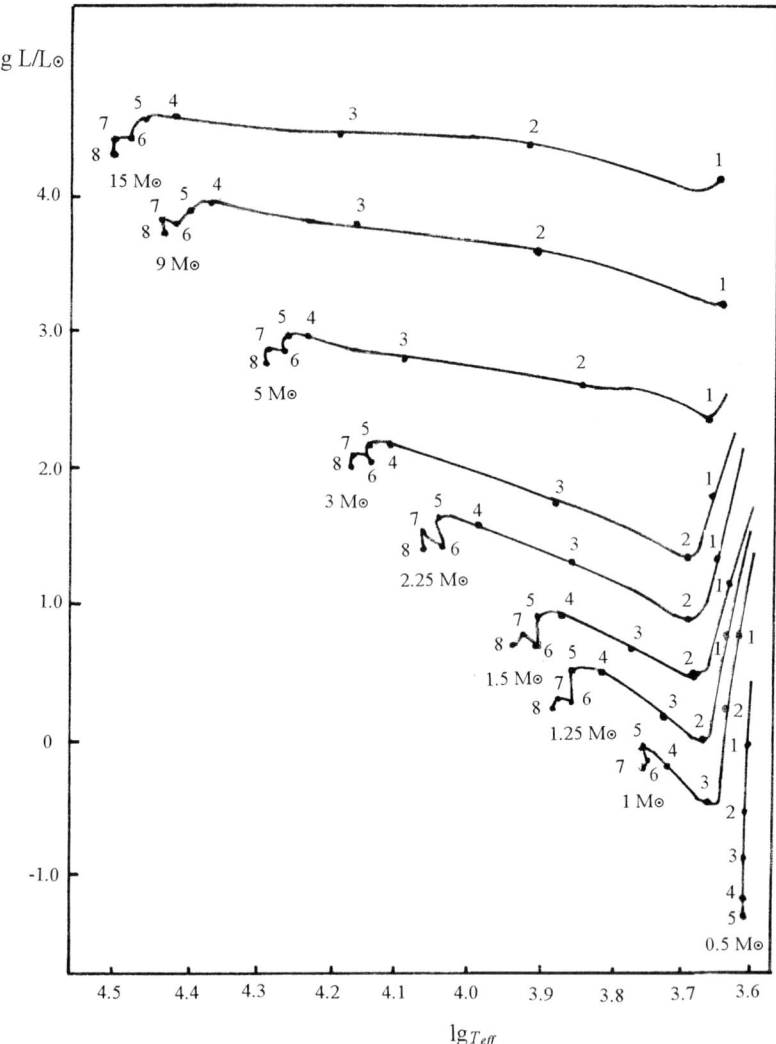

Fig. 8.11 The evolutionary tracks in the HR diagram for stars with masses 0.5, ..., 15M⊙ during the pre main sequence stage (according to Iben 1965). The numbers on the evolutionary tracks are the ages (see Table 8.1).

stars will be in the lower right region. Considering members of the same stellar cluster (i.e., stars formed at the same time), calculations show that the low mass stars are still under contraction and evolving towards the main sequence when massive stars have already evolved to the main sequence. This has been confirmed by observations. In some young stellar clusters, such as NGC2264, only those stars with luminosities L greater than a certain value (i.e., the massive stars) are positioned on the main sequence, while those with smaller L are positioned to the right of the main sequence and are still undergoing the pre main sequence contraction.

Table 8.1 Evolutionary age* (according to Iben 1965).

Points on the evolutionary track	mass ($M_\odot$)								
	15.0	9.0	5.0	3.0	2.25	1.5	1.25	1.0	0.5
1	6.740 (2)	1.443 (3)	2.936 (4)	3.420 (4)	7.862 (4)	2.347 (5)	4.508 (5)	1.189 (5)	3.195 (5)
2	3.766 (3)	1.473 (4)	1.069 (5)	2.078 (5)	5.940 (5)	2.363 (6)	3.957 (6)	1.508 (6)	1.786 (6)
3	9.350 (3)	3.645 (4)	2.001 (5)	7.663 (5)	1.883 (6)	5.801 (6)	8.800 (6)	8.910 (6)	8.711 (6)
4	2.203 (4)	6.987 (4)	2.860 (5)	1.135 (6)	2.505 (6)	7.584 (6)	1.155 (7)	1.821 (7)	3.092 (7)
5	2.657 (4)	7.992 (4)	3.137 (5)	1.250 (6)	2.818 (6)	8.620 (6)	1.404 (7)	2.529 (7)	1.550 (8)
6	3.984 (4)	1.019 (5)	3.880 (5)	1.465 (5)	3.319 (6)	1.043 (7)	1.755 (7)	3.418 (7)	–
7	4.585 (4)	1.195 (5)	4.559 (5)	1.741 (6)	3.993 (6)	1.339 (7)	2.796 (7)	5.016 (7)	–
8	6.170 (4)	1.505 (5)	5.759 (5)	2.514 (6)	5.855 (6)	1.821 (7)	2.954 (7)	–	–

*Numbers in brackets represents powers of 10, e.g., 2.347(5) $\equiv$ 2.347 $\times$ 10^5.

We should also point out that thermonuclear reactions through deuterium (^{2}H) and lithium (Li) burning may occur before hydrogen burning commences during the pre main sequence stage where the star contracts and the internal temperature rises continually. It is known that the minimum temperature for p-p reaction of hydrogen is 8×10^6 K, but that for deuterium burning is 10^6 K, 3×10^6 K for Li, 5×10^6 K for Be and 6×10^6 K for B. It is usually supposed that the deuterium inside a star with $M \leq 2M_\odot$ can be exhausted through the ^{2}H(p, γ)^{3}He reaction in the pre main sequence stage. At the same time, the reactions for Li in the envelope during the pre main sequence stage will exhaust all Li. However, the abundance for these elements are extremely small in a star, so the reactions will not alter the scenario that the radiation is mainly supported by energy released from contraction during the pre main sequence stage. Nevertheless, these nuclear reactions will change the abundance of these elements in the star which in turn change the abundance ratios of these elements with other elements, e.g. the Li/Be ratio. According to the observations by Goldberg *et al.* (1960), the Li/Be ratio is 1/30 at the solar surface, 6 in meteorites and 5–15 on Earth. To explain the different values on the solar surface and inside the solar system, Greenstein and Richardson (1951) proposed that Li is exhausted since the temperature in the convective layer in the envelope has reached the temperature for Li burning. From observations of Bonsack and Greenstein (1960) and Bonsack (1961), the abundance ratios of Li and other metals in T Taurus type stars are greater than those on the solar surface by a factor of about 100, and are close to the values obtained inside the solar system. This has led people to think that T Taurus type stars are likely stars during the pre main sequence contraction.

8.2.3 Properties of Contraction for the Pre main sequence Star

The stars in the pre main sequence stage have low internal temperatures so they cannot have thermonuclear reactions. At this time, the output energy is mainly supported by the released gravitational energy, and the star is under homogenous contraction. The internal structure of such a star can be studied by the polytropic model (see §7.8). We can derive the characteristics of changes in physical quantities inside the pre main sequence star through the known relations from the polytropic model.

From Eq. (7.138), we know

$$r = \frac{1}{A} Z \tag{8.81}$$

$$R = \frac{1}{A} Z_n \tag{8.82}$$

where Z_n is the Z value at the stellar surface, R is the stellar radius. When the polytropic index n is fixed, Z_n is a constant. From Eqs. (8.81) and (8.82), we get

$$r = R \frac{Z}{Z_n} \tag{8.83}$$

Employing Eq. (8.83), we can study the characteristics of changes in r for a certain Z value (corresponding to a certain M_r value inside the star). From Eq. (8.83), for a fixed Z, we have

$$\frac{\dot{r}}{r} = \frac{\dot{R}}{R} = const \tag{8.84}$$

which is valid for any Z value inside the star. Therefore, when the star contracts, the relative changes of the radius for all layers inside the star are the same, and are equal to the relative change of the radius for the entire star.

Further considering changes in other physical quantities, we can write Eq. (8.84) in the form

$$\frac{\partial}{\partial M_r}\left(\frac{\dot{r}}{r}\right) = 0 \tag{8.85}$$

so

$$\frac{\partial}{\partial M_r}\left(\frac{\dot{r}}{r}\right) = \frac{\partial}{\partial M_r}\left(\frac{\partial \ln r}{\partial t}\right) = \frac{\partial}{\partial t}\left(\frac{\partial \ln r}{\partial M_r}\right)$$

$$= \frac{\partial}{\partial t}\left(\frac{1}{r}\frac{1}{4\pi r^2 \rho}\right)$$

$$= \frac{1}{4\pi r^2 \rho}\left(-3\frac{\dot{r}}{r} - \frac{\dot{\rho}}{\rho}\right) = 0 \tag{8.86}$$

from which we get

$$\frac{\dot{\rho}}{\rho} = -3\frac{\dot{r}}{r} = const \tag{8.87}$$

From this we know the relative changes in the density for all layers inside the star are also the same, and the relation between the relative changes in the density and the relative changes in the radius fulfil Eq. (8.87).

According to $P = \int_{M_r}^{M} \frac{GM_r}{4\pi r^4}\,dM_r$, we have

$$\dot{P} = \int_{M_r}^{M} \frac{\partial}{\partial t}\left(\frac{1}{r^4}\right)\frac{GM_r}{4\pi}\,dM_r$$

$$= \int_{M_r}^{M} \left(-\frac{4\dot{r}}{r}\right)\frac{GM_r}{4\pi r^4}\,dM_r \tag{8.88}$$

Since $\frac{\dot{r}}{r} = const$, $\frac{GM_r}{4\pi r^4} = -\frac{dP}{dM_r}$, the integration of Eq. (8.88) gives

$$\frac{\dot{P}}{P} = -\frac{4\dot{r}}{r} = const \tag{8.89}$$

Therefore, the relative changes in the pressure for all layers inside the star are also the same, and the relation between the relative changes in the pressure and the relative changes in the radius fulfil Eq. (8.89). From the equation of state for an ideal gas $T \sim P/\rho$, we obtain

$$\frac{\dot{T}}{T} \approx \frac{\dot{P}}{P} - \frac{\dot{\rho}}{\rho} = \frac{\dot{r}}{r}(-4 + 3) \tag{8.90}$$

so

$$\frac{\dot{T}}{T} \approx -\frac{\dot{r}}{r} = const \tag{8.91}$$

Further considering the variation in ε_g during contraction of the star, we have from Eq. (7.12)

$$\varepsilon_g = -C_P \dot{T} + \frac{\delta}{\rho} \dot{P} \tag{8.92}$$

Since $\nabla_{ad} = (\delta P / C_P \rho T) = 0.4$, the above equation can be written as

$$\begin{aligned}
\varepsilon_g &= -C_P T \left(\frac{\dot{T}}{T} - \nabla_{ad} \frac{\dot{P}}{P} \right) = -C_P T \left(-\frac{\dot{R}}{R} + 4 \nabla_{ad} \frac{\dot{R}}{R} \right) \\
&= -\frac{3}{5} C_P T \frac{\dot{R}}{R}
\end{aligned} \tag{8.93}$$

which shows that $\varepsilon_g > 0$ for $\dot{R} < 0$ (contraction), i.e., the thermal energy increases through contraction and $\varepsilon_g \sim T$ so the internal temperature increases during the contraction.

We proceed to study the characteristics of change in the luminosity L:

$$L_r = \int_0^{M_r} \varepsilon_g dM_r = -\frac{3}{5} C_P \frac{\dot{R}}{R} \int_0^{M_r} T dM_r \tag{8.94}$$

For the stellar surface, we have

$$L = \int_0^{M} \varepsilon_g dM_r = -\frac{3}{5} C_P \frac{\dot{R}}{R} \int_0^{M} T dM_r \tag{8.95}$$

From Eqs. (8.94) and (8.95), we have

$$\frac{\dot{L}_r}{L_r} = \frac{\dot{L}}{L} = const \qquad (8.96)$$

that is, when the star contracts, the relative changes in the luminosity for all layers inside the star are also the same, and equal to the relative change in the luminosity for the entire star.

8.3 MAIN SEQUENCE

8.3.1 Zero Age Main Sequence

Astronomical observations have shown that most stars fall onto a strip in the HR diagram stretching from the top left to bottom right, which is called the main sequence strip. Furthermore, the mass and temperature of the stars decrease from the top left to bottom right of the main sequence strip. There are empirical relationships between the mass M and the luminosity L, and between M and the radius R for main sequence stars, known as the mass–luminosity relation and the mass–radius relation, respectively (see §1.5).

The physical meaning behind the main sequence strip is revealed by the development of theories on stellar structure and evolution. According to discussions of §8.2, the internal temperature of pre main sequence stars are relatively low so there are no thermonuclear reactions. The radiation from the stars are supported by the gravitational energy released during contraction, part of which will support the external radiation while another part of which will increase the internal thermal energy or the temperature. When the central temperature rises to a point where thermonuclear reactions can take place, the star will reach the left edge of the main sequence strip in the HR diagram. This line at the left edge of the main sequence strip is known as the zero age main sequence. Thus, the zero age main sequence is the position in the HR diagram representing stars with same chemical compositions but different masses, and with hydrogen burning ignited in their central regions.

The stars evolve with the Kelvin–Helmholtz time scale during pre main sequence contraction. Starting from the zero age main sequence, there are thermonuclear reactions inside and, in particular, the speed of thermonuclear reactions for hydrogen is extremely slow, so the stars evolve with the nuclear time scale. The time spent for a star with any mass in the pre main sequence stage is much smaller than the time spent in the main sequence and post main sequence stage. In this way, the pre main sequence stage can almost be "neglected", and the age of the star is counted since there are thermonuclear reactions. This is the reason why the left edge of the main sequence is referred to as the "zero age" main sequence.

The position of the zero age main sequence is dependent on the chemical composition of the star. Figure 8.12 gives the position of the zero age main sequence in the HR diagram for stars with different heavy element abundances.

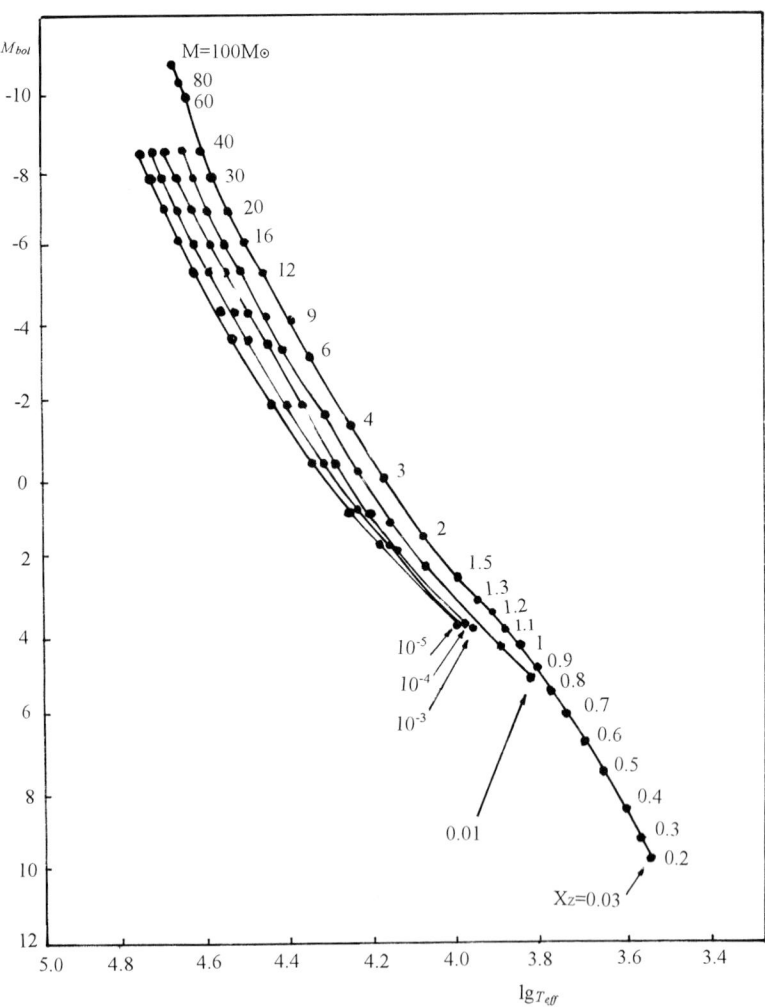

Fig. 8.12 The zero age main sequence in the HR diagram for stars with different heavy element abundances (according to Vanderlinden 1982).

The rate of pre main sequence evolution and the landing position on the zero age main sequence will be different for stars with different masses. Massive stars will evolve faster and land on the upper part of the zero age main sequence, while low-mass stars will evolve slower and land on the lower part of the zero age main sequence. Figure 8.12 also gives the distribution of stars with different masses on the zero age main sequence.

8.3.2 Properties of Main Sequence Stars

It is relatively easy to compute main sequence star models under hydrostatic equilibrium with uniform chemical compositions and hydrogen burning. Some properties of zero age main sequence stars can be obtained from numerical model calculations. For example, the model calculations of stars with the chemical composition $X_H = 0.685$, $X_{He} = 0.294$ and $X_Z = 0.021$ and different masses give the relations between the mass M and the luminosity L and between the mass M and the radius R as shown by the solid lines in Figs. 1.10 and 1.11. From the curve in Fig. 1.10, the relation between the mass M and the luminosity L for main sequence stars (or mass–luminosity relation in short) can be expressed as

$$L \approx M^\alpha \tag{8.97}$$

and the average value of α for the whole mass range is about 3.2. For $M = 1.0, ..., 10M_\odot$ the average value of α is 3.88, while in the larger range $M = 1.0, ..., 40M_\odot$ it is 3.35. From the curve in Fig. 1.11, the relation between the mass M and the radius R for main sequence stars (or mass–radius relation in short) can be expressed as

$$R \approx M^\beta \tag{8.98}$$

and the average value of β for stars in the upper half of the main sequence is obtained to be $\beta = 0.57$ while that for stars in the lower half is $\beta = 0.8$.

Figure 8.13 gives the relations between the core temperature T_c and the core density ρ_c for different masses of zero age main sequence stars. The relations show that with the decrease of the mass of the main sequence star, the core temperature T_c of the star decreases at a rate four times faster while the core density increases at a rate 100 times faster. For a higher temperature, the energy generation rate from

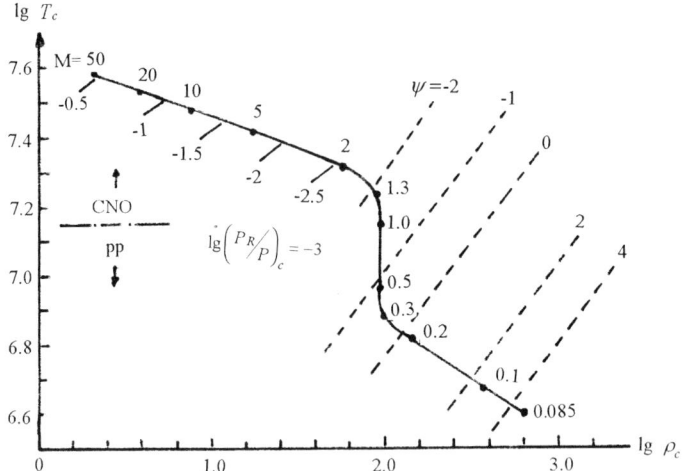

Fig. 8.13 The relation between the core temperature T_c (K) and the core density ρ_c (gcm^{-3}) for different masses of zero age main sequence stars. The figure also shows the temperature separating the CNO cycle reaction and the pp reaction chain, the ratio between the radiation pressure and the total pressure, and the electron gas degeneracy represented by the broken lines (ψ = constant) (adapted from Kippenhahn and Weigert 1991).

hydrogen burning will be dominated by the energy generation rate ε_{CNO} from CNO cycle reactions (since $\varepsilon_{CNO} \sim T^{20}$, and $\varepsilon_{pp} \sim T^5$). Figure 8.13 gives the temperature separating the energy generation dominated by the ε_{CNO} and ε_{pp}. Figure 8.14 shows the ratios between the energy generation rate ε_{CNO} through the CNO cycle reaction and the total energy generation rate ε in the core of some main sequence stars. From Fig. 8.14, it can be seen that ε_{CNO} dominates for main sequence stars with mass greater than 1.5M$_\odot$ while ε_{pp} dominates for main sequence stars with mass smaller than 1.3M$_\odot$.

When the core density ρ_c increases quickly with the decrease of the mass M of low mass main sequence stars, the central electron gas will become degenerate, and the degeneracy grows with the value of ρ_c. The broken line in Fig. 8.13 gives the regions with different degeneracy parameters ψ, and also gives the change of the ratio between the radiation pressure P_R and the total pressure P with respect to the mass. Since the radiation pressure $P_R \sim T^4$, the contribution of the radiation pressure to the total pressure in low mass stars is extremely small. For a 1.3M$_\odot$ star, the radiation pressure is only one-thousandth of the total pressure.

Figure 8.15 shows the distribution of convective regions inside zero age main sequence stars of different masses. It can be seen that there are convective regions in the core of stars with M > 1M$_\odot$ while the outer region is under radiative equilibrium. For stars with M < 1M$_\odot$, the core is under radiative equilibrium while the outer region has convective regions. The results in Fig. 8.15 are obtained using the Schwarschild criterion ($\nabla_R > \nabla_{ad}$) as the criterion for convective instabilities. Figure 8.16 shows the variation of ∇_R, ∇_{ad} and the real temperature gradient ∇ with temperature for a 1M$_\odot$ star (Fig. 8.16a)

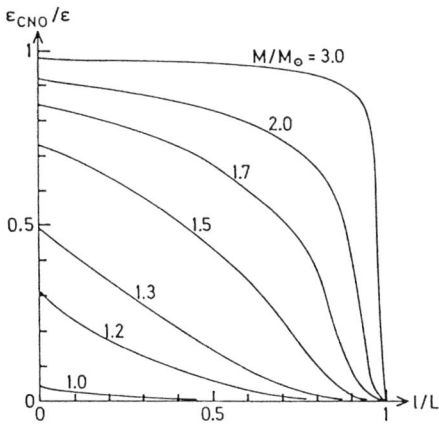

Fig. 8.14 The ratios between the energy generation rate ε_{CNO} through the *CNO* cycle reaction and the total energy generation rate ε in the core of some main sequence stars, where l is the local luminosity (according to Kippenhahn and Weigert 1991).

and a $10M_\odot$ star (Fig. 8.16b). In massive stars, there are *CNO* cycle reactions in the core and $\varepsilon_{CNO} \sim T^{20}$, so the energy generation rate and L_r are very large, and thus $\nabla_R \left(\sim \frac{\bar{\kappa} L_r}{4\pi r^2}\right)$ is very large and is greater than ∇_{ad}. Therefore, the core is convective in massive main sequence stars. In low mass stars, the *pp* reactions dominate in the core and $\varepsilon_{pp} \sim T^5$, so the energy generation rate and L_r are relatively small, and thus $\nabla_R < \nabla_{ad}$. Therefore, the core is not convective for stars with mass smaller than $1M_\odot$. Nonetheless, the above boundary of $1M_\odot$ is

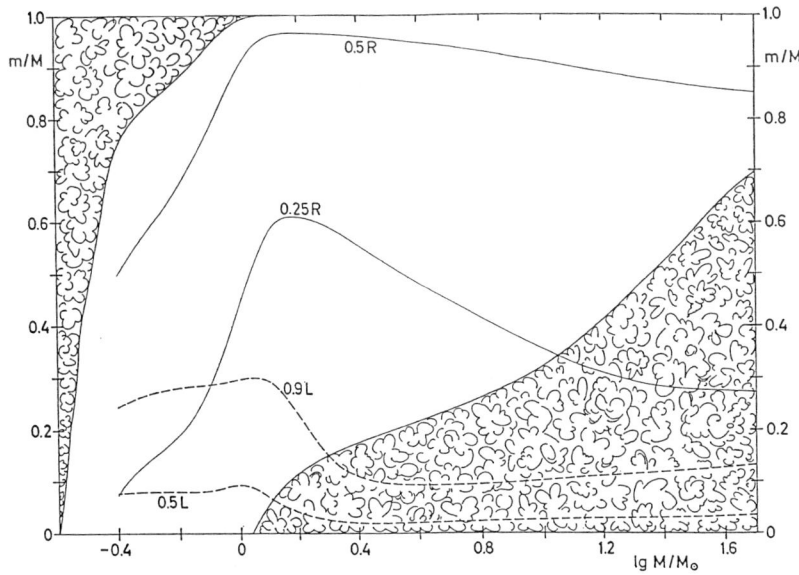

Fig. 8.15 The convective regions (regions with "clouds") inside zero age main sequence stars of different masses. The distributions inside the stars with 0.5R, 0.25R, 0.9L and 0.25L are also shown (according to Kippenhahn and Weigert 1991).

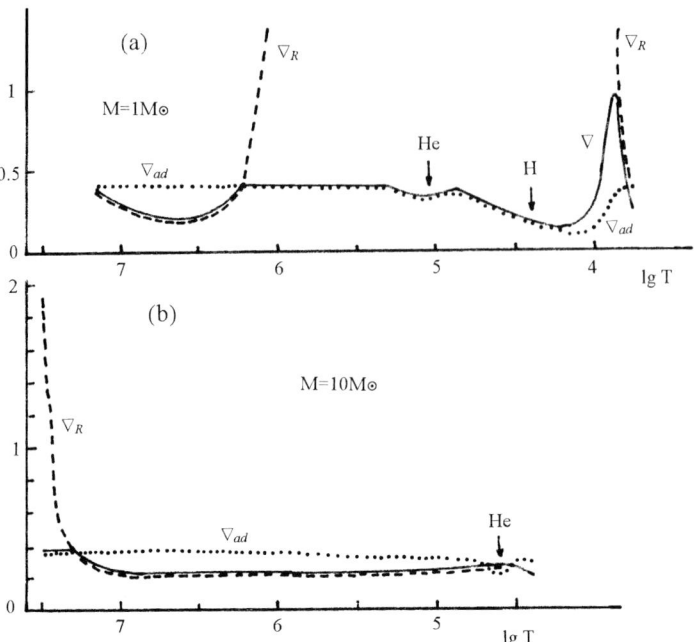

Fig. 8.16 The variation of some temperature gradients with temperature for zero age main sequence stars: (a) a $1M_\odot$ star (b) a $10M_\odot$ star. The arrows show the ionization regions for the corresponding elements (adapted from Kippenhahn and Weigert 1991).

not absolute, since the boundary will also be dependent on the chemical composition of the star and, in particular, how we handle the convection and whether we consider convective overshoot.

8.3.3 Main Sequence Strip

Hydrogen burning in the core starts from the zero age main sequence. At this time, the energy source of the star is mainly the nuclear energy released in the fusion of hydrogen into helium. At the same time, the star will leave the zero age main sequence and moves rightward to the Hayashi line. When hydrogen is completely burnt in the core of the star, the evolutionary stage in the main sequence is completed. Therefore, the main sequence has a certain width in the HR diagram which is determined by the positions of the stars with different masses when their hydrogen burning is completed (see Fig. 8.17).

Since hydrogen is the most abundant element in stars, and the fusion of hydrogen is extremely slow, the stars will spend more than 90% of their time for core hydrogen burning, i.e., on the main sequence evolutionary stage. This also explains why most stars are observed to be on the main sequence. Moreover, main sequence stars are stable and fulfil hydrostatic equilibrium, so they undergo quasi-static evolution.

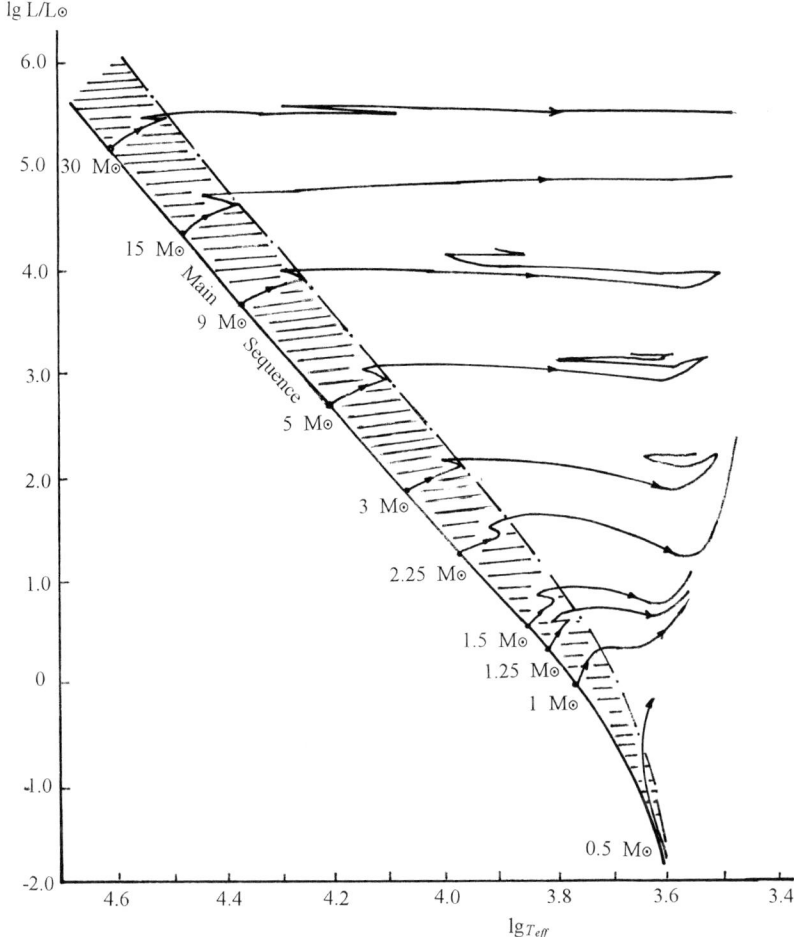

Fig. 8.17 The main sequence strip in the HR diagram.

From the mass–luminosity relation $M \sim L^{3.2}$ for main sequence stars, we see that hydrogen burning is faster for more massive stars so they leave the main sequence towards the red giants at a faster rate. Representing the total mass of hydrogen in the core of the star as M_{Hc} and the time needed for the complete burning of hydrogen in the core as τ_H, we have

$$\tau_H \sim -\frac{M_{Hc}}{dM_{Hc}/dt} \qquad (8.99)$$

Apparently, M_{Hc} is proportional to the total mass M of the star, i.e., $M_{Hc} \sim M$, and $-(dM_{Hc}/dt)$ is proportional to the total luminosity of the star, i.e., $-(dM_{Hc}/dt) \sim L \sim M^{3.2}$, from Eq. (8.99), we have

$$\tau_H \sim \frac{M}{M^{3.2}} = M^{-2.2} \tag{8.100}$$

This implies that the time τ_H needed for the complete burning of hydrogen in the core of more massive stars is shorter or they leave the main sequence earlier. Therefore, for the same evolutionary age, the distribution of stars with the same chemical composition but different masses on the HR diagram is like that in Fig. 8.18, which is known as an "isochron". We know that stellar clusters in the galaxy are composed of stars with the same chemical composition and evolutionary age but different masses. Therefore, their distributions in the HR diagram resemble the isochron in Fig. 8.18. In other words, starting from certain points on the main sequence (called turn off points), brighter stars (or the more massive stars) will leave the main sequence earlier to the right of the main sequence, while stars with lower masses will still stay on the main sequence. The altitude of turn off points will depend on the age of the stellar clusters. For a larger age, stars with lower masses can leave the main sequence so the turn off point will move downhill the main sequence. The above can satisfactorily explain the HR diagrams observed for some galactic stellar clusters. Figure 1.8 gives the HR diagram for some galactic stellar clusters constructed by Hayashi *et al.* according to the information provided by Sandage (1962). From the different turn off points for these stellar clusters, we know they have different ages. NGC2362 and Perseus $h + \chi$ are relatively young stellar clusters, while NGC188 and M67 are relatively old stellar clusters. Figure 1.9 gives the

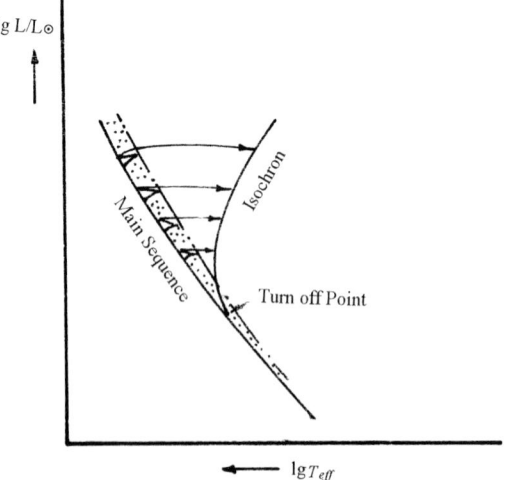

Fig. 8.18 An isochron in the HR diagram.

HR diagram for some globular clusters with extremely poor metal abundance constructed using the information of Sandage (1962), in which all the stars are population II stars. It can be seen that the isochrons for these globular clusters are very similar, which shows that the age of these clusters are also similar.

It should be pointed out that the theoretical width of the main sequence strip, the shape of the isochrons and the positions of the turn off points are all related to whether we have considered stellar wind loss and convective overshoot in the model calculations. The stellar structure and evolution model, taking into account the mass loss due to stellar wind, will have a lower core temperature and thus a lower hydrogen fusion reaction rate in the core, a longer period for hydrogen burning and a wider main sequence strip. This effect is more prominent for massive stars ($M > 10M_\odot$). For intermediate and low mass stars, this effect is less prominent due to weaker stellar wind in the main sequence stage.

Convective overshoot will have significant effects on the stellar structure and evolution of massive and intermediate massive stars. It enlarges the central convective core to enable more hydrogen to participate in the fusion reactions, so the main sequence stage will be lengthened and the main sequence strip becomes wider.

8.3.4 Other Main Sequences

In the earlier section, the positions on the HR diagram of stellar models with uniform chemical compositions but different masses and when core hydrogen burning starts have been defined as the main sequence or the hydrogen main sequence (H-MS). This concept can be extended to positions on the HR diagram of stellar models with uniform chemical compositions but different masses when core helium burning starts which are defined as the helium main sequence or He-MS, and similarly to positions on the HR diagram of stellar models when core carbon burning starts which are defined as the carbon main sequence or C-MS. The hydrogen main sequence is the most important since hydrogen is the most abundant element and also the temperature of hydrogen burning is the lowest (1.5–10×10^7 K) so the reaction rate is slow and the time spent by the star on hydrogen burning is the longest. Observations have in fact shown that more than 90% of stars are on the hydrogen main sequence. The helium main sequence and the carbon main sequence have practical meanings only for understanding of the final stage evolution of the stars.

A massive star becomes a helium star when hydrogen is exhausted in the core and the envelope is completely lost due to a strong stellar wind. When core helium burning starts, the star will fall on the helium main sequence on the HR diagram. For a low mass star, almost the entire star is convective so a C-O star is formed after hydrogen burning. When core carbon burning starts, the star will fall on the carbon main sequence on the HR diagram. Since helium burns at a higher temperature ($1–2 \times 10^8$ K), the helium reaction rate is much greater than the hydrogen reaction rate. Therefore, a star spends much less time on the helium main sequence than on the hydrogen main sequence. Nevertheless, the time spent on the helium main sequence is still long enough to be observed. For example, the Wolf–Rayet (WR) star and the low-mass stars on the horizontal branch are helium main sequence stars. Carbon main sequence stars are formed after helium burning completes in the stars. However, carbon burns at a higher temperature ($>7 \times 10^8$ K), and the reaction is even faster, so the time spent by a star on the carbon main sequence stage is extremely small and carbon main sequence stars are difficult to observe.

On the HR diagram, helium main sequence stars and carbon main sequence stars are situated on the left of the hydrogen main sequence stars, where the temperature is higher (Fig. 8.19). If stars with similar masses on the hydrogen main sequence, helium main sequence and carbon main sequence are compared, it can be found that the luminosities of the helium main sequence stars are higher than those

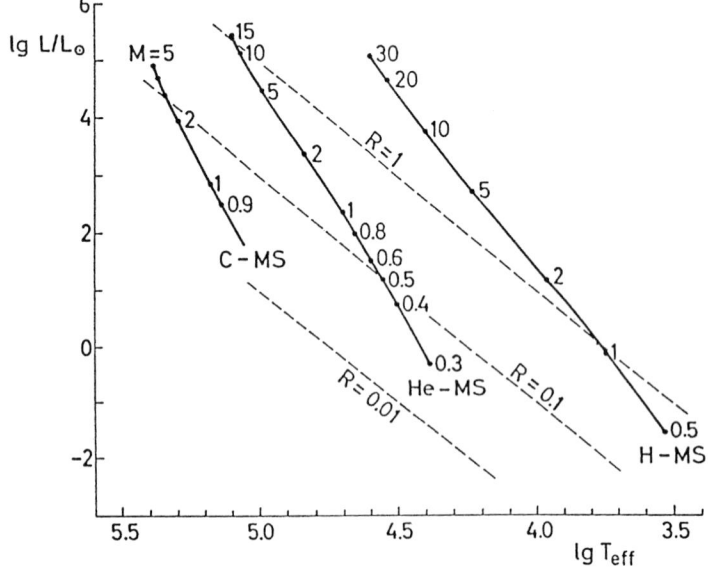

Fig. 8.19 The distribution of the hydrogen main sequence (H-MS, $X_H = 0.685$, $X_{He} = 0.294$), helium main sequence (He-MS, $X_H = 0$, $X_{He} = 0.979$), and the carbon main sequence (C-MS, $X_H = X_{He} = 0$, $X_C = X_O = 0.497$) on the HR diagram. The numbers on the solid lines are the masses of the stars (in units of $M_\odot$) and the broken lines refer to the constant radius R. (adapted from Kippenhahn and Weigert 1991).

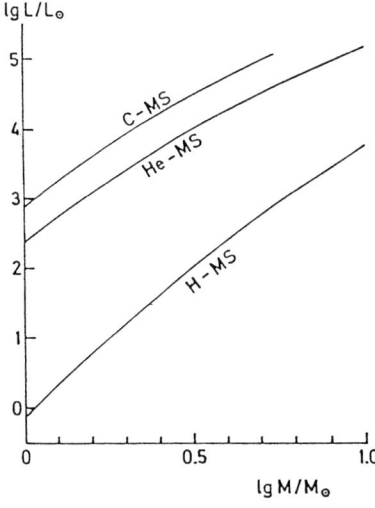

Fig. 8.20 The mass–luminosity relation for the hydrogen main sequence, helium main sequence and the carbon main sequence (according to Kippenhahn and Weigert 1991).

of the hydrogen main sequence stars but lower than those of the carbon main sequence stars (Figs. 8.19 and 8.20). At the same time, the radii of helium main sequence stars are smaller than those of hydrogen main sequence stars but larger than those of carbon main sequence stars (Fig. 8.19). Figure 8.21 gives the relation between core temperature and core density for stars in the hydrogen main sequence, helium main sequence and the carbon main sequence. It can be seen that the core temperature T_c and the core density ρ_c for helium main sequence stars are far greater than those of hydrogen main sequence stars, but far smaller than those of carbon main sequence stars.

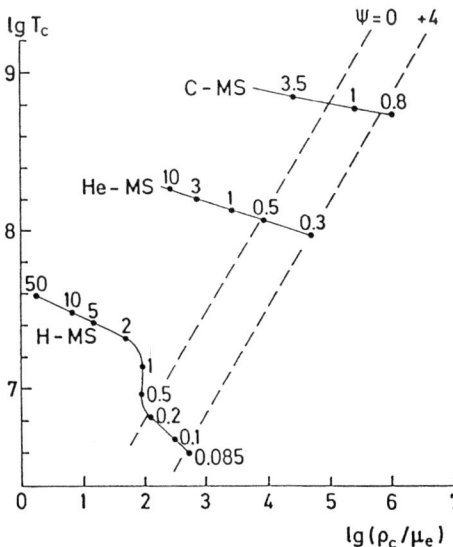

Fig. 8.21 The relation between the core temperature and the core density for stars in the hydrogen main sequence, helium main sequence and the carbon main sequence. The numbers on the solid lines are the masses of the stars (in units of $M_\odot$) and the broken lines refer to constant degeneracy parameters ($\psi =$ constant) (according to Kippenhahn and Weigert 1991).

CHAPTER NINE

EVOLUTION FROM THE MAIN SEQUENCE

The discussions of stellar evolution in this chapter will start from the zero age main sequence where there are thermonuclear reactions inside the star. The huge amount of energy released from fusion is the main energy source for stellar radiation. The thermonuclear reactions will gradually lead to non-uniform chemical compositions inside the star, and to a more complicated internal structure for the entire star. In this way, the evolution of stars with thermonuclear reactions and non-uniform chemical compositions will be very different from that of pre main sequence stars with uniform chemical compositions and without thermonuclear reactions.

Stars of different masses will have different internal structures and evolution. They can be separated into three different classes according to their initial masses, namely, low mass stars ($M < 2.3M_\odot$), intermediate massive stars ($2.3M_\odot < M < 9M_\odot$) and massive stars ($M > 9M_\odot$). Low mass stars have rather high core densities and the helium core formed at the end of hydrogen burning is already degenerate. Intermediate massive stars have a bit lower core densities and the helium core formed at the end of hydrogen burning is not yet degenerate but the C-O core formed at the end of helium burning is degenerate. Massive stars have even lower core densities so the C-O core formed at the end of helium burning is also not degenerate.

The degeneracy in the core of the star will significantly affect whether the temperature increase during contraction of the core can lead to new nuclear reactions, and thus significantly affect evolutionary properties and the evolutionary fate.

Besides, stars with different masses will also have other different properties. For example, massive stars have strong mass loss due to stellar winds. The cores of massive stars are convective but the envelopes are under radiative equilibrium whilst on the main sequence stage. Low mass stars have very small mass loss due to stellar winds and have smaller or no convective cores. But the envelope of low mass stars are convective whilst on the main sequence stage. These different features will also significantly affect their structures and evolution properties.

It should be noted that the boundaries of the three classes of stars are not very clear cut, because these will be largely influenced by how we treat convection in stellar models, in particular, whether we consider convective overshoot. Convective overshoot not only enlarges the convective core and enables more matter to participate in nuclear reactions, but also affects the degeneracy in the core. For example, when we consider convective overshoot, and keep the boundary to be whether degeneracy occurs in the helium core and the C-O core, the boundaries for low mass stars will change to $M < 1.6$–$1.8M_\odot$, those for intermediate massive stars to 1.6–$1.8M_\odot < M < 6M_\odot$ and those for massive stars to $M > 6M_\odot$

9.1 EVOLUTION OF INTERMEDIATE MASSIVE STARS

We use a $5M_\odot$ star as our example to explain the evolution of intermediate massive stars from the main sequence. Figure 9.1 shows the evolutionary track of a $5M_\odot$ star ($X_H = 0.602$, $X_Z = 0.044$) on the HR diagram calculated by Kippenhahn, Thomas and Weigert (1965) and Weigert (1966). In the figure, the evolution has been separated into important stages: main sequence stage, crossing the Hertzsprung gap, helium burning stage and the AGB stage. Finer evolution stages are labeled A, B, C, ..., M. Figure 9.2 shows the temporal variation of the internal structure for a $5M_\odot$ star from their calculations. The different evolutionary stages will be separately discussed in the following subsections.

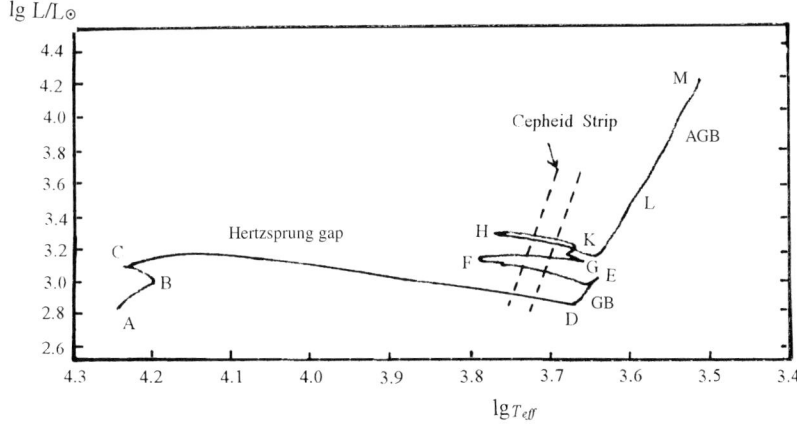

Fig. 9.1 The evolutionary track of a 5M$_\odot$ star on the HR diagram (Kippenhahn, Thomas and Weigert 1965, and Weigert 1966).

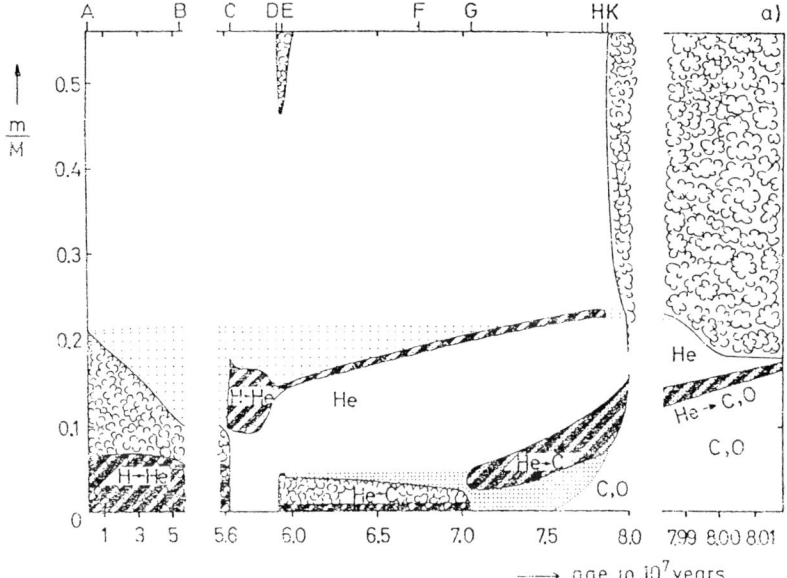

Fig. 9.2 The temporal variation of the internal structure for a 5M$_\odot$ star (Kippenhahn, Thomas and Weigert 1965, and Weigert 1966).

9.1.1. Main Sequence Stage

The main sequence stage in which core hydrogen burning occurs is represented by the path from point A to point C in Fig. 9.1. The core hydrogen burning is dominated by CNO cycle reactions and the core is completely convective. It takes 5.37×10^7 y to go from A to B, when the hydrogen abundance decreases from $X_H = 0.602$ to $X_H = 0.054$, and the core mass decreases from $M_c = 0.22 M_\odot$ to $M_c = 0.10 M_\odot$ due to contraction of the convective region.

From point B to C, hydrogen is gradually exhausted in the core and convection gradually comes to a stop. When hydrogen becomes gradually exhausted, hydrogen burning will be ignited at the external boundary of the core. At point C (with an age of 5.63×10^7 y), hydrogen is completely exhausted in the core, convection is completely stopped and the core has become a helium core. At the same time, there is a shell hydrogen burning at the boundary of the core. At this time, the energy of the star relies totally on the shell hydrogen burning. There are no nuclear reactions in the helium core interior to the shell. Its temperature will decrease and will finally match that of the shell source; it then becomes an isothermal core.

Figure 9.3 gives the change in the hydrogen abundance of a $5M_\odot$ star whilst on the main sequence. The solid curve refers to the distribution of hydrogen abundance at point C. The broken line with label "0" refers to the distribution of hydrogen abundance at point A, while those with labels "1" and "2" refer to the distributions for intermediate states.

9.1.2 Crossing the Hertzsprung Gap

The star evolves at a faster rate from point C to D across the Hertzsprung gap in the HR diagram. The evolution time scale is the Kelvin–Helmholtz time scale. During this stage, there is an unburnt isothermal helium core, and a shell hydrogen burning at the outer boundary of the helium core. The helium core interior to the shell source will quickly contract inwards while the envelope external to the shell burning will quickly expand outwards, so a "mirror reflection" phenomenon will emerge with the shell burning acting as a mirror (see Fig. 9.4). Since the envelope of the star quickly expands outwards, the radius of the star increases, the surface temperature

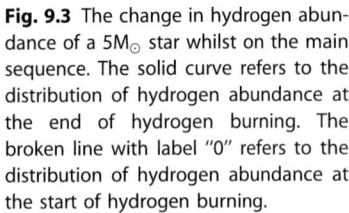

Fig. 9.3 The change in hydrogen abundance of a $5M_\odot$ star whilst on the main sequence. The solid curve refers to the distribution of hydrogen abundance at the end of hydrogen burning. The broken line with label "0" refers to the distribution of hydrogen abundance at the start of hydrogen burning.

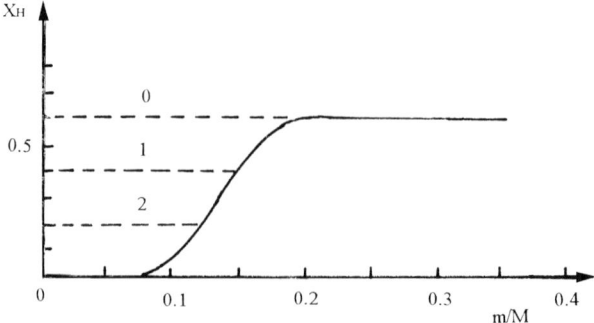

decreases and the color is reddened, so the star evolves quickly to the red giant branch on the HR diagram. The star evolves from point C to D on a Kelvin–Helmholtz time scale, so the time spent by the star in this stage is very short. Thus, it is difficult to observe stars in this region, and this explains why the region is called the Hertzsprung gap.

At this stage, the main energy source comes from shell hydrogen burning. The shell hydrogen burning keeps moving outwards and at the same time "engulfing" outer hydrogen, turning it into helium and passing the helium to the core. In this way, the mass of the helium core is also increasing. When the shell hydrogen burning extends outwards, its thickness will become thinner. In the stage from point C to D, there is no convective regions inside the star.

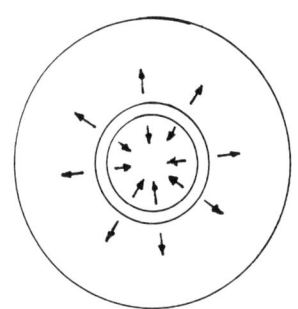

Fig. 9.4 The "mirror reflection" phenomenon with the shell hydrogen burning as the mirror.

The reason behind the contraction of the isothermal helium core will be explored in §9.3. Along with the contraction of the helium core, the core temperature will increase. At point D, the core temperature has already reached 10^8 K and helium burning starts.

9.1.3 Helium Burning Stage

The helium burning stage refers to the processes D $\rightarrow$ E $\rightarrow$ F $\rightarrow$ G, which takes about 1.1×10^7 y and corresponds to 1/5 of the time spent on the main sequence. The main nuclear reactions for helium burning are ^{4}He$(2\alpha, \gamma)^{12}$C and ^{12}C$(\alpha, \gamma)^{16}$O, i.e., the fusion of ^{4}He atoms to form ^{12}C and ^{16}O.

In the stage from point D to E, the energy provided by helium burning increases. However, the size of the helium core is only $0.05\ M$, and is much smaller than the hydrogen burning core. Moreover, the energy generation rate per unit mass is only 1/10 that for hydrogen burning. Therefore, the total energy provided by helium burning contributes only a small part to the total energy of the star. For example, it contributes only 6% at point E. Most of the energy of the star comes from shell hydrogen burning. The helium core is convective. In the stage from point D to E, the star is near the Hayashi line on the HR diagram. An important feature of its internal structure is the inward extension of the external convective layer. When point E is reached, the external convective layer has penetrated to the largest depth, i.e., $m/M = 0.46$. In other words, it has reached the region with changes in chemical compositions produced in the main sequence stage. It will move the more abundant helium from this region to the entire envelope and the stellar surface, which leads to a small change in the chemical composition in the entire envelope.

Since the helium core is also convective, the entire star is convective at point E.

In the stage E $\rightarrow$ F $\rightarrow$ G, the energy generated by core helium burning increases. At point F, the energy generated by core helium burning reaches 20% of the total energy. At the same time, the helium abundance in the core decreases to $X_{He} = 0.25$ so 75% of the core is made of ^{12}C and ^{16}O. On reaching point G, the energy generated by core helium burning has reached 48% of the total energy, while the helium abundance in the core decreases to $X_{He} < 0.1$ so the core is now made of ^{12}C and ^{16}O. At the outer boundary of the C-O core, there is shell helium burning. Therefore, there are two shell burning sources in the star, one being shell helium burning outside the C-O core and the other being shell hydrogen burning outside shell helium burning.

In the stage E $\rightarrow$ F, the star leaves the Hayashi line and turns left in the HR diagram. After reaching point F, it turns right again towards the Hayashi line. It is important to point out that the star passes the Cepheid pulsation strip twice when it goes from right to left for E $\rightarrow$ F, and from left to right for F $\rightarrow$ G (called loops) (Fig. 9.1). Entering the Cepheid pulsation strip, an oscillation mechanism (called the κ-mechanism) will be excited in the ionization regions of hydrogen and helium in the envelope, and radial oscillation happens in the star and the star becomes a Cepheid variable. When the evolutionary track of the star leaves the Cepheid pulsation strip, the star will resume to normal.

9.1.4 Evolution after Helium Burning

After point G, there are two shell burning sources inside the star which have a special mirror reflection phenomenon. The internal C-O core contracts inwards, the helium region between the two shell burning sources expands, and the envelope of the star outside shell hydrogen burning contracts. Therefore, the star will move away from the Hayashi line in the HR diagram.

When point H is reached, the shell hydrogen burning has moved outwards to a lower temperature region so hydrogen burning has ceased and there is only one shell helium burning left. The mirror reflection phenomenon for one shell burning emerges. The C-O core within the helium shell burning continues to contract inwards, and the envelope of the star outside the shell helium burning expands outwards. In this way, the evolutionary track of the star turns right again towards the Hayashi line. The continuous contraction of the C-O core has raised the core density to $\log\rho_c = 5.5$ and the core becomes degenerate. In the stage H → K, the external convective layer extends inwards. On reaching K, the external convective layer has extended into the original shell hydrogen burning region, and for the second time moves the helium formed from this region to the entire envelope, so the helium abundance at the stellar surface is increased slightly again.

9.1.5 AGB Stage

After point K, the external convective layer continues to extend inwards, and passes through the original shell hydrogen burning region and approaches the shell helium burning. When the external convective layer extends to the region $M_r/M = 0.1717$, the temperature has reached $T \approx 2 \times 10^8$ K which makes the hydrogen burn. Therefore, there are again two shell burning sources in the star, i.e., a shell hydrogen burning emerges outside the shell helium burning. At this time, the energy of the star is mainly supported by shell hydrogen burning. The C-O core interior to the shell helium burning is degenerate. With the increase in the mass of the C-O core, the luminosity of the star increases quickly, and the star undergoes K → L → M and moves along the so-called asymptotic giant branch (abbreviated as the AGB). The star is said to have entered the AGB evolutionary stage.

The stars will have some peculiar phenomena in the AGB stage. When point M is reached, periodic unstable thermal reactions occur in the shell helium burning, i.e., the helium burning inside the shell is not static but is explosive. In other words, a large amount of energy is released within a very short time, so the luminosity L_r of the shell burning can instantly exceed the total luminosity of the star by many times (see Fig. 9.5). Nevertheless, the large amount of energy instantly released by the shell helium burning cannot be transferred to the stellar surface instantly and affect the total luminosity of the star. The energy can only lead to an abrupt increase in the temperature of the shell burning, to an expansion in the volume and a decrease in the density (see Fig. 9.5). After all, the helium inside the shell source has been exhausted in a relatively short time and this type of unstable thermonuclear reaction will disappear. Only when sufficient helium has been accumulated again in the shell source (which takes about several thousand years) will explosive thermonuclear reactions happen again. This type of periodic explosive thermonuclear reaction is called

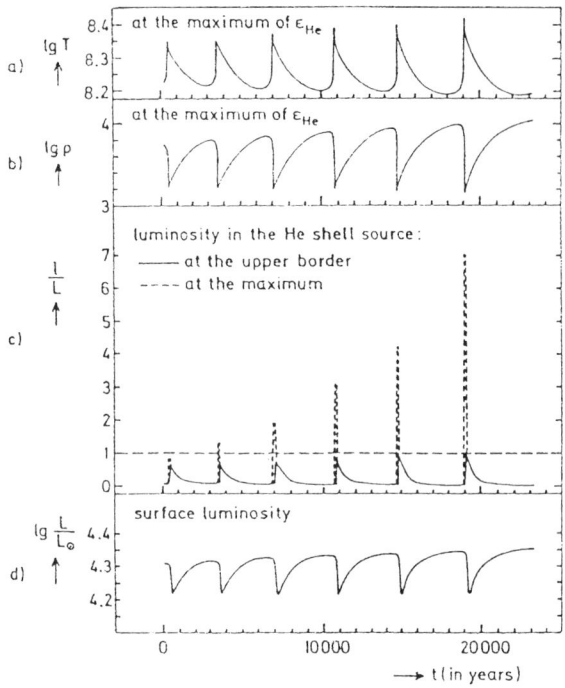

Fig. 9.5 The change in temperature, density and luminosity during the thermal pulse of the shell helium burning (according to Weigert 1966).

the "thermal pulse". From Fig. 9.5, we can see the change in the period of thermal pulse from 3200 y to 4300 y.

The next peculiar phenomenon of a star in the AGB stage is the production of oscillations with large amplitudes, which is excited by the κ-mechanism in hydrogen and helium ionization regions in the envelope of the star.

At the late stage of AGB evolution, the star will produce very strong stellar wind. From observations, the rate of mass loss due to stellar wind can reach the order of 10^{-5} to $10^{-6}\mathrm{M}_\odot/y$. This is also called the superwind. Because of this, the star can lose a large amount of matter in the envelope.

During the AGB evolutionary stage, the central region of the degenerate C-O core will generate a lot of neutrinos which carry vast amounts of energy when they escape from the star and lower the temperature of the central region of the degenerate C-O core.

9.1.6 Fate of Evolution

On reaching the AGB stage, the stars will have a degenerate C-O core, outside which are a shell helium burning and a shell hydrogen burning.

The ^{12}C and ^{16}O produced in shell helium burning is again compressed into the C-O core so the mass M_c of the C-O core is increased. At the same time, a very strong superwind appears on the stellar surface which leads to the loss of a vast amount of matter. The peculiar features of the AGB stars will significantly affect the fate of the stellar evolution.

For a degenerate C-O core, if the mass M_c is smaller than a critical mass M_{ch}, the core temperature will not increase with the increase in the core density as a result of contraction of the core, and so there will be no new nuclear reactions. This is because the equation of state is $P = k\rho^\gamma$, i.e., the density is independent of the temperature.

However, if there is still shell helium burning outside the degenerate C-O core, and the shell helium burning can increase the mass of the C-O core, when the mass M_c of the C-O core increases to the critical mass M_{ch}, the contraction of the central core can raise the core temperature (see Fig. 9.3) so new nuclear reactions can take place.

The capability of shell helium burning to increase the mass M_c of the C-O core to the critical mass M_{ch} will be determined by the mass of the envelope exterior to the shell helium burning, which is related to two factors. The first is the initial mass of the star, or the stellar mass whilst on the main sequence. If the initial mass is large, the mass of the envelope outside the shell source should also be large. This is the crucial factor. The second is the mass loss due to superwind. If the mass loss is significant, the mass of the envelope will be smaller.

From the above, we see that the degeneracy of the C-O core will lead to two possible evolutionary fates for the star. If shell helium burning cannot increase the mass of the degenerate C-O core to the critical mass, there will be no new nuclear reactions in the core of the star, and the star will change into a planetary nebula from the AGB stage, and will finally become a white dwarf. If shell helium burning can increase the mass of the degenerate C-O core to the critical mass, there will be new nuclear reactions in the core of the star. The carbon burning in the degenerate C-O core is unstable and explosive and a large amount of energy will be released instantly. The unstable burning in a degenerate core is called a "flash".

For stars with initial masses of 6–8$M_\odot$, the mass M_c of the degenerate C-O core formed after helium burning is smaller than the critical mass M_{ch} and the shell helium burning cannot increase M_c to M_{ch} so they will evolve to C-O white dwarfs. These stars constitute the majority of the stars. Stars with initial masses greater than 8$M_\odot$ are likely to evolve as supernovae, but can also evolve as white dwarfs, depending on the mass loss due to superwind at the final stage of the AGB evolution.

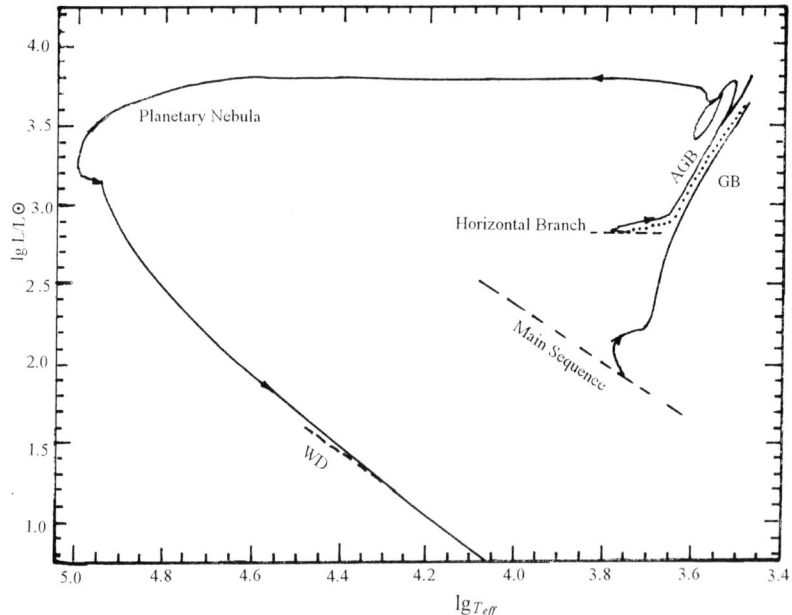

Fig. 9.6 The evolutionary track of a low mass star on the HR diagram.

9.2 EVOLUTION OF LOW MASS STARS

For stars with masses $M < 2.3 M_\odot$, the helium core formed after hydrogen burning is degenerate. The evolutionary properties and fates of these stars are different from those of massive and intermediate massive stars. Furthermore, these stars are located at the lower right portion of the main sequence and are relatively close to the Hayashi line. They have very small or even no convective regions in their cores whilst on the main sequence, which has obvious influence on the evolutionary properties and will lead to tracks in the HR diagram different from those of massive and intermediate massive stars.

Figure 9.6 gives the evolutionary track of a low mass star on the HR diagram. From the figure, the evolution can be broadly divided into several stages, namely, main sequence, red giant branch, horizontal branch, AGB stage and the white dwarf. However, not all low mass stars will experience all the above evolutionary stages. Generally speaking, the smaller the stellar mass, the more likely it will experience all the above stages.

9.2.1 Main Sequence Stage

We take a population II star ($X_H = 0.90$, $X_{He} = 0.090$, $X_Z = 0.001$) with a mass of $1.3M_\odot$ calculated by Thomas (1967) as our example to explain the evolution of a low mass star from the main sequence. Figure 9.7 is the track of this star on the HR diagram. The alphabets A, B, C, ... denote some meaningful evolutionary stages. Figure 9.8 shows the change in the internal structure of the star with the age (10^9 y). The hatched area in the figure represents the regions with energy generation rates greater than 10 erg/g.s; the dotted areas represent regions with nuclear reactions which also lead to changes in the hydrogen abundance greater than 0.01; and the areas with clouds are convective regions.

The stages A $\rightarrow$ C in Fig. 9.7 stand for the core hydrogen burning stage, i.e., the main sequence stage. At point A (zero age main sequence), the core temperature $T_c = 1.48 \times 10^7$ K and there are *pp* reactions inside the core, within which a small region with a mass 0.043 M is convective. The convective region disappears before the exhaust of hydrogen in the same region. The outside part of the star has a convective layer which starts from the photosphere and extends inwards to $r = 0.95$ R. Since the core hydrogen burning takes place in a non-convective core, the change in hydrogen abundance is that shown in Fig. 9.9. It is significantly different from the change in hydrogen abundance of an intermediate massive star whilst in the main sequence stage (Fig. 9.3). At the end of core hydrogen burning of a low mass star,

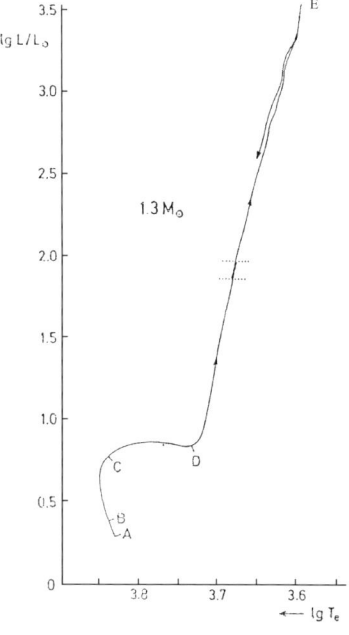

Fig. 9.7 The evolutionary track on the HR diagram of a star with $X_H = 0.90$, $X_{He} = 0.090$, $X_Z = 0.001$, and a mass of $1.3M_\odot$ (according to Thomas 1967).

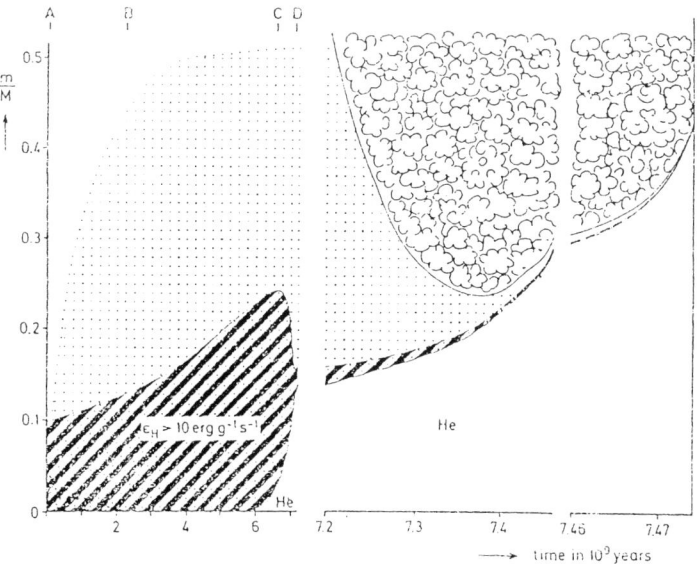

Fig. 9.8 The change in the internal structure of the $1.3M_\odot$ star with age (according to Thomas 1967).

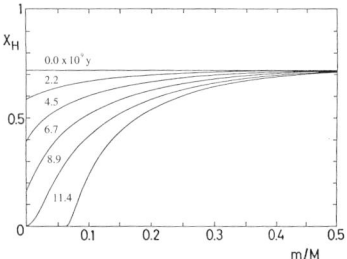

Fig. 9.9 The change in hydrogen abundance in the main sequence stage for a $1.3M_\odot$ star.

a core with varying helium abundance is formed; while at the end of core hydrogen burning of an intermediate massive star, a core with uniform helium abundance is formed which is not continuous in the chemical composition with the envelope.

9.2.2 Transition to the Red Giant Branch

When the star evolves to point C, hydrogen in the core has been exhausted and a helium core is formed. At the outer boundary of this helium core, shell hydrogen burning occurs (see Fig. 9.8). During the evolution of C → D, the helium core contracts inwards, and the envelope outside the shell hydrogen burning expands outwards. In the HR diagram, the star turns right towards the Hayashi line. When the evolutionary tracks of a low mass star and an intermediate massive star in the HR diagram are compared, we will see that the one for the intermediate massive star cannot be smooth, while that for the low mass star can be smooth. This is because the convective regions inside low mass stars are very small, and will disappear quickly, so core hydrogen burning does not take place in a large area convective region and will not be extinguished over a large area at the same time. The situation is opposite for intermediate massive stars. For these stars, the hydrogen burning takes place and extinguishes over a large convective region. These explain why the evolutionary tracks of low mass stars can be smooth. Moreover, after completion of core hydrogen burning, the tracks of low mass stars can move a very short distance after turning right on the HR diagram, which is the distance between the Hayashi line and the main sequence and becomes very short for very low values of the luminosity L.

9.2.3 Red Giant Branch Stage

During D → E, the helium core inside the star continues to contract inwards, the envelope outside the shell hydrogen burning continues to expand outwards, and the star becomes a red giant. In the HR diagram, the star moves along the red giant branch and goes upwards quickly. The convective region in the outer part of the star continues to extend into the inside of the star, and can reach the central region with varying chemical compositions formed at the hydrogen burning stage, and move the helium outwards to the entire envelope and the stellar surface.

The shell hydrogen burning continues to move outwards and at the same time "engulfs" the hydrogen in the envelope turning it into helium. Thus, the mass of the helium core increases continuously and, at the same time, the radius decreases continuously, so the density also increases continuously. In this way, the gravitational acceleration at the surface of the helium core increases rapidly. For example, for a helium core with $M_c = 0.35 M_\odot$, radius $R_c = (1/40) R_\odot$ and $g_c = 1.4 \times 10^7$ dyn.cm^2, the gravitational acceleration is greater than that at the solar surface by a factor of 500, which is close to that at the surface of a white dwarf. We can view the low mass star reaching the red giant branch as an object with a white dwarf at the center surrounded by a large and hydrogen rich envelope. Due to the high density inside the helium core, the electrons are already degenerate. The helium core is isothermal due to the high thermal conductivity of the degenerate electrons. In the central region of the helium core, a very strong neutrino radiation is produced, which decreases the temperature near the center.

There is still another important feature in the D $\rightarrow$ E stage: the luminosity L_H and the temperature of shell hydrogen burning are only dependent on the mass M_c and the radius R_c of the core, and are independent of the total mass of the star. Refsdal and Weigert (1970) gave the relations among L_H, T, M_c and R_c as

$$L_H \approx M_c^7 R_c^{-16/3} \qquad (9.1)$$

$$T \approx M_c R_c \qquad (9.2)$$

From Eqs. (9.1) and (9.2), when the core mass increases and the core radius decreases, the temperature T and the luminosity L_H of the shell burning increases. The luminosity L of the star is given by $L = L_H$ so the luminosity L of a low mass star is only dependent on the core mass M_c and the core radius R_c at this stage, and is independent of the total mass of the star. Therefore, it does not obey the mass–luminosity relation for main sequence stars.

9.2.4 From Helium Flash to Zero Age Horizontal Branch

When a low mass star evolves to the red giant branch, the internal helium core is degenerate. If the mass M_c of the degenerate helium core is smaller than the critical mass $M_{cr}(\text{He})$ ($= 0.45 M_\odot$), the contraction of the helium core will not lead to an increase in the core temperature, and thus not to helium burning. However, for a star with an initial mass of $1.3 M_\odot$, there is shell hydrogen burning outside the helium core.

The shell hydrogen burning produces ^{12}C and ^{16}O and can increase the core mass M_c continuously, which will finally reach $M_{cr}(He)$. When the core mass reaches $M_{cr}(He)$, the temperature of the helium core will rise during contraction (see §9.3.3).When the star has evolved to point E in the HR diagram (Fig. 9.7), the temperature of the helium core has already reached $\sim 10^8$ K which leads to helium burning. Since the helium core is degenerate, helium burning is unstable and explosive, which is called the helium flash. The time for helium flash is very short, ranging from several seconds to several minutes in general. The vast amount of energy released in such a short time cannot be transferred instantly to the stellar surface to be observed.

The degenerate state will disappear automatically several seconds or several minutes after the helium flash, i.e., the gas in the helium core again fulfils the equation of state for ideal gas. In this way, a rise in the temperature will increase the pressure and the volume, and the volume expansion will decrease the temperature, so thermonuclear reactions resume the stable nuclear reactions. The discussions on the flash in the degenerate helium core, the automatic disappearance of the degenerate state after a short time, and the restoration of the normal stable nuclear reactions will be presented in §9.3.4.

When the star has evolved to point E in the HR diagram (Fig. 9.7), a large amount of neutrinos will be produced in the central region of the helium core. These neutrinos can penetrate through the entire star and escape without any obstruction, carrying away a large amount of energy. Due to the energy loss by neutrinos which results in a relatively low temperature in the central region of the core, helium burning will not start from the central region but starts from regions away from the center, which then extends to the entire helium core.

After the helium flash, the star has moved to some point in the lower left of the HR diagram (Figs. 9.6 and 9.7). To study this position, Faulkner (1966) constructed a series of stellar models with a helium core mass $M_c = 0.5 M_\odot$ and a shell hydrogen burning outside the core. In the envelopes of these models, the hydrogen abundance is $X_H = 0.65$, and the abundance of C, N and O are $X_{CNO} = 10^{-2}$. The different models differ in their total masses, which are 1.25, 0.75 and $0.60 M_\odot$, respectively. The calculations of Faulkner (1966) show that these stellar models fall on the so-called "zero age horizontal branch" (shown in Fig. 9.10) in the HR diagram. The model with $M = 0.6 M_\odot$ is on the left-most region, while those for $M = 1.25 M_\odot$ and $M = 0.75 M_\odot$ are close to the Hayashi line. Furthermore, the positions of the zero age horizontal branch is also dependent on the abundance X_{CNO}. If X_{CNO} decreases, e.g., $X_{CNO} = 10^{-5}$, the zero age horizontal branch will move

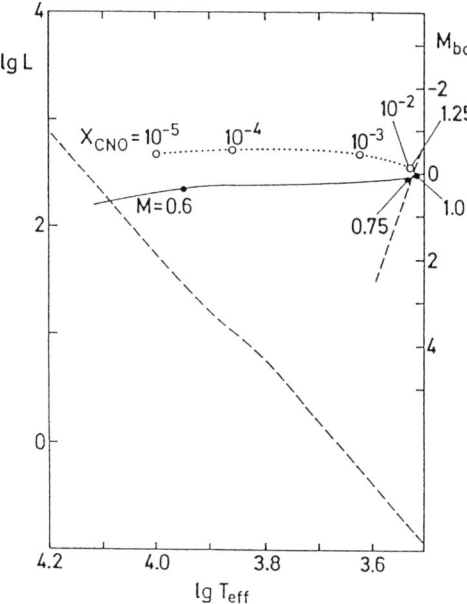

Fig. 9.10 The position in the HR diagram of stellar models with the same helium core mass M_c, the same abundance X_H and X_{CNO}, but different total masses. The solid line and the dotted lines are the zero-age horizontal branches for $X_{CNO} = 10^{-2}$ and $X_{CNO} = 10^{-5}$, respectively. The broken lines are the main sequence and the Hayashi line (according to Faulkner 1966).

upwards (the dotted line in Fig. 9.10). To understand these results, people suppose low mass stars have a relatively large amount of mass loss during the red giant stage and fall onto different positions on the zero age horizontal branch according to the mass loss after the helium flash. These have been justified by astronomical observations. Firstly, astronomical observations show that red giants have stronger stellar wind. Secondly, they have shown the existence of the horizontal branch for globular clusters in the HR diagram (Fig. 1.5).

To understand the abrupt evolution of the stellar model in the HR diagram from point E on the red giant branch to some point on the horizontal branch, we have to note the following facts. On the red giant branch, the helium core is degenerate, and the star is under equilibrium. During helium flash, the helium core is unstable. After the helium flash, the star has reached a new equilibrium state, which differs from the previous equilibrium in that the helium core is now non-degenerate so that the positions on the HR diagram are also different.

9.2.5 AGB Stage

There are two energy sources in the stellar model on the zero age horizontal branch. One is the burning of the helium core (the helium core is degenerate) and the other is the shell hydrogen burning at the bottom of the envelope of the star. Kippenhahn *et al.* (1968), Strom *et al.* (1970) and Iben and Renzini (1983) studied the evolution for this model. Figure 9.6 shows the evolutionary track of this model in the HR diagram. The star starts to move towards the top right region from the zero age horizontal branch. When helium in the helium core is exhausted, the core becomes a C-O core made of ^{12}C and ^{16}O, and shell helium burning appears outside the boundary of the C-O core, and shell hydrogen burning outside the shell helium burning. The C-O core contracts inwards rapidly and forms a degenerate C-O core. The envelope outside the shell hydrogen burning rapidly expands to turn the star into a giant. The evolutionary track of the star moves upwards along an asymptotic giant branch (AGB). AGB is different from the red giant branch in that there is still no helium burning inside the stars on the red giant branch.

There are many peculiar features for stars in the AGB stage. First, thermal pulses will appear in the shell helium burning, i.e., shell helium burning takes place in form of an explosion, and the vast amount of energy will be released in an extremely short time. Second, oscillation with very large amplitudes can occur in hydrogen and helium ionization regions in the envelope due to the κ-mechanism. At the late AGB stage, there will be very large mass loss due to stellar wind, reaching the order of 10^{-5} to $10^{-4} M_\odot/y$ and being called the superwind. The superwind can lead to a large amount of mass loss from the envelope.

9.2.6 Fate of Evolution

For a low mass star with $M < 0.5 M_\odot$, the helium core formed after hydrogen burning is degenerate, and the mass M_c of the helium core is given by $M_c < M_{cr}(\text{He})$ ($= 0.45 M_\odot$). Since the mass of the envelope outside the shell hydrogen is very small, shell hydrogen burning cannot increase the mass of the helium core to reach $M_{cr}(\text{He})$. In this way, when the degenerate helium core contracts, the temperature will not increase and there will be no helium burning. Therefore, stars with $M < 0.5 M_\odot$ will evolve to become helium white dwarfs.

For low mass stars with mass $0.5 M_\odot < M < 2.3 M_\odot$, the C-O core formed after core helium burning is degenerate, and the mass M_c of the C-O core is less than the critical mass M_{ch}. The mass of the envelope outside the shell helium burning is not very large, in particular, there can be a vast amount of mass loss from the surface due to the superwind in the late AGB stage. Therefore, it is not easy for the shell helium burning to increase the mass M_c of the degenerate C-O core to reach the critical mass M_{ch}. Thus, the contraction of the C-O core will not raise the temperature to have carbon burning. In this way, for a star with a mass $0.5M_\odot < M < 2.3M_\odot$, a large amount of envelope mass will be lost after the AGB stage, and the shell burning source will rapidly approach the surface and disappear, so the evolutionary track of the star will turn left. The star becomes a planetary nebula and finally evolves to a C-O white dwarf.

9.3 SOME PHYSICAL PROCESSES DURING EVOLUTION

In the previous two sections, we have discussed the evolution and internal structure of intermediate massive and low mass stars starting from the main sequence. However, some important phenomena have not been studied in detail, e.g., the renewed contraction, temperature rise and core burning as a result of hydrogen or helium burning in the core, the unstable thermonuclear burning that occurs in the degenerate core or in the shell (or simply called flash or thermal pulse), etc. This section will discuss these important phenomena and the associated physical processes during the above evolution.

9.3.1 Schönberg–Chandrasekhar Limit

In the discussions of the evolution of a $5M_\odot$ star in §9.1, we have pointed out that an isothermal helium core is formed after the completion of hydrogen burning and that the helium core will contract during shell hydrogen burning. The contraction will lead to a release of gravitational energy. According to the Virial theorem, part of the released gravitational energy will be used for compensation of the outward radiation, while the remaining will be used to increase the internal energy of the core, thereby increasing the core temperature to cause helium burning in the helium core.

It is important to ask why an isothermal helium core will contract inwards, and what are the physical conditions for the contraction of an isothermal helium core, because different kinds of isothermal cores will emerge during the evolution of different types of stars, e.g., isothermal hydrogen core, isothermal helium core, isothermal carbon-oxygen core, etc.

There is an envelope outside an isothermal core, so the surface pressure of the core is non-zero. We have discussed the expression for the Virial theorem for a non-zero surface pressure in §7.6, which is valid for the study of the isothermal core. We suppose the mass of the core is M_c, radius is R_c, temperature is T_c and the pressure at the surface of the core is P_c. From Eq. (7.101), when hydrostatic equilibrium is fulfilled, P_c should satisfy the relation

$$P_c = A\frac{T_c M_c}{R_c^3} - B\frac{M_c^2}{R_c^4} \tag{9.3}$$

where

$$A = \frac{3\Re}{4\pi\mu} > 0,\, B = \frac{C_1 G}{4\pi} > 0 \tag{9.4}$$

The first term on the right-hand side of Eq. (9.3) represents the gas pressure inside the core ($\sim\bar{\rho}T_c$), and the second term represents the pressure caused by self-gravitation ($\sim R_c\,\bar{\rho}\bar{g}$) since $\bar{\rho} \sim M_c R_c^{-3}$ and $\bar{g} \sim M_c R_c^{-2}$. Thus, P_c is the difference in the two terms. Apparently, P_c is a function of R_c for fixed M_c and T_c, and P_c is maximum for some R_c. We will now determine the position of this maximum. From the condition for producing the maximum,

$$\frac{\partial P_c}{\partial R_c} = -3A\frac{T_c M_c}{R_c^4} + 4B\frac{M_c^2}{R_c^5} = 0 \tag{9.5}$$

which gives

$$R_c = \frac{4B}{3A}\frac{M_c}{T_c} \tag{9.6}$$

Substituting Eq. (9.6) into Eq. (9.3), we obtain

$$P_{cmax} \sim \frac{T_c^4}{M_c^2} \tag{9.7}$$

which says that P_{cmax} decreases rapidly as the core mass M_c increases.

We further consider the pressure outside the boundary of the central core, i.e., the pressure P_h at the bottom of the envelope. This can be estimated through the hydrostatic equilibrium condition in Eq. (7.5). Supposing the pressure at the stellar surface is P_0, we can deduce from Eq. (7.5)

$$\frac{P_0 - P_h}{R} \cong -\frac{GM\bar{\rho}}{R^2} = -\frac{GM^2}{4R^5} \tag{9.8}$$

Since $P_o \ll P_h$, we have from Eq. (9.8)

$$P_h \approx \frac{M^2}{R^4} \qquad (9.9)$$

Form the equation of state for an ideal gas ($P_h \sim \bar{\rho} T_c$) and Eq. (9.9), we get

$$T_c \approx \frac{M}{R} \qquad (9.10)$$

Substituting Eq. (9.10) into Eq. (9.9), we have

$$P_h \approx \frac{T_c^4}{M^2} \qquad (9.11)$$

From Eqs. (9.11) and (9.7), we get

$$\frac{P_h}{P_{cmax}} \approx \left(\frac{M_c}{M}\right)^2 \qquad (9.12)$$

We now discuss when the central core can maintain hydrostatic equilibrium. To achieve hydrostatic equilibrium, it has to fulfil

$$P_c = P_h \qquad (9.13)$$

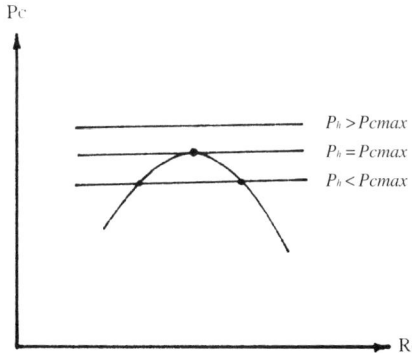

Fig. 9.11 The curve shows schematically the pressure P_c at the surface of the isothermal core as a function of the core radius R_c. Horizontal lines indicate the pressure P_h at the bottom of the envelope. The stable solution is $P_h < P_{cmax}$.

at the boundary of the core. From Fig. 9.11, we see that this can be achieved only when $P_h < P_{cmax}$. In this way, the condition for the central core to have hydrostatic equilibrium is

$$\frac{P_h}{P_{cmax}} < 1 \qquad (9.14)$$

From Eqs. (9.12) and (9.14), we know the condition for a stable core is that (M_c/M) should be less than a certain value (known as the critical value). In other words, if (M_c/M) exceeds this critical value, the core cannot reach a stable equilibrium. Schönberg–Chandrasekhar have calculated this limit which is about

$$\left(\frac{M_c}{M}\right)_{sc} \cong 0.11 \tag{9.15}$$

and is known as the Schönberg–Chandrasekhar limit. If $(M_c/M) > (M_c/M)_{SC}$, we have $P_h > P_{cmax}$ so the pressure outside the core is too large and will compress the core. This explains the previous finding that the isothermal helium core of a $5M_\odot$ star will contract inwards when the mass M_c of the core is larger than $0.11\ M$.

9.3.2 Properties of Contraction of an Isothermal Core

We now discuss the change in the internal physical state of an isothermal core during its contraction. The properties of the contraction of an isothermal core without thermonuclear reactions are the same as those for the pre main sequence stellar model as described in §8.2. From Eqs. (8.87) and (8.89), we have

$$\frac{\dot{\rho}}{\rho} = -3\frac{\dot{r}}{r} \tag{9.16}$$

$$\frac{\dot{P}}{P} = -4\frac{\dot{r}}{r} \tag{9.17}$$

The equation of state for the matter inside the isothermal core can be written as

$$\rho \approx P^\alpha T^{-\delta} \tag{9.18}$$

from which we get

$$\frac{\dot{\rho}}{\rho} = \alpha\frac{\dot{P}}{P} - \delta\frac{\dot{T}}{T} \tag{9.19}$$

Eliminating $(\dot{P}/P)$ and $(\dot{r}/r)$ from Eqs. (9.16), (9.17) and (9.18), we have

$$\frac{\dot{T}}{T} = \frac{\frac{4}{3}\alpha - 1}{\delta}\frac{\dot{\rho}}{\rho} \tag{9.20}$$

which gives the change of temperature with the density during contraction of the isothermal core. There are two cases for the contraction, which will be discussed in the following.

(A) Contraction of a non-degenerate core

Suppose the matter inside the isothermal core fulfils the equation of state for an ideal gas, so $\alpha = \delta = 1$. From Eq. (9.20), we have

$$\frac{\dot{T}}{T} = \frac{1}{3}\frac{\dot{\rho}}{\rho} \qquad (9.21)$$

Therefore, when the matter inside the core is a non-degenerate ideal gas, contraction of the core will increase the density and raise the temperature. During contraction of the non-degenerate ideal gas core, the temperature density curve should be a straight line with a slope 1/3 on the $\log T - \log \rho$ plane (in Fig. 9.12, the slope of the temperature density curve should be a straight line with a slope 1/3 in the ideal gas region).

(B) Contraction of a degenerate core

If the isothermal core is degenerate, the values of α and δ will change with the degree of degeneracy (Ψ). For example, when the electrons are completely non-relativistic degenerate, $P \sim \rho^{5/3}$, so $\alpha = 3/5$ and $\delta = 0$. When the electrons are completely relativistic degenerate, $P \sim \rho^{4/3}$, so $\alpha = 3/4$ and $\delta = 0$. In Fig. 9.12, the curve with $\psi = 0$ represents the boundary between the ideal gas case and degenerate case; and the dot-dashed line represents the boundary between the non-relativistic degenerate and relativistic degenerate cases.

During contraction of the degenerate core, with the increase in the density, the degree of degeneracy increases, the values of α and δ change, and the slope of the $\log T - \log \rho$ curve changes according to Eq. (9.20). When the core contracts to a stage where the electrons become relativistic degenerate, the core contraction will lead to a decrease in the temperature (see the directions of the arrows in the relativistic degenerate region in Fig. 9.12). It can be seen that contraction of the degenerate core will not raise the core temperature and will not lead to new nuclear reactions.

The initial temperatures and densities of the isothermal cores for stars with different masses are different. When the cores contract, their temperatures and densities will change along different curves.

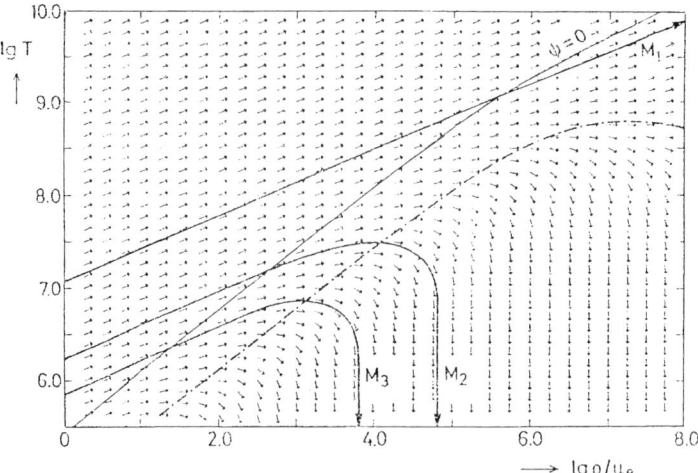

Fig. 9.12 The change in temperature with density during contraction of an isothermal core. The arrows give the direction of change of temperature with density (according to Kippenhahn and Weigert 1991).

Figure 9.12 gives the temperature density curves for contraction of the cores of three stellar models with masses M_1, M_2 and M_3 ($M_1 > M_2 > M_3$) at the same time. From Fig. 9.12, it can be seen that the cores for lower mass stars will become degenerate at lower temperatures and will have lower maximum temperatures. For example, stars with masses less than $0.08 M_\odot$ will become degenerate during the pre main sequence contraction when the core temperature is lower than 6×10^6 K, and the core maximum temperature reached is lower than 6×10^6 K, i.e., lower than the temperature for hydrogen burning. Therefore, the cores for stars with mass less than $0.08 M_\odot$ will not have hydrogen burning. This is the lower limit for the zero age main sequence. It is common to refer stars with masses less than $0.08 M_\odot$ with no core hydrogen burning as brown dwarfs, which belong to the family of white dwarfs.

The cores for stars with larger masses will become degenerate at higher temperatures. For example, the cores for stars with masses greater than $10 M_\odot$ will not become degenerate.

Figure 9.13 shows the evolution of core temperature and density for stellar models with different masses calculated for no mass loss and for no convective overshoot. The solid lines are evolutionary curves of the core temperature and density for stars with different masses. The numbers attached to the curves represent the stellar masses (in units of $M_\odot$). The broken line ($\alpha = 3/4$, $\delta = 0$) is the boundary separating the $\lg T_c - \lg \rho_c$ plane into degenerate and non-degenerate regions. The three dot-dashed lines give the ignition for hydrogen burning, helium burning and carbon burning, respectively. From Fig. 9.13, it can be seen that the helium cores formed after hydrogen burning in the interior of stars with mass less than $2.3 M_\odot$ are degenerate; the carbon-oxygen cores formed

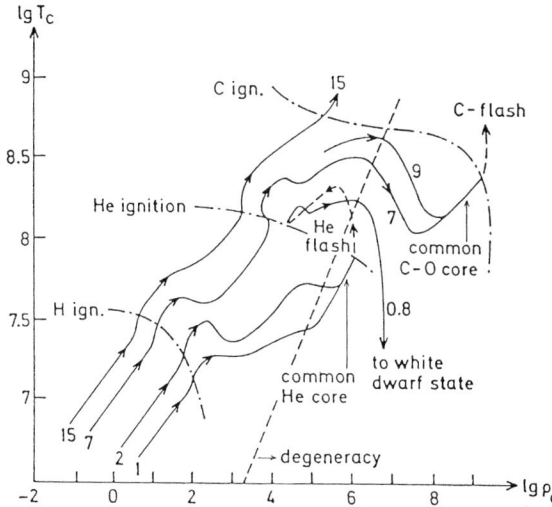

Fig. 9.13 The change in the core temperature and density for stars with different masses. The numbers attached to the curves represent stellar masses ($M_\odot$). The broken line is the boundary separating the degenerate and non-degenerate cases. The dot-dashed lines give the ignition points for different nuclear reactions (according to Iben 1974).

after helium burning in the interior of intermediate massive stars with masses $2.3 M_\odot < M < 9 M_\odot$ are degenerate; and the cores of massive stars with masses larger than $9 M_\odot$ after the start of carbon burning will not yet become degenerate.

The abovementioned features that the temperature of a degenerate core will not increase on contraction and will not lead to new nuclear reactions are not always true. They realize only when the mass M_c of the degenerate core is less than a critical mass M_{cr}. If M_c reaches or exceeds M_{cr}, the temperature can increase on contraction and lead to new nuclear reactions. These will be discussed in detail in the following section.

9.3.3 Critical Mass for a Degenerate Core to have Temperature Increase on Contraction

Under certain circumstances, the core mass of stars will increase during evolution. For example, the shell hydrogen burning or the shell helium burning can increase the mass of the helium core or the carbon-oxygen core. In this section, we will study the case in which the mass of the degenerate core has reached or exceeded the critical mass M_{cr} so the temperature of the degenerate core can rise on contraction and lead to new nuclear reactions.

For a degenerate core, the internal pressure P_c is mainly the electron pressure so it can be approximated by

$$P_c \approx P_e = \frac{\mathfrak{R}}{\mu_e} \rho_c T_c + K_\gamma \left(\frac{\rho_c}{\mu_e}\right)^\gamma \tag{9.22}$$

where ρ_c and T_c are the density and temperature of the core, and μ_e is the mean molecular weight for electrons. The first term on the right-hand side of Eq. (9.22) is the non-degenerate electron pressure, and the second term is the degenerate electron pressure. Here, γ is not a constant and can change from $\gamma = 5/3$ (when $\rho \ll 10^6$ gcm^{-3}) to $\gamma = 4/3$ (when $\rho \gg 10^6$ gcm^{-3}) when the degenerate electrons change from non-relativistic to relativistic. The corresponding coefficient K_γ will change from the expression in Eq. (4.119) to that in Eq. (4.129). To illustrate the changes in γ and K_γ, we can write them as $\gamma = (4/3) + h$ and $K_\gamma = K_{(4/3+h)}$, where h is a variable. When the electrons become relativistic degenerate, $h \to 0$. Therefore, Eq. (9.22) can be written as

$$P_c \approx P_e = \frac{\mathfrak{R}}{\mu_e} \rho_c T_c + K_{\left(\frac{4}{3}+h\right)} \left(\frac{\rho_c}{\mu_e}\right)^{\left(\frac{4}{3}+h\right)} \tag{9.23}$$

On the other hand, we can employ the hydrostatic equilibrium equation (7.7) to estimate the core pressure P_c as

$$P_c \approx \frac{GM_c\bar{\rho}}{R_c} = fGM_c^{2/3}\rho_c^{4/3} \tag{9.24}$$

where

$$\bar{\rho} = \frac{3M_c}{4\pi R_c^3} \ , \ f = \bar{\rho}/\rho_c \tag{9.25}$$

and f can be treated as a constant during core contraction.

Substituting Eq. (9.24) into Eq. (9.23), we have

$$\frac{\mathfrak{R}}{\mu_e} T_c = \left[fGM_c^{2/3} - K_{\left(\frac{4}{3}+h\right)} \mu_e^{-\left(\frac{4}{3}+h\right)} \rho_c^h \right] \cdot \rho_c^{1/3} \tag{9.26}$$

If the core is a non-degenerate ideal gas, the second term inside the bracket on the right-hand side of Eq. (9.26) or the degenerate electron pressure should be equal to zero. At this time, we obtain $T_c \sim \rho_c^{1/3}$ for a fixed M_c. If the core is non-relativistic degenerate, $h \neq 0$, so the signs of the two terms inside the bracket on the right-hand side of Eq. (9.26) are opposite. It is known that for a given M_c, the T_c-ρ_c curve will have a maximum T_{cmax} at some ρ_{co}, and electrons will become relativistic degenerate when ρ_c continues to increase so $h \to 0$. At this time, if the mass M_c of the degenerate core is larger than the critical mass M_{cr}, i.e., M_{cr} fulfils the relation

$$M_c > M_{cr} = \left(\frac{K_{4/3}}{fG}\right)^{3/2} \mu_e^{-2} \tag{9.27}$$

T_c can increase on contraction of the degenerate core and $T_c \sim \rho_c^{1/3}$. The critical mass can be obtained from Eqs. (9.27) and (9.25). If the core is a helium core, the critical mass $M_{cr}(\text{He}) \approx 0.45 M_\odot$; if the core is a carbon-oxygen core, the critical mass $M_{cr}(\text{C-O}) \approx 1.4 M_\odot$, where $M_{cr}(\text{C-O})$ is also known as the Chandrasekhar mass limit and abbreviated as M_{ch}.

From Eqs. (9.25) and (9.27), we see that $M_{cr}(\text{He})$ and $M_{cr}(\text{C-O})$ are closely related to the chemical composition, degree of degeneracy and the stellar models (whether convective overshoot and mass loss are considered). Figure 9.14 gives the temperature density curves for contraction of the cores of three stars with masses M_1, M_2 and M_3 ($M_1 > M_2 > M_3$) at the same time. The broken line ($\psi = 0$) is the boundary between ideal gas and non-relativistic degeneracy. The dot-dashed line ($\alpha = 3/4$, $\delta = 0$) is the boundary separating the relativistic degenerate and non-relativistic degenerate regions. The cores of the stars with masses M_1 and M_2 are beyond the critical mass, while the core of the star with mass M_3 is below the critical mass. From Fig. 9.14, when the core is a non-degenerate ideal gas, $T_c \sim \rho^{1/3}$ during core contraction. After the core has become degenerate, for the non-relativistic case, T_c has a maximum value T_{cmax} at ρ_{co} and the curve then descends. After

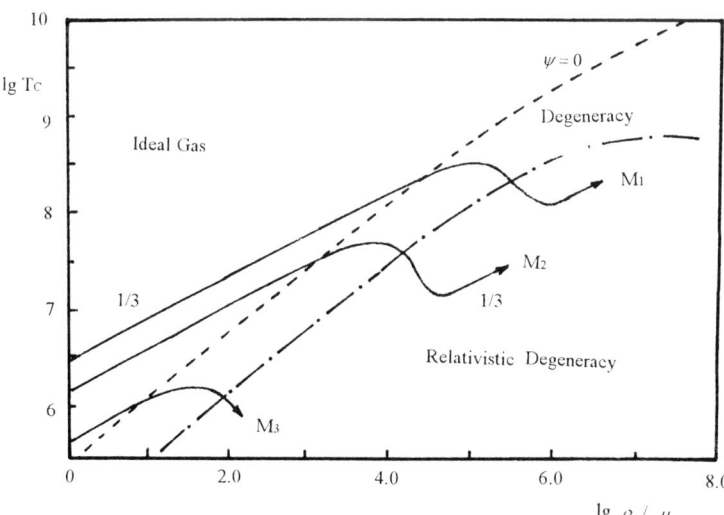

Fig. 9.14 The temperature-density curves for contraction of the cores of three stars with masses M_1, M_2 and M_3 ($M_1 > M_2 > M_3$) at the same time. The broken line ($\psi = 0$) is the boundary between ideal gas and non-relativistic degeneracy. The dot-dashed line ($\alpha = 3/4$, $\delta = 0$) is the boundary separating the relativistic degenerate and non-relativistic degenerate regions. The cores of the stars with masses M_1 and M_2 are beyond the critical mass, while the core of the star with mass M_3 is below the critical mass.

entering the relativistic degenerate region, for the case where the mass of the degenerate core is greater than the critical mass, T_c increases again according to $T_c \sim \rho_c^{1/3}$ (see the degenerate cores of the stars M_1 and M_2 in the figure); for the case where the mass of the degenerate core is smaller than the critical mass, T_c decreases (see the degenerate core of the star M_3 in the figure).

From the above discussions, if the mass of the degenerate core is greater than or equal to the critical mass, or the condition in Eq. (9.27) is fulfilled, the temperature T_c can increase according to $T_c \sim \rho_c^{1/3}$ during contraction of the degenerate core and lead to new nuclear reactions. Conversely, if the mass of the degenerate core is smaller than the critical mass, or the condition in Eq. (9.27) is not fulfilled, the temperature T_c will not increase according to $T_c \sim \rho_c^{1/3}$ during contraction of the degenerate core.

9.3.4 Instabilities of Thermonuclear Reactions

If the temperature of the degenerate core rises and leads to nuclear reactions, the nuclear reactions will be unstable and take place in form of explosions, i.e., a vast amount of energy is released within an extremely short time. It is common to call this kind of unstable violent thermonuclear reactions in the degenerate core the "flash". Similar unstable thermonuclear reactions will also occur during shell burning. For example, when a $5M_\odot$ star has evolved to the AGB stage, the shell helium will have violent explosive type burning (see Fig. 9.5). People refer unstable violent nuclear reactions in the shell layer as the "thermal pulse". It is worth noting that the necessary condition for a flash to occur in the core is that the electrons in the core should already be degenerate while the thermal pulse takes place in the shell burning when the electrons are non-degenerate. This section will focus on the reasons behind the instabilities in thermonuclear reactions and the features of unstable thermonuclear reactions. The discussions will also be divided into cases for core burning and shell burning (see Kippenhahn and Weigert 1991; Han 1987).

(A) Flash in a degenerate core

Suppose there is a perturbation in the core at a certain instance, which changes the nuclear energy generation rate by a small quantity ε_{N1} (assuming ε_{N1} is positive). If the star can internally and automatically regulate and decrease the energy generation rate to offset ε_{N1}, the thermonuclear reactions are stable inside the core. Conversely, if the changes inside the star can somehow increase the energy generation rate inside the core, the chain reaction will cause a rapid increase in the energy generation rate within an extremely short time, and the violent explosive "flash" will occur. At this time, thermonuclear reactions in the core are unstable. We adopt the energy equation (7.13)

$$\frac{dL_r}{dM_r} = \varepsilon_N - C_P \dot{T} + \frac{\delta}{\rho} \dot{P} \qquad (9.28)$$

to study this problem, where $\dot{T}$ and $\dot{P}$ are time differentials, and the term $-\varepsilon_\nu$ has been absorbed into $-\varepsilon_N$ here.

Suppose there is a perturbation at some instance, which not only changes the energy generation rate by ε_{N1}, but also slightly changes P, T, ρ, ..., i.e.,

$$
\begin{aligned}
T &= T_0 + T_1 \\
P &= P_0 + P_1 \\
\rho &= \rho_0 + \rho_1 \\
\varepsilon_N &= \varepsilon_{N0} + \varepsilon_{N1} \\
L_r &= L_{r0} + L_{r1}
\end{aligned}
\qquad (9.29)
$$

$$\vdots$$

where the subscripts "1" refer to the perturbation terms, and subscripts "0" refer to the unperturbed terms. For simplicity, we can assume the core has been under equilibrium before the perturbation, so the time differentials are zero, i.e.,

$$\dot{T}_0 = \dot{P}_0 = 0 \qquad (9.30)$$

Since the perturbation takes place in an extremely short time, the energy change cannot be transferred out in this short time, so we have

$$L_{r1} = 0 \qquad (9.31)$$

Substituting Eqs. (9.29), (9.30) and (9.31) into Eq. (9.28), we get

$$
\begin{aligned}
\varepsilon_{N1} &= C_P \dot{T}_1 - \frac{\delta}{\rho} \dot{P}_1 \\
&= C_P \dot{T}_1 \left(1 - \frac{\delta P}{C_P \rho T} \frac{\dot{P}_1/P}{\dot{T}_1/T} \right)
\end{aligned}
\qquad (9.32)
$$

Using Eq. (3.34)

$$\nabla_{ad} \equiv \left(\frac{dlnT}{dlnP} \right)_{ad} = \frac{\delta P}{C_P \rho T}$$

Eq. (9.32) can be written as

$$\varepsilon_{N1} = C_P \dot{T}_1 \left[1 - \nabla_{ad} = \frac{\dot{P}_1/P}{\dot{T}_1/T} \right] \tag{9.33}$$

The $(\dot{P}_1/P)$ term in the bracket on the right-hand side is a function of $(\dot{T}_1/T)$. To obtain this functional relationship, we have to rely on the equation of state

$$\rho \approx P^\alpha T^{-\delta} \tag{9.34}$$

Taking the differential of logarithm of Eq. (9.34), and substituting Eq. (9.29), we get

$$\frac{\dot{\rho}_1}{\rho} = \alpha \frac{\dot{P}_1}{P} - \delta \frac{\dot{T}_1}{T} \tag{9.35}$$

Furthermore, the mass M_r of the core will not change no matter whether it expands or contracts, i.e.,

$$M_r \approx \rho r^3 = const \tag{9.36}$$

Taking the differential of logarithm of Eq. (9.36), and substituting Eq. (9.29), we get

$$\frac{\dot{\rho}_1}{\rho} = -3 \frac{\dot{r}_1}{r} \tag{9.37}$$

Since the pressure can be expressed as

$$P = \int_{M_r}^{M} \frac{GM_r}{4\pi r^4} dM_r \tag{9.38}$$

and considering Eq. (9.36), we have

$$\dot{P} = \int_{M_r}^{M} \frac{GM_r}{4\pi} \frac{\partial}{\partial t} \left(\frac{1}{r^4} \right) dM_r$$

$$= -\int_{M_r}^{M} 4\frac{\dot{r}}{r}\frac{GM_r}{4\pi r^4} dM_r = -4\frac{\dot{r}}{r}P \tag{9.39}$$

Substituting Eq. (9.29), we get

$$\frac{\dot{P_1}}{P} = 4\frac{\dot{r_1}}{r} \tag{9.40}$$

From Eqs. (9.35), (9.37) and (9.40), and eliminating $(\dot{\rho_1}/\rho)$ and $(\dot{r_1}/r)$, we have

$$\frac{\dot{P_1}}{P} = \frac{4\delta}{4\alpha - 3}\frac{\dot{T_1}}{T} \tag{9.41}$$

Substituting this into Eq. (9.31), we obtain

$$\dot{T_1} = \frac{\varepsilon_{N1}}{C_P}\frac{1}{1 - \nabla_{ad}\left[\frac{4\delta}{4\alpha-3}\right]} \tag{9.42}$$

We now discuss two different cases using Eq. (9.42).

(1) When the core is a non-degenerate ideal gas
If the matter inside the core is a non-degenerate ideal gas, $\alpha = \delta = 1$, $\nabla_{ad} = 2/5$, so from Eq. (9.42) we have

$$\dot{T_1} = -\frac{5}{3}\frac{\varepsilon_{N1}}{C_P} \tag{9.43}$$

Since C_P is always greater than zero, when the perturbation of the energy generation rate $\varepsilon_{N1} > 0$, $\dot{T_1} < 0$, which leads to a decrease in the energy generation rate ε_N. In other words, a positive perturbation in the energy generation rate will decrease the temperature which will in turn decrease the energy generation rate. Therefore the thermonuclear reactions inside the core are stable. Thus, when the matter inside the core is a non-degenerate ideal gas, the nuclear reactions are stable and there will be no violent explosive flash phenomena.

(2) When the core is degenerate
For simplicity, we will discuss the complete non-relativistic degenerate case. Here $P \sim \rho^{5/3}$, $\alpha = 3/5$ and $\delta = 0$. From Eq. (9.42), we have

$$\dot{T} = \frac{\varepsilon_{N1}}{C_P} \qquad (9.44)$$

From Eq. (9.44), for a degenerate core, a perturbation in the energy generation rate $\varepsilon_{N1} > 0$ will lead to $\dot{T}_1 > 0$ or a rise in the core temperature. Since the energy generation rate is given by $\varepsilon_N \sim T^\nu$ (ν is a relatively large positive number), the rise in the core temperature will lead to an even larger energy generation rate. The outcome of this chain reaction is an avalanche type increase in the energy generation rate. Therefore, the chain reaction in the degenerate core is unstable and there are so-called "flash" phenomena.

We now study whether the energy generation rate will increase indefinitely after the occurrence of the unstable nuclear reaction (i.e., the flash) in the degenerate core. Suppose the electrons were relativistic degenerate in the core before the flash, which corresponds to point N in Fig. 9.15. When the flash takes place, the temperature increases instantly. However, according to Eq. (9.41) and $\delta = 0$, the pressure inside the degenerate core remains unchanged. From Eqs. (9.37) and (9.40), we further know that the density also remains unchanged. Thus, the state of the core will move vertically upwards in Fig. 9.15. Nevertheless, when the temperature reaches a certain point, the core has entered the non-degenerate ideal gas region, so the electrons are now non-degenerate. Here Eq. (9.41) is no longer valid. According to the equation of state for an ideal gas, the rise in temperature will lead to a rise in the pressure, which will in turn increase the volume, decrease the density and decrease the temperature. Therefore, the flash will stop automatically. It can be seen that the flash will not increase the energy generation rate indefinitely. When the energy generation rate increases to a certain extent, i.e., when the core

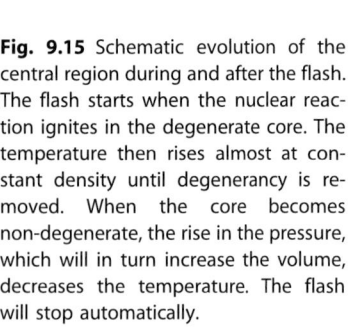

Fig. 9.15 Schematic evolution of the central region during and after the flash. The flash starts when the nuclear reaction ignites in the degenerate core. The temperature then rises almost at constant density until degeneracy is removed. When the core becomes non-degenerate, the rise in the pressure, which will in turn increase the volume, decreases the temperature. The flash will stop automatically.

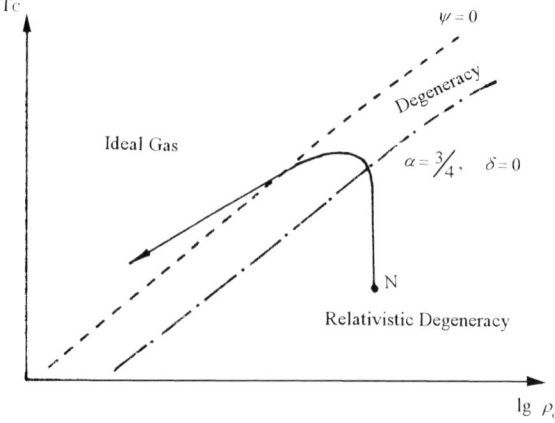

temperature reaches a certain point, the core will become non-degenerate and the flash will stop automatically. Only when the core becomes degenerate and there are nuclear reactions again will there be another flash.

(B) Thermal pulse of the shell burning

We consider a shell layer with inner radius r_i and thickness of Δ (Fig. 9.16). Assuming that the mass within the shell layer remains unchanged instantly, we have

$$M_s = \frac{4}{3}\pi\left[(r_i + \Delta)^3 - r_i^3\right]\rho = const \tag{9.45}$$

We further assume that the shell burning always changes the outside only and leave the inner side unchanged, i.e., $r_i = $ constant. Then, differentiating Eq. (9.45) with respect to time, substituting Eq. (9.29), and considering $r = r_i + \Delta = r_o + r_1$, we have

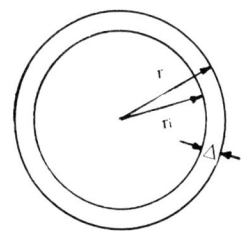

Fig. 9.16 A shell layer with inner radius r_i and thickness Δ.

$$\frac{\dot{\rho}_1}{\rho} = -\left(1 + \frac{3r\Delta - \Delta^2}{3r^2 - 3r\Delta + \Delta^2}\right)\frac{r}{\Delta}\cdot\frac{\dot{r}_1}{r}$$

$$= A \cdot \frac{\dot{r}_1}{r} \tag{9.46}$$

where

$$A \equiv \left(1 + \frac{3r\Delta - \Delta^2}{3r^2 - 3r\Delta + \Delta^2}\right)\frac{r}{\Delta} \tag{9.47}$$

Comparing Eq. (9.46) with Eq. (9.37) for the core, we see that the two equations only differ in the coefficients, i.e., Eq. (9.37) will become Eq. (9.46) if the coefficient is replaced by A. Besides, the two equations (9.35) and (9.39) obtained for the core have general applications, and are valid for shell burning. Using Eqs. (9.35), (9.39) and (9.46), and eliminating $(\dot{\rho}_1/\rho)$ and $(\dot{r}_1/r)$, we have

$$\frac{\dot{P}_1}{P} = \frac{4\delta}{4\alpha - A}\cdot\frac{\dot{T}_1}{T} \tag{9.48}$$

Equation (9.33) for the core has similar general applications and is valid for shell burning. Substituting Eq. (9.48) into Eq. (9.33), we get

$$\dot{T}_1 = \frac{\varepsilon_{N1}}{C_P}\frac{1}{1 - \frac{4\delta\nabla_{ad}}{4\alpha - A}} \tag{9.49}$$

From Eq. (9.49), we get the condition for the occurrence of instabilities in shell burning to be $1 - \frac{4\delta\nabla_{ad}}{4\alpha - A} > 0$, i.e.,

$$\frac{4\delta\nabla_{ad}}{4\alpha - A} < 1 \qquad (9.50)$$

The inequality (9.50) is valid only under two conditions, i.e., $(4\alpha - A) > 4\delta\nabla_{ad}$ or $4\alpha - A < 0$. Through simple physical analysis, we know that $(4\alpha - A) > 4\delta\nabla_{ad}$ does not hold, so only when $4\alpha - A < 0$ will the shell burning have instabilities. If the matter within the shell layer is a non-degenerate ideal gas, $\alpha = \delta = 1$ and $\nabla_{ad} = 2/5$, so from the condition $4\alpha - A < 0$, we obtain $(\Delta/r) < 0.37$. In other words, when the ratio between the thickness and the radius of the shell layer is smaller than 0.37, instabilities will occur in thermonuclear reactions. This kind of unstable burning in the shell layer is called the "thermal pulse". For example, when a $5M_\odot$ star evolves to the AGB stage, the thermal pulse in the shell helium burning arises from unstable thermonuclear reactions.

From the above discussions, we know that electron degeneracy is the reason behind the production of unstable thermonuclear reactions in the core, and the change in geometry is the reason behind the production of unstable nuclear reactions in the shell layer. After the flash in the core, the rise in temperature without changes in the density will eliminate the electron degeneracy and stop the flash automatically. Thus, the flash in the core will not burn up all the matter inside the core instantly. On the other hand, when there is a thermal pulse in the shell layer, the unstable thermonuclear reactions will not be eliminated automatically, so the thermal pulse will not stop until all matter inside the shell layer is burnt out.

9.3.5 Instabilities and Strong Stellar Wind on the Surface of AGB and RGB Stars

When intermediate massive stars and low mass stars evolve to become AGB and RGB stars, their envelopes are convective. If the sum of the outward radiation pressure acceleration and the turbulent pressure acceleration exceeds the inward gravitational acceleration at some place inside the convective envelope, i.e.,

$$g_R + g_t - g > 0 \qquad (9.51)$$

where the inward gravitational acceleration is

$$g = \frac{GM_r}{r^2} \tag{9.52}$$

the outward radiation pressure acceleration is

$$g_R = -\frac{1}{\rho}\frac{dP_R}{dr} \tag{9.53}$$

and the turbulent pressure acceleration is

$$g_t = -\frac{1}{\rho}\frac{dP_t}{dr} \tag{9.54}$$

the region is no longer at hydrostatic equilibrium and starts to move, and the motion will satisfy the equation of motion

$$\rho\frac{dv}{dt} + \frac{dP_g}{dr} = -\rho g - \frac{dP_R}{dr} - \frac{dP_t}{dr} \tag{9.55}$$

Where P_g is the gas pressure. At this time, dynamic instabilities have emerged.

However, the entire star will be made to fulfil hydrostatic equilibrium during computations for the stellar model, i.e., the time dependent term in Eq. (9.55) will be neglected, and the model will fulfil

$$\frac{dP_g}{dr} = -\rho g - \frac{dP_R}{dr} - \frac{dP_t}{dr} \tag{9.56}$$

In this way, in regions with dynamic instabilities, from Eqs. (9.56) and (9.51), we should have

$$\frac{dP_g}{dr} > 0 \tag{9.57}$$

i.e., the gas pressure increases outwards. Since the gas pressure is given by $P_g = (\mu/\Re)\rho T$ and $(dT/dr) < 0$ for radiative equilibrium, from Eq. (9.57), we have

$$\frac{d\rho}{dr} > 0 \tag{9.58}$$

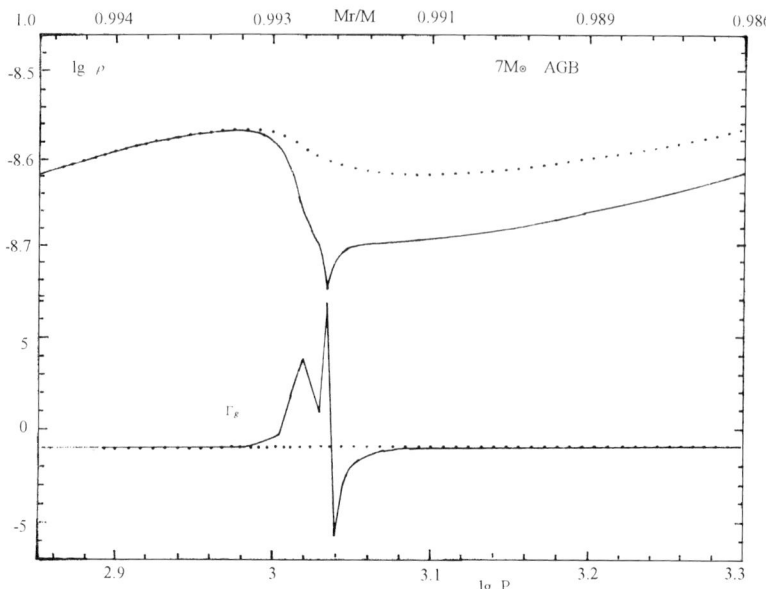

Fig. 9.17 The distribution in the density ρ and the function $\Gamma_g \equiv g_R + g_t - g$ in the convective envelope of a $7M_\odot$ AGB star (Jiang and Huang 1997b).

i.e., the density increases outwards in regions with dynamic instabilities, which is known as the density inversion.

From the above, if the entire star fulfils the hydrostatic equilibrium equation when computing the stellar structure and evolution models, and if the function $\Gamma_g \equiv g_R + g_t - g > 0$ and if there is density inversion $(d\rho/dr) > 0$ in some regions inside the convective envelope, we can say that the region has dynamic instabilities.

Jiang and Huang (1997a, b) studied the evolution of a $7M_\odot$ star from the main sequence to the AGB stage. Figure 9.17 shows the change in the density ρ and the function Γ_g in the envelope at the late AGB stage for the $7M_\odot$ star. The solid line refers to the case which has considered the turbulent pressure. The dotted line refers to the case which has not considered the turbulent pressure. From Fig. 9.17, it can clearly be seen that at the late AGB stage of a $7M_\odot$ star, the convective envelope does have density inversion and $\Gamma_g > 0$, and dynamic instabilities occur in regions close to the stellar surface. These results show that the production of superwind in AGB stars is due to dynamic instabilities which occur in the convective envelope near the surface and which causes mass outflow. At the same time, Fig. 9.17 clearly shows that $\Gamma_g > 0$ will not occur without the consideration of the turbulent pressure (see the dotted line in Fig. 9.17), and the density inversion is substantially weakened, so there will be no dynamic instabilities. This shows that the crucial factor for dynamic instabilities is the turbulent pressure.

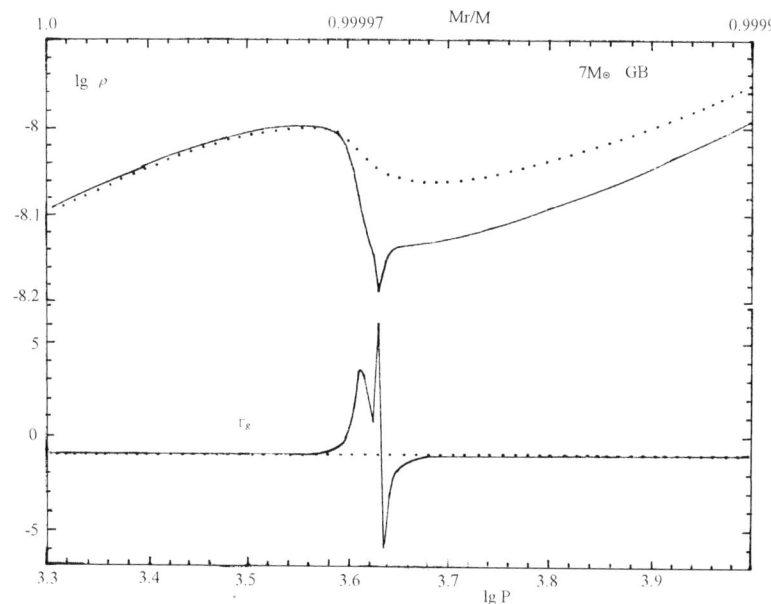

Fig. 9.18 The distribution in the density ρ and the function $\Gamma_g \equiv g_R + g_t - g$ in the convective envelope of a $7M_\odot$ giant star (Jiang and Huang 1997b).

Jiang and Huang (1997a, b) also studied the evolution of a $7M_\odot$ star to the red gaint stage. They found that when turbulent pressure is considered in the model calculations, the region in the convective envelope close to the surface has also density inversion $(d\rho/dr) > 0$ and $\Gamma_g > 0$, or dynamic instabilities (see Fig. 9.18) which lead to mass outflow and strong stellar wind. However, when Figs. 9.17 and 9.18 are compared, the region for dynamic instabilities in the envelope during the red giant star stage is much smaller than that during the AGB stage. Therefore, we know that the rate of mass loss during the red giant stage is much smaller than that during the AGB stage.

In the discussions of instabilities in the convective envelopes in AGB and RGB stars, we have shown that stellar wind can also be substantial or even superwind can occur in AGB and RGB stars with low surface temperature and thus low radiation pressure. The physical reason is the occurrence of dynamic instabilities in regions close to the surface under the turbulent pressure, which leads to mass outflow.

9.3.6 Effects of Convective Overshoot on Stellar Evolution

In stellar model calculations, the methods in determination of the boundary of the convective region and in calculation of the convective energy transfer will have important effects on the stellar structure and evolution. The size of the central convective core is related to the amount of matter participating in thermonuclear reactions, i.e., the energy source of

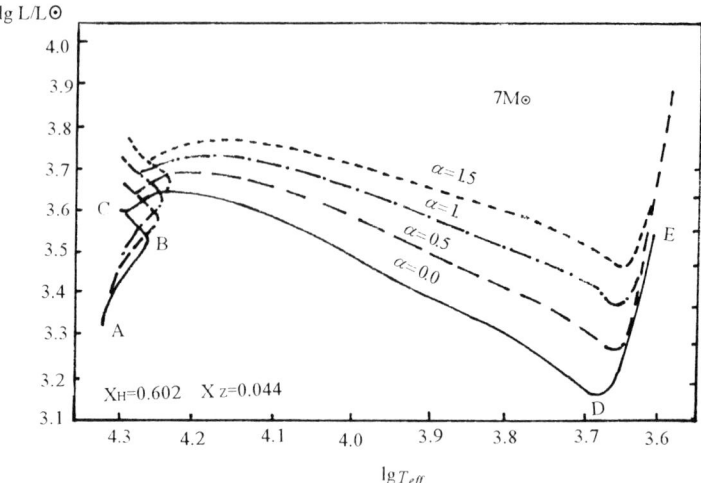

Fig. 9.19 The evolutionary tracks of a 7M$_\odot$ star in the HR diagram before helium burning for different cases of overshoot (Matraka et al. 1981).

the star. The convective energy transfer will affect the temperature distribution inside the star, the core temperature and the distribution of other physical parameters such as the pressure and density.

Matraka et al. (1981) used the Schwarzschild criterion and convective overshoot to handle the convection problem. They studied the evolution of a 7M$_\odot$ star ($X_H = 0.602$, $X_Z = 0.044$). Figure 9.19 shows the evolutionary tracks of this 7M$_\odot$ star in the HR diagram before helium burning. The parameter α attached to the evolutionary tracks is related to the overshoot. The case of $\alpha = 0$ refers to no overshoot, and larger values of α refer to larger overshoot. From Fig. 9.19, in the hydrogen burning stage (from point A to C in the figure), the overshoot has moved the evolutionary track to the upper right region, so the main sequence is widened. This is easily comprehensible because convective overshoot will enlarge the convective core so that more hydrogen can participate in nuclear reactions, and the time to the exhaust of hydrogen is prolonged, and thus the main sequence is widened. In the shell hydrogen burning stage (from point C to D in the figure), the overshoot has moved the evolutionary track upwards, i.e., towards regions with higher luminosities. Figure 9.20 shows the evolutionary tracks of a 7M$_\odot$ star in the HR diagram during helium burning for different cases of overshoot. The evolutionary tracks of intermediate massive stars during the helium burning stage have a peculiar feature, namely the loops in the HR diagram. The overshoot will diminish the amplitude of the loops. From Fig. 9.20, we see that for larger overshoot, the bluest point F of the loop of the evolutionary track migrates towards lower effective temperature, i.e., towards the red side.

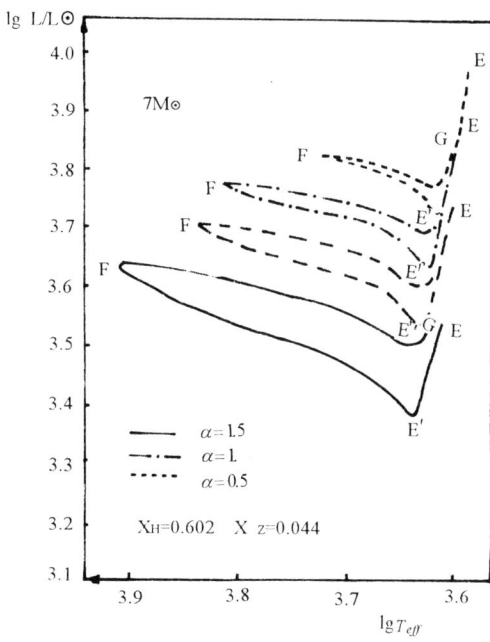

Fig. 9.20 The evolutionary tracks of a $7M_\odot$ star in the HR diagram during helium burning for different cases of overshoot (Matraka et al. 1981).

On the other hand, the envelope is also convective during the helium burning stage of the star. Matraka *et al.* (1981) showed that the ratio of the mixing length and the pressure scale height, $\alpha_o = l/H_P$, for the external convective layer has important effects on the properties of loops in the evolutionary tracks. For smaller values of α_o, the bluest point F will move towards the blue side or higher effective temperatures.

From the above, we know that the amplitude of loops in evolutionary tracks in the HR diagram during helium burning is determined by the overshoot in the internal convective layer (i.e., magnitude of the parameter α) and the ratio of mixing length to pressure scale height in the external convective layer (i.e., magnitude of α_o). However, theories on convection nowadays still cannot provide the values of these two parameters. Huang and Weigert (1983) developed a method to determine these two parameters based on the fact that a star will become a Cepheid variable if it passes the Cepheid pulsation strip while looping on the HR diagram during hydrogen burning, and based on the observation that the mass of Cepheid variables ranges from 5 to $14M_\odot$. They calculated an evolution model for a $5M_\odot$ star and found that the star will pass the Cepheid pulsation strip and become a Cepheid variable only when $\alpha = 0.5$ and $\alpha_o = 0.5$ (see Fig. 9.21), thereby determining the α and α_o values for the $5M_\odot$ star.

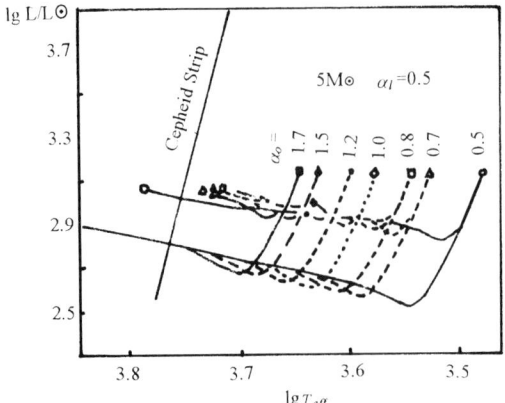

Fig. 9.21 The bluest point for the loops of a $5M_\odot$ star in the HR diagram for $\alpha = 0.5$ and for different values of α_o (Huang and Weigert 1983).

9.4 EVOLUTION OF MASSIVE STARS

Massive stars are those stars with masses greater than $9M_\odot$ whilst on the zero age main sequence. They are usually O type and B type stars with luminosity classes III (giant stars), II (bright giant stars) and I (supergiant stars), and are the brightest stellar objects in the galaxy. These stars constitute about 10% of the stars in the galaxy but have special properties and meanings.

Different from the cases for intermediate massive and low mass stars, the carbon and oxygen cores formed in massive stars after helium burning are non-degenerate and there are no violent flash phenomena in the core during carbon burning. Therefore, the outcome of massive stars is very likely the supernova explosion.

Through ultraviolet observations, people have found that massive stars have strong stellar winds, so that most or all of the envelope of a massive star will be lost during its evolution, and the core is left behind exposed. Moreover, the mass loss due to stellar wind will also greatly affect the internal structure of the star, and leads to peculiar changes in the evolutionary track of a massive star in the HR diagram.

The internal temperatures of massive stars are very high and thermonuclear reactions are vigorous in larger regions. Therefore, radiation alone cannot transfer the vast amount of energy, and large convective regions occur inside the star. Convection becomes a major component of energy transfer. Therefore, the way we treat convection, e.g., determination of the size of convective regions and the efficiency of convective energy transfer etc., will have important effects on the structure and evolution of massive stars.

Since massive stars lose mass continuously into space during evolution, in particular, they throw out heavy elements formed through thermonuclear reactions at the late evolutionary stage, they are also important sources for the interstellar medium, especially sources of heavy elements. Through observations, people can understand the products of thermonuclear reactions inside massive stars and the processes for their changes.

9.4.1 The HR Diagram of Massive Stars

Figure 9.22 is the HR diagram of massive stars in our galaxy (Humphreys and Davidson 1979). From the HR diagram, two important features can be observed. First, there is an upper limit in the luminosities of massive stars in the HR diagram, which is known as the Humphreys and Davidson limit or abbreviated as the HD limit. There are extremely few massive stars above the HD limit. Near the HD limit, there are some super giant variables, such as the luminous blue variables (LBV), Hubble–Sandage variables, P Cygni and η Carnae variables etc. The surface of these super giant variables are all unstable, which have large amounts of mass outflow. The HD limit is the highest in the O type star region, reaching about $M_{bol} = -12$. With the decrease of effective temperature, the position of the HD limit also descends. After entering the region for cool giant stars, the position of the HD limit remains unchanged. From the position of HD limit for cool giant stars, we know there will be no red super giants with a luminosity of 10^6 $L_\odot$ and a temperature lower than 15000 K.

The presence of HD limit in the HR diagram for massive stars in fact contradicts the results of the mass conservative stellar evolution theory, which shows that the evolutionary track for stars with mass greater than $60M_\odot$ can go beyond the HD limit, or in other words, the HD limit should not exist.

Besides, from Fig. 9.22, most massive stars are positioned in the very wide region from $O3$ type to Ao type stars in the HR diagram, and a small number of massive stars are positioned in the red giant region on the right of the HR diagram. There are only a very limited number of stars in the gap region between the above two regions. This shows that the main sequence for massive stars is very wide which extends to the Ao type. The observed width of the main sequence for massive stars also contradicts the results from the mass conservative evolution model calculations which show that the width of the main sequence

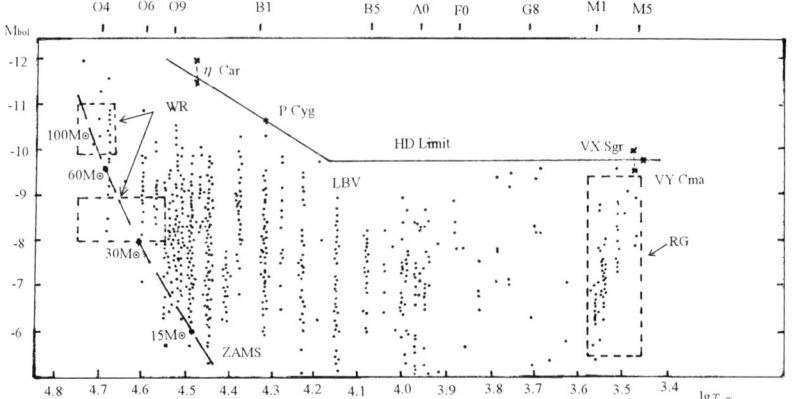

Fig. 9.22 The HR diagram of massive stars in our galaxy. The solid line is the Humphreys and Davidson limit abbreviated as the HD limit. The broken line is the zero-age main sequence. The positions for Wolf–Rayet stars (WR), luminous blue variables (LBV), and red giants (RG) have been plotted (Humphreys and Davidson 1979).

should only extend to the O9 to B0 type, which is much narrower than the observed main sequence.

9.4.2 Effects of Mass Loss due to Stellar Wind on Evolution of Massive Stars

An important feature for massive stars is the strong mass loss due to stellar wind. From observations, the mass loss rate due to stellar wind ranges from $10^{-11} M_\odot yr^{-1}$ (a 9$M_\odot$ star on the zero age main sequence) to $10^{-4} M_\odot yr^{-1}$ (a star close to the HD limit or a 40$M_\odot$ red giant). The velocity of the stellar wind ranges from 10–100 kms^{-1} (K, M type red giants) to 600–3500 kms^{-1} (O, B type early stars). Figure 9.23 shows the curves for the same mass loss rate for different types of stars in the HR diagram given by de Jager *et al.* (1983). From the figure, we see that the mass loss rate due to stellar wind is closely related to the luminosity and the effective temperature of the star. The mass loss rate due to stellar wind will be higher for stars with higher luminosities L. It can reach the order of $10^{-4} M_\odot yr^{-1}$ for supergiants close to the HD limit, and has a value of 10^{-5}–$10^{-6} M_\odot yr^{-1}$ for WR stars and 10^{-8}–$10^{-10} M_\odot yr^{-1}$ for Be stars.

For the mass loss due to stellar wind for early type stars, Castor *et al.* (1975) and Abbott (1982) proposed the radiation pressure driven mechanism, and obtained the relation among the mass loss rate, the stellar luminosity L, the effective temperature T_{eff}, the mass M and the abundance of heavy elements X_Z as

$$\dot{M} = 1.4 \times 10^{-15} \left(\frac{L}{L_\odot}\right)^{1.98} \left(\frac{X_Z}{X_{Z\odot}}\right)^{0.94} \left(\frac{M_{eff}}{M_\odot}\right)^{-1.03} \left(\frac{T_{eff}}{10^4\,K}\right)^{-0.02} (M_\odot yr^{-1})$$

(9.59)

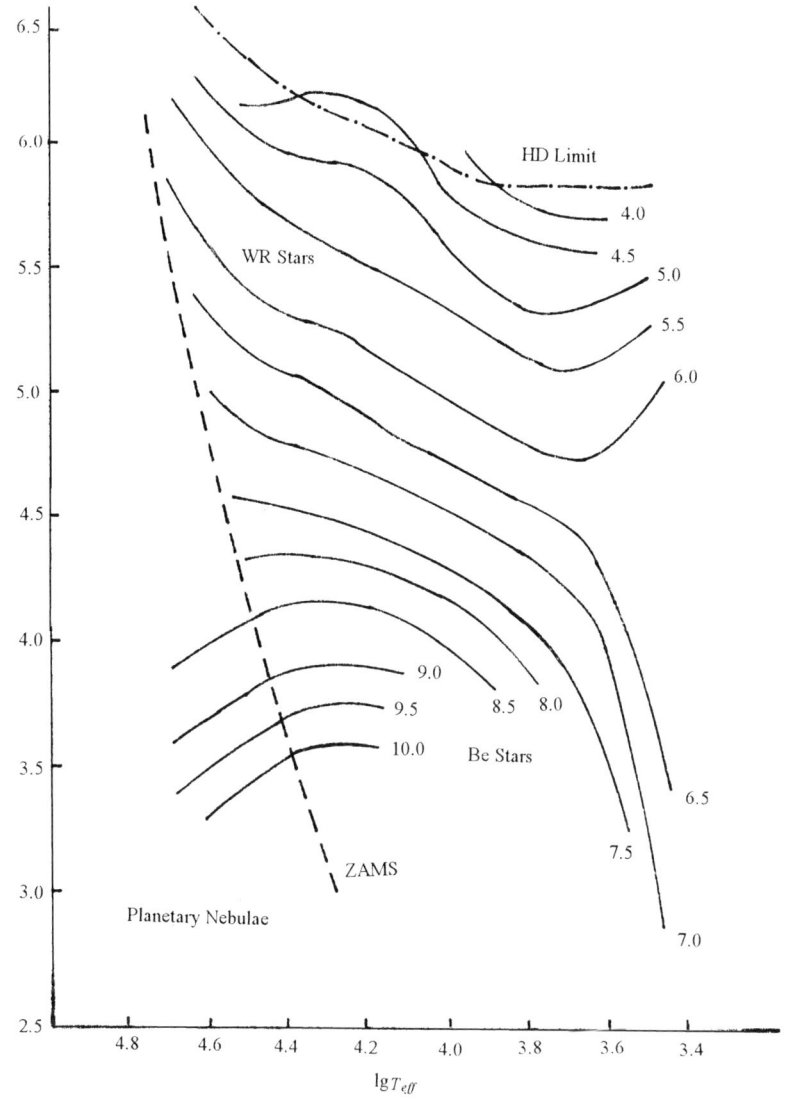

6.5

Fig. 9.23 The curves for the same mass loss rate for different types of stars in the HR diagram given by de Jager *et al.* (1983). The numbers attached to the curves are values of −log $\dot{M}$.

where

$$M_{eff} = M(1 - \Gamma) \,, \quad \Gamma = \sigma_e L/4\pi cGM \qquad (9.60)$$

and σ_e is the scattering absorption coefficient.

However, the mass loss rate due to stellar wind calculated for early type stars using Eq. (9.59) is always less than the observed value. Therefore, people tend to use the empirical formula obtained from observations (see Chiosi and Maeder 1986), for example, the semi-empirical formula given by de Jager *et al.* (1988):

$$log\,\dot{M} = 1.24\,log\left(\frac{L}{L_\odot}\right) + 0.16log\left(\frac{M}{M_\odot}\right) + 0.81log\left(\frac{R}{R_\odot}\right) - 14.02$$

$$(9.61)$$

or written as (Nieuwenhuijzen and de Jager 1990):

$$log\,\dot{M} = 1.64log\left(\frac{L}{L_\odot}\right) - 1.61logT_{eff} + 0.16log\left(\frac{M}{M_\odot}\right) - 7.93 \quad (9.62)$$

where $\dot{M}$ is in units of $M_\odot yr^{-1}$

There are still no unified theories on the mass loss mechanism due to stellar wind for late type giant stars and red giants. Reimers (1975) has provided an empirical formula as

$$\dot{M} = 4 \times 10^{-13}\eta\left(\frac{L}{gR}\right) \quad (\eta = 0.3 - 3) \quad (9.63)$$

where L and R are in solar units $L_\odot$ and $R_\odot$, and $\dot{M}$ is in units of $M_\odot yr^{-1}$.

Through model evolution calculations, we can summarize the effects of mass loss due to stellar wind on the structure and evolution of massive stars as follows.

(1) Hydrogen burning stage

When mass loss due to stellar wind is considered, the core temperature will be lowered. Since the energy generation rate in the CNO cycle reaction is very sensitive to the temperature ($\varepsilon_{CNO} \sim T^{20}$), the decrease in core temperature will lower the energy generation rate ε_{CNO} thus lowering the luminosity and prolongs the time to completion of hydrogen burning.

In the HR diagram for massive stars, the lowering of luminosity of stars due to mass loss of stellar wind will turn the evolutionary track toward the low luminosity region. Figure 9.24 shows the evolutionary tracks of stars with masses from $20M_\odot$ to $50M_\odot$ in the HR diagram calculated by De Loore *et al.* (1977). It can clearly be seen that the evolutionary tracks turn to low luminosity regions with mass loss, and the extent of turn is related to the mass loss rate due to stellar wind adopted for the model calculations.

The mass loss due to stellar wind has also prolonged the time for completion of hydrogen burning, which is equivalent to the broadening of the main sequence. This explains why the observed main sequence is very wide and extends down to the $A0$ spectral type.

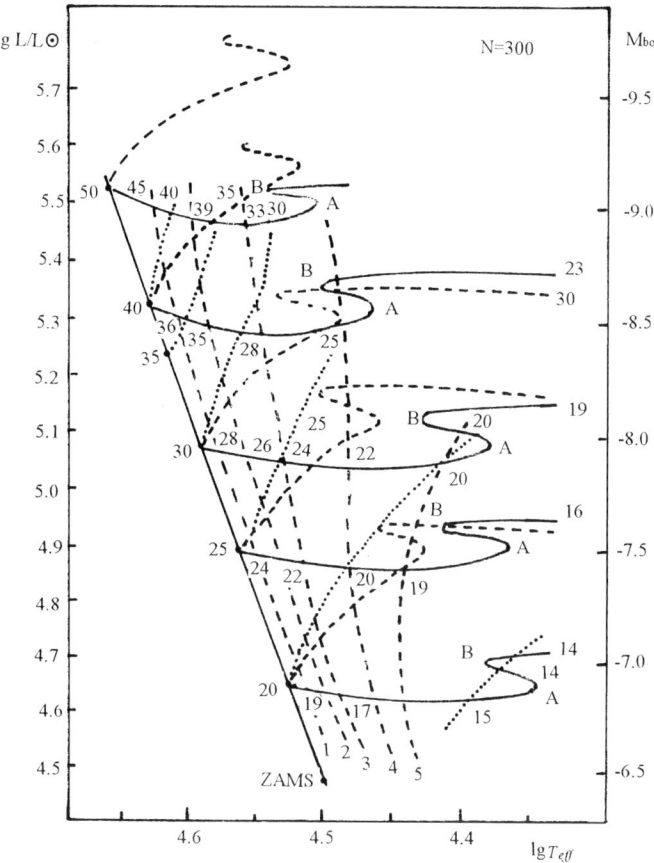

Fig. 9.24 The evolutionary tracks of stars with different masses in the HR diagram. The broken lines correspond to evolutionary tracks with constant mass while the solid lines correspond to evolutionary tracks with mass loss due to stellar wind. The numbers attached to the tracks are masses ($M_\odot$) of the stars, and the alphabets A and B represent the points at which core hydrogen burning ends and shell helium burning starts, respectively. The dotted-lines are iso-mass curves, and the dashed-lines are isochrons (De Loore *et al.* 1977).

(2) Helium burning stage

With mass loss due to stellar wind, the helium core formed after hydrogen burning will be smaller, and the start of helium burning is related to both the initial mass and rate of mass loss. Figure 9.25 gives the evolutionary tracks of massive stars with different masses and with mass loss in the HR diagram. The figure gives the regions for hydrogen burning and helium burning. It is noticed that for massive stars with different masses, some of the start points for helium burning are in the blue super giant (BSG) stage, some in the yellow super giant (YSG) stage, and some are in the red super giant (RSG) stage, while those for very massive stars are in the region of the Wolf–Rayet stars (WN or WC).

(3) Inversion of evolutionary tracks

With mass loss due to stellar wind, massive stars will lose most or all of the envelope during hydrogen or helium burning. If the star loses

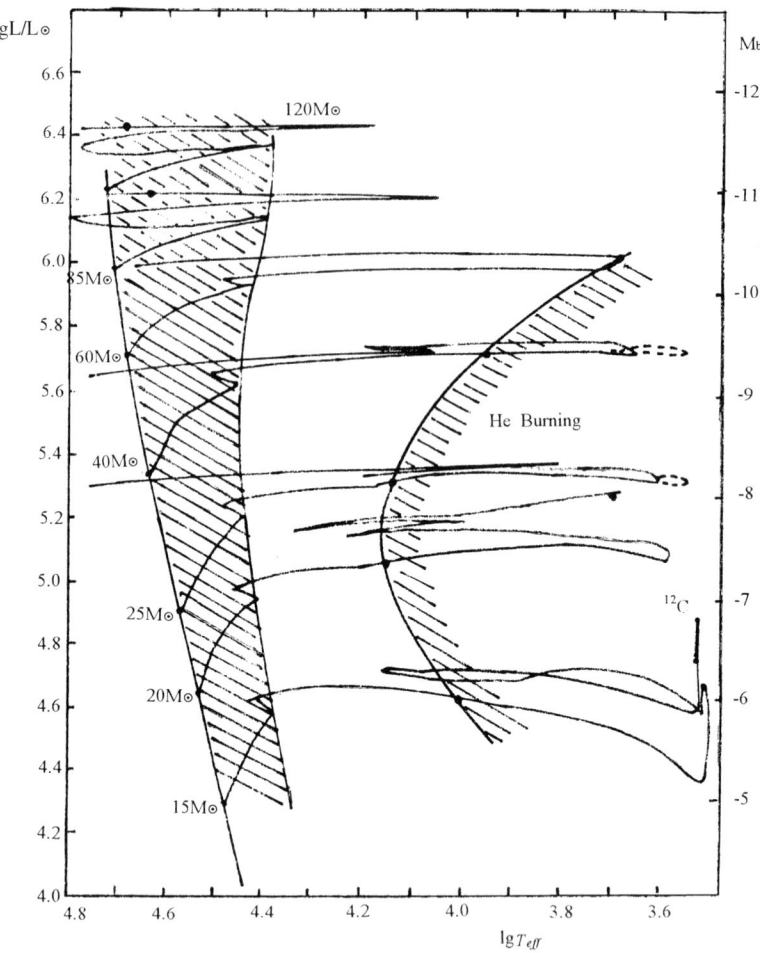

Fig. 9.25 The evolutionary tracks of massive stars with different masses and with mass loss due to stellar wind in the HR diagram. The hatched region on the left is the hydrogen burning region while the hatched region on the right is the helium burning region. The first dots on the evolutionary tracks represent the start point for helium burning. Some evolutionary tracks have a second dot which represents the start point for carbon burning (Chiosi and Maeder 1986).

the entire envelope, its evolutionary track in the HR diagram will be inverted, and move towards the high temperature blue direction. This explains the presence of the Humphreys–Davidson limit in the HR diagram for massive stars, since all the envelope has been lost when the star reaches the HD limit so the evolutionary track will be inverted and move towards the blue direction.

It should be pointed out that the envelope cannot be completely lost before reaching the HD limit by considering mass loss due to stellar wind using either the theoretical rate formula (9.59) or the semi-empirical formulae in Eqs. (9.61) and (9.62). It is believed that there are other physical mechanisms which cause instabilities in the envelope when the star evolves to near the HD limit, and produces a large amount of mass outflow.

9.4.3 Effects of Convection, Metal Abundance and Mixing on Evolution of Massive Stars

Besides mass loss, convection, metal abundance and mixing also have important effects on the evolution of massive stars. Nevertheless, up to now, there are no certain theoretical or observational evidence on how the above physical effects should be handled.

(A) Convection

In §9.3.6, we have pointed out that the treatments of convection, i.e., methods in determining the boundary of the convective region and in calculating the convective energy transfer, will have important effects on the structure and evolution of different types of stars, especially massive stars. This is because the size of the convective core will directly determine the amount of matter participating in nuclear reactions, i.e., the energy source of the star. There are many methods in handling convection in the studies of massive stars nowadays, e.g., the use of the Schwarzchild criterion, the Ledoux criterion, and the Schwarzchild criterion together with convective overshoot, etc., for determining the boundary of the convective core. Most studies are based on the last method and the size of the convective overshoot region is adopted to be 0.2–0.4 H_P (see Maeder and Conti 1994).

The effects of handling convection by use of the Schwarzchild criterion together with convective overshoot on massive stars in the hydrogen burning stage are similar to those mentioned in §9.3.6. On one hand, the evolutionary track will move towards the upper right region in the HR diagram, i.e., luminosity increases and effective temperature decreases (see Fig. 9.19). On the other hand, the time for completion of hydrogen burning is prolonged, so the main sequence is broadened. During the shell hydrogen burning stage, the evolutionary track moves upwards noticeably, i.e., towards the high luminosity region.

The observed main sequence in the HR diagram is very wide for massive stars. Even when mass loss due to stellar wind is taken into account, the theoretically obtained main sequence cannot be that wide. Therefore, convective overshoot should be considered at the same time.

(B) Metal abundance

The internal temperatures for massive stars are relatively high, so the opacity is mainly determined by electron scattering. In this way, the

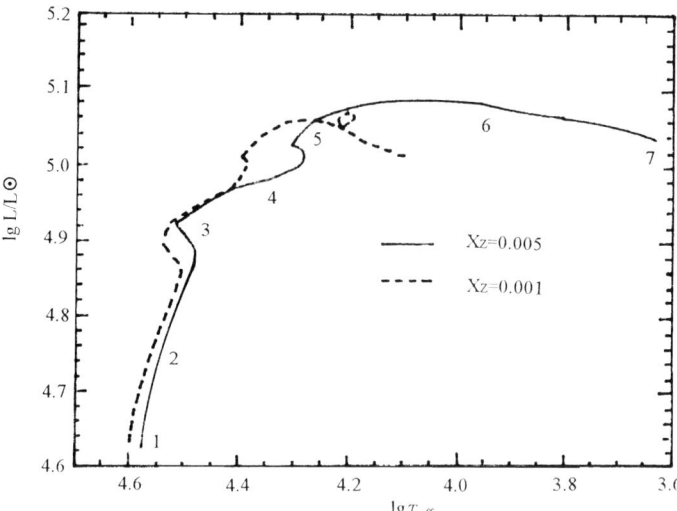

Fig. 9.26 The evolutionary tracks for a 20M$_\odot$ star with mass loss due to stellar wind with different metal abundance X_Z (from Weiss 1989).

metal abundance in massive stars will have lesser influence on opacity. Nevertheless, the difference in metal abundance will directly affect the rate of CNO cycle reactions, and thus the energy generation rate. Besides, the reaction rate for shell hydrogen burning stage is closely related to metal abundance. If the metal abundance X_Z is reduced, the shell hydrogen burning will become more violent, and the star will remain in the blue super giant region in the HR diagram during helium burning and will not evolve to the red super giant region. Furthermore, the mass loss rate due to stellar wind is related to the metal abundance X_Z [see Eq. (9.59)]. Figure 9.26 depicts the evolutionary tracks for a 20M$_\odot$ star with mass loss due to stellar wind with different metal abundance X_Z. From the figure, we see that the model with $X_Z = 0.005$ can evolve to become a red super giant, and the model with $X_Z = 0.001$ can only evolve to become a blue super giant.

The effect of metal abundance should be considered for massive stars because the precursor of the observed supernova 1987A (SN1987A) is only a blue super giant but not a red super giant in the Large Megellanic Cloud (LMC), and at the same time, spectral analyses show that the metal abundance for the LMC is smaller than the solar value by a factor of 2 to 3, and the value for the envelope of SN1987A is smaller than the solar value by a factor of 3 to 4.

(C) Mixing

If there is mixing of mass across different layers inside a massive star, e.g., mixing caused by rotation, the chemical composition in different layers will be changed. If the mixing is uniform, the helium formed by hydrogen burning can be mixed into the envelope very quickly, and the star will evolve to become a WR star whilst in the main sequence stage. If the mixing is not very uniform, the helium abundance will increase in the envelope, which favors the evolution of the star to a blue super giant. Saio *et al.* (1988) have manually increased the helium abundance in the envelope of a red super giant, and found that this star would evolve immediately from a red super giant to a blue super giant.

Mixing should be considered for massive stars because the precursor of SN1987A is a blue super giant and spectral analyses show that its envelope has relatively high values of He/H. Moreover, the helium abundance for OB super giants are also observed to be relatively high.

9.4.4 Evolution of Massive Stars with Different Initial Masses

(A) Stars with $M > 60M_\odot$

Massive stars with initial masses larger than $60M_\odot$ will have a very large mass loss rate due to stellar wind, and can lose most of the envelope during hydrogen burning and early shell hydrogen burning. When atmospheric hydrogen abundance decreases to 0.3, the evolutionary track of the star in the HR diagram will turn left. Therefore, this type of stars will not evolve to become red super giants. They are always situated in the left half of the HR diagram (Fig. 9.27). When the star loses most of its envelope and after the evolutionary track turns left, the star will enter the LBV region in the HR diagram and becomes a LBV star (see Fig. 9.22). LBV stars are also-called the Hubble–Sandage objects, and have large mass loss rate which can reach the order of $10^{-3}M_\odot yr^{-1}$ (see Laemrs 1985). Therefore, the star will lose all the remaining envelope during the LBV stage and is left with a bare core and becomes a Wolf–Rayet star. From the Wolf–Rayet stage, the star continues to evolve towards the supernova.

Therefore, the evolution of a massive star with mass greater than $60M_\odot$ will follow the following route:

O type star $\rightarrow$ Of type star $\rightarrow$ blue super giant (BSG) $\rightarrow$ luminous blue variable (LBV) $\rightarrow$ WN type Wolf–Rayet star $\rightarrow$ WC type Wolf–Rayet star $\rightarrow$ supernova (SN)

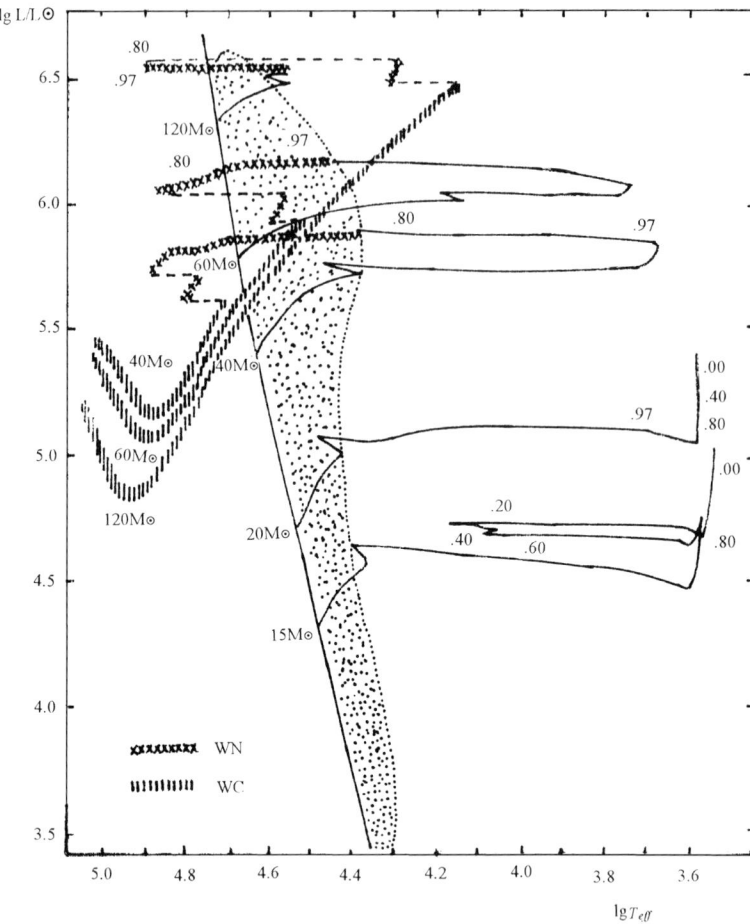

Fig. 9.27 The evolutionary tracks of stars with masses from 15 to 120M$_\odot$ ($Z = 0.02$). The numbers attached to the curves are the core helium abundances (according to de Jager *et al.* 1988).

(B) Stars with mass 25M$_\odot$ < M < 60M$_\odot$

For stars with mass in this range, the mass loss rate due to stellar wind is not very large, so the star will not lose the entire envelope during or near the main sequence. Thus, the star will rapidly evolve to become a red super giant and spend a considerable fraction of the helium burning stage within the red super giant region. For stars with masses in the upper part of the range, e.g., between 40M$_\odot$ and 60M$_\odot$, since the mass loss rate due to stellar wind is very large, they can lose the entire envelope relatively quickly, and their evolutionary tracks will move towards the direction of blue super giants. These stars which become blue super giants again will have chemical compositions different from the blue super giants evolved originally from the main sequence. From the blue super giant stage, the star will evolve to become a Wolf–Rayet star and ultimately becomes a supernova.

For stars with masses in the lower range, e.g., between 25 to $40 M_\odot$, since the mass loss rate due to stellar wind is relatively small, the envelope will not be lost entirely. The star will evolve directly from the red super giant to the supernova.

Therefore, the evolution of stars with mass between 25 and $60 M_\odot$ will follow the following routes:

O type star $\rightarrow$ blue super giant (BSG) $\rightarrow$ yellow super giant (YSG)
$\rightarrow$ red super giant (RSG) $\rightarrow$ blue super giant (BSG)
$\rightarrow$ WN type Wolf–Rayet star $\rightarrow$ supernova (SN)

$$(\text{for } 40 M_\odot < M < 60 M_\odot)$$

O type star $\rightarrow$ blue super giant (BSG) $\rightarrow$ yellow super giant (YSG)
$\rightarrow$ red super giant (RSG) $\rightarrow$ supernova (SN)

$$(\text{for } 25 M_\odot < M < 40 M_\odot)$$

(C) Stars with mass $M < 25 M_\odot$

For stars with mass less than $25 M_\odot$, the mass loss rate due to stellar wind is small, so the star will not lose the entire envelope during the main sequence or the red super giant stages. For stars with larger masses between $20 M_\odot$ and $25 M_\odot$, the evolution route starts from the main sequence, passes through the blue super giants and red super giants regions, and then evolves directly to form supernovae. For stars with masses lower than $20 M_\odot$, the evolution route goes directly from the main sequence to red super giants region, then resembles that for intermediate massive stars, i.e., there are loops in the HR diagram (see Fig. 9.27) for which the evolutionary tracks go from the red super giants region back to the blue region, passing the Cepheid pulsation strip (the stars become Cepheid variables here), and then moves back to the red super giants region, and finally reaches the supernova.

CHAPTER TEN

FINAL STAGE OF STELLAR EVOLUTION

10.1 FINAL STAGES OF STARS WITH DIFFERENT INITIAL MASSES

Stellar evolution is mainly determined by internal thermonuclear burning, which may get through evolutionary processes shown in Fig. 10.1.

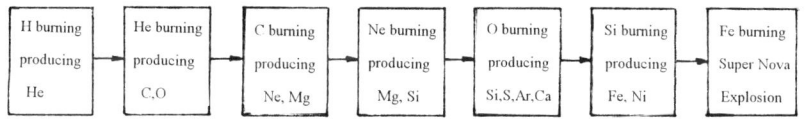

Fig. 10.1 Possible nuclear burning processes and products in the core of a star in the evolution.

If a star can go through all nuclear burning processes shown in Fig. 10.1, the final stage will be a supernova, with the explosion possibly leaving behind a neutron star or a black hole. If the star cannot go through all processes, the star will lose its envelope due to some reasons (stellar wind, surface instabilities or other causes) and evolve to a white dwarf. The main factors determining whether a star can go through all the processes include:

(A) Initial mass

In §9.3.2, it has been pointed out that if the core of a star is not degenerate, contraction of the core will lead to a rise in temperature and a new nuclear burning which is stable. On the contrary, if the core is degenerate, contraction of the core will not generally lead to a rise in temperature and new thermonuclear burning. Only when the mass of the degenerate core reaches a critical mass M_{cr} will the contraction lead to a rise in temperature and thermonuclear burning (see §9.3.3). Whether degeneracy occurs in the core of a star is determined by the core density, which is closely related to the initial mass of the star. The smaller the initial mass, the larger the core density. Therefore, for a star with an initial mass smaller than $2.3M_\odot$, the helium core formed after hydrogen burning is already degenerate. For a star with an initial mass between $2.3M_\odot$ and $9M_\odot$, the C-O core formed after helium burning is already degenerate (see Fig. 9.13). Furthermore, the masses of these degenerate cores are in general smaller than the critical mass M_{cr}. To reach the critical mass, shell burning external to the degenerate core should "engulf" the mass of the envelope, so that the product from shell burning can increase the mass of the degenerate core. In this way, the capability for the core to further contract to raise the temperature is closely related to the mass of the envelope, which is again determined by the initial mass of the star.

Generally speaking, a massive star with an initial mass greater than $10M_\odot$ can go through all nuclear burning processes in Fig. 10.1. Therefore, the final evolution stage of these massive stars is a supernova, which will leave a neutron star or black hole after explosion. These massive stars constitute only about 10% of the total number of stars. Intermediate massive and low mass stars with initial masses less than $8M_\odot$ will not go through all processes in Fig. 10.1, so their final evolution stage will be the white dwarf. These intermediate massive and low mass stars constitute about 90% of the total number of stars.

(B) Mass loss and accretion of mass

The mass loss from the stellar surface due to different mechanisms can affect the total mass of the star as well as the mass of the stellar envelope. The total mass of the star will directly influence the core density, whether the core will be degenerate, and the degree of degeneracy. Therefore, the mass loss from the stellar surface will have significant effects on whether the core will become degenerate, and whether the mass of the degenerate core will increase to the critical mass M_{cr}. In particular, the strong stellar wind with significant loss rate for massive stars and the superwind for intermediate massive and low mass stars during the later AGB stage will determine whether the star will evolve as a white dwarf or a supernova in its final stage of evolution. Although it has been mentioned before that massive stars will evolve as supernovae while intermediate massive and low mass stars will evolve as white dwarfs, the boundary has yet to be set by the mass loss from the stellar surface.

The stars can also accrete mass. In particular, members in binary systems can accrete mass from their companions. Mass accretion can change the total mass of stars as well as the mass of stellar envelopes. Therefore, intermediate massive and low mass stars, or even white dwarfs can be converted into supernovae. From §10.4, it can be seen that after mass accretion, the C-O white dwarf in a binary system can become a type I supernova without leaving compact objects after explosion.

10.2 WHITE DWARFS

White dwarfs are products of the final evolutionary stage of intermediate massive and low mass stars. They have two distinct features when compared to normal stars. First, they have a small radius ($10^{-2}R_\odot$), high density ($\sim 10^6$ gcm^{-3}) and strong surface gravitational acceleration. Second, their interiors do not have thermonuclear burning so their evolution rely on the dissipation of their own thermal energy.

Research on white dwarfs has three directions. The first one is on the structure of white dwarfs including the distribution of internal pressure and density, the mass-radius relationship and the mass limit of white dwarfs etc. The second is on thermal properties including problems on radiation and evolution. The third is on physical processes which occur inside the non-degenerate envelopes of white dwarfs, e.g., oscillation and diffusion etc. In this chapter, only the first two directions will be discussed. In Chapter 12, the oscillation in white dwarfs will be studied.

10.2.1 Structure of White Dwarfs–Chandrasekhar's Theory

The structure of white dwarfs is built on the following two assumptions. Electrons are all degenerate, and the degenerate electrons will contribute all the pressure. Ions are non-degenerate and they contribute almost the entire mass. Based on these two assumptions, and by employing the equation of hydrostatic equilibrium, the equation for the structure of white dwarfs can be obtained. The form of this equation is very similar to the Emden differential equation (7.139) of the polytropic model. By using boundary conditions, the equation can be solved to give the internal structure of white dwarfs, such as the distribution of density and mass, the mass-radius relationship and the upper limit of mass of white dwarfs etc.

Although the electrons inside white dwarfs are all degenerate, the motions of electrons are closely related to the density of white dwarfs. For white dwarfs with lower mass, the mean density is $\rho \leq 10^6$ gcm^{-3}, and the degenerate electrons have velocities far smaller than the velocity of light or are non-relativistic. For white dwarfs with larger mass, the mean density is greater than 10^6 gcm^{-3}, and the degenerate electrons have velocities close to the velocity of light or are relativistic. Therefore, the velocities of degenerate electrons inside white dwarfs are determined by the mean densities. In §4.3.3, a parameter $\xi = p_F/m_e c$ has been introduced to indicate the degree of relativistic degeneracy of electrons [see Eq. (4.123)]. According to Eqs. (4.124) and (4.122), the pressure of degenerate electrons can be written as

$$P_e = C_1 f(\xi) \tag{10.1}$$

where

$$f(\xi) = \frac{8}{m_e^5 c^5} \int_0^{p_F} \frac{p^4 dp}{(1 + p^2/m_e^2 c^2)^{1/2}} = 8 \int_0^\xi y^3 d[(1 + y^2)^{1/2}] \tag{10.2}$$

$$C_1 = \frac{\pi m_e^4 c^5}{3h^3}, \ y = p/m_e c \tag{10.3}$$

According to Eq. (4.115), the density can be further expressed as

$$\rho = C_2 \xi^3 \tag{10.4}$$

where

$$C_2 = \frac{8\pi m_e^3 c^3 m_P \mu_e}{3h^3} \tag{10.5}$$

Based on the assumption that the degenerate electrons constitute the entire pressure, the equation of hydrostatic equilibrium can be written as

$$dP_e + g_r \rho dr = 0 \tag{10.6}$$

Using $g_r = \frac{G}{r^2} \int_0^r 4\pi r^2 \rho dr$, Eq. (10.6) can be written as

$$\frac{r^2}{\rho}\frac{dP_e}{dr} + G \int_0^r 4\pi r^2 \rho dr = 0 \qquad (10.7)$$

Differentiating Eq. (10.7), we have

$$\frac{1}{r^2}\frac{d}{dr}\left(\frac{r^2}{\rho}\frac{dP_e}{dr}\right) + 4\pi G\rho = 0 \qquad (10.8)$$

Substituting Eqs. (10.1) and (10.4) into Eq.(10.8), we have

$$\frac{C_1}{C_2} \cdot \frac{1}{r^2}\frac{d}{dr}\left(\frac{r^2}{\xi^3}\frac{df(\xi)}{dr}\right) + 4\pi G C_2 \xi^3 = 0 \qquad (10.9)$$

From Eq. (10.2), we have

$$\frac{1}{\xi^3}\frac{df(\xi)}{dr} = 8\frac{d}{dr}[(1+\xi^2)^{1/2}] \qquad (10.10)$$

Introducing the parameter

$$x^2 \equiv 1 + \xi^2 \qquad (10.11)$$

and substituting Eqs. (10.10) and (10.11) into Eq. (10.9), we get

$$\frac{1}{r^2}\frac{d}{dr}\left(r^2\frac{dx}{dr}\right) + \frac{\pi G C_2^2}{2C_1}(x^2-1)^{3/2} = 0 \qquad (10.12)$$

Further introducing two dimensionless quantities Z and U to replace r and x:

$$Z \equiv r/A , \quad A = \sqrt{\frac{2C_1}{\pi G}}\frac{1}{C_2 x_c} \qquad (10.13)$$

$$U \equiv x/x_c \qquad (10.14)$$

where x_c represents the value of x at the center. Therefore Eq. (10.12) can be rewritten as

$$\frac{1}{Z^2}\frac{d}{dZ}\left(Z^2\frac{dU}{dZ}\right) + \left(U^2 - \frac{1}{x_c^2}\right)^{3/2} = 0 \qquad (10.15)$$

Equation (10.15) is the differential equation describing the structure of white dwarfs, and was first derived by Chandrasekhar. The form of this equation is very similar to the Emden differential equation (7.139) of the polytropic model. In fact, when $x \to \infty$ (i.e., $\xi \to \infty$, the fully relativistic condition) and $x \to 1$ (i.e., $\xi \to 0$, the fully non-relativistic condition), Eq. (10.15) becomes the Emden differential equation with $n = 3$ and $n = 1.5$, respectively. The boundary conditions at the center for the differential equation (10.15) are:

At the center $(Z = 0):$ $U = 1, U' = 0$ (10.16)

Using the boundary conditions at the center and by providing the x_c value (i.e., providing the parameter ξ_c), integration of Eq. (10.15) can be performed from the center outwards. The Z value from the integration out to the surface is Z_s. Since the density at the surface is $\rho = 0$, according to Eqs. (10.4) and (10.11), we can get some parameters at the surface as:

When $Z = Z_s$: $\xi_s = 0,$ $x_s = 1,$ $U_s = 1/x_c$ (10.17)

From Eq. (10.4), the distribution of density for white dwarfs is

$$\rho = C_2 \xi^3 = C_2 (x^2 - 1)^{3/2} = C_2 x_c^3 \left(U^2 - \frac{1}{x_c^2} \right)^{3/2}$$ (10.18)

The radius R of a white dwarf is

$$R = AZ_s = \sqrt{\frac{2C_1}{\pi G}} \cdot \frac{1}{C_2 x_c} Z_s$$ (10.19)

The mass M of a white dwarf can be obtained from Eqs. (10.14) and (10.19) through elimination of r and U, i.e.,

$$
\begin{aligned}
M &= \int_0^R 4\pi r^2 \rho\, dr = 4\pi A^3 C_2 x_c^3 \int_0^{U_s} Z^2 \left(U^2 - \frac{1}{x_c^2} \right)^{3/2} dZ \\
&= 4\pi A^3 C_2 x_c^3 \left(-Z^2 \frac{dU}{dZ} \right)_s \\
&= \frac{4\pi}{C_2^2} \left(\frac{2C_1}{\pi G} \right)^{3/2} \left(-Z^2 \frac{dU}{dZ} \right)_s
\end{aligned}
$$ (10.20)

Table 10.1 The central density, mass and radius for white dwarfs corresponding to different values of x_c (taken from Cox and Giuli 1968).

$1/x_c^2$	ξ_c	Z_s	$\left(-Z^2\frac{dU}{dZ}\right)_s$	ρ_c/μ_e (gcm^{-3})	$\mu_e^2 M$ (M$_\odot$)	$\mu_e R$ (km)
0	∞	6.8968	2.0182	∞	5.84	0
0.01	9.95	5.3571	1.9321	9.48×10^8	5.60	4.170
0.02	7	4.9857	1.8652	3.31×10^8	5.41	5.500
0.05	4.36	4.4601	1.7096	7.98×10^8	4.95	7.760
0.1	3	4.0690	1.5186	2.59×10^7	4.40	10.000
0.2	2	3.7271	1.2430	7.70×10^6	3.60	13.000
0.3	1.53	3.5803	1.0337	3.43×10^6	2.99	16.000
0.5	1	3.5330	0.7070	9.63×10^5	2.04	19.500
0.8	0.5	4.0446	0.3091	1.21×10^5	0.89	28.200
1.0	0	∞	0	0	0	∞

Taking values between $x_c = \infty$ (fully relativistic) and $x_c = 1$ (fully non-relativistic) and integrating Eq. (10.15), different Z_s values at the surface can be obtained. On providing the value of μ_e (which determines the value of C_2), we can use Eqs. (10.18), (10.19) and (10.20) to calculate the corresponding central density ρ_c, radius R and mass M for white dwarfs (see Table 10.1).

From Table 10.1, the mass-radius relationship for white dwarfs can be obtained. It can also be observed that, when $x_c = \infty$, i.e., the fully relativistic condition, $\rho_c = \infty$, $R = 0$, the mass M (written as M_{ch}) should be the upper limit for the mass of white dwarfs. If $M > M_{ch}$, the gravitational force will be greater than the electron pressure and the white dwarf will collapse. However, R is already zero at this time so no more collapse can occur. Therefore, M_{ch} should be the limiting mass for white dwarfs. From Table 10.1, we have

$$M_{ch} = 5.84/\mu_e^2 \qquad (10.21)$$

so the limiting mass is related to the chemical composition μ_e. According to Eq. (4.39), $\mu_e = 2/(1 + X_H)$. There is almost no hydrogen in white dwarfs so $\mu_e \approx 2$, so

$$M_{ch} = 1.44 M_\odot \qquad (10.22)$$

The limiting mass obtained above, together with the observation from Table 10.1 that the radius of white dwarfs decreases with the increase of mass, do not agree with the normal conception of a star that the mass should grow with the radius. The discrepancy arises because the need for white dwarfs to fulfil hydrostatic equilibrium and the internal gas pressure is mainly constituted by the degenerate electron pressure. We have the following discussions for these points.

The equation of hydrostatic equilibrium is:

$$\frac{dP}{dr} = -\rho \frac{GM_r}{r^2} \qquad (10.23)$$

Under complete non-relativistic degeneracy

$$P \sim \rho^{5/3} \approx \frac{M^{5/3}}{R^5} \qquad (10.24)$$

Therefore, the left-hand side of Eq. (10.23) can be written as

$$\frac{dP}{dr} \cong \frac{M^{5/3}}{R^6} \qquad (10.25)$$

Under complete relativistic degeneracy

$$P \sim \rho^{4/3} \approx \frac{M^{4/3}}{R^4} \qquad (10.26)$$

Therefore, the left-hand side of Eq. (10.23) can be written as

$$\frac{dP}{dr} \cong \frac{M^{4/3}}{R^5} \qquad (10.27)$$

On the other hand, the right-hand side of Eq. (10.23) can be written as

$$\rho \frac{GM_r}{r^2} \cong \frac{M^2}{R^5} \qquad (10.28)$$

From Eqs. (10.25) and (10.28), we obtain

$$M^{1/3}R = \text{const} \tag{10.29}$$

which means that when the mass of a white dwarf is not very large, the density is not very high, and when the electrons are non-relativistic, the mass is inversely proportional to the cube of the radius. In other words, the mass gets larger when the radius gets smaller. From Eqs. (10.27) and (10.28), we obtain

$$C\frac{M^{4/3}}{R^5} = \frac{M^2}{R^5}$$

$$\text{or} \quad M = C^{-3/2} \tag{10.30}$$

which means when the mass of the white dwarf grows to the mass limit, the density has increased to make the electrons relativistic and the mass becomes a constant and is no longer related to the radius. This corresponds to the upper limit for the mass of white dwarfs.

In Chandrasekhar's theoretical derivations presented above, Eqs. (10.1) and (10.4) have been employed, which together give the equation of state $P = f(\rho)$ (assuming that the electron pressure is the most important). It is worth pointing out that this equation of state has been obtained under the assumption of no interactions between particles, i.e., under the condition that the interior of a white dwarf is treated as an ideal gas. However, when the internal density of the white dwarf is very high, the gas particles have Coulomb interactions, e.g., Coulomb interactions between ions and ions, electrons and electrons, and ions and electrons. If Coulomb interactions for a non-ideal gas under a high density is taken into account, the obtained total pressure and thus the upper limit for the mass of white dwarfs will be decreased a little bit. In order to demonstrate the relationship between the decrease of the total pressure caused by Coulomb interactions and the density of the white dwarf, we have the following derivations. Assume there are n nearest ions around an ion, and the mean distances between the nearest ions are r_i, so that the ion-ion Coulomb interaction can be approximately represented as

$$E_I \sim \frac{1}{2}n^2 N_i \frac{(Ze)^2}{r_i} \tag{10.31}$$

where N_i is the total number of ions in the white dwarf, Z is the number of nearest electrons around an ion. Assuming that the mean distance between an ion and its nearest electrons is r_e, since the volume occupied by an ion is larger than that occupied by an electron by a factor of Z, we have

$$r_e \cong Z^{-1/3} r_i \tag{10.32}$$

The energy of electron-electron Coulomb interaction can be approximately written as

$$
\begin{aligned}
E_e &\sim \frac{1}{2} (Zn)^2 \frac{e^2}{r_e} N_i \\
&\cong Z^{1/3} \left(\frac{1}{2} n^2 N_i \frac{(Ze)^2}{r_i} \right) = Z^{1/3} E_I
\end{aligned}
\tag{10.33}
$$

Finally, the energy of ion-electron Coulomb interaction can be approximately written as

$$
\begin{aligned}
E_{Ie} &\sim -n(Zn) N_i \frac{(Ze)e}{r_{ie}} \\
&\cong -n^2 N_i \frac{(Ze)^2}{r_i} Z^{1/3} = -2 Z^{1/3} E_I
\end{aligned}
\tag{10.34}
$$

In the above equations, we have assumed $r_{ie} \approx r_e$. Therefore, the total energy of Coulomb interaction between particles is

$$E_{Coul} = E_I + E_e + E_{Ie} \cong \frac{1}{2} n^2 N_i \frac{(Ze)^2}{r_i} (1 - Z^{1/3}) \tag{10.35}$$

Considering that the volume occupied by an ion is $V_i = V/N_i$ (V is the volume of the white dwarf), we have $r_i \sim V_i^{1/3} \sim (V/N_i)^{1/3}$, so

$$E_{Coul} \cong \frac{1}{2} n^2 N_i^{4/3} \frac{(Ze)^2}{V^{1/3}} (1 - Z^{1/3}) \tag{10.36}$$

Based on the Maxwell relationship, through partial differentiating Eq. (10.36) with respect to V (keeping N_i unchanged), we can show the change of the total pressure when Coulomb interactions are considered:

$$
\begin{aligned}
\Delta P = \Delta P_{Coul} &= \left(\frac{\partial E_{Coul}}{\partial V} \right)_{N_i} = \frac{1}{6} n^2 (Ze)^2 (1 - Z^{1/3}) \left(\frac{N_i}{V} \right)^{4/3} \\
&= n^2 \frac{(Ze)^2 (1 - Z^{1/3})}{6 A^{4/3} m_p^{4/3}} \rho^{4/3} \\
&\approx 1.94 \times 10^{12} \frac{Z^2 (1 - Z^{1/3})}{A^{4/3}} \rho^{4/3}
\end{aligned}
\tag{10.37}
$$

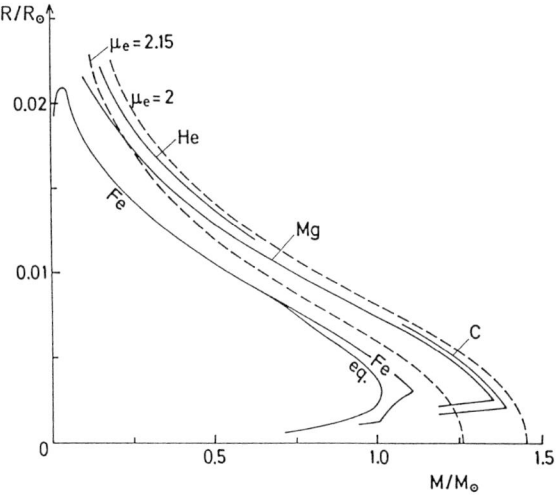

Fig. 10.2 The mass–radius relationship for white dwarfs. The broken lines refer to the conditions $\mu_e = 2$ (^{4}He, ^{12}C, ^{16}O, ...) and $\mu_e = 2.15$ (^{56}Fe), and are computed based on Chandrasekhar's theory. The other solid lines are obtained with the consideration of Coulomb interactions and correspond to white dwarfs composed of He, C, Mg and Fe (according to Hamada and Salpeter 1961).

Equation (10.37) gives the relationship between the change in total pressure with the density when Coulomb interactions are considered. In the derivations, we have employed $\rho = (N_i A m_p)/V$. When the total pressure changes, there are some changes in the structural equation obtained from the equation of hydrostatic equilibrium in Eq. (10.6) and the mass-radius relationship obtained from the structural equation, and also in the upper limits of mass of white dwarfs. Figure 10.2 shows the mass-radius relationship for white dwarfs, where the broken lines refer to results obtained by the integration of Eq. (10.15) under the conditions $\mu_e = 2$ and $\mu_e = 2.15$, and through Eqs. (10.18), (10.19) and (10.20), while the other solid lines are the mass-radius relationship obtained with the consideration of Coulomb interactions and correspond to white dwarfs composed of He, C, Mg and Fe.

10.2.2. Structure of the Envelope

From observations, the surface temperature of white dwarfs is of the order of 10^4 K, and the density is smaller than 10^2 gcm^{-3}. From these, we know that the envelopes of white dwarfs are composed of non-degenerate, or very lowly degenerate ideal gas. However, inside the white dwarfs, the electrons are completely degenerate. Since degenerate electrons have very high thermal conductivity, and since the luminosities L of white dwarfs are comparatively small, the inside of white dwarfs should have small temperature gradients and can be approximated to have a uniform temperature. Therefore, a white dwarf consists of a core with a uniform temperature and degenerate electrons, and an envelope made of an ideal gas.

The fact that a white dwarf has an envelope composed of a non-degenerate ideal gas has important implications. Inside such an envelope, the energy transfer is mainly through radiation and convection. However, the transfer efficiencies of radiation and convection are far smaller than the heat conduction of completely degenerate electrons. Therefore, the presence of the envelope will greatly reduce the outward energy transfer, thereby greatly slowing down the cooling process and the evolution of white dwarfs.

The thickness and mass of the envelope, the distribution of pressure, temperature and density in the envelope are all interesting research topics.

Assume that the core with constant temperature intersects with the envelope at a certain boundary, i.e., the part inside this boundary is made of a gas with completely degenerate electrons while the part outside is made of a non-degenerate ideal gas. The temperature and density at this boundary can be obtained by equating the pressures for degenerate electrons and for an ideal gas. From the pressure of degenerate electrons given in Eq. (4.118),

$$\frac{\Re}{\mu} \rho_0 T_0 = \frac{h^2}{20 m_e m_P} \left(\frac{3}{\pi m_P}\right)^{2/3} \left(\frac{\rho_0}{\mu_e}\right)^{5/3} \tag{10.38}$$

where the subscript "o" refers to the quantities at the boundary. From Eq. (10.38),we have

$$\rho_0 = C_1 T_0^{3/2} \tag{10.39}$$

and

$$C_1 = 2.21 \times 10^{-8} \frac{\mu_e^{5/2}}{\mu^{3/2}} \tag{10.40}$$

Since the envelope is at hydrostatic equilibrium, energy transfer is mainly through radiation. Therefore, the equation of hydrostatic equilibrium and the radiative energy transfer equation can be applied in the envelope:

$$\frac{dP}{dr} = -g_r \rho = -\frac{GM}{r^2} \rho \tag{10.41}$$

$$\frac{dT}{dr} = -\frac{3\kappa\rho}{16\pi ac}\frac{L}{r^2 T^3} \tag{10.42}$$

In the above equations, we assume $M_r \approx M$, $g_r \approx g$, $L_r \approx L$. If the opacity κ can be represented by the Kramers equation, i.e., $\kappa = \kappa_0 \rho T^{-3.5}$ [see Eq. (5.91)], dividing Eq. (10.41) by Eq. (10.42) gives dP/dT and subsequent integration will give

$$P = \left(\frac{16\pi acG}{12.75\kappa_0} \cdot \frac{M}{L}\right)^{1/2} T^{4.25} \tag{10.43}$$

In the above integration, the zero boundary condition $T = 0$ at $P = 0$ has been employed. Replacing P in Eq.(10.43) by $\Re T\rho/\mu$, we have

$$\rho = C_2 \cdot T^{3.25} \tag{10.44}$$

where

$$C_2 = \left(\frac{16\pi acG}{12.75\kappa_0} \cdot \frac{M}{L}\right)^{1/2}\frac{\mu}{\Re} \tag{10.45}$$

Applying Eq. (10.44) at the boundary, we get

$$\rho_0 = C_2 \cdot T_0^{3.25} \tag{10.46}$$

Using Eqs. (10.39) and (10.46) to eliminate ρ_0, we obtain

$$T_0^{3.5} = \left(\frac{C_1}{C_2}\right)^2 = B\frac{L/L_\odot}{M/M_\odot} \tag{10.47}$$

From Eq. (10.47), we arrive at

$$T_0 = B^{2/7}\left(\frac{L/L_\odot}{M/M_\odot}\right)^{2/7} \cong 5.9 \times 10^7 \left(\frac{L/L_\odot}{M/M_\odot}\right)^{2/7} K \tag{10.48}$$

Observational results for white dwarfs include: $M \approx M_\odot$, $L/L_\odot \approx 10^{-4}$–$10^{-2}$. Substituting these into Eq. (10.48), we find the temperature at the boundary to be $T_0 \approx 4.2 - 16 \times 10^6$ K. Since the temperature inside the boundary is constant, we know the core temperature of the white dwarf is also T_0. If $T_0 \approx 10^6$ K is adopted and put into Eq. (10.39), the density at the boundary can be obtained to be $\rho_0 \approx 10^3$ gcm^{-3}.

The thickness l of the envelope of the white dwarf can be approximated as

$$l \approx P/\rho g \cong \frac{kT}{\mu g} \tag{10.49}$$

where $g = GM/R^2$ and $g \sim 5 \times 10^8$ cm/s^2 for a typical white dwarf, temperature $T \leq 10^6$ K, so $l \leq 10^{-3}$R $\approx$ 1–10 km. After knowing l and ρ_0, the mass of the envelope M_s can be estimated to be

$$M_s \leq 4\pi R^2 l \rho_0 \cong 2 \times 10^{-4} M_\odot \tag{10.50}$$

From this, it can be seen that the mass of the envelope is very small, which justifies our previous assumption of $M_r = M$, $g_r = g$, $L_r = L$ when we used the equation of hydrostatic equilibrium and the radiative transfer equation.

When Eqs. (10.43), (10.44) and the ideal gas law are further substituted into Eq. (10.41) and differentiation is then performed, we have

$$T = \frac{2}{8.5} \frac{\mu}{k} MG\left(\frac{1}{r} - \frac{1}{R}\right) \tag{10.51}$$

which gives the temperature distribution in the envelope of white dwarfs. Knowing the temperature distribution, the pressure distribution can be obtained from Eq. (10.43). From the distribution of T and P, the density distribution can be determined from the ideal gas law.

10.2.3 Cooling of White Dwarfs

Since thermonuclear reactions have ceased in white dwarfs, the external radiation relies entirely on the dissipation of its own thermal energy. With the loss of thermal energy, the white dwarf becomes dim when it is cooled. When all the thermal energy is lost and thus $T \to 0$, the white dwarf becomes a black dwarf. Therefore, the evolution of a white dwarf is in fact its cooling. Our interest lies in the length of time elapsed before the white dwarf has lost all its thermal energy and becomes a black dwarf, which is, in other words, the lifetime of the white dwarf.

Integrating the energy equation (7.11), we have

$$L = \int_0^M (\varepsilon_n - \varepsilon_\nu + \varepsilon_g) dm \tag{10.52}$$

Since thermonuclear reactions have ceased in the white dwarf, and since not a large amount of neutrinos are produced due to conditions of temperature and density inside the white dwarf, $\varepsilon_n = \varepsilon_v = 0$ in Eq. (10.52). If T and ρ are taken as independent variables, from the first law of thermodynamics, we have

$$dQ = C_V dT + P d\left(\frac{1}{\rho}\right) = C_V dT - \frac{P}{\rho^2} d\rho \qquad (10.53)$$

so that

$$\varepsilon_g = -\frac{dQ}{dt} = -C_V \dot{T} + \frac{P}{\rho^2} \dot{\rho} \qquad (10.54)$$

Since a white dwarf is a compact stellar object with a high degree of degeneracy, its compression term, i.e., the second term on the right of Eq. (10.54), can be neglected. Therefore, Eq. (10.52) becomes

$$L = -\int_0^M C_V \dot{T} dm \qquad (10.55)$$

The envelope of the white dwarf is composed of ions and electrons (assuming part of them are degenerate). For such a system, the theorem of equipartition of energy is not applicable. In this way, the internal energy density can be written as

$$U = \frac{3}{5} n_e E_F + \frac{\pi^2}{4} n_e \left(\frac{kT}{E_F}\right) kT + \frac{3}{2}\left(\frac{\rho}{\mu m_p}\right) kT \qquad (10.56)$$

On the right of Eq. (10.56), the last term is the thermal energy of ions with a temperature T. The first term is the minimum energy for degenerate electrons at $T = 0$ where E_F is the Fermi energy. The Fermi energy can be visualized in the following way. When the density of a star is high and the electrons become degenerate, the distance between electrons become comparable to the Compton wavelength so the properties of matter are determined by the uncertainty principle in quantum mechanics, that is, the momentum p and the linear distance λ of electrons are required to fulfil the uncertainty relationship $p\lambda \geq \hbar$. Since the volume of the electron is $V_o \sim \lambda^3$, we have $p \geq \hbar(V_o^{1/3})$, so the electron energy is

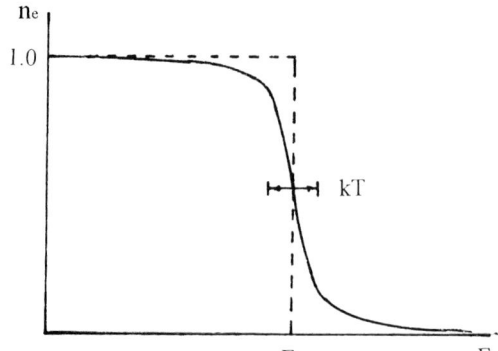

Fig. 10.3 The relationship between the number density n_e and the electron energy E. For a complete degenerate system at $T = 0$, the distribution is given by the broken line. For $kT << E_F$, the distribution is given by the solid line.

$$E_0 \approx \frac{p^2}{2m_e} \geq \frac{\hbar^2}{2m_e} V_0^{-2/3} = \frac{\hbar^2}{2m_e} \left(\frac{N_e}{V}\right)^{2/3} = \frac{\hbar^2}{2m_e} n_e^{2/3} \equiv E_F \qquad (10.57)$$

which shows that the electron energy will be E_F but not zero when the temperature drops to zero.

The second term on the right of Eq. (10.56) is the thermal energy of electrons at temperature T. This can be visualized as follows. When $T = 0$, all electrons are at the lowest energy states and the electron energy is E_F. The relationship between the number density n_e and the electron energy E is shown by the broken line in Fig. 10.3.

However, under the temperature ($kT << E_F$) of the envelope of the white dwarf, the n_e distribution is given by the solid line in Fig. 10.3, which deviates from that for $T = 0$ only within a range of kT around the energy E_F. In this way, only a small amount (kT/E_F) of the electrons will have a mean energy of kT, so that the mean thermal energy of electrons with a temperature T is proportional to $n_e(kT/E_F)kT$. Out of the three terms on the right of Eq. (10.56), the third term (thermal energy of ions) is the largest and is much larger than the thermal energy of electrons ($kT/E_F << 1$). Therefore,

$$C_V = \left(\frac{\partial U}{\partial T}\right)_V \cong \frac{3}{2}\frac{k\rho}{\mu m_p} = \frac{3}{2}\frac{k}{A m_p} \qquad (10.58)$$

Substituting Eq. (10.58) into Eq. (10.55), and considering the constant temperature inside the white dwarf, $dT/dt = dT_c/dt = dT_0/dt$ (where T_c is the core temperature of the white dwarf which is equal to the temperature T_0 at the boundary between the core and the envelope). Moving C_V and dT_c/dt out of the integration sign, we finally arrive at

$$L = -\frac{3}{2}\frac{k}{Am_P}\frac{dT_c}{dt}\int_0^M dm = -\frac{3}{2}\frac{kM}{Am_P}\frac{dT_c}{dt} \qquad (10.59)$$

Putting Eq. (10.59) into Eq. (10.47), eliminating L and rewriting T_o as T_c, we have

$$\frac{dT_c}{dt} = -\frac{L_\odot}{M_\odot}\frac{2Am_P}{3}\frac{1}{k}\frac{1}{B}T_c^{3.5} \qquad (10.60)$$

Integrating Eq. (10.60) with respect to time, denoting the lifetime of the white dwarf as τ, assuming $T_c(0) \gg T_c(\tau)$, and after integration, replacing T_c by L through Eq. (10.48), we have

$$\begin{aligned}
\tau &= \frac{2}{5}\left(\frac{M_\odot}{L_\odot}B\right)^{2/7}\frac{3}{2}\frac{k}{Am_P}\left(\frac{M}{L}\right)^{5/7} \\
&\simeq \frac{4.7\times 10^7}{A}\left(\frac{M/M_\odot}{L/L_\odot}\right)^{5/7} \quad \text{(yr)}
\end{aligned} \qquad (10.61)$$

When $A = 4$, $M = M_\odot$, $L/L_\odot = 10^{-3}$, we get $\tau \approx 10^9\,y$, which is comparable to the age of the universe.

The evolutionary track in the HR diagram when the white dwarf is cooled can be determined by the mass-radius relationship, $M \sim R^{-3}$,

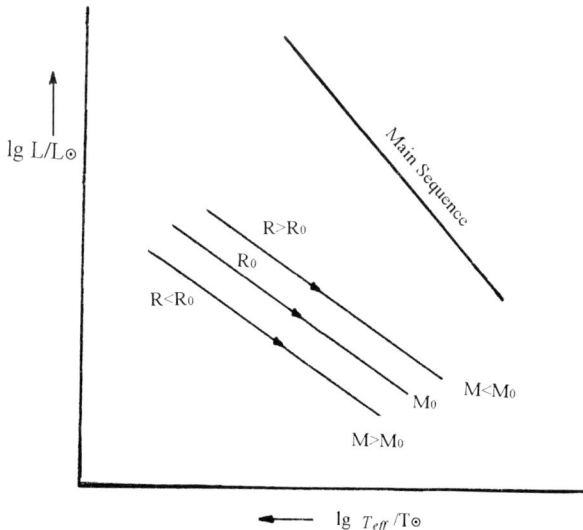

Fig. 10.4 Evolutionary tracks of white dwarfs in the HR diagram.

given by Eq. (10.29), and the definition of the effective temperature $L \sim R^2 T_{eff}^4$, through the elimination of the radius R, i.e.,

$$\log\left(\frac{L}{L_\odot}\right) = 4\log\left(\frac{T_{eff}}{T_\odot}\right) - \frac{2}{3}\log\left(\frac{M}{M_\odot}\right) + C \qquad (10.62)$$

Figure 10.4 gives the evolutionary tracks of white dwarfs in the HR diagram. It is observed that the evolutionary track is at a lower position when the mass of the white dwarf is larger.

10.3 WEAK INTERACTION PROCESSES

A massive star with an initial mass greater than $10M_\odot$ can go through all nuclear burning processes shown in Fig. 10.1, and the final stage is a supernova. However, starting from carbon burning, the core temperature exceeds 0.5×10^9 K, and weak interactions will occur. For example, during the carbon, neon, oxygen and silicon burning processes, different types of neutrinos-producing weak interactions become important. On the completion of the core silicon burning, the temperature has exceeded 5×10^9 K and the density has reached $\sim 10^9$ gcm^{-3}. Under these conditions, photodisintegration and electron capture will occur. These weak interactions have important effects in the final evolution of stars. For example, different types of neutrinos-producing weak interactions can greatly accelerate the carbon, neon, oxygen and silicon burning, so that the time taken to go through these processes is greatly shortened. On the other hand, the photodisintegration and electron capture will cause the collapse of the iron core of stars and finally lead to supernova explosion.

10.3.1 Neutrino Loss

Starting from carbon burning, the core temperature of the star has exceeded 0.5×10^9 K. Under such a high temperature, different types of neutrinos-producing weak interactions will take place, e.g., electron pair annihilation neutrino process, photoneutrino process, plasmaneutrino process and bremsstrahlung neutrino process (see §6.15 and Fig. 6.7). These weak interactions produce large amounts of neutrinos in the core of the star, which can escape through the star without obstruction and take away large amounts of energy. During carbon, neon, oxygen

Table 10.2 The ratios between the neutrino luminosity L_ν and the surface luminosity L for a $15M_\odot$ star and a $25M_\odot$ star during carbon, neon, oxygen and silicon burning and the lifetime for each nuclear burning process (Weaver *et al.* 1978).

Nuclear burning process	$M = 15M_\odot$ $(L \cong 10^4 L_\odot)$		$M = 25M_\odot$ $(L \cong 3 \times 10^5 L_\odot)$	
	L_ν/L	$\tau(y)$	L_ν/L	$\tau(y)$
C	1.0	6.3×10^3	8.3	1.7×10^2
Ne	1.8×10^3	7	6.5×10^3	1.2
O	2.1×10^4	1.7	1.9×10^4	0.51
Si	9.2×10^5	0.017	3.2×10^6	0.004

and silicon burning, the energy loss due to neutrinos far exceeds the energy radiated from the stellar surface. Weaver *et al.* (1978) studied the energy loss due to neutrinos for a $15M_\odot$ star and a $25M_\odot$ star during carbon, neon, oxygen and silicon burning. Their results showed that during nuclear burning processes towards the end of the evolution, the neutrino luminosity L_ν (energy loss due to neutrinos per second) is greater than the total energy L radiated from the stellar surface by a factor which can reach 10^6 (see Table 10.2). Because of this, the lifetime for nuclear burning will be greatly shortened. Table 10.2 lists the lifetime for each nuclear burning process. It can be seen that neon burning only lasts for a few years while silicon burning lasts for a few days. In conclusion, energy loss by neutrinos greatly shortens the late stage evolution of stars.

10.3.2 Photodisintegration

If the core temperature of a star exceeds 5×10^9 K, thermal energies of photons have exceeded the binding energies of nuclei of heavy elements including those of iron. In such conditions, the photons can disintegrate different atomic nuclei in the core, which is called the photodisintegration of heavy atomic nuclei. For example, the photo-disintegration of ^{20}Ne is

$$^{20}\text{Ne} + \gamma \;\rightleftarrows\; ^{16}\text{O} + ^4\text{He} \qquad (10.63)$$

This process is similar to ionization and recombination of atoms. At equilibrium, the number densities for Ne, O and He fulfil an equation similar to the Saha equation (4.10), i.e.,

$$\frac{n_O n_{He}}{n_{Ne}} = \frac{g_O g_{He}}{g_{Ne}} \frac{(2\pi m_O m_{He} kT)^{3/2}}{h^3} e^{-Q_{Ne}/kT} \qquad (10.64)$$

where n_O, n_{He} and n_{Ne} are the number densities of ^{16}O, ^4He and ^{20}Ne, g_O, g_{He} and g_{Ne} are the corresponding weighting factors, Q_{Ne} is the binding energy for the Ne atom, i.e.,

$$Q_{Ne} = (m_O + m_{He} - m_{Ne})c^2 \qquad (10.65)$$

In general, when an atomic nucleus with nucleon number A and electric charge Z undergoes photodisintegration, there are two possible outcomes, the first producing ^4He and the second producing a neutron:

$$(Z, A) + \gamma \rightleftarrows (Z - 2, A - 4) + ^4He \qquad (10.66)$$

$$(Z, A) + \gamma \rightleftarrows (Z, A - 1) + n \qquad (10.67)$$

For example, a ^{56}Fe nucleus can go through 13 ^4He-producing photodisintegrations and four neutron-producing photodisintegrations to form 13 ^4He nuclei and four neutrons, i.e.,

$$^{56}Fe + \gamma \rightleftarrows 13 ^4He + 4n \qquad (10.68)$$

From Eq. (10.64), n_{Fe}, n_{He} and n_n fulfil the relationship

$$\frac{n_{He}^{13} n_n^4}{n_{Fe}} = \frac{g_{He}^{13} g_n^4}{g_{Fe}} \left(\frac{2\pi kT}{h^2}\right)^{24} \left(\frac{m_{He}^{13} m_n^4}{m_{Fe}}\right)^{3/2} e^{-Q_{Fe}/kT} \qquad (10.69)$$

where

$$Q_{Fe} = (13 m_{He} + 4 m_n - m_{Fe})c^2 \qquad (10.70)$$

The ^4He formed from ^{56}Fe can be further photodisintegrated as

$$^4He + \gamma \rightleftarrows 2 ^1H + 2n \qquad (10.71)$$

Photodisintegration is an endothermic process. For example, if an iron core with a mass of $1.4 M_\odot$ undergoes photodisintegration to form ^4He

and neutrons, 4×10^{51} erg should be absorbed; if ^{4}He is further photodisintegrated, another 1×10^{52} erg should be absorbed. Therefore the total energy required to convert the iron core into neutrons and protons is $E_{photon} = 1.41 \times 10^{52}$ erg.

When the core temperature of a star is sufficiently high, all heavy atoms will undergo photodisintegrations, and the products can also undergo photodisintegrations. The number density of atoms in every photodisintegration processes should fulfil equations similar to Eq. (10.64), so a series of "Saha equations" can be written down. However, these equations alone do not suffice to give the number densities of different atoms at statistical equilibrium for a fixed temperature T and density ρ. We still need the relationship between the number densities of different atoms and the density ρ, and the abundance ratios between different particles and neutrons. For example, when a core constituted mainly by iron undergoes photodisintegration, the density ρ can be written as

$$\rho = (56n_{Fe} + 4n_{He} + n_n)m_u \qquad (10.72)$$

where m_u is the atomic mass unit ($1\ m_u = 1.6603 \times 10^{-24}$ g). For the iron core, we can assume that the abundance ratio between ^{4}He and neutrons to be $n_{He}/n_n = 13/15$. From this, the left-hand side of Eq.(10.69) can be expressed as

$$\left(\frac{4}{13}\right)^4 \frac{n_{He}^{17}}{n_{Fe}} \qquad (10.73)$$

In this way, from the three equations (10.69), (10.72) and (10.73), (it is already known that $g_{He} = 1$, $g_{Fe} = 1$ and $g_n = 2$), n_{Fe} and n_{He} can be solved for a given temperature T and density ρ. Figure 10.5 gives the distribution of ^{56}Fe and ^{4}He on the T–ρ plane for $n_{He}/n_n = 13/15$. It can be seen that ^{56}Fe dominates in the lower temperature region while ^{4}He dominates in the higher temperature region. Figure 10.6 gives the equilibrium distribution of different atoms on the T–ρ plane for $n_{He}/n_n = 1$. It is observed that ^{56}Ni dominates in the lower temperature region, ^{54}Fe + 2p dominates in the region with a higher temperature, while ^{4}He dominates in the region with an even higher temperature.

10.3.3 Electron Capture

A neutron can undergo a β decay to form a proton, an electron and an antineutrino, i.e.,

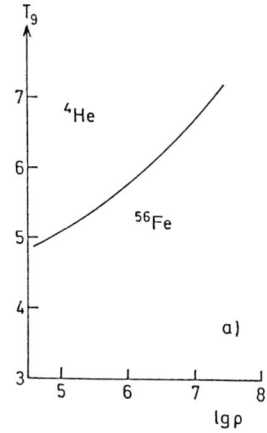

Fig. 10.5 The distribution of ^{56}Fe and ^{4}He on the T–ρ plane for $n_{He}/n_n = 13/15$ (according to Kippenhahn and Weigert 1991).

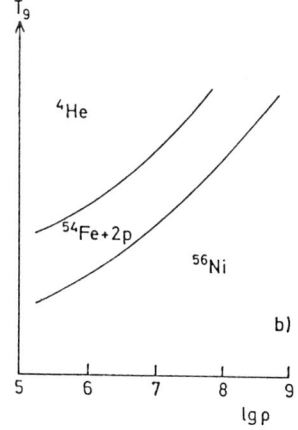

Fig. 10.6 The distribution of different atoms on the T–ρ plane for $n_{He}/n_n = 1$ (according to Kippenhahn and Weigert 1991).

$$n \rightarrow p + \bar{e} + \bar{\nu}_e \qquad (10.74)$$

The electron and the antineutrino should possess an energy of 1.3 MeV which is equivalent to the difference between the neutron mass and the proton mass. In other words, if the electron and antineutrino with such high energies cannot be produced, the neutron will not undergo β decay. The reverse of the β decay is the electron capture. We consider a neutron formed in the degenerate electrons in a certain density state, and the degenerate electrons have occupied the energy states up to 1.3 MeV. The electron gas density can be given by the Fermi energy of degenerate electrons as [see Eq. (10.57)]

$$\varepsilon_E = \frac{\hbar^2}{2m_e} n_e^{2/3} + m_e c^2 = 1.3 MeV + m_e c^2 \qquad (10.75)$$

If this electron gas density is exceeded, the electron energy will exceed $1.3 MeV + m_e c^2$, and the electron can be captured by a proton to form a neutron and a neutrino, i.e., the reverse of the β decay :

$$\bar{e} + p \rightarrow n + \nu_e \qquad (10.76)$$

The proton which captures the electron does not necessarily need to be a free proton, and can be bound to the nucleus of a heavy atom, which is the case for the cores of massive stars. When the iron core contracts so that its density rises to 1.1×10^9 gcm^{-3}, the electron energy exceeds 3.7 MeV $+ m_e c^2$, so the protons bound to the ^{56}Fe nuclei can capture electrons with very high energies to form ^{56}Mn and neutrinos, i.e., the inverse β decay

$$\bar{e} + {}^{56}Fe \rightarrow {}^{56}Mn + \nu_e \qquad (10.77)$$

and ^{56}Mn can further undergo inverse β decay at a density of 1.5×10^{10} gcm^{-3} to form ^{56}Cr and neutrinos:

$$\bar{e} + {}^{56}Mn \rightarrow {}^{56}C_r + \nu_e \qquad (10.78)$$

If the core of a star collapses to increase its density to exceed 10^{11} gcm^{-3}, all nuclei of heavy atoms can undergo inverse β decays. At this time, inverse β decay processes are very fast, which produce a large amount of neutrinos.

Electron capture is an endothermic process. For an iron core with an initial mass of $1.4M_\odot$, it is not difficult to calculate the absorbed energy through electron capture. This iron core consists of about 10^{57} electrons which can be captured by heavy atoms to form 10^{57} neutrinos. If the energy of each neutrino is about 10 MeV, the total energy absorbed through electron capture should be

$$E_{cap} \cong 10^{57} \times (10 \times 1.6 \times 10^{-6}) = 1.6 \times 10^{52} \text{ (erg)} \qquad (10.79)$$

which is the energy to be carried away by neutrinos.

10.4 SUPERNOVA

Supernova explosion is the catastrophic phenomena which occurs towards the end of the evolution of massive stars. During the entire evolution of a star, very violent unstable nuclear burning will occur, e.g., flash in the core or thermal pulse of shell burning, but these unstable nuclear burning will not lead to the destruction of the whole star which occurs after a supernova explosion. A possible outcome of supernova explosion is that no remnant is left behind and the entire star has shattered into space. Another possible outcome is that only the envelope of the star is shattered into space leaving behind the compact core to form a neutron star or a black hole.

10.4.1 Types of Supernovae

Supernovae are classified according to their spectra and features of their light curves. Those with hydrogen lines in their spectra are called type II supernova (SNII), while those without hydrogen lines are called type I supernova (SNI). The common feature for the light curves of supernovae is that they have a steep rise at the beginning, and their brightness reaches the peak values within a very short time and then decreases. On average, the absolute magnitude for type II supernovae at maximum brightness is brighter than that for type I supernovae by one magnitude. Type II supernovae have a broader brightness peak in the light curve, and the decrease rate of the brightness after reaching the peak values, which is about $0.^m03$ to $0.^m05$ per day in the first 30 to 100 days, is slower than that for type I supernovae, which is about $0.^m1$ per day in the first 20 to 30 days.

Type II supernovae can be further classified into two subgroups according to the descending part of their light curves. Those with

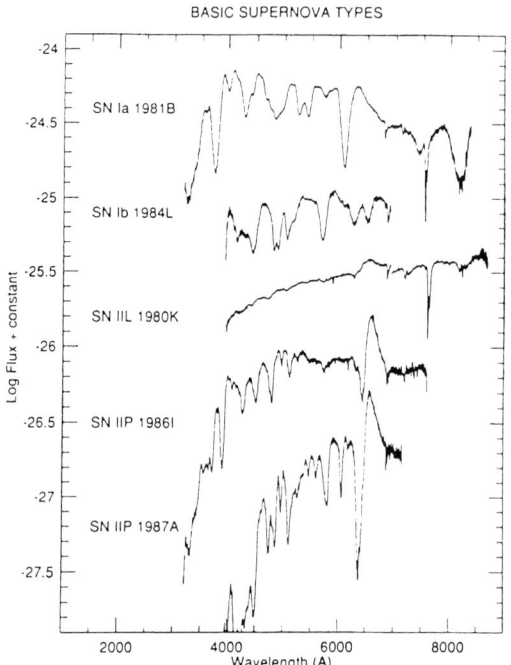

Fig. 10.7 The spectra of different types of supernovae several weeks after explosion (Wheeler and Harkmess 1992).

smallest decrease rates, with an average of $0.^m03$ per day, and with a plateau after 30 days of decrease are called type II-P (SNII-P). Those with somewhat larger decrease rates, with an average of $0.^m05$ per day, and with no plateau are called type II-L (SNII-L).

Type I supernovae can also be further classified into two subgroups according to the spectral features when their brightness reach the peak values. Those with strong SiII absorption lines ($\lambda 6355$Å) and no HeI absorption lines ($\lambda 6150$Å) are called type Ia (SNIa). Those with strong HeI absorption lines and no SiII absorption lines are called type Ib (SNIb). Figure 10.7 gives the spectra of different types of supernovae several weeks after explosion while Fig. 10.8 gives the light curves of different types of supernovae. Table 10.3 lists the characteristics of the spectra and the light curves of different types of supernovae.

Type II supernovae do not occur in elliptical galaxies and only occur in early type spiral galaxies, and in regions closely connected to HII regions in the spiral arms. These regions are supposed to have continuous star formation. Therefore, type II supernovae are assumed to be young massive population I stars. Type I supernovae occur in all galaxies, including spiral galaxies and elliptical galaxies, and in the halo or spiral arms. These regions do not have young stars, so type I supernovae are assumed to be old low-mass population II stars, such as white dwarfs.

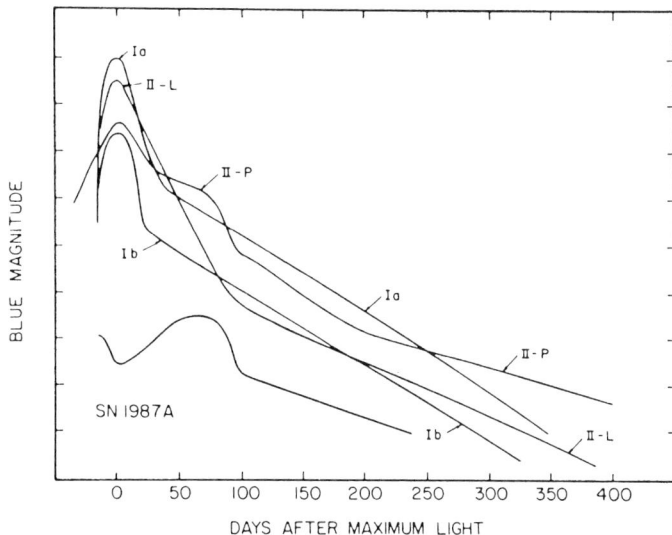

Fig. 10.8 The light curves of different types of supernovae (Wheeler and Harkmess 1992).

Table 10.3 Characteristics of spectra and light curves of different types of supernovae.

Characteristics	SNIa	SNIb	SNII-L	SNII-P
Criterion	no hydrogen	no hydrogen	with hydrogen	with hydrogen
Location	various galaxies; halo or spiral arms	late type galaxies; near HII region	early type galaxies; near HII region	early type galaxies; near HII region
Optical spectrum	P Cygnus type line SiIIλ6355 Å absorption line	P Cygnus type line HeI λ6150 Å line	P Cygnus type line complicated spectrum	P Cygnus type line complicated spectrum
Ejection velocity	$\geq$ 10000 km/s	$\geq$ 10000 km/s	$\leq$ 10000 km/s	$\leq$ 10000 km/s
Absolute magnitude	-20	-18; -18.5	≤ -18	≤ -18
Light curve shape	steep rise; decrease by $\sim$0.^{m}1 per day for 30 days; then decrease by $\sim$0.^{m}02 per day	steep rise; decrease by $\sim$0.^{m}1 per day for 30 days; then decrease by $<$ 0.^{m}01 per day	steep rise; decrease by $\sim$0.^{m}05 per day for 100 days; then decrease by $<$ 0.^{m}01 per day	steep rise; decrease by $\sim$0.^{m}03 per day for 30 days; then a plateau occurs until the 80th day; then decrease by $\sim$0.^{m}05 per day until the 125th day; then decrease by $\sim$0.^{m}006 per day

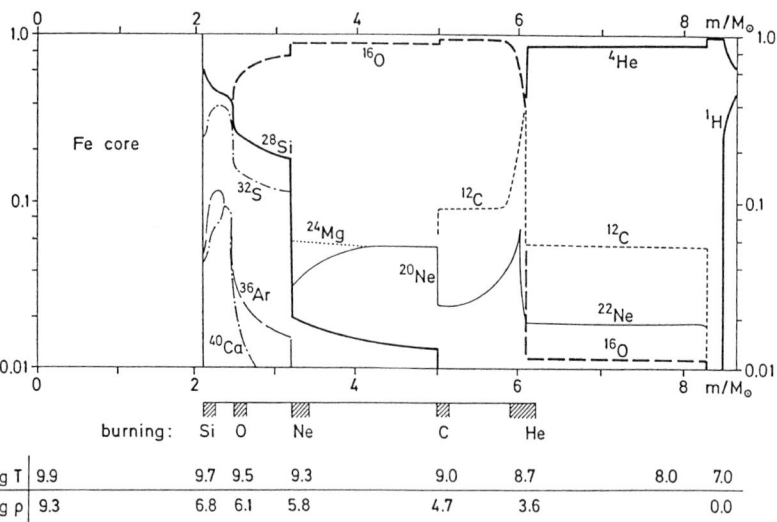

Fig. 10.9 The distribution of internal chemical composition, core temperature and core density for a star with a mass $25M_\odot$ after silicon burning (Woosley and Weaver 1986c).

10.4.2 Type II Supernovae

It is commonly supposed that type II supernovae are massive stars which have evolved to post-silicon burning, collapse of the iron core and the release of a large amount of gravitational energy to explode the whole star. Details of theoretical models of type II supernovae can be found in Woosley and Weaver (1986a,b), Hillebraudt (1985) and Arnett *et al.* (1989). The formation and the explosion for type II supernovae will be briefly described in the following sections.

(A) Presupernova

For a massive star, the core neon, oxygen and silicon burning subsequent to carbon burning happens without degeneracy, so the star can undergo all these burning processes to form a pre-supernova. For example, according to the calculations of Woosley and Weaver (1986c), for a star with a mass of $25_\odot$, the distribution of internal chemical composition, core temperature and core density after silicon burning and before becoming a supernova, are shown in Fig. 10.9. From Fig. 10.9, it can be seen that the innermost part of the star is an iron core (^{54}Fe, ^{56}Fe) with a mass of about $2.05M_\odot$. Exterior to the iron core is a shield with a mass of about $4M_\odot$ made of ^{16}O, ^{28}Si, ^{32}S, ^{20}Ne, ^{24}Mg. The envelope made of helium is mainly exterior to this shield. Inside the star, there are a number of shells in which Si, O, Ne, C and He burning are separately taking place.

(B) Collapse of the iron core

When the iron core contracts, the gravitational energy released raises the temperature and density of the iron core continuously. When the temperature reaches 5×10^9 K and the density reaches $\sim 10^{10}$ gcm^{-3}, weak interactions of photodisintegration and electron capture will take place in the iron core, which are strongly endothermic processes and are main processes leading to the collapse of the iron core. In §10.3.2 and 10.3.3, calculations have been performed for an iron core with a mass of 1.4M$_\odot$, which should absorb 1.4×10^{52} erg for the photodisintegation of ^{56}Fe and 1.6×10^{52} erg for electron capture. Therefore, a large amount of energy will be absorbed within a short time if the iron core undergoes these two processes, which will consume vast amount of kinetic energy of all types of particles (mainly electrons) in the iron core, implying that the pressure will be drastically reduced. At the same time, the pressure inside the iron core is mainly constituted by the degenerate electron pressure. When a large amount of electrons are captured by the heavy atoms, the pressure will be drastically reduced. In this way, hydrostatic equilibrium will be destroyed and the iron core will collapse due to its self gravitation. The time taken for the collapse of the iron core is extremely short, and it is carried out on the free falling timescale. For a core with a density of 10^{10} gcm^{-3}, this timescale is $\tau_{ff} = (G\bar{\rho})^{-1/2} \cong 1$ ms.

As the iron core starts to collapse, matter falls towards the center quickly. The central density and pressure suddenly increases so the falling speed of matter near the center is reduced. On the other hand, the matter in the outer regions still has a velocity close to the free falling velocity. Therefore, there is a transition point (referred to as the sonic point) in the iron core, outside of which matter will fall with supersonic speeds, while inside of which matter will fall with subsonic speeds. There is also a region near the center of the iron core in which matter will fall with uniform speed, the mass of which is the Chandrasekhar mass limit M_{ch}. The equivalence of this mass limit to the mass of the region can be rationalized as follows. We refer to the previous discussion of the structure of white dwarfs without macroscopic motion in §10.2.1. The pressure in white dwarfs is mainly constituted by the degenerate electron pressure. The structural equation for white dwarfs can be obtained by the hydrostatic equilibrium condition when the electron degenerate pressure and the gravitational force reach an equilibrium. Under the condition that degenerate electrons are relativistic, the mass limit obtained for white dwarfs is the Chandrasekhar mass limit. For a collapsing iron core which has macroscopic motion, the central density is extremely high and the electrons are relativistic degenerate. At this time, the condition for the degenerate electron pressure and the gravitational force to reach equilibrium leads to the uniform falling speed of matter. Therefore, the mass of the region in which matter falls with uniform speed is equal to M_{ch}. From Eq. (10.21), $M_{ch} \sim \mu_e^{-2}$ (μ_e is the mean molecular weight for electron). During the collapse, a large amount of electrons are captured so μ_e increases continuously which decreases M_{ch} and, thus, the region for uniform falling speed continuously. Apparently, the decrease of M_{ch} is related to the rate of electron capture, the neutrino transportation and the equation of state.

(C) Reflection and shock wave

The collapse rapidly raises the density inside the innermost part of the iron core, i.e., the region with mass M_{ch}, to approach the nuclear density, which can be obtained from the following equations. The radius R for an atom with nucleon number A is

$$R = r_o A^{1/3} \tag{10.80}$$

where $r_o = 1.2 \times 10^{-13}$ cm. The nuclear density ρ_{nuc} can be written as

$$\rho_{nuc} = \frac{3 A m_N}{4\pi R^3} = \frac{3 m_N}{4\pi r_o^3} = 2.3 \times 10^{14} \text{ g cm}^{-3} \tag{10.81}$$

where m_N is the mass of a nucleon. If the density of the innermost part of the iron core approaches the nuclear density $\rho_{nuc} = 2.3 \times 10^{14}$ gcm^{-3}, the nuclear force becomes very important. At this time, the neutrons are degenerate, and matter is incompressible. When the iron core center becomes incompressible, the collapse at the center ceases. However, the matter exterior to M_{ch} still falls with supersonic speeds, so a strong shock wave is produced near M_{ch}. When the collapse in the innermost part ceases, according to the law of conservation of momentum, a reflection should occur which propagates the shock wave outwards.

(D) Neutrino effects

The collapse can raise the temperature of the inner core to exceed 7×10^9 K and the density to reach 10^{11} to 10^{14} gcm^{-3}. Under such a high temperature and high density, photodisintegration and electron capture processes described in §10.3.2 and §10.3.3 will produce a lot of neutrons and neutrinos. The majority of the gravitational energy released from the collapse will be converted into neutrino energy. If we treat the inner core of the iron core as a neutron star, and the iron core at the start of the collapse as a white dwarf, the gravitational energy released during the collapse process is

$$E \cong GM_e^2 \left(\frac{1}{R_n} - \frac{1}{R_W} \right) \cong \frac{GM_c^2}{R_n} \cong 3 \times 10^{53} \quad \text{erg} \qquad (10.82)$$

where M_c is the mass of the collapsing iron core, R_n and R_W are typical radii of a neutron star and a white dwarf. The upper limit of the energy required to blow off the uncollapsed envelope can be approximately written as

$$E_e = \int_{M_W}^{M} \frac{Gmdm}{r} << \frac{GM^2}{R_W} = 3 \times 10^{52} \quad \text{erg} \qquad (10.83)$$

In the above calculations, we have taken $M = 10M_\odot$. The actual energy required to blow off the envelope is far less than this value; it is only of the order 10^{50} erg. From observations, we can also estimate the kinetic energy of matter ejected by the supernova explosion to be about 10^{51} erg, and the electromagnetic energy radiated by the explosion is of the order 10^{49} erg.

From these, it can be seen that the total energy required to blow off the envelope, to eject matter and for the electromagnetic radiation is only a small part of the gravitational energy released during the collapse of the iron core; the majority of the gravitation energy has been converted into neutrino energy.

If the large amount of neutrinos produced during the collapse can instantly penetrate through the star and escape, the inner part of the iron core will be completely composed of neutrons. However, not all the neutrinos will escape since the mean free path for neutrinos $l_\nu \approx 2 \times 10^{20}/\rho$ cm [see Eq.(6.291)] is related to the density ρ. When the density $\rho \geq 3 \times 10^{11}$ gcm^{-3}, the neutrinos can no longer escape during the free falling timescale. There is a region to trap neutrinos inside the

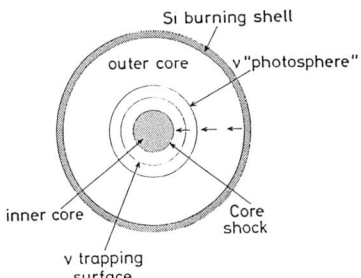

Si burning shell

outer core

ν "photosphere"

inner core

Core shock

ν trapping surface

Fig. 10.10 A schematic diagram showing the collapse and reflection of the star.

iron core, or called the neutrino trapping region (see Fig. 10.10). The size of the neutrino trapping region can be determined by equating the outward diffusion rate of neutrino to the falling speed of matter.

When the neutrino density in the inner core increases, the inverse β decay process in which heavy atoms capture electrons to form neutrinos, or called the neutronization process will slow down gradually.

(E) Prompt explosion

When the reflection propagates the shock wave out, the energy stored in the shock wave will directly lead to explosion of the star, which is referred to as the prompt explosion. This mechanism will be discussed in the following.

Numerical calculations show that the energy stored in the shock wave, or the binding energy in the inner core, is about

$$E_{\text{shock wave}} \approx E^{ic}_{\text{binding energy}} \approx (4-8) \times 10^{51} \text{ erg} \qquad (10.84)$$

The places reached by the shock wave on its way out will have photodisintegration. From §10.3.2, we know that photodisintegration is strongly endothermic and will consume the energy stored in the shock wave. This energy stored is about $(4-8) \times 10^{51}$ erg and can only support $0.5 M_\odot$ of heavy atoms for photodisintegration.

On the other hand, $\mu_e \approx 2.89$ for the inner core when reflection occurs. Therefore the mass of the inner core can be estimated to be $M^{ic} = 5.84/\mu_e^2 = 0.70 M_\odot$ [see Eq.(10.21)]. In this way, the prompt explosion can at most explode a star with a mass $0.7 M_\odot + 0.5 M_\odot = 1.2 M_\odot$, leaving behind a neutron star of $0.7 M_\odot$. According to model calculations for stellar evolution, only stars with initial masses of 8–$12 M_\odot$ (maximum is $15 M_\odot$) can evolve as pre-supernovae with core masses $\leq 1.2 M_\odot$. Therefore, the prompt explosion mechanism can explain the supernova explosion for stars with initial masses smaller than $15 M_\odot$. However, it is known that SN1987A is a star with an initial mass of $20 M_\odot$. There are difficulties for the prompt explosion mechanism for supernovae with initial masses greater than $15 M_\odot$.

(F) Delayed explosion

Since the prompt explosion mechanism cannot explain a supernova explosion with a core mass larger than $1.2 M_\odot$, Wilson (1985), Bethe and Wilson (1985) proposed the delayed explosion mechanism. This mechanism suggests that after several hundred ms from the reflection, the energy transferred by neutrinos can form another shock wave further away from the inner core. When this shock wave propagates out, it can blow off the external matter. This solves the problem of the prompt explosion mechanism that the shock wave can only propagate out for $0.5 M_\odot$.

10.4.3 Type I Supernovae

Type I supernovae, or more precisely type Ia supernovae, display no hydrogen in their spectra. At the same time, SNIa only occurs in elliptical galaxies or in regions between spiral arms in spiral galaxies, where young stars are absent. Therefore, we know that SNIa should be low mass white dwarfs before explosion. Current theories propose that SNIa is a C-O white dwarf and is a member of a binary system which undergoes unstable thermonuclear reactions and explodes after accreting mass from the companion. Therefore, the physical mechanisms involve various physical processes of unstable thermonuclear reactions in the degenerate C-O core.

(A) Detonation or deflagration of degenerate C-O core burning

From §9.3.4, it is known that when a degenerate core undergoes nuclear reactions, the reactions are unstable, i.e., the nuclear energy generation rate and the temperature will increase tremendously within a very short time but the pressure remains unchanged. When the temperature rises to a certain extent, electron degeneracy is removed and the non-degenerate ideal gas state is resumed. At this time, the pressure is again related to the temperature. As the temperature rises, the pressure rises and the volume increases which will decrease the temperature (see Fig. 9.15). The unstable nuclear reaction which happens during carbon burning of the degenerate C-O core is called the carbon flash, which is a more violent burning. The whole process from a sharp rise in temperature to an increase in the pressure and then to an increase in the volume will be completed within microseconds.

We proceed to discuss the nuclear burning in a degenerate C-O core starting from some local position such as the center. We denote the time taken for this local ignition point to raise its temperature, increase its pressure and then increase its volume as τ_{cc}. Since the energy instantly released by nuclear reactions is completely used to raise the temperature of matter, τ_{cc} can be written as

$$\tau_{cc} = \frac{C_p T}{\varepsilon_{cc}} \tag{10.85}$$

where ε_{cc} is the energy generation rate from carbon burning and C_P is the specific heat at constant pressure.

The pressure increase at the ignition point should also affect other non-ignited points inside the core through the timescale of hydrodynamics. From Eq.(7.106), the timescale of hydrodynamics τ_{dyn} can be written as

$$\tau_{dyn} = \left(\frac{2}{3}\pi G\bar{\rho}\right)^{-1/2} \tag{10.86}$$

where $\bar{\rho}$ is the mean density of the C-O core.

If $\tau_{cc} \gg \tau_{dyn}$, the pressure increase at the ignition point is relatively slow and the transportation of the pressure perturbation is very rapid so that the pressure increase due to burning will not produce compression in the surrounding matter. Conversely, if $\tau_{cc} \ll \tau_{dyn}$, the pressure increase at the ignition point is rapid and the transportation of the pressure perturbation is slow so that the pressure increase due to burning will produce a compression in the surrounding matter. If the pressure increases drastically, the compression becomes a shock front which propagates with supersonic speed in the core.

On the other hand, the degenerate C-O core is in general convective. The convection can transfer the released energy outwards and at the same time replenish the fuel for burning. The characteristic timescale τ_{con} for convection can be written as

$$\tau_{con} \approx l/C_s \tag{10.87}$$

where l is the mixed length and C_s is the speed of sound.

If $\tau_{cc} \gg \tau_{con}$, the convection can immediately take away the energy released from burning. Conversely, if $\tau_{cc} \ll \tau_{con}$, the convection cannot immediately take away the energy, so the energy released at the ignition point can only raise the temperature and thus increase the pressure of the matter at that point, which leads to the formation and outward propagation of the compression.

Assuming a density of $\rho > 10^8$ gcm^{-3} and a temperature $T \approx 2 - 3 \times 10^9$ K for a degenerate C-O core, we can calculate τ_{dyn} and τ_{con} to be about 0.1s. Since τ_{cc} is 10^{-6} s, we are certain that $\tau_{cc} \ll \tau_{dyn}$ and $\tau_{cc} \ll \tau_{con}$. In other words, if an unstable carbon burning (carbon flash) occurs at a certain point in the degenerate C-O core, a strong shock front with almost zero surface thickness will be produced and propagates out from the ignition point. This shock front has a lot of energy. When it propagates outwards, it will raise the temperature of the unburnt matter at places through which it passes to initiate burning. In this way, it has produced a burning front with almost zero surface thickness to propagate outwards with and behind the shock front. The burning front can be divided into two categories. First, when the shock front reaches a particular place, the matter is immediately burnt because of the high temperature and high pressure. Therefore, the burning front overlaps with the shock front, and both travel outwards at a supersonic speed. This burning front is called the detonation front.

Second, the matter is not immediately burnt when the shock front reaches it. When the shock front has passed by, the temperature of the matter is raised to ignition point through energy transfer (convective or conductive). The burning front propagates outwards at a subsonic speed, and is called a deflagration front. Apparently, the speed of the deflagration front is determined by the heat conduction rate in the core, and the temperature difference between the deflagration front and its foreground.

In order to study the different features between the detonation front and the deflagration front, we can analyze the state of matter following the shock front. For the shock wave, the conservation of mass can be written as

$$\rho_1 D = \rho_2 (D - v_2) \tag{10.88}$$

and the conservation of momentum as

$$P_1 + \rho_1 D^2 = P_2 + \rho_2 (D - v_2)^2 \tag{10.89}$$

where ρ and P are the density and pressure of the fluid, D is the front speed, the subscripts "1" and "2" represent the states ahead of and behind the front respectively, v is the speed of matter, and we have assumed $v_1 = 0$. Denoting $j = \rho_1 D$ as the mass flowing into the front, from Eq. (10.88), we have

$$D = jV_1 \tag{10.90a}$$

$$D - v_2 = jV_2 \tag{10.90b}$$

where $V = 1/\rho$ is the specific volume. Substituting Eq. (10.90) into Eq. (10.89), we have

$$P_1 + j^2 V_1 = P_2 + j^2 V_2$$

or

$$j^2 = \frac{D^2}{V_1^2} = -\frac{P_2 - P_1}{V_2 - V_1} \qquad (10.91)$$

which shows that the mass flow j and the front speed D are both determined by the ratio of the pressure change to the specific volume change. Since $j^2 > 0$, the pressure change and the specific volume change have opposite directions behind the front, i.e., either both pressure and density increase or both decrease. The former corresponds to the case for the detonation front while the latter corresponds to the case for the deflagration front. For a detonation front, both pressure and density increase when crossing to the region behind the shock front. For a deflagration front, both pressure and density decrease in the region behind the shock front (i.e., the region ahead of the deflagration front). These features can also be shown numerically. Figure 10.11 shows the change in conditions for a detonation front in the C-O core. The seven curves in the figure represent the seven different stages of the C-O core during the outward propagation of the detonation front from the center. The condition of the center of the C-O core are represented by dots. Curves 1 to 5 show the case that the C-O core temperature quickly increases but the core surface has not changed, which means the detonation front has not reached the surface. If we view inwards from the core surface, the density and the temperature will rise across the shock front region. When the detonation front reaches the surface (curve 6 in Fig. 10.11),

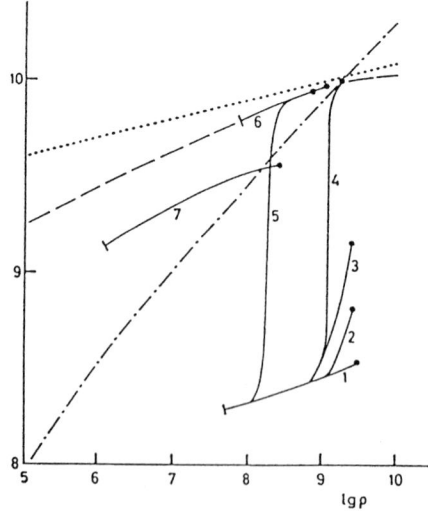

Fig. 10.11 The change in conditions for a detonation front in the C-O core. The dot-dashed curve separates the log$T$–logρ plane into degenerate and nondegenerate areas. The broken line shows the temperature change when all the energy released from carbon burning is used to raise the temperature of matter (after Arnett, 1969).

the temperature of the entire C-O core will rise to exceed 5×10^9 K. Under such a high temperature, the whole core will become an iron core.

Figure 10.12 shows the change in conditions for a deflagration front in the degenerate C-O core. From curves 1 to 3, it can be seen that the C-O core temperature increases abruptly but the core surface has not changed. When the deflagration front reaches the core surface (curve 4 in Fig. 10.12), the temperature and density of the surface will decrease. If we view inwards from the core surface, the density and temperature will decrease across the shock front region. As a consequence of the C-O core burning, only the temperature of the innermost part of the C-O core (inner core) can reach 5×10^9 K to form an iron core. The temperature of the outer regions of the C-O core will be far less than 5×10^9 K due to expansion.

(B) Some possible fates of the C-O white dwarf after accretion of mass

The outcome of a C-O white dwarf in a binary system after accretion of mass from its companion is closely related to the accretion rate and the initial mass of the white dwarf. Figure 10.13 shows some possible thermonuclear explosions of the C-O white dwarf after accretion of mass given by Nomoto *et al.* (1985). From Fig. 10.13, it can be observed that, if the mass accretion rate is larger than $10^{-6} M_\odot yr^{-1}$, i.e., larger than $\dot{M}_{RG}$, a very thick envelope will occur in the outer part of the white dwarf and turn the white dwarf into a red giant. If the mass accretion rate falls into the range $4 \times 10^{-8} M_\odot yr^{-1} \leq \dot{M} \leq \dot{M}_{RG}$, relatively weak unstable thermonuclear reactions will take place in the helium shell at the surface of the white dwarf. The C and O produced from the shell helium burning will increase the mass of the C-O white dwarf. When the mass of the C-O white dwarf reaches the critical mass, M_{ch}, carbon flash will occur in the C-O core. A subsonic propagating deflagration front is also produced which converts an inner core of about $0.6 M_\odot$ into ^{56}Ni and converts the outer part into intermediate elements such as oxygen, calcium, etc. The total energy released from these thermonuclear reactions which take place within a short time can not only explode the entire white dwarf, but also provides a kinetic energy of 10^{51} erg after the explosion. Moreover, the observed features of light curves for type Ia supernovae can be explained by the decay characteristics of ^{56}Ni.

If the mass accretion rate falls in the range $1.5 \times 10^{-9} M_\odot yr^{-1} \leq \dot{M} \leq 4 \times 10^{-8} M_\odot yr^{-1}$, the helium shell at the surface of the white dwarf will have strong unstable thermonuclear reactions. A supersonic propagating detonation front is also produced which can completely burn the entire helium shell. The detonation front can also propagate inwards to the

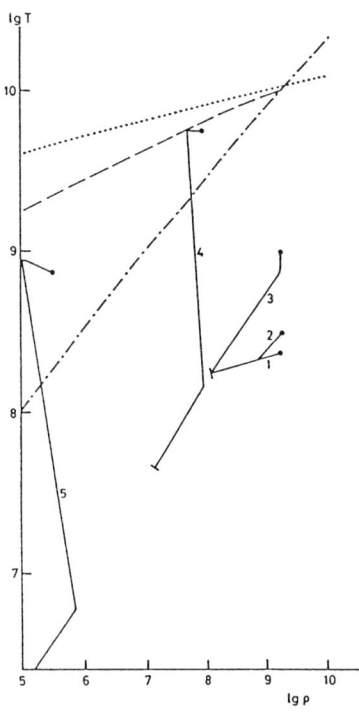

Fig. 10.12 Change in conditions for a deflagration front in the degenerate C-O core. The meanings of the dot-dashed curve and the broken curve are the same as those in Fig. 10.11 (adapted from Nomoto *et al.* 1976).

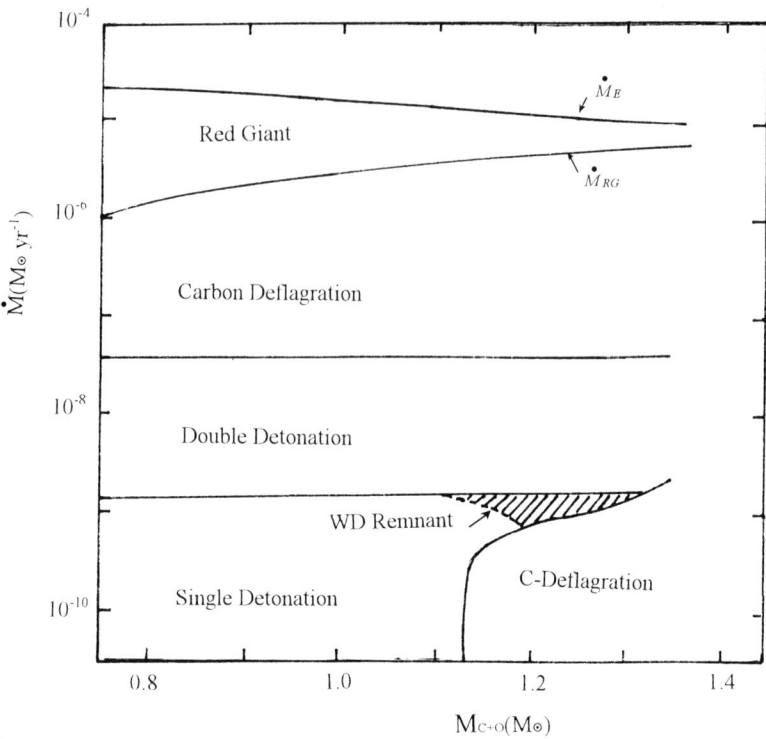

Fig. 10.13 Some possible nuclear explosions for the C-O white dwarf after accretion of mass (after Nomoto *et al.* 1985).

degenerate C-O core which leads to a flash in the C-O core and produces a new powerful detonation front to burn the entire C-O white dwarf. As a result the whole white dwarf explodes. This is the case for a double detonation.

If the mass accretion rate $\dot{M} < 1.5 \times 10^{-9} M_\odot \text{yr}^{-1}$, the consequence is closely related to the initial mass of the white dwarf. For a large initial mass (about $1.12 M_\odot$), the white dwarf will ultimately have carbon flash and deflagration. For a smaller initial mass (less than $1.12 M_\odot$), the helium shell at the surface of the white dwarf will have unstable burning and detonation. The detonation can explode the whole white dwarf, either leaving behind a WR star (hatched area in Fig. 10.13) or nothing.

10.5 NEUTRON STAR

Neutron star is the core left behind a supernova explosion, which is a stellar object with very high density. In general, its mass is 1.5 to 2.5 times the solar mass, but the radius is only about 10 km. Therefore, the internal density for a neutron star can exceed 10^{14} gcm^{-3}. Similar to white dwarfs, neutron stars do not have thermonuclear reactions, and their energy source in maintaining the external radiation is the dissipation of their internal heat energy.

Neutron stars have important impacts on many fields in contemporary astrophysics. For example, observationally discovered radio pulsars are rapidly spinning neutron stars with strong magnetic fields, and many X-ray sources and γ-ray sources arise from the neutron star in a binary system accreting mass from its companion.

Although neutron stars are related to many important astrophysical phenomena, it is very difficult to thoroughly understand the internal structure and properties of neutron stars. First, when the density reaches or even exceeds the nuclear density ($\rho_{nuc} = 2.4 \times 10^{14}$ gcm^{-3}), the strong interaction (nuclear force) can no longer be neglected. The equation of state with strong interaction is still not very clear, but the structure and properties of the neutron star is very sensitive to the equation of state. Second, in the case of extremely high densities, the effects of general relativity become very important. In the following, we briefly study the internal composition of a neutron star.

At the instant of supernova explosion, the temperature of the neutron star can reach 10^{11} to 10^{12} K. Due to neutrino loss, the temperature decreases quickly. The temperature can decrease to 10^9 K one day after explosion, and to 10^8 K one hundred years after explosion. The temperature of 10^8 K is extremely high for the earth and the sun, but is extremely low for a neutron star with a high density. Since the particles under a temperature 10^8 K have a thermal energy $kT \approx 10$ keV, but the Fermi energy of degenerate neutrons under the density $\rho = 10^{14}$ gcm^{-3} is $E_F \approx 1000$ MeV which is higher than the thermal energy of the particles by a factor of 10^5, so we can treat the internal matter of a neutron star as having $T \approx 0$, i.e., the internal degenerate electrons, protons and neutrons of a neutron star are at the lowest energy state. This inhibits β decay of neutrons, i.e.,

$$n \to p + \bar{e} + \bar{\nu}_e \tag{10.92}$$

because this process requires the energy of the created electron and neutrino to attain the large energy corresponding to the difference in the mass of the proton and the neutron, but the electron is at the lowest energy state and cannot have such high energy.

The internal density of a neutron star is very high. The inverse β decay process through capturing an electron by a proton in a nucleus to form a neutron, or the neutronization process, is enhanced with the increase of the density. At the same time, the β decay process which eliminates neutrons cannot occur. In this way, the inner part of a neutron star should be mainly composed of neutrons. Regions with higher densities should have more neutrons. We can study the chemical composition of different parts of a neutron star according to the densities.

In regions with densities less than 10^{11} gcm^{-3}, due to capture of electrons by atomic nuclei to form atomic nuclei with more neutrons, the matter is mainly composed of heavy atoms and electrons, with a small amount of protons.

In regions with densities reaching 4×10^{11} gcm^{-3}, the atomic nucleus with the most abundant neutrons, ^{118}Kr, starts to release neutrons. This process is called neutron drip. Therefore, the matter in these regions is mainly composed of atomic nuclei, electrons, protons and a small amount of neutrons. In these regions, the pressure is still mainly contributed by the degenerate electron pressure, i.e., $P \approx P_e >> P_n$.

The neutron drip process becomes more violent with the increase of density. In regions with densities reaching the nuclear density ($\rho_{nuc} = 2.4 \times 10^{14}$ gcm^{-3}), the atomic nuclei vanish, and the matter is mainly composed of degenerate neutrons (liquid) and a small amount of electrons and protons.

In regions with densities close to 10^{15} gcm^{-3}, since the energies of the degenerate electrons and degenerate neutrons have exceeded the energy corresponding to the rest mass of the proton, the high energy processes producing the hyperons Λ and Σ will occur, i.e.,

$$e^- + p \rightarrow \Lambda + \nu \tag{10.93}$$

$$e^- + n \rightarrow \Sigma^- + \nu \tag{10.94}$$

Therefore, the matter in these regions is mainly composed of the hyperons Λ and Σ. Figure 10.14 shows the density distribution and the corresponding matter composition inside a neutron star.

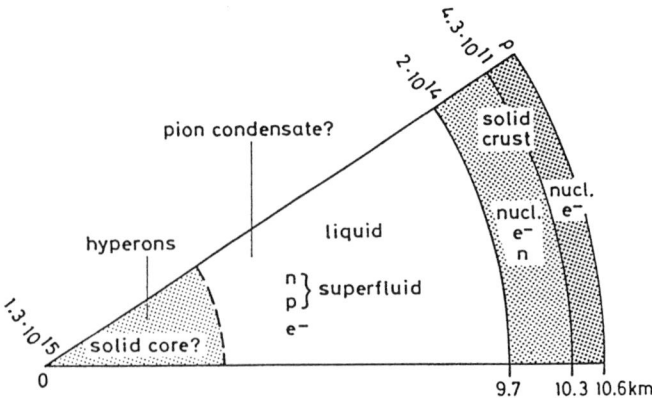

Fig. 10.14 The density distribution and the corresponding matter composition inside a neutron star.

CHAPTER ELEVEN

EVOLUTION OF INTERACTING BINARY SYSTEMS

Approximately 50% of stars are binaries. If the two stars in a binary system are widely separated, their interaction will be very small so their properties and evolution should be similar to those of single stars. However, in a great number of binary systems, the distance between the two components is small. Here, each component is subjected to stronger effects of the gravitational field and the radiation field of the companion. Because of interaction, the rotation of the two components are synchronized with the orbital motion; and at certain evolutionary stages, there is mass transfer between the two components. These processes cause a significant difference between properties and evolution of a component of the binary and a single star. These binary systems are categorized as interacting binary systems. A lot of special astronomical phenomena, such as Type Ia supernova, nova, X-ray sources etc., are in fact due to binary systems containing compact stellar objects. The study of binary systems has become an important area in contemporary astrophysics.

11.1 ROCHE MODEL

11.1.1 Roche Equipotential Surfaces

The Roche model treats the masses of the two stars as concentrated at their centers, and investigates the equipotential surfaces produced by the gravitational force and the rotation of the system. Suppose the distance between the two stars of masses M_1 and M_2 is A. The coordinate system (x, y, z) is chosen such that the origin is at the center of the primary, the x-axis coincides with the line joining the centers of the primary and the secondary, the y-axis is inside the orbital plane and perpendicular to the x-axis, and the z-axis is parallel to the axis of co-rotation (see Fig. 11.1). We now consider the potential ϕ on an arbitrary point $P(x, y, z)$ in space, which can be written as:

$$\phi = -\frac{GM_1}{r_1} - \frac{GM_2}{r_2} - \frac{\omega^2}{2}s^2 \qquad (11.1)$$

where

$$\left.\begin{aligned}
r_1^2 &= x^2 + y^2 + z^2 \\
r_2^2 &= (A - x)^2 + y^2 + z^2 \\
s^2 &= \left(x - \frac{M_2 A}{M_1 + M_2}\right)^2 + y^2 \\
\omega^2 &= G\frac{M_1 + M_2}{A^3}
\end{aligned}\right\} \qquad (11.2)$$

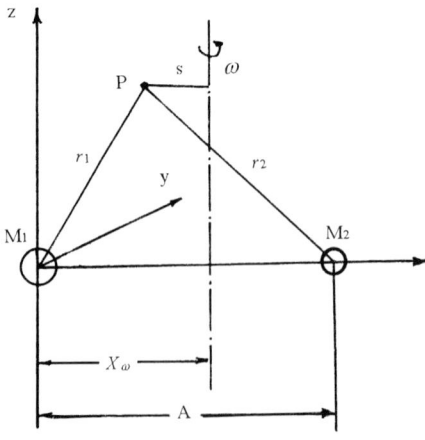

Fig. 11.1 The geometry for calculation of the Roche potential.

The 1st and 2nd terms on the right of Eq. (11.1) are potentials due to gravitational forces from the stars with masses M_1 and M_2, while the 3rd term is the potential due to centrifugal force caused by rotation of the point P around the co-rotation axis. In Eq. (11.1), r_1 and r_2 are the distances from P to centers of the primary M_1 and the secondary M_2, s is the distance from P to the co-rotation axis, and ω is the angular velocity of the co-rotation. Substituting Eq. (11.2) into (11.1), and introducing new variables:

$$\xi = \frac{x}{A}, \quad \eta = \frac{y}{A}, \quad R_1 = \frac{r_1}{A}, \quad R_2 = \frac{r_2}{A} \tag{11.3}$$

Eq. (11.1) can be expressed as:

$$-\frac{2}{1+q} \cdot \frac{\phi A}{GM_1} = \frac{2}{1+q} \cdot \frac{1}{R_1} + \frac{2q}{1+q} \cdot \frac{1}{R_2} + \left[\left(\xi - \frac{q}{1+q} \right)^2 + \eta^2 \right] \tag{11.4}$$

where $q = \dfrac{M_2}{M_1}$. \hfill (11.5)

In the derivation of Eq. (11.4), we have set all z values to zero, i.e., we only consider the cross section of the potential surface on the xy-plane. If we introduce the variable

$$\psi \equiv -\frac{2}{1+q} \cdot \frac{\phi A}{GM_1} \tag{11.6}$$

Eq. (11.4) can be written as

$$\psi \equiv \frac{2}{1+q} \cdot \frac{1}{R_1} + \frac{2q}{1+q} \cdot \frac{1}{R_2} + \left[\left(\xi - \frac{q}{1+q} \right)^2 + \eta^2 \right] \tag{11.7}$$

From Eq. (11.7), the equipotential surface given by $\psi = $ constant in the $\xi\eta$-plane can be constructed, which is known as the Roche equipotential surface. Figure 11.2 is the Roche equipotential surface in the orbital plane when $q = 0.215$. In Fig. 11.2, L_1 is the inner Lagrangian point, while L_2 and L_3 are outer Lagrangian points. The equipotential surface passing through L_1 is called the Roche critical equipotential surface. From Eq. (11.7), it is easy to see that when $R_1 << R_2$, $\psi \approx \frac{2}{1+q} \cdot \frac{1}{R_1}$ and the equipotential surfaces are a set of circles with the center at the center of the primary; when $R_2 << R_1$, $\psi \approx \frac{2q}{1+q} \cdot \frac{1}{R_2}$ and the equipotential surfaces are a set of circles with the center at the center of the secondary.

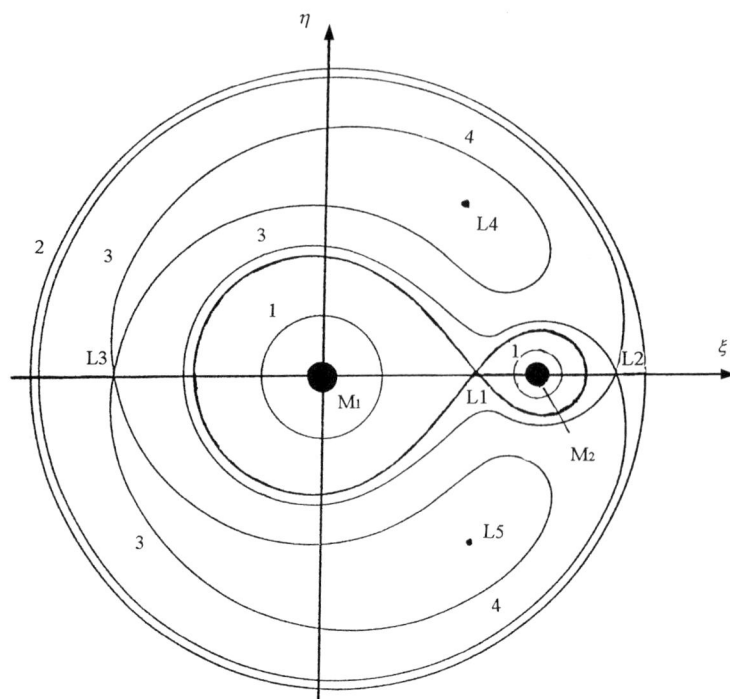

Fig. 11.2 Roche equipotential surfaces. L_i are the Lagrangian points; L_1: inner Lagrangian point; L_2, L_3: outer Lagrangian points. The equipotential surface passing through L_1 is called the Roche critical equipotential surface.

11.1.2 Critical Radius of the Roche Lobe

The Roche critical equipotential surface, or briefly, the Roche lobe has a significant meaning. The volume enclosed by this surface is called the Roche critical volume. When a test particle is inside the Roche lobe, the major force acting on it comes from the star within this surface. In other words, when a star has a volume smaller than the Roche critical volume, this star can be treated as an individual body experiencing relatively small effects from the gravitational field of the companion star, which cannot accrete surface matter from this star. However, if this star expands during evolution to fill up its Roche lobe, the gravitational effect of the companion star can accrete its surface matter through L_1, thus starting the mass transfer, so the star will lose a considerable part of its matter within a relatively short time, and its radius will decrease. Therefore, the Roche critical volume is the maximum volume to which a component of a binary system can expand. From Fig. 11.2, it can be seen that the Roche lobe is not spherical. Nevertheless, we can equate the Roche critical volume to that of a sphere, the radius of which is referred to as the critical radius of the Roche lobe and expressed as R_{cr1} or R_{cr2}.

Precise calculations of R_{cr1} or R_{cr2} are required by model evolution calculations of binary systems. According to Kopal (1959), the spherical polar coordinates (r, φ, θ) are introduced with the origin at the center of the primary:

$$x = r \cos \varphi \sin \theta = r\lambda$$
$$y = r \sin \varphi \sin \theta = r\mu \qquad (11.8)$$
$$z = r \cos \theta = r\nu$$

Substituting Eq. (11.8) into (11.1), and using new variables:

$$R \equiv r/A \qquad (11.9)$$

$$\zeta \equiv -\frac{A\phi}{GM_1} - \frac{M_2^2}{2M_1(M_1 + M_2)} \qquad (11.10)$$

we obtain from Eq. (11.1)

$$\zeta = \frac{1}{R} + q\left(\frac{1}{\sqrt{1 - 2\lambda R + R^2}} - \lambda R\right) + \frac{1 + q}{2} R^2(1 - \nu^2) \qquad (11.11)$$

For fixed values of ζ and q, Eq. (11.11) is a function of $R(\lambda, \nu)$. When the term $(1 - 2\lambda R + R^2)^{-1/2}$ of Eq. (11.11) is expanded by Leggendre polynomial $P_n(\lambda)$, noting that $P_0(\lambda) = 1$, $P_1(\lambda) = \lambda$, $P_2(\lambda) = \frac{1}{2}(3\lambda^2 - 1)$, $P_3(\lambda) = \frac{1}{2}(5\lambda^3 - 3\lambda)$, $P_4(\lambda) = \frac{1}{8}(35\lambda^4 - 30\lambda^2 + 3)$, Eq. (11.11) becomes

$$(\zeta - q)R = 1 + q\sum_{n=3}^{\infty} R^{n+1}P_n(\lambda) + \frac{1 + q}{2} R^3(1 - \nu^2) + \frac{q}{2} R^3(3\lambda^2 - 1)$$

$$(11.12)$$

We only consider the case where the radius of the Roche lobe is smaller than the distance A between the two stars, i.e., $R < 1$. When the precision is required up to R^3, and if only the orbital plane is considered, i.e., $\theta = 90°$ and $\nu = 0$, we obtain from Eq. (11.12):

$$R = \frac{1 + \frac{q}{2}(3\cos^2\varphi - 1)R^3 + \frac{1+q}{2} R^3}{\zeta - q} \qquad (11.13)$$

Substituting ζ in Eq. (11.13) by the ζ_1 value on the Roche lobe, we can obtain the formula for the Roche lobe as

$$R_{cr1}^* = \frac{1 + \frac{q}{2}(3\cos^2\varphi - 1)R_{cr1}^{*3} + \frac{1+q}{2} R_{cr1}^{*3}}{\zeta_1 - q} \qquad (11.14)$$

From Eqs. (11.1) and (11.10), ζ_1 can be obtained from

$$\zeta_1 = \frac{1}{x_1} + \frac{q}{1-x_1} + \frac{1}{2}(1+q)\left[x_1 - \frac{q}{1+q}\right]^2 - \frac{q^2}{2(1+q)} \tag{11.15}$$

where x_1 is the x-axis component of the inner Lagrangian point $L_1(x_1, 0, 0)$, which can in turn be obtained from

$$\frac{1}{x_1^2} - \frac{q}{(1-x_1)^2} - (1-q)x_1 + q = 0 \tag{11.16}$$

Using Eq. (11.14), and for fixed φ, R^*_{cr1} can be obtained numerically. The Roche critical volume can then be evaluated through integration:

$$\begin{aligned}
V_{cr1} &= \int_0^\pi \pi R^{*2}_{cr1} \sin^2\varphi \cdot R^*_{cr1} \sin\varphi \, d\varphi \\
&= \pi \int_0^\pi R^{*3}_{cr1} \sin^3\varphi \, d\varphi
\end{aligned} \tag{11.17}$$

and the critical radius of the Roche lobe R_{cr1} is given by

$$R_{cr1} = \left(\frac{3V_{cr1}}{4\pi}\right)^{1/3} \tag{11.18}$$

where R_{cr1} is in units of A.

To solve for the critical radius R_{cr2} for the Roche lobe of the secondary, it only needs to replace x_1 by $(1-x_1)$ and q by $(\frac{1}{q})$ in Eq. (11.14) and (11.15), i.e.

$$\left.\begin{aligned}
\zeta_2 &= \frac{1}{1-x_1} + \frac{1}{qx_1} + \frac{1+q}{2q}\left(1-x_1 - \frac{1}{1+q}\right)^2 - \frac{1}{2q(1+q)} \\
R^*_{cr2} &= \frac{1 + \frac{3}{2q}R^{*3}_{cr2}\cos^2\varphi + \frac{1}{2}\left(1+\frac{1}{q}\right)R^{*3}_{cr2}}{\zeta_2 - \frac{1}{q}} \\
V_{cr2} &= \pi \int_0^\pi R^{*3}_{cr2} \sin^3\varphi \, d\varphi \\
R_{cr2} &= \left(\frac{3V_{cr2}}{4\pi}\right)^{1/3}
\end{aligned}\right\} \tag{11.19}$$

The above steps illustrate the detailed calculations of the critical radii of the Roche lobe. In some cases, approximations can be used. The approximations given by Eggleton (1983) in calculating the critical radius of the Roche lobe is ($q = M_1/M_2$)

$$R_{cr1}/A = \frac{0.49q^{2/3}}{0.6q^{2/3} + \ln(1 + q^{1/3})} \qquad (11.20)$$

which is valid for all q values. The value of R_{cr2}/A is obtained by replacing q with $1/q$. The approximations given by Paczynski (1967) are

$$R_{cr1}/A = \begin{cases} 0.38 + 0.2\log q & 0.5 \leq q < 20 \\ 0.462\left(\frac{q}{1+q}\right)^{1/3} & 0 < q < 0.5 \end{cases} \qquad (11.21)$$

and R_{cr2}/A is obtained by replacing q with $1/q$.

11.1.3 Classification of Interacting Binary Systems

Interacting binary systems can be classified into three types according to whether the stars have filled the Roche lobes.

(1) When both stars are smaller than the corresponding Roche lobes (Fig. 11.3), the system is called a detached binary system.
(2) When one star is smaller than the Roche lobe, and the other star has filled its Roche lobe, the system is called a semi-detached binary system.
(3) When both stars have filled the Roche lobes, the system is called a contact binary system.

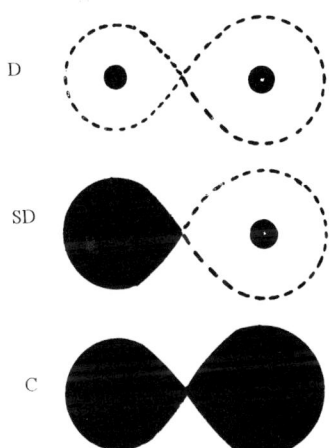

Fig. 11.3 Classification of interacting binary systems. D: detached; SD: semi-detached; C: contact.

11.2 CONSERVATIVE EVOLUTION OF BINARY SYSTEMS

11.2.1 Mass Transfer

The more massive star (primary) in a binary system evolves more quickly. When the primary has filled its Roche lobe, mass transfer starts. The gas on the surface of the primary will flow through the inner Lagrangian point L_1 to the companion star (secondary). During mass

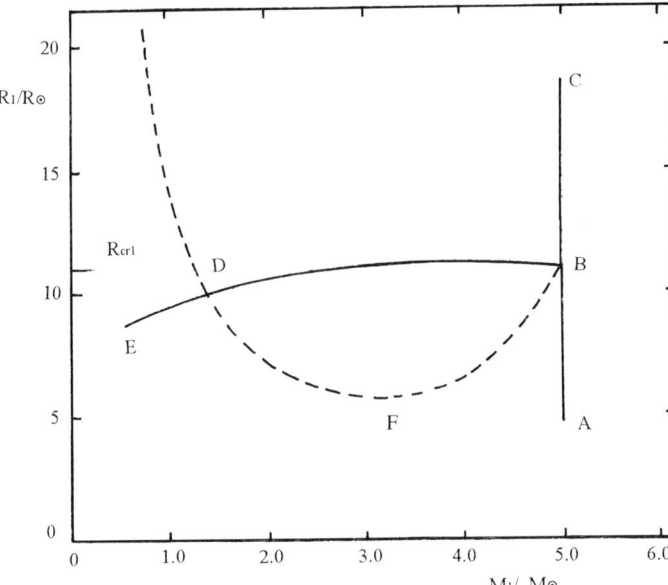

Fig. 11.4 The variation of the radius of the primary and the corresponding critical radius of the Roche lobe during mass transfer for a $5M_\odot + 2M_\odot$ binary system.

transfer, the radii of the primary and secondary, and the corresponding critical radii of Roche lobes will change accordingly. Figure 11.4 shows the variation of the radius of the primary and the corresponding critical radius of Roche lobe during mass transfer for a $5M_\odot + 2M_\odot$ binary system. If the primary with a mass of $5M_\odot$ is a single star, its radius R_1 will vary along the solid line ABC in Fig. 11.4. However, in a binary system, the radius of the primary can only expand to the critical radius of the Roche lobe, i.e. $R_1 = R_{cr1}$, corresponding to point B in Fig. 11.4. At that time, mass transfer begins. If the primary has a mass loss rate in such a way that the variation of the radius is small (varies along the solid line BDE in Fig. 11.4), the critical radius R_{cr1} of the Roche lobe will vary along the broken line BFD. In the BF portion, the difference between the critical radius of the Roche lobe and the radius of the primary increases. At point F, the masses of the two stars are equal, i.e., $M_1 = M_2$. In the FD portion, the mass of the secondary becomes greater than that of the primary. At the same time, the difference between the critical radius of the Roche lobe and the radius of the primary decreases. After point D, the critical radius of the Roche lobe is greater than the radius of the primary so the mass transfer ceases. During the processes from B to D, the primary loses a large amount of mass to the secondary. When compared to the nuclear timescale for evolution of the primary, the mass transfer is a fast process. Calculations of evolutionary models show that if the envelope of the primary is at radiative equilibrium at the start of mass transfer, the mass transfer will take place on the Kelvin–Helmholtz timescale; if the envelope of the primary is convective, the mass transfer will take place on the hydrodynamical timescale.

It should be pointed out that the mass transfer rate has been chosen arbitrarily for Fig. 11.4 for demonstrating the variation characteristics of the radius of the primary and the critical radius of the Roche lobe during mass transfer. This arbitrary mass transfer rate can cause a drastic difference between the variations of the radius of the primary and the critical radius of the Roche lobe. On the contrary, the realistic mass transfer rate should be chosen in such a way that the changes of the two radii are approximately equal, i.e. $R_1 \approx R_{cr1}$. When mass transfer starts, if the envelope of the primary is at radiative equilibrium, it will take place on the Kelvin–Helmholtz timescale. From Eq. (7.118),

$$t_K \approx 2 \times 10^7 \frac{(M/M_\odot)^2}{(L/L_\odot)(R/R_\odot)} \qquad \text{(yr)} \qquad (11.22)$$

so the mass transfer rate is

$$\dot{M} = \frac{M}{t_K} = 5 \times 10^{-8} \frac{(R/R_\odot)(L/L_\odot)}{(M/M_\odot)} \qquad (M_\odot/\text{yr}). \qquad (11.23)$$

When mass transfer starts, if the envelope of the primary is convective, it will take place on the hydrodynamical timescale. Under such a condition, it is difficult to achieve synchronous variations of the radius of the primary and the critical radius of the Roche lobe.

11.2.2 Angular Momentum Transfer

Suppose the rotation and the orbital motion of the components in a binary system are synchronized. The total angular momentum of the system can be written as

$$J = J_0 + (I_1 + I_2)\omega, \qquad (11.24)$$

where J_0 is the orbital angular momentum of the two stars. From Fig. 11.1, it is easy to see that J_0 can be expressed as

$$J_0 = [M_1 X_\omega^2 + M_2(A - X_\omega)^2]\omega. \qquad (11.25)$$

Since $X_\omega = \frac{M_2 A}{M_1 + M_2}$ and $\omega^2 = G\frac{M_1 + M_2}{A^3}$, Eq. (11.25) can also be written as

$$J_0 = (GA)^{1/2} \frac{M_1 M_2}{(M_1 + M_2)^{1/2}}. \qquad (11.26)$$

The term $(I_1 + I_2)\omega$ in Eq. (11.24) is the rotational angular momenta of the two stars, where I_1 and I_2 are moments of inertia of the two stars, which can be expressed as

$$I_i = \frac{8\pi}{3} \int_0^{R_i} r^4 \rho\, dr \qquad (i = 1, 2) \qquad (11.27)$$

There are two forms of angular momentum exchange during the evolution of the binary system. The first one is the conversion of orbital angular momentum into rotational angular momentum. Since the volume of the stars increase during evolution, their moments of inertia and thus rotational angular momenta will increase. Under conservation of total angular momentum, the orbital angular momentum should decrease. The second form occurs during mass transfer between the two stars. Accompanying the transfer of mass, there should also be a transfer of angular momentum from one star to the other.

Huang *et al.* (1992) have studied the evolution of a $8M_\odot + 5.5M_\odot$ binary system. It has been found that the radius of the primary increases by a factor of 26 when it evolves from a main sequence star to a red giant, and the rotational angular momentum increases by a factor of 10^3 accordingly. This decreases the orbital angular momentum significantly and shortens the period. The results also proved that the rotational angular momenta should not be ignored, i.e., the $(I_1 + I_2)\omega$ term in Eq. (11.24) should be introduced. If only the orbital angular momentum is taken into account in calculations for the evolution of a binary system, the conversion from orbital angular momentum into rotational angular momenta cannot be considered, and the determined orbital period will be significantly different.

11.2.3 Variation of the Orbital Period

The transfer of mass and angular momentum during evolution of a binary system will certainly lead to a change in the orbital period of the system. Suppose a mass ΔM flows from the primary to the secondary in a certain time interval during mass transfer. The mass ratio $q(= M_2/M_1)$ of the two stars will change to

$$q = \frac{M_2 - \Delta M}{M_1 + \Delta M}, \quad (\Delta M < 0) \tag{11.28}$$

From Eq. (11.24), the total angular momentum of the binary system is

$$J = J_0 + (I_1 + I_2)\omega \tag{11.29}$$

Introducing two dimensionless parameters a and b, where

$$a \equiv \frac{(I_1 + I_2)\omega}{J_0} \tag{11.30}$$

$$b \equiv \frac{(I_1 + I_2)\omega}{J} = \frac{1}{1 + 1/a} \tag{11.31}$$

Eq. (11.29) can be written as

$$J = J_0(1 + a) \tag{11.32}$$

Differentiating the logarithms of the above equation, and considering the conservation of total angular momentum, we have

$$0 = \frac{\Delta J_0}{J_0} + \frac{\Delta a}{1 + a} = \frac{\Delta J_0}{J_0} + \frac{\Delta a}{a} b \tag{11.33}$$

From Eqs. (11.30), (11.26) and $\omega^2 = G\frac{M_1 + M_2}{A^3}$, and considering the conservation of mass in the system, i.e., $M_1 + M_2 = \text{const}$, $\Delta M = \Delta M_1 = -\Delta M_2$, we have

$$\frac{\Delta a}{a} = \frac{\Delta \omega}{\omega} + \frac{\Delta(I_1 + I_2)}{I_1 + I_2} - \frac{\Delta J_0}{J_0} \tag{11.34}$$

$$\frac{\Delta J_0}{J_0} = \frac{1}{2}\frac{\Delta A}{A} + \frac{\Delta M}{M_1} - \frac{\Delta M}{M_2} \tag{11.35}$$

$$\frac{\Delta \omega}{\omega} = -\frac{3}{2}\frac{\Delta A}{A} \tag{11.36}$$

Substituting Eqs. (11.34),(11.35) and (11.36) into (11.33), we obtain

$$\Delta A = \frac{A}{2b - 1/2}\left[\frac{\Delta M}{M_1}(1 - b) - \frac{\Delta M}{M_2}(1 - b) + \frac{\Delta(I_1 + I_2)}{I_1 + I_2}b\right] \tag{11.37}$$

which gives the change in the distance A between the two stars due to transfer of mass and angular momentum. If the change in the moments of inertia of the stars due to transfer of mass ΔM is negligible, i.e., $\frac{\Delta(I_1+I_2)}{I_1+I_2} \approx 0$, the following can be known from Eq. (11.37): when mass flows from the more massive star to the less massive star (if $M_1 > M_2$), ΔA is negative so the distance between the two stars decreases; when mass flows from the less massive star to the more massive star (if $M_1 < M_2$), ΔA is positive so the distance between the two stars increases. At $q = 1$, ΔA is equal to zero, so A has a minimum. It can also be seen from Eq. (11.37) that the volume expansion of stars can change the moments of inertia to change A even when there is no mass transfer ($\Delta M = 0$).

Once the variation in A is known, the orbital period of the binary system can be determined from

$$P = \frac{2\pi}{\omega} = \frac{2\pi A^{3/2}}{G^{1/2}(M_1 + M_2)^{1/2}} \tag{11.38}$$

Furthermore, once the variations in q and A are known, the new critical radii of Roche lobes can be calculated from Eqs. (11.18) and (11.19), or Eqs. (11.20) and (11.21).

11.2.4 Evolution of Binary Systems under Different Conditions

The evolution of a binary system is closely related to the evolutionary stage of the primary when mass transfer starts. To illustrate this, we study the variation of the radius of a $5M_\odot$ primary during evolution. From Fig. 11.5, it can be seen that the change in the radius of this star can be separated into three stages, viz, the core hydrogen burning stage, shell hydrogen burning stage and the core helium burning stage. The mass transfer can occur at different stages mentioned above, and there are three different cases for evolution of the binary system.

Case A: mass transfer occurs at core hydrogen burning stage of the primary.

Case B: mass transfer occurs at shell hydrogen burning stage of the primary.

Case C: mass transfer occurs at core helium burning stage of the primary.

The evolution of the binary system for these three cases will be discussed in the following.

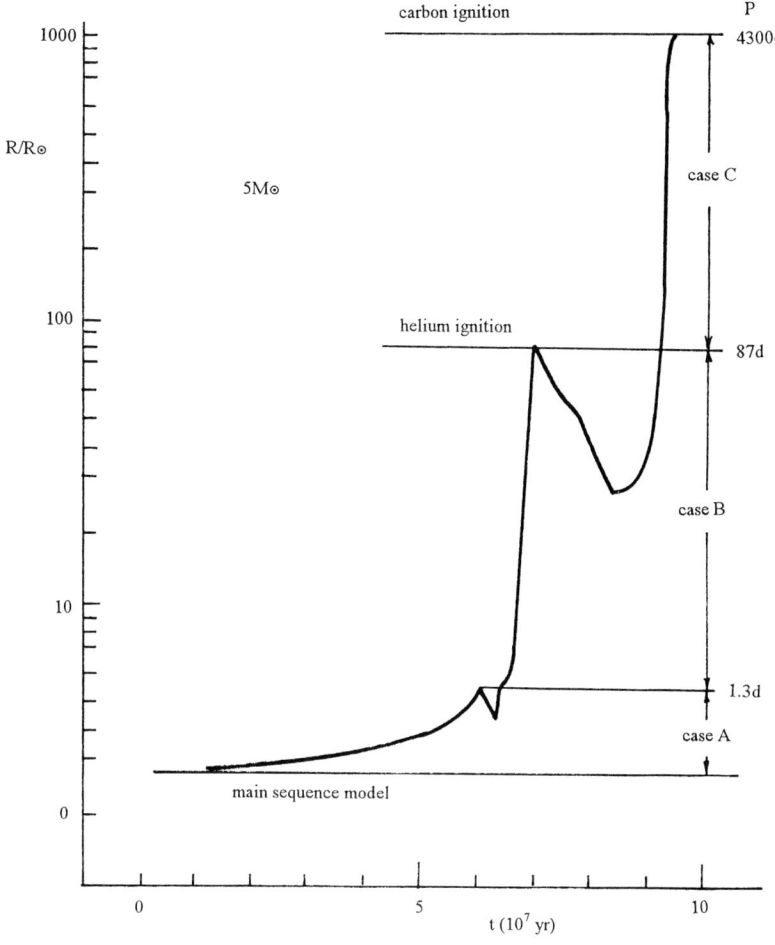

Fig. 11.5 The temporal variation of the radius of a 5M$_\odot$ primary.

(A) The Evolution for Case A

The 9M$_\odot$ + 5M$_\odot$ binary system employed by Kippenhahn and Weigert (1967) for calculations is used here as an example to illustrate the evolution for the case where mass transfer occurs at the core hydrogen burning stage of the primary. The initial distance between the two stars in this system is $A = 13.2$R$_\odot$, and the chemical composition of the two stars is $X_H = 0.602$ and $X_Z = 0.044$. Since the primary evolves much faster than the secondary, only the evolution for the primary is calculated. Figure 11.6 gives the evolutionary track of the primary in the H-R diagram. At the beginning, the primary is a zero age main sequence star. After 12.5×10^6 yrs, about half of the hydrogen in the core has been burnt. At this time, the primary fills up the Roche lobe and mass transfer starts (point a in Fig. 11.6). The mass transfer from

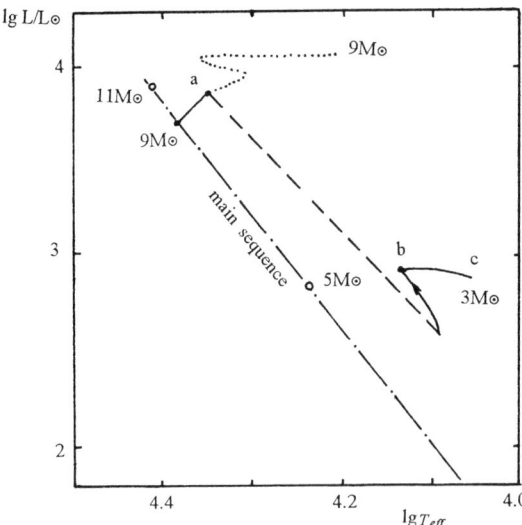

Fig. 11.6 The evolutionary track of the primary in a $9M_\odot + 5M_\odot$ binary system in the H–R diagram (Kippenhahn and Weigert 1969).

point a to b in Fig. 11.6 takes place on the Kelvin–Helmholtz timescale, which is a fast process and takes 6×10^4 yrs. During this stage, the primary loses $5.27M_\odot$ and becomes a star with $M_1 = 3.73M_\odot$, while the mass of the secondary changes from $5M_\odot$ to $10.27M_\odot$. Therefore, the roles of the primary and the secondary have interchanged, i.e., the original more massive primary now becomes the less massive secondary, and the secondary becomes the primary. During the fast process of mass transfer, the luminosity and the effective temperature of the primary will decrease quickly. The reasons are as follows: The surface of the primary loses matter quickly and continuously. In order to retain hydrodynamical equilibrium, the matter below surface is made to expand outwards, which requires energy and causes a decrease in the effective temperature and the luminosity.

After the fast process of mass transfer, the original primary (M_1) still has about half of the unburnt hydrogen in the core. It should thus still expand and fill up the Roche lobe, and mass transfer occurs. At this time, the mass transfer takes place on a nuclear timescale, which is a slow process of mass transfer (the stage from point b to c in Fig. 11.6). During this process, M_1 continues to lose mass of $0.7M_\odot$. Table 11.1 gives some parameters of the binary system at different stages of the evolution, where a, b and c are identical to the points a, b and c in Fig. 11.6. After the slow process of mass transfer, the original primary with a mass of $9M_\odot$ becomes a secondary of $3.02M_\odot$ which fills up the Roche lobe. Its position in the H-R diagram has moved to the right of

Table 11.1 Some parameters for case A's evolution of a $9M_\odot + 5M_\odot$ binary system.

	a	b	c
M_1 ($M_\odot$)	9	3.73	3.02
M_2 ($M_\odot$)	5	10.27	10.78
$\log A$ (cm)	11.965	12.11	12.34
$\log R_{cr1}$ (cm)	11.60	11.57	11.67
$\log R_1$ (cm)	11.60	11.57	11.67

the main sequence, and its luminosity is greater than those of main sequence stars with the same mass.

On the other hand, the original secondary (M_2) with a mass of $5M_\odot$ has now become the primary with a mass of $10.78M_\odot$. Its position on the H-R diagram is still on the main sequence, but has moved along the main sequence towards the upper left direction. This is because only a very small fraction of hydrogen in its core has been burnt during the whole evolution process. At this time, the binary system is semi-detached, and is a typical Algol type binary system. Algol type binary systems have the following features: the more massive primary is a main sequence star, which obeys the mass–luminosity relationship and the mass–radius relationship of main sequence stars; the less massive secondary is on the right of the main sequence in the H-R diagram, which does not obey the mass-luminosity relationship of main sequence stars and is much brighter than main sequence stars with the same mass. These features of Algol type binary systems are contradicting the evolution model of single stars. According to the theory of evolution of single stars, the more massive primary evolves more quickly and should be on the right of the main sequence, while the less massive secondary evolves more slowly and should be on the main sequence. The contradiction is also known as the Algol paradox. Obviously, the Algol paradox arises because there is mass transfer during evolution of the binary system. This also proves that observed Algol type binary systems are those which have undergone mass transfer.

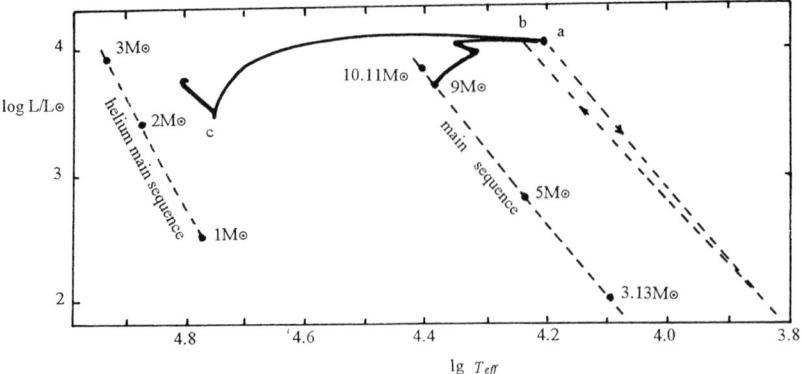

Fig. 11.7 The evolutionary track of the primary in the 9M$_\odot$ + 3.13M$_\odot$ binary system in the H–R diagram (Kippenhahn and Weigert 1967).

(B) The Evolution for Case B

(I). The case for intermediate massive stars

The 9M$_\odot$ + 3.13M$_\odot$ binary system employed by Kippenhahn and Weigert (1967) for calculations is used here as an example to illustrate the evolution for the case where mass transfer occurs at the shell hydrogen burning stage of the primary. The initial distance between the two stars in this system is $A = 29.5$R$_\odot$, and the chemical composition of the two stars is $X_H = 0.602$ and $X_Z = 0.044$. Figure 11.7 shows the evolutionary track of the primary in the H-R diagram. At point a, the primary fills up the Roche lobe and mass transfer starts. The core hydrogen burning of the primary has been completed, but there is shell hydrogen burning at the outer boundary of the core. The helium core is contracting inwards while the outer portion of the shell hydrogen burning is expanding outwards. After mass transfer starts, the evolutionary track of the primary in the H-R diagram follows the broken line from point a to b in Fig. 11.7. After about 4×10^4 yrs, the primary has lost a total mass of 7M$_\odot$ and has effectively lost all the mass in the envelope, and becomes a helium star with a mass of 2.02M$_\odot$. On the other hand, the mass of the original secondary (M_2) changes from 3.13M$_\odot$ to 10.11M$_\odot$. At this time, the roles of the primary and secondary have interchanged. After completion of the fast process of mass transfer (point b in Fig. 11.7), the helium star with a mass of only 2.02M$_\odot$ can no longer fill up the Roche lobe. Its volume continues to decrease, which is equivalent to the evolution from point b to c in the H-R diagram, and it takes a total time of 3×10^5 yrs to reach point c, i.e., to reach the helium main sequence. At point c, the helium in its core starts to burn, and the evolution of the star starts to take place on a slow nuclear timescale. Its radius is much smaller than the critical

Table 11.2 Some parameters for case B's evolution of a $9M_\odot + 3.13M_\odot$ binary system.

	a	b	c
M_1 ($M_\odot$)	9	2.02	2.02
M_2 ($M_\odot$)	3.13	10.11	10.11
$\log A$ (cm)	12.313	12.589	12.589
$\log R_{cr1}$ (cm)	11.989	11.987	11.987
$\log R_1$ (cm)	11.989	11.987	10.865

radius of the Roche lobe. Therefore, the binary system now belongs to a detached system, and is also an Algol type binary system. Table 11.2 gives some parameters of this binary system at different evolutionary stages.

(2) The case for low mass stars

Refsdal and Weigert (1969) calculated case B's evolution of a $1.4M_\odot + 1.1M_\odot$ low mass binary system. The initial distance between the two stars is $A = 8.6R_\odot$, and the chemical composition of the two stars is $X_H = 0.602$ and $X_Z = 0.044$. Figure 11.8 shows the evolutionary track of the primary of this system in the H-R diagram. In Fig. 11.8, point a is the zero age main sequence, c refers to the completion of core hydrogen burning and the start of shell hydrogen burning, and point d refers to filling up of the Roche lobe and start of mass transfer. The primary takes 2×10^9 years to evolve from point a to d. If mass transfer does not occur at point d, the primary will move along the broken line in Fig. 11.8. The fast process of mass transfer starting from point d causes the primary to lose $0.5M_\odot$ within 2×10^6 yrs to reach point f (note when the mass of a star decreases, the corresponding Hayashi line will also move to the right). The slow process of mass transfer starts from point f. At this stage (about 3×10^8 yrs), the helium core inside the primary continues to contract inwards and the helium core becomes degenerate. The envelope external to the shell hydrogen burning expands outwards. Although the radius of the primary does not exceed the critical radius of the Roche lobe, since the critical radius

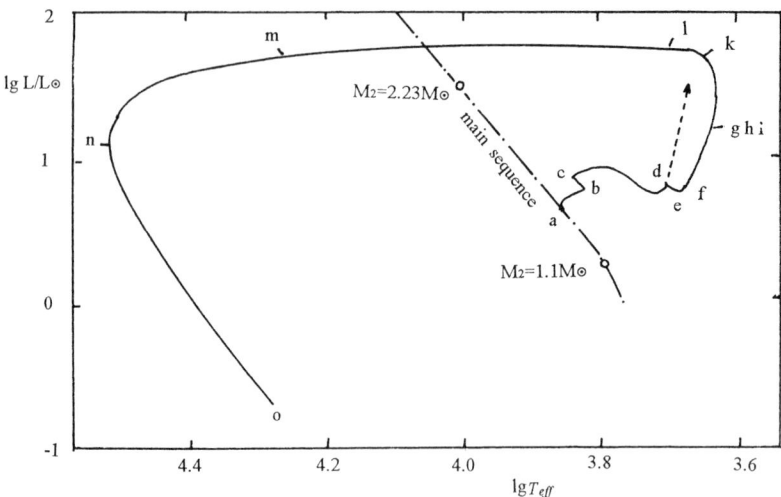

Fig. 11.8 The evolutionary track of the primary in a 1.4M$_\odot$ + 1.1M$_\odot$ binary system in the H–R diagram (Refsdal and Weigert 1969).

of the Roche lobe grows continuously, the radius of the primary keeps increasing and the primary becomes a red giant, and its luminosity also increases. The shell hydrogen burning also keeps moving outwards. When the primary evolves to point k in Fig. 11.8, the slow process of mass transfer ceases. At this time, the mass of the primary is only 0.26M$_\odot$, 96% of which is concentrated in the degenerate helium core and the rest constituting a very thick envelope. When viewed externally, the primary is a red giant. When the primary evolves to a point not far beyond point k in Fig. 11.8, the shell hydrogen burning has moved to a place near the surface, where the temperature is very low, and disappears. In this way, the evolutionary track of the primary moves to the left, passes through most regions in the H-R diagram and finally reaches the white dwarf branch. The mass of the original secondary (M_2) has now increased to 2.24M$_\odot$, but it is still a main sequence star. Therefore, the final binary system is a detached system composed of a more massive (2.24M$_\odot$) main sequence star and a less massive (0.26M$_\odot$) white dwarf. The distance between the two stars has changed to $A = 55$R$_\odot$ and the corresponding orbital period is 31 days.

(C) The evolution for case C

Lauterborn (1970) calculated the case C evolution of a 5M$_\odot$ + 2M$_\odot$ binary system. The initial distance between the two stars is $A = 301.9$R$_\odot$, and the chemical composition of the two stars is $X_H = 0.602$ and $X_Z = 0.044$. Figure 11.9 shows the evolutionary track of the primary of this binary system in the H-R diagram. Figure 11.9(a)

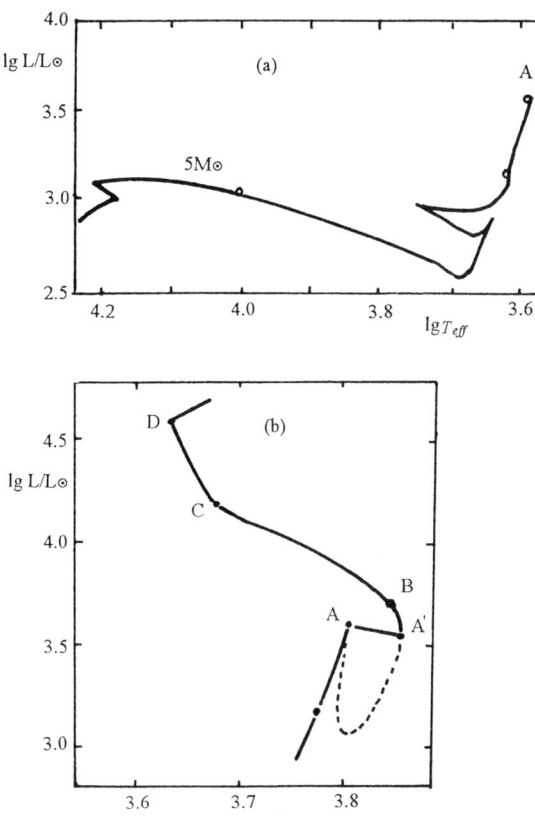

Fig. 11.9 The evolutionary track of the primary in a 5M$_\odot$ + 2M$_\odot$ binary system in the H–R diagram.

gives the evolutionary track of the primary with mass 5M$_\odot$ in the H-R diagram. Point A refers to the completion of core helium burning of the primary, at which the core has a mass of 0.5M$_\odot$ consisting of carbon and oxygen. External to the core of carbon and oxygen, there is a shell helium burning with a mass of 0.45M$_\odot$. The envelope external to the shell helium burning has a mass of 4.05M$_\odot$. Figure 11.9(b) shows the evolutionary track of the primary in the H-R diagram beyond point A. The primary fills up the Roche lobe at A and mass transfer starts. In Fig. 11.9(b), the mass transfer along A $\rightarrow$ A' $\rightarrow$ B is a fast process and takes 2×10^5 yrs. During this period, the primary loses a total mass of 3.98M$_\odot$ and becomes a star of only 1.02M$_\odot$ Therefore, the roles of the primary and the secondary have interchanged. During the evolutionary stages B $\rightarrow$ C $\rightarrow$ D, the radius of the star M_1 is less than the critical radius of the Roche lobe, so the binary system is now a detached system. During the stage B $\rightarrow$ C, the outer convective region of the star M_1 extends inwards, and mixes the inner helium with the envelope, so the helium abundance of the envelope changes from $X_{He} = 0.359$ to $X_{He} = 0.745$. The evolutionary track in the HR diagram therefore changes to the left. During the stage C → D, the core of the star M_1 has a compact C-O core, and a shell helium burning external to that core which continuously moves outwards. Since the luminosity and the temperature of the shell helium burning increase with mass M_c of the compact core, the temperature of the shell helium burning can reach 2×10^8 K which is higher than the temperature for hydrogen burning. When the shell helium burning moves outwards to a hydrogen rich region, hydrogen burning will start again (starting from point C in Fig. 11.9(b)). Therefore, during the stage C → D, the star M_1 has two shells burning, i.e. there is shell hydrogen burning external to shell helium burning. The stage BCD takes a total period of 6.3×10^5 yrs. When it has evolved to point D, the star M_1 has a total mass of $1.02 M_\odot$, 95% of which is concentrated in the very high density core of carbon and oxygen. There are two shells burning external to the core. The envelope has a mass of $0.046 M_\odot$, which contributes to 99.9% of the radius of M_1. The whole star M_1 has a radius of 192 $R_\odot$ and is a red giant. At point D, the star M_1 fills up the Roche lobe again and the slow process of mass transfer starts. However, the calculation of Lauterborn (1970) were up to point D. It is expected that, during the slow process of mass transfer, M_1 will lose the entire envelope of $0.046 M_\odot$ and becomes a C-O white dwarf with about $1 M_\odot$. The final binary system may consist of a $6 M_\odot$ main sequence star and a $1 M_\odot$ white dwarf, and is a detached system.

11.2.5 Fate of Evolution

Of the three evolution cases mentioned above, case A is relatively rare amounting to less than 15%, while case B is more common. In §11.2.4, we have introduced that intermediate massive stars evolve through case B to form a detached binary system consisting of a less massive helium star and a more massive main sequence star. But what will be the fate of the binary system if it evolves further? Delgado and Thomas (1981) determined the evolution of a $9 M_\odot + 6 M_\odot$ binary system with an initial period of 2.4 days, which underwent case B's evolution and became a detached system consisting of a $2 M_\odot$ helium star and a $13 M_\odot$ main sequence star and having a period of 21.5 days. Fig. 11.10 shows the evolutionary track of the $2 M_\odot$ helium star (M_1) in the H-R diagram from the helium main sequence. After 3.27×10^6 yrs from the helium main sequence, the volume of the $2 M_\odot$ helium star has increased to fill up the Roche lobe, and case BB mass transfer starts. Case BB evolution refers to the situation in which a second time mass transfer happens

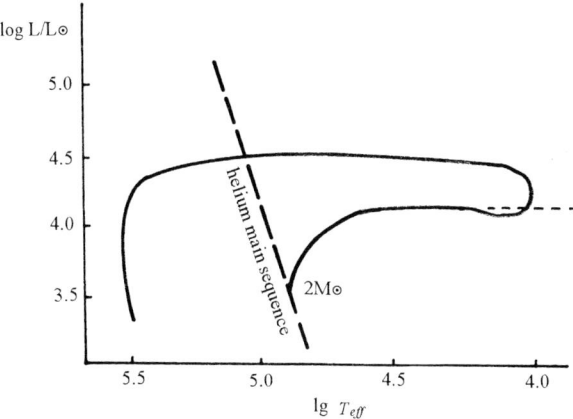

Fig. 11.10 The evolutionary track of the $2M_\odot$ helium star of the $2M_\odot + 13M_\odot$ binary system in the H–R diagram.

when core helium burning of the helium star has completed (the carbon-oxygen core has formed) and when shell helium burning is continuously moving outwards. Through case BB's mass transfer, the binary system becomes a detached system consisting of a $0.993M_\odot$ C-O white dwarf and a $14.007M_\odot$ main sequence star, and the orbital period is 140.45 days.

Generally speaking, helium stars with a mass less than $2M_\odot$ can evolve to C-O white dwarfs through case BB mass transfer. Helium stars with a mass of 2–$3M_\odot$ can evolve through case BB's mass transfer processes to a white dwarf of mass 1.2–$1.4M_\odot$ composed of O, Ne and Mg. If the helium star has a mass greater than $4M_\odot$, it can finally form a supernova (Miyaji *et al*; 1980; Sygimoto and Namoto 1980). It is therefore seen that the fate of evolution of a binary system is closely related to the initial mass of the primary and the orbital period of the system. Figure 11.11 gives the relationship between the evolutionary fate of the primary and its initial mass and its orbital period when filling up the Roche lobe (mass ratio is 1).

11.3 NON-CONSERVATIVE EVOLUTION OF MASSIVE BINARY SYSTEMS

11.3.1 Introduction

A binary system is called a massive binary system if at least one component is a massive star with a mass greater than $10M_\odot$. From §9.4, it is known that massive stars have a lot of special features, e.g., they have strong stellar wind and thus great mass loss, high surface

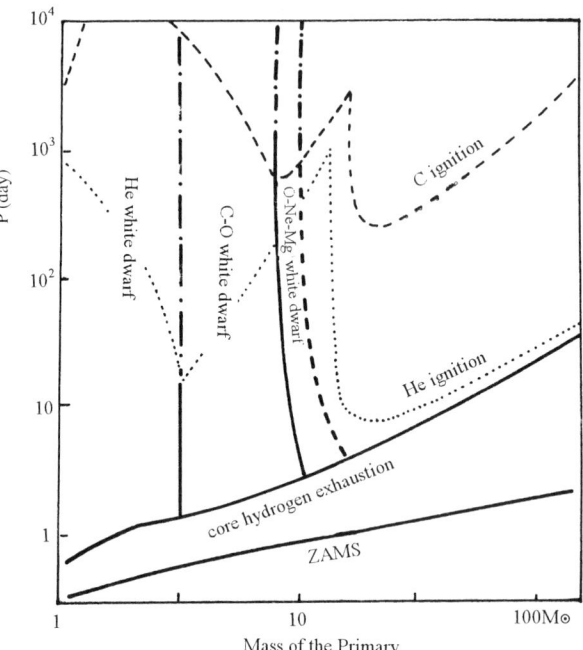

Fig. 11.11 The relationship between the evolutionary fate of the primary and its initial mass and its orbital period when filling up the Roche lobe (mass ratio is 1).

temperature and strong radiation pressure, and will evolve to supernova explosion. If a massive star is at the same time a component of a binary system, it will also be subjected to the effects of gravitational and radiational field of its companion star. Therefore, a massive binary system will encounter a series of very sophisticated situations during its evolution:

(a) Since a massive star is an early type hot star, according to the line driven wind model proposed by Caster *et al.* (1975), the rate of mass loss due to stellar wind is related to the luminosity of the star. The irradiation of the two stars in a massive binary system can increase the luminosities of the stars, thus increasing the rate of mass loss of the massive star. Furthermore, according to Friend and Castor (1982), when the gravitational acceleration on the surface of the star decreases, the rate of mass loss will increase. For binary systems, the tidal effect and the effect of rotation of the system can decrease the gravitational acceleration on the surface of the star. Therefore, for massive binary systems, we have to know the effects of tide, rotation and irradiation on the mass loss rate.

(b) The matter escaping from the stellar surface in form of stellar wind will take away angular momentum at the same time. Therefore, we have to know the relationship between mass loss and angular momentum loss.

(c) It is interesting to know how the orbital period of the binary system changes when both stars have mass loss and angular momentum loss and at the same time mass transfer occurs.

(d) We need to know, in the case that the stars have very high surface temperature and very strong radiation pressure, how their radiation affects the Roche lobe and the mass transfer.

(e) When both stars have stellar wind, the collision of stellar winds will produce a shock front. The shape, position and effects of the shock front should be known.

(f) If a component evolves to a supernova, the explosion being another way for significant loss of mass and angular momentum, the questions arise whether the explosion will destroy the binary system and what the criteria are for this to occur.

(g) Under certain conditions, the two stars can evolve to a system with a common envelope which is then lost through the outer Lagrangian point. This is yet another way of significant loss of mass and angular momentum in binary systems. After the loss of the common envelope, the binary system becomes one with a very short period. It is important to determine the criteria for the two stars to evolve to have a common envelope, and the change in the orbital period of the binary system after loss of the common envelope.

(h) Massive binary systems may evolve to massive X-ray binary systems or low mass X-ray systems. Accretion of matter is the main mechanism of generating X-rays. It is of interest to investigate how accretion disks are formed around compact stellar objects and how X-rays are produced, or how compact stellar objects accrete stellar winds to produce X-rays.

The above problems will be encountered when discussing non-conservative evolution of massive binary systems. Therefore, these problems will first be discussed in the following sections.

11.3.2 Effects of Tide and Rotation on the Mass Loss Rate

From Fig. 11.12 and Eq. (11.1), we can obtain the Roche equipotential on the orbital plane of the binary system:

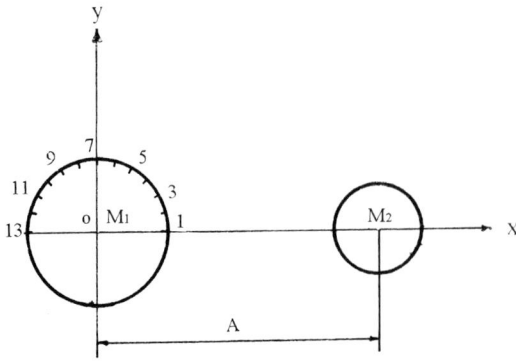

Fig. 11.12 The coordinate system for calculation of effects of tide and rotation.

$$\Omega \equiv \frac{\varphi}{GM_1} = \frac{1}{(x^2 + y^2)^{1/2}} + \frac{q}{[(A - x)^2 + y^2]^{1/2}} + \frac{1+q}{2A^3}\left[\left(x - \frac{qA}{1+q}\right)^2 + y^2\right] \tag{11.39}$$

in which the 2nd and the 3rd terms on the right are contributions to the Roche equipotential from the gravitational force of the companion star (M_2) and the centrifugal force of rotation of the binary system, respectively. Differentiating Eq. (11.39) with respect to x and y, we have

$$\frac{d\Omega}{dx} = -\frac{x}{(x^2 + y^2)^{3/2}} + \frac{q(A - x)}{[(A - x)^2 + y^2]^{3/2}} + \frac{1+q}{A^3}\left(x - \frac{qA}{1+q}\right) \tag{11.40}$$

$$\frac{d\Omega}{dy} = -\frac{y}{(x^2 + y^2)^{3/2}} - \frac{qy}{[(A - x)^2 + y^2]^{3/2}} + \frac{(1+q)y}{A^3} \tag{11.41}$$

The effective gravitational acceleration g_d at point $r(x, y)$ on the orbital plane is

$$g_d = \sqrt{\left(\frac{d\Omega}{dx}\right)^2 + \left(\frac{d\Omega}{dy}\right)^2}. \tag{11.42}$$

Using Eq. (11.42), the effective gravitational acceleration of the points 1,2,3 ... on the surface of M_1 can be determined (see Fig. 11.12), and the average value $\bar{g}_d$ can then be calculated.

If M_1 is a single star, the 2nd and the 3rd terms on the right of Eq. (11.39) are zero. Using the same method as above, the gravitational acceleration g_s of a single star can be obtained as

$$g_s = \frac{1}{x^2 + y^2} \qquad (11.43)$$

In this way, the ratio $g_s/\bar{g}_d$ gives the factor f_1 for the decrease of effective gravitational force on the surface of the star M_1 due to effects of tide and rotation and is also the mass loss enhancement factor on the primary, i.e.,

$$f_1 = g_s/\bar{g}_d \qquad (11.44)$$

11.3.3 Effects of Irradiation on the Mass Loss Rate

The loss rate of stellar wind from the primary M_1 is closely related to its surface luminosity L which can be increased by irradiation from the companion M_2. It is noticed from Fig. 11.13 that the total energy flow from the secondary per second to the surface of the primary is:

$$\ell = L_2 \cdot \frac{dF}{4\pi d^2} = \frac{L_2}{2}(1 - \cos\theta) = \frac{L_2}{2}\left(1 - \frac{\sqrt{A^2 - R_1^2}}{A}\right) \qquad (11.45)$$

where L_2 is the luminosity of the secondary and R_1 is the radius of the primary. A portion of this energy will be absorbed by the primary and the luminosity, L_1, will increase by an amount ΔL_1 given by

$$\Delta L_1 = f_2 \ell \qquad (11.46)$$

where f_2 is the absorption coefficient. In general, $f_2 \approx 0.1$–0.5 can be assumed. Huang and Taam (1990) calculated case A and case B non-

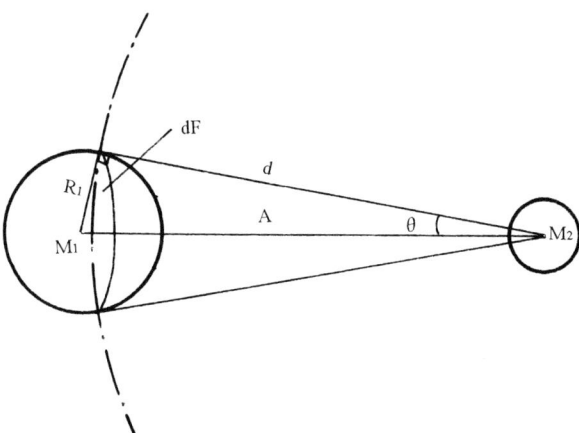

Fig. 11.13 Schematic diagram used in calculations of the effect of irradiation.

conservative evolution of a $40M_\odot + 25M_\odot$ binary system. It was proved that tide, rotation and irradiation affect the increase of mass loss rate for case A's evolution only, but not very significantly, and have insignificant effects for case B's evolution.

11.3.4 Angular Momentum Loss due to Stellar Wind

Suppose the rotations and the orbital motions of the two stars in a binary system are synchronized by tidal effect, and the angular velocity is ω. We further assume that, within a certain time interval, the mass losses from the two stars due to stellar wind ΔM_{1w} and ΔM_{2w} are distributed within spherical shells on the surfaces of the two stars (see Fig. 11.1) and the shells are rigid. The angular momentum taken away by the lost mass is given by Huang and Xie (1985) as:

$$\Delta J = \left(\frac{2}{3}R_1^2 + X_\omega^2\right)\Delta M_{1w}\cdot\omega + \left[\frac{2}{3}R_2^2 + (A - X_\omega)^2\right]\Delta M_{2w}\cdot\omega \quad (11.47)$$

where R_1 and R_2 are the radii of the two stars, $X_\omega = \frac{AM_2}{M_1+M_2}$ the distance between the rotational axis of the system and the origin of the coordinate system (Fig. 11.1), $\left(\frac{2}{3}R_1^2 + X_\omega^2\right)\Delta M_{1w}$ and $\left[\frac{2}{3}R_2^2 + (A - X_\omega)^2\right]\Delta M_{2w}$ are, respectively, moments of inertia of the shells on the surface of the two stars. Considering $\omega^2 = \frac{G(M_1+M_2)}{A^3}$, Eq. (11.47) can be written as

$$\Delta J = \frac{G^{1/2}(M_1 + M_2)^{1/2}}{A^{3/2}}\left[\frac{2}{3}(\Delta M_{1w}R_1^2 + \Delta M_{2w}R_2^2) + \frac{A^2 M_2^2}{(M_1 + M_2)^2}\right.$$
$$\left.(\Delta M_{1w} + \Delta M_{2w}) + \Delta M_{2w}A^2\left(1 - \frac{2M_2}{M_1 + M_2}\right)\right] \quad (11.48)$$

which gives the angular momentum loss of the system.

11.3.5 Change in Orbital Period

If the two stars in a binary system have mass loss due to stellar wind as well as mass transfer, the change in the mass ratio q is

$$q = \frac{M_2 + \Delta M_{2w} - \Delta M}{M_1 + \Delta M_{1w} + \Delta M} \quad (11.49)$$

where ΔM_{1w} and ΔM_{2w} are the mass losses due to stellar wind (ΔM_{1w}, $\Delta M_{2w} < 0$) within a certain time interval, and ΔM is the amount of mass transfer ($\Delta M < 0$) between the two stars within the same time interval.

The change in the orbital period for the non-conservative case can be determined using exactly the same method as in §11.2.3. From Eqs. (11.29) and (11.32), the total angular momentum of the binary system can be written as

$$J = J_0(1 + a) \tag{11.50}$$

For the non-conservative case, from Eq. (11.50), the relative change of angular momentum of the system is

$$\Delta J / J = \Delta J_0 / J_0 + \frac{\Delta a}{a} b \tag{11.51}$$

where the definitions of J_0, a and b are the same as those in Eqs. (11.26), (11.30) and (11.31). When the total mass ($M_1 + M_2$) of the system is non-conservative, from Eqs. (11.26), (11.30) and the definition of ω, we have

$$\Delta J_0 / J_0 = \frac{1}{2} \frac{\Delta A}{A} + \frac{\Delta M_1}{M_1} + \frac{\Delta M_2}{M_2} - \frac{1}{2} \frac{\Delta M_1 + \Delta M_2}{M_1 + M_2} \tag{11.52}$$

$$\Delta a / a = \frac{\Delta \omega}{\omega} + \frac{\Delta I_1 + \Delta I_2}{I_1 + I_2} - \Delta J_0 / J_0 \tag{11.53}$$

$$\Delta \omega / \omega = \frac{1}{2} \cdot \frac{\Delta M_1 + \Delta M_2}{M_1 + M_2} - \frac{3}{2} \frac{\Delta A}{A} \tag{11.54}$$

where I_1 and I_2 are moments of inertia of the two stars [see Eq. (11.27)].

Substituting Eqs. (11.52) to (11.54) into (11.51), we can obtain the change in distance ΔA between the two stars as

$$\Delta A = \frac{A}{\frac{1}{2} - 2b} \left[\frac{\Delta J}{J} - \frac{\Delta M_1}{M_1}(1 - b) - \frac{\Delta M_2}{M_2}(1 - b) - \frac{\Delta M_1 + \Delta M_2}{M_1 + M_2}(b - 1/2) \right. $$
$$\left. - \frac{\Delta I_1 + \Delta I_2}{I_1 + I_2} b \right] \tag{11.55}$$

Once the change in distance A between the two stars is known, the orbital period P of the binary system can be obtained from Eq. (11.38). Furthermore, if changes in q and A are known, the new critical radius of the Roche lobe can be calculated from Eqs. (11.18) and (11.19), or Eqs. (11.20) and (11.21).

11.3.6 Effects of Radiation on the Roche Lobe and on Mass Transfer

When the components in a binary system are massive hot stars and have high radiation pressure, the force of radiation from the primary and secondary acting on a point P in space (see Fig. 11.1) can be expressed as:

$$(F_1)_{rad} = \frac{a_1 GM_1}{r_1^2} \tag{11.56}$$

and

$$(F_2)_{rad} = \frac{a_2 GM_2}{r_2^2} \tag{11.57}$$

where $a_1 = \dfrac{\kappa_e L_i}{4\pi GM_i c} = 1.5306 \times 10^{-5}(1+X_{Hi})\dfrac{L_i/L_\odot}{M_i/M_\odot}, \ (c=1,2),$ (11.58)

L_i and M_i are, respectively, luminosities and masses of the two components, X_{Hi} are abundances of hydrogen on the surface of the components, and $\kappa_e = 0.2(1 + X_{Hi})$ is the absorption coefficient of electron scattering. Since the force acting on point P due to radiation has a direction opposite to the gravitational force, the parameter a_i is introduced such that the radiative force and the gravitational force can be expressed by the same form. In this way, the Roche equipotential can be written as

$$\varphi = -\frac{G(1-a_1)M_1}{r_1} - \frac{G(1-a_2)M_2}{r_2} - \frac{\omega^2}{2}\left[\left(x - \frac{M_2 A}{M_1 + M_2}\right)^2 + y^2\right] \tag{11.59}$$

Using the method as in §11.1.1, Eq. (11.59) becomes

$$\psi \equiv \frac{2}{1+q}\cdot\frac{(1-a_1)}{R_1} + \frac{2q}{1+q}\cdot\frac{(1-a_2)}{R_2} + \left[\left(\xi - \frac{q}{1+q}\right)^2 + \eta^2\right] \tag{11.60}$$

where

$$\left.\begin{aligned} \psi &\equiv -\frac{2}{1+q}\frac{\varphi A}{GM_1} \\ \xi &= \frac{x}{A} \\ \eta &= \frac{y}{A} \\ q &= M_2/M_1 \end{aligned}\right\} \tag{11.61}$$

The Roche lobe can be constructed using Eq. (11.60). From Eq. (11.60), it is obvious that the shape of the Roche lobe depends on q as well as the radiation parameters a_1 and a_2. Figure 11.14 shows the shape of the Roche lobe under influence of radiation.

Figure 11.14(a) is the Roche lobe for $a_1 = a_2 = 0$, i.e., no radiative effects. Its shape should be identical to that shown in Fig. 2. Figure 11.14(b) is the Roche lobe for $a_1 = 0.3$ and $a_2 = 0.2$, i.e., under influence of radiation. It is observed that under influence of radiation, the right portion of the Roche lobe has opened up, which has once led people to think that radiative effects can enormously complicate the mass transfer in binary systems. However, a detailed investigation of the different nature of radiative and gravitational forces will reveal that radiation will in fact not affect the mass transfer between the two stars

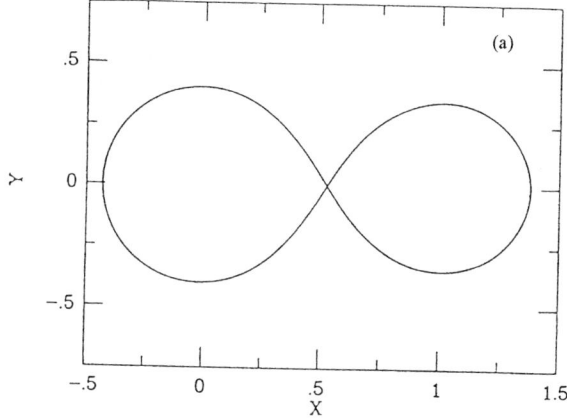

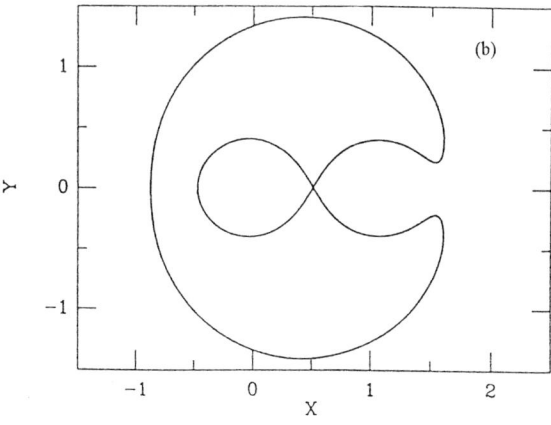

Fig. 11.14 The Roche critical equipotential surface under influence of radiation (Huang and Taam, 1990).

in the system. This is because the criterion for radiation of the star to be effective at any point in space is that there should be no matter between the star and the point to absorb the radiation. If this condition is not fulfilled, the radiation from the stars is absorbed and will not have any effects on that point. On the contrary, the gravitational force on any point are not affected by the intervening matter. Huang and Taam (1990) calculated the optical thickness τ within the Roche critical volume of the secondary in a $40M_\odot + 25M_\odot$ binary system when the primary fills up the Roche lobe and transfers mass to the secondary through the inner Lagrangian point at a rate of $10^{-4}M_\odot$ yr^{-1}. The τ value is found to be able to reach ≈ 100, i.e., the space is opaque, so the radiation from the two stars will be totally absorbed and will not affect the Roche lobe. Therefore, the Roche lobe shown in Fig. 11.14(b) is replaced by that shown in Fig. 11.14(a). In other words, the radiation will not influence the mass transfer between the two stars.

11.3.7 Shock Front of Colliding Winds and X-ray Sources

When both components in a binary system have stellar winds with hypersonic velocities, the collision of stellar winds will produce a shock front. Kudritzki and Reimers (1978) were the first to observe spectroscopically FeII emission lines from the shock front of colliding winds in the α-Sco binary system. Subsequent to these observations, Shore and Brown (1988), Shore and Cororan (1992), Sahade and Brandi (1991) have proved the existence of shock fronts of colliding winds in massive binary systems. Recently, X-ray observations of the ROSAT satellite have revealed that X-rays are coming from shock fronts of colliding winds in many massive binary systems, especially WR binary systems [*IAU Symp.* No. 163(1994)]. Since astronomical X-ray sources are usually found in binary systems with compact objects, the X-rays observed to be coming from shock fronts of colliding winds in massive binary systems have attracted much interest in the astrophysical community. Huang and Weigert (1982) were the first to study the hydrodynamics of the shock front of colliding winds, and proposed a simple method to calculate the position and the shape of the shock front. Their work was based on the following two basic assumptions. Firstly, since the mean free path of particles in the shock front is negligible compared to the distance A between the two stars ($l/A \approx 10^{-5}$), we can assume that the shock front is extremely thin. Secondly, it is assumed that

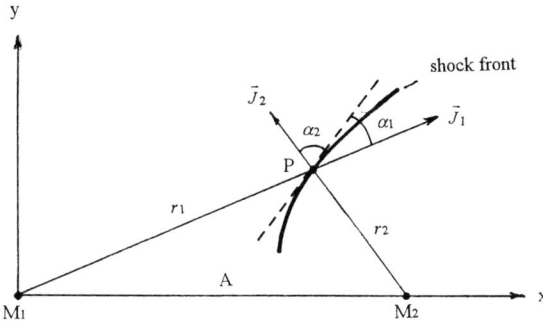

Fig. 11.15 The x-y plane containing a portion of the shock front. The broken line is tangential to the point P on the shock front. $\vec{J}_1$ and $\vec{J}_2$ are the momentum flux of stellar wind from the primary and the secondary.

the velocity v_i of stellar wind is much larger than the velocity v_{orb} of orbital motion of the components, i.e., the approximation $v_{orb} \approx 0$ is taken.

Introducing a coordinate system with the origin at the center of the primary and the x-axis passing through the center of the secondary (see Fig. 11.15), and according to earlier assumptions, the presence of a stable shock front should fulfil the following conditions:

(1) The forces along the direction perpendicular to the shock front at point P should be at equilibrium, i.e.,

$$J_1 \sin^2 \alpha_1 - J_2 \sin^2 \alpha_2 = \frac{\rho_F v^2}{R} \tag{11.62}$$

where the terms $J_i = \frac{\dot{m}_i v_i}{4\pi r_i^2}$ $(i = 1, 2)$ on the left are the momentum flux of stellar wind from the two stars at point P ($\dot{m}_i$ are rates of mass loss by stellar winds, v_i the velocities of stellar winds and r_i the distances between the stars and point P), $J_i \sin^2 \alpha_i$ are the components of momentum flux along the normal direction and the term $\frac{\rho_F v^2}{R}$ on the right is the centrifugal force produced when matter flows along the shock front (ρ_F is the density of the shock front, v is the flow velocity of matter along the shock front and R is the radius of curvature of the shock front, i.e. $R = \left| \frac{(1+y'^2)^{3/2}}{y''} \right|$).

(2) The resultant force along the tangent to the shock front should be equal to the variation of momentum flux of matter on the shock front, i.e.,

$$J_1 \sin \alpha_1 \cdot \cos \alpha_1 + J_2 \sin \alpha_2 \cdot \cos \alpha_2 = \frac{d(\rho_F v^2)}{ds} \tag{11.63}$$

Introducing new variables

$$\xi \equiv \frac{x}{A}, \quad \eta \equiv \frac{y}{A}, \quad \zeta \equiv \frac{d\eta}{d\xi}, \quad \varphi \equiv \frac{\rho_F v^2 A}{\dot{m}_1 v_1} \tag{11.64}$$

Eqs. (11.62) and (11.63) can be changed into three linear differential equations with ξ as the independent variable:

$$\eta' = \zeta \tag{11.65}$$

$$\zeta' = -\frac{1}{\varphi}(1+\zeta^2)^{1/2}\left\{\frac{(\xi\zeta-\eta)^2}{(\xi^2+\eta^2)^2} - j\frac{(\eta+\zeta-\xi\zeta)^2}{(1-2\xi+\xi^2+\eta^2)^2}\right\} \tag{11.66}$$

$$\varphi' = (1+\zeta^2)^{-1/2}\left\{\frac{(\xi\zeta-\eta)(\zeta\eta+\xi)}{(\xi^2+\eta^2)^2} + j\frac{(\zeta-\zeta\xi+\eta)(\zeta\eta+\xi-1)}{(1-2\xi+\xi^2+\eta^2)^2}\right\} \tag{11.67}$$

These equations have determined the shape and position of the shock front. It can also be seen that there is only one variable in these equations, namely

$$j = \frac{\dot{m}_2 v_2}{\dot{m}_1 v_1} \tag{11.68}$$

which means that the variable j alone determines the shape and the position of the shock front. When a value of j is chosen, Eqs. (11.65) to (11.67) can be integrated after choosing suitable initial values. Huang and Weigert (1982) gave the initial values as

$$\xi_1 = \frac{1}{1+j^{1/2}} + \mu \tag{11.69}$$

$$\eta_1 = \left[\left(\frac{3j^{1/2}}{1-j}\right)\mu - \frac{3}{5}\left(\frac{1+j-5j^{1/2}}{1+j-2j^{1/2}}\right)\mu^2\right]^{1/2} \tag{11.70}$$

$$\zeta_1 = \left[\frac{3j^{1/2}}{1-j} - \frac{6}{5}\left(\frac{1+j-5j^{1/2}}{1+j-2j^{1/2}}\right)\mu\right]/2\eta_1 \tag{11.71}$$

$$\varphi_1 = \frac{\mu}{(1+\zeta_1^2)^{1/2}}\left[\frac{(\xi_1\zeta_1-\eta_1)(\eta_1\zeta_1+\xi_1)}{(\xi_1^2+\eta_1^2)^2} + j\frac{(\zeta_1-\xi_1\zeta_1+\eta_1)(\eta_1\zeta_1-1+\xi_1)}{(1-2\xi_1+\xi_1^2+\eta_1^2)^2}\right] \tag{11.72}$$

where μ is an arbitrarily small number. Figure 11.16 gives the shape and the position of the shock front in the $\xi - \eta$ plane according to different

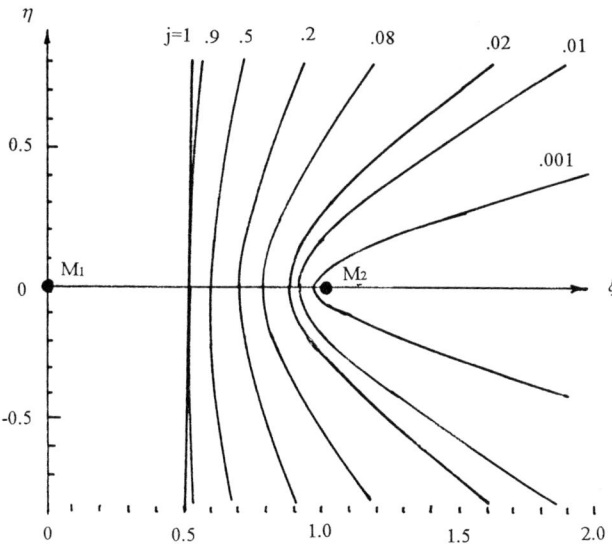

Fig. 11.16 The shape and position of the shock front in the $\xi - \eta$ plane according to different values of j (Huang and Weigert 1982).

values of j. The real shock front should be the surface generated by rotation around the ξ-axis.

11.3.8 Conditions for the Binary System not to be destroyed by Supernova Explosion

To simplify our discussions, it is assumed that the orbits of two stars are circular. Figure 11.17 gives schematic diagrams of the binary system before and after the supernova explosion. Assuming the masses of the primary and secondary are M_1 and M_2, respectively, before the explosion, and the distance between them is A, the total energy E_{tot} (the sum of kinetic and potential energies) is

$$E_{tot} = \frac{1}{2}\frac{M_1 M_2}{M_1 + M_2}v^2 - \frac{GM_1 M_2}{A} \tag{11.73}$$

Since $\omega^2 = \frac{G(M_1 + M_2)}{A^3}$ we have

$$v^2 = \frac{G(M_1 + M_2)}{A} \tag{11.74}$$

Substituting (11.74) into (11.73),

$$E_{tot} = -\frac{1}{2}\frac{GM_1 M_2}{A} \tag{11.75}$$

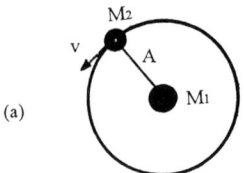

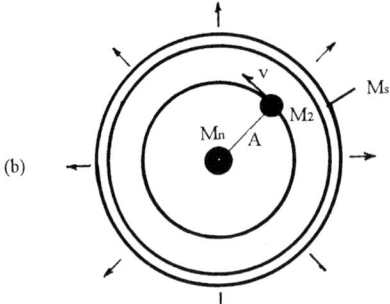

Fig. 11.17 Schematic diagrams of the binary system: (a) before supernova explosion (b) after supernova explosion.

Suppose the neutron star left over by supernova explosion has a mass M_n, the envelope ejected during the explosion has a mass $M_s = M_1 - M_n$. In addition, if the binary system is not affected at the moment of supernova explosion, i.e., the distance between the two stars remains at A and the velocity of the star M_2 remains at v, the total energy E'_{tot} of the binary system after explosion is

$$E'_{tot} = \frac{1}{2} \frac{M_n M_2}{M_n + M_2} v^2 - \frac{G M_n M_2}{A} \tag{11.76}$$

Substituting (11.74) into (11.76), we get

$$E'_{tot} = \frac{G M_n M_2}{2A} \left(\frac{M_1 + M_2}{M_n + M_2} - 2 \right) \tag{11.77}$$

The condition for the binary system to remain intact after explosion is that the total energy E'_{tot} should be negative. From Eq. (11.77), this condition can be written as

$$M_n > \frac{1}{2}(M_1 - M_2) \tag{11.78}$$

or

$$M_s < \frac{1}{2}(M_1 + M_2) \tag{11.79}$$

11.3.9 Conditions for the System to have Common Envelope and Period after Loss of the Envelope

The following outlines the conditions for a binary system to evolve to one having a common envelope:

(a) It has been pointed out by Darwin (1909) and Counselman (1973) that if the secondary has a mass so small that the orbital angular momentum J_{orb} and the rotational angular momentum of the primary J_{rot} satisfy the relationship

$$J_{orb} \leq 3J_{rot} \qquad (11.80)$$

the secondary will spiral into the primary in such a way that the two stars have a common envelope.

The reasons for the condition set in Eq. (11.80) are as follows: if the rotation of a component slows down, which is due to an increase in the volume, the tidal effect will speed up the rotation and will dissipate the orbital angular momentum, i.e., converting part of the orbital angular momentum into the rotational angular momentum. However, the decrease in the orbital angular momentum will shorten the distance between the two stars and speed up the orbital rotation. These will further speed up the rotation of the stars through the tidal effect, and once again convert a portion of the orbital angular momentum into the rotational angular momentum. The above will be repeated to speed up the rotation and shorten the distance between the two stars continuously. In this way, a small perturbation to slow down the rotation of a component will make the two stars get closer continuously, i.e., the two stars will spiral-in to form a system with a common envelope. However, such a spiral-in process can only occur when the orbital angular momentum of the binary system is not large enough (to satisfy Eq. (11.80)). If the orbital angular momentum of the binary system is sufficiently large, the part of orbital angular momentum converted into rotational angular momentum due to tidal effect can only lead to a very small relative change in the orbital angular momentum. Under such a condition, the process for speeding up the rotation will stop at some point at which the rotation of the stars and the orbital rotation are stably synchronized to a certain velocity.

(b) If the mass ratio $q = M_2/M_1$ of the two stars is too small, e.g., when the mass of the primary is larger than that of the secondary by a factor greater than 2.5, the thermal timescale of the envelope of the primary is far less than that of the secondary. During mass transfer, the primary loses mass on a thermal timescale. The secondary receives a large amount of mass from the primary in a short time interval, but cannot carry out thermal adjustment, because the timescale of the envelope of the secondary is much greater than that of the primary. Therefore, the secondary will expand quickly and fill up the Roche lobe, so the binary system becomes one with a common envelope.

(c) If the envelope of the primary is convective during mass transfer, and loses mass on a hydrodynamical timescale, the secondary will also receive a large amount of matter in a short time interval and will not have sufficient time to carry out thermal adjustment. The secondary will thus expand quickly and fill up the Roche lobe, so the binary system becomes one with a common envelope.

The two former conditions are the main ones for the evolution of a massive binary system to become one having a common envelope. A binary system with a common envelope can lose some envelope mass through the outer Lagrangian point and become a system with a very short period. To determine the period change, we introduce the parameter (according to Livio and Soker (1988), van den Heuvel (1992))

$$\alpha_{ce} = \Delta E_b / \Delta E_{orb} \tag{11.81}$$

where ΔE_b is the binding energy of the ejected matter from the envelope, and ΔE_{orb} is the decrease in the orbital energy when the two stars spiral in to form a system with a common envelope. The parameter α_{ce} indicates the portion of decrease in orbital energy used for ejection of mass during the spiral-in process. Therefore, the parameter is an efficiency parameter with a value close to 1. We denote M_{1c} and M_{1e} to be the masses of the core and the envelope of the primary, and M_2 to be the mass of the secondary. Before the spiral-in, the primary fills up the Roche lobe (see Fig. 11.18a). If the Roche critical radius is R_{cr1} (in units of A_1), the radius of the primary is $R_1 = A_1 R_{cr1}$, where A_1 and A_2 are the distances between the two stars respectively before the spiral-in and after the loss of the common envelope. We therefore have

$$\Delta E_b = \frac{G(M_{1c} + M_{1e})M_{1e}}{\lambda A_1 R_{cr1}} \tag{11.82}$$

where λ is a weighting factor determined by the density of the envelope and is of the order of 1, and

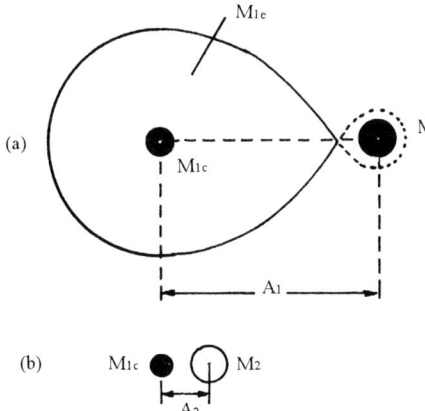

(a)

(b)

Fig. 11.18 The change of parameters of the binary system through evolution of the common envelope. (a) Before having a common envelope, the star M_1 fills up the Roche lobe; M_{1c} and M_{1e} are the masses of the core and the envelope of the star M_1. (b) The binary system after loss of the common envelope.

$$\Delta E_{orb} = \frac{GM_{1c}M_2}{2A_2} - \frac{G(M_{1c}+M_{1e})M_2}{2A_1} \qquad (11.83)$$

Substituting Eqs. (11.82) and (11.83) into (11.81), we obtain

$$\frac{A_2}{A_1} = \frac{\alpha_{ce}M_{1c}M_2}{(M_{1c}+M_{1e})}/(\alpha_{ce}M_2 + 2M_{1e}/\lambda R_{cr1}) \qquad (11.84)$$

Using Eq. (11.84), we can calculate the change in the distance between the two stars after loss of the common envelope.

11.3.10 Accretion in Binary Systems and X-ray Emission

The X-ray emission in binary systems with compact objects is due to accretion. There are in general two cases of accretion. First, during mass transfer, the accreted matter will first flow through the inner Lagrangian point and form an accretion disk around the compact object. Second, the accreted matter is stellar wind and will not form an accretion disk around the compact object.

(A) Accretion disk and its luminosity L_{dis}

During mass transfer in a binary system, matter flows out from the surface of the star M_1 which has filled up the Roche lobe, and passes through the inner Lagrangian point L_1. Since the matter possesses angular momentum, it will not flow to the companion star M_2 in a straight line and will instead move stably in a circular orbit. The radius of this circular orbit is usually smaller than the critical radius R_{cr2} of

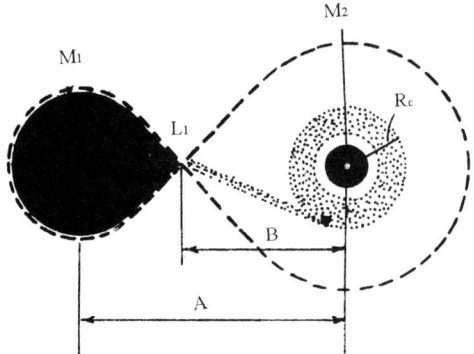

Fig. 11.19 A binary system with an accretion disk. The star M_1 fills up the Roche lobe, and the star M_2 is a compact object (white dwarf or neutron star).

the Roche lobe of the companion star M_2, but larger than the radius R_2 of the star M_2. In other words, a disk is formed around the secondary, which is called the accretion disk. Due to viscosity of matter within the disk, the gas moving in circular orbits will convert its mechanical energy (kinetic energy and potential energy) into thermal energy, and drops to circular orbits with a lower angular momenta. The accreted matter continues to drop and finally into the central compact star M_2. Accompanied by the continuous inward movement of the accreted matter, there is a radial and outward transportation of angular momentum.

Consider a coordinate system rotating with the two stars (see Fig. 11.19). The velocity of the gas flowing out from the inner Lagrangian point L_1 can be resolved into two components, viz., one parallel to the line joining the two stars, $v_{\parallel}$, and one perpendicular to the line, $v_{\perp}$. We have also

$$v_{\parallel} \le C_s \tag{11.85}$$

and
$$v_{\perp} = B\omega \tag{11.86}$$

where C_s is the velocity of sound in the envelope of the star M_1 (when the temperature of the envelope of a star $\le 10^5$ K, $C_s \sim 10$ kms^{-1}), and B is the distance between the inner Lagrangian point L_1 and the star M_2. According to Plavec and Kratochvil (1964), B can be approximated by

$$B = (0.50 - 0.227 \log q)A \tag{11.87}$$

where $q = M_2/M_1$.

The most stable orbit of a gas with a fixed angular momentum is a circular one because its energy is the lowest. Therefore, the gas flowing out from L_1 will move stably in a circular orbit after a certain time. Assuming the radius of the stable circular orbit to be R_c, the movement of a gas particle in the circular orbit will fulfil

$$v(R_c) = \left(\frac{GM_2}{R_c}\right)^{1/2} \tag{11.88}$$

and
$$R_c \cdot v(R_c) = B^2 \omega \tag{11.89}$$

Substituting Eq. (11.88) into (11.89), and considering $\omega = 2\pi/P$, we have

$$R_c \,/\, A = \frac{4\pi^2}{GM_2}\left(\frac{A^3}{P^2}\right)\left(\frac{B}{A}\right)^4 \tag{11.90}$$

Employing the Kepler equation $A^3/P^2 = G(M_1 + M_2)/4\pi^2$, we have

$$R_c/A = (1+q)[0.50 - 0.227\log q]^4 \tag{11.91}$$

Using Eqs. (11.91) and (11.21), it is straightforward to prove that, except for very small q values ($q \leq 0.05$), R_{cr2} is greater than R_c by a factor of 2–3. In other words, the stable circular orbit of the gas particle is inside the Roche lobe of the star M_2.

On the other hand, using the Kepler law $A^3 = G(M_1 + M_2)P^2/4\pi^2$, Eq. (11.91) can be written as

$$R_c \cong 4(1+q)^{4/3}[0.50 - 0.227\log q]^4 P_{day}^{2/3} \quad (R_\odot) \tag{11.92}$$

From Eq. (11.92), we observe that R_c and $R_\odot$ have the same order of magnitude, e.g.,

$$\text{when } q = 0.3 \,, \ R_c \cong 1.2 P_{day}^{2/3} \quad (R_\odot) \tag{11.93}$$

$$\text{when } q > 0.5 \,, \ R_c \cong 0.6 P_{day}^{2/3} \quad (R_\odot) \tag{11.94}$$

Therefore, R_c can be approximated by

$$R_c \geq 3.5 \times 10^9 \, P_{hr}^{2/3} \quad (\text{cm}) \tag{11.95}$$

Equation (11.95) illustrates the following: if the star M_2 is a compact object, i.e., $R_2 < 10^9$ cm and the period is generally of the order of hours, we have $R_c > R_2$; if M_2 is a main sequence star, i.e., $R_2 > 10^{10}$ cm and the period is of the order of days or longer, we have $R_c < R_2$. In other words, if the star M_2 is a compact object, the gas flowing out from L_1 will move in a stable circular orbit of which the radius is smaller than the critical radius of the Roche lobe and larger than the radius of the compact object, so that a disk is formed; if the star M_2 is a main sequence star, the gas flowing out from L_1 will go straight to the surface of the star M_2 because the radius of the stable circular orbit is smaller than the radius of the star M_2.

At equilibrium, if the mass accreted per second is $\dot{M}$, the gravitational energy released is

$$L_{acc} = \frac{GM_2\dot{M}}{R_2} \tag{11.96}$$

of which about half is used for radiation and the rest converted to the kinetic energy for the movement of gas on the accretion disk. Therefore, the accretion luminosity is expressed as

$$L_{dis} = \frac{1}{2}\frac{GM_2\dot{M}}{R_2} \tag{11.97}$$

(B) Accretion of stellar wind

If a component of the binary system is a massive early type star, has great mass loss due to stellar wind and has a volume less than the Roche lobe while the companion is a compact neutron star, the neutron star can accrete the stellar wind from the massive star and produce X-ray emission.

We assume that the velocity of stellar wind v_w of the massive star M_1 can be approximately represented by the escape velocity v_{esc}, i.e.,

$$v_w \approx v_{esc} = \left(\frac{2GM_1}{R_1}\right)^{1/2} \tag{11.98}$$

Denoting the orbital velocity of the neutron star as v_{orb}, the resultant velocity of v_{orb} and v_w, i.e., the velocity of the stellar wind relative to the neutron star v_{rel} (see Fig. 11.20), is given by

$$\left.\begin{array}{l} v_{rel} \cong (v_{orb}^2 + v_w^2)^{1/2} \\ \beta = \tan^{-1}(v_{orb}/v_w) \end{array}\right\} \tag{11.99}$$

The cylinder with the direction of $\vec{v}_{rel}$ as the axis and r_{acc} as the radius is called the accretion cylinder, where r_{acc} is called the accretion radius and is calculated by equalizing the kinetic energy and the potential energy of the stellar wind, i.e.,

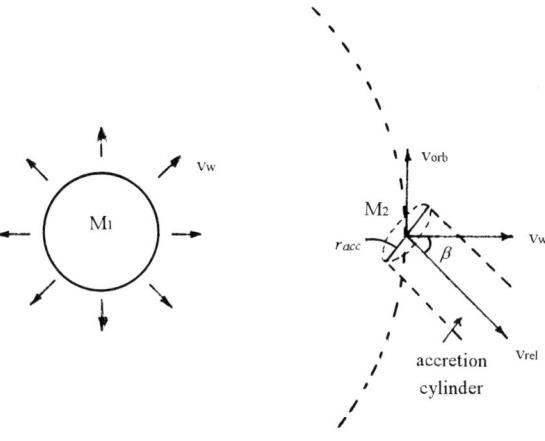

Fig. 11.20 The stellar wind of the primary M_1 accreted by the secondary M_2 (neutron star).

$$\frac{v_{rel}^2}{2} = \frac{GM_2}{r_{acc}} \qquad (11.100)$$

Only the stellar wind entering the accretion cylinder will be accreted by the neutron star M_2.

In most cases, the velocity of stellar wind v_w of the massive star is very large and is much larger than the orbital velocity of the neutron star. Under such a condition, $v_w \approx v_{rel}$ and $\beta = 0$, and the ratio between $\dot{M}$ (the mass flowing into the accretion cylinder per second) and $\dot{M}_w$ (the rate of mass loss due to stellar wind) can be written as

$$\frac{\dot{M}}{\dot{M}_w} \simeq \frac{\pi r_{acc}^2 v_{rel}}{4\pi A^2 v_w} \qquad (11.101)$$

where A is the distance between the two stars.

Substituting Eqs. (11.98) and (11.100), and $v_w \approx v_{rel}$ into Eq. (11.101), we obtain

$$\frac{\dot{M}}{\dot{M}_w} \simeq \frac{G^2 M_2^2}{A^2 v_w^4} = \frac{1}{4}\left(\frac{M_2}{M_1}\right)^2 \left(\frac{R_1}{A}\right)^2 \qquad (11.102)$$

If typical parameters for a massive X-ray binary system are employed, the ratio is $\dot{M}/\dot{M}_w \approx 10^{-4}$ to 10^{-3}, i.e., the matter flowing into the accretion cylinder is only 10^{-4} to 10^{-3} of the total amount of stellar wind. The accretion luminosity is

$$L_{acc} = \frac{GM_2\dot{M}}{R_2} \qquad (11.103)$$

11.3.11 Evolution to a WR + O Type Binary System

(A) Case A evolution

We use the $40M_\odot + 25M_\odot$ binary system studied by Huang and Taam (1990) as an example to illustrate the non-conservative evolution of a massive binary system in which mass transfer occurs at the stage of core hydrogen burning of the primary. The initial distance between the two components is $40.73R_\odot$, and the chemical composition is $X_H = 0.602$ and $X_Z = 0.044$. The evolution of the two components are followed simultaneously. Figure 11.21 shows the evolutionary tracks of the two components in the H-R diagram. In Fig. 11.21, the solid line represents the track for no mass transfer between the two components while the broken line represents the track with mass transfer. The points a and a′ correspond to zero age main sequence models of the primary and secondary; points b and b′ correspond to start of the first mass transfer of the primary and secondary; points c and c′ correspond to the end of the first mass transfer; points d and d′ the start of the second mass transfer; points e and e′ the end of the second mass transfer.

Table 11.3 gives some parameters of the binary system at different stages. From Table 11.3, we can see that after 2.6×10^6 yrs from the zero age main sequence (point a), the volume of the primary expands to fill the Roche lobe and the first mass transfer starts (point b). At this time, due to mass loss of stellar wind, the primary and secondary have lost $2.93M_\odot$ and $0.07M_\odot$, respectively. Because of loss of mass and angular

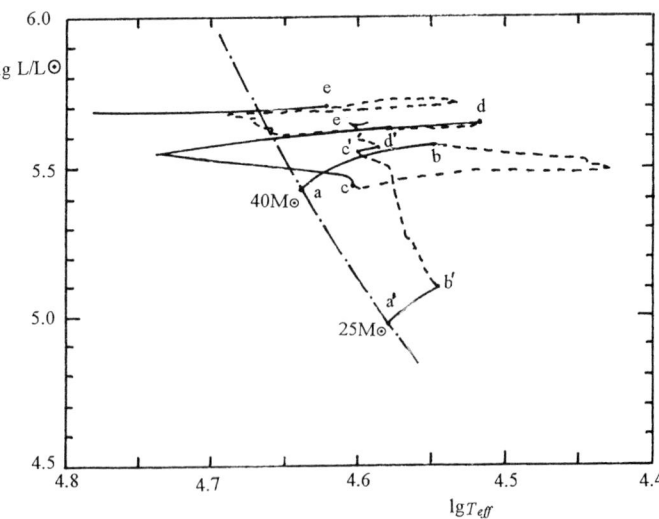

Fig. 11.21 The evolutionary tracks of the two components in a $40M_\odot + 25M_\odot$ binary system on the H–R diagram for case A's non-conservative evolution. The solid line is the track for no mass transfer while the broken line is the track with mass transfer (Huang and Taam 1990).

Table 11.3 Some parameters of a $40M_\odot + 25M_\odot$ binary system at different evolutionary stages.

	$\tau(10^6 y)$	$A(R_\odot)$	$P(d)$	$M_1(M_\odot)$	$M_2(M_\odot)$	$R_1(R_\odot)$	$R_2(R_\odot)$	$\dot{M}_1(M_\odot/y)$	$\dot{M}_2(M_\odot/y)$
a	0.00	40.73	3.74	40.00	25.00	9.12	7.12	5.76×10^{-7}	1.21×10^{-7}
b	2.598	40.57	3.82	37.07	24.53	16.41	9.45	2.63×10^{-6}	2.76×10^{-7}
c	3.219	36.78	3.41	19.19	38.17	11.06	12.64	9.70×10^{-7}	1.47×10^{-6}
d	3.350	36.82	3.43	19.07	37.96	20.29	13.57	8.01×10^{-6}	1.78×10^{-6}
e	3.357	48.47	5.18	15.11	41.88	11.23	13.69	1.70×10^{-6}	2.06×10^{-6}

momentum, the distance between the two components in the binary system has changed to $40.57R_\odot$ and the period has increased from $3.74d$ to $3.82d$. The first mass transfer has taken a duration of 6×10^5 yrs. Within this period, the primary has given up mass of $14.1M_\odot$ to the secondary. At the end of the first mass transfer (point c), the radius of the primary has become less than the critical radius of the Roche lobe and the binary system becomes detached. However, after 1.4×10^5 yrs, due to core hydrogen burning of the primary, its volume will expand to fill the Roche lobe again and the second mass transfer starts (point d). The duration of the second mass transfer is very short which lasts only 7×10^3 yrs. During this period, the primary continues to lose a mass of $3.9M_\odot$ to the secondary. The end of the second mass transfer is 3.4×10^6 yrs from the zero age main sequence. At this time, the primary has a mass of only $15.1M_\odot$ but the secondary has a mass increased to $41.88M_\odot$, and the orbital period of the binary system has increased to 5.18 days. Since the original primary has lost a large amount of mass, the hydrogen abundance on its surface has decreased to 0.1. Its evolutionary track in the H-R diagram moves quickly to the left and reaches the helium main sequence. The ultimate binary system obtained consists of a Wolf–Rayet star with a mass $15.11M_\odot$ and an O5I super giant with a mass of $41.88M_\odot$ (i.e., WR + O5I binary system) and the orbital period is 5.18d.

(B) Case B evolution

Huang and Taam (1990) calculated case B's non-conservative evolution of a $40M_\odot + 25M_\odot$ binary system. The initial distance between the two stars is $113.75R_\odot$ (the corresponding orbital period is 17.44 days), and the chemical compositions of the two stars are $X_H = 0.602$ and $X_Z = 0.044$. Figure 11.22 gives the evolutionary tracks of the two components in the H-R diagram. In the figure, the solid lines

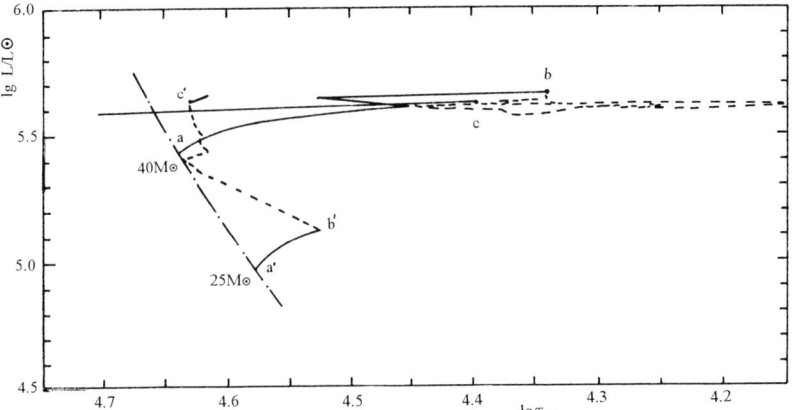

Fig. 11.22 The evolutionary tracks of the two components in the H–R diagram for case B's non-conservative evolution of a 40M$_\odot$ + 25M$_\odot$ binary system. The solid lines correspond to evolutionary tracks for no mass transfer while the broken lines correspond to evolutionary tracks with mass transfer (Huang and Taam, 1990).

correspond to the evolutionary tracks for no mass transfer while the broken lines correspond to evolutionary tracks with mass transfer. After 3.29×10^6 yrs from the zero age main sequence (point a), the primary expands to fill the Roche lobe and the first mass transfer starts (point b). At this time, core hydrogen burning of the primary has completed and the stage of shell hydrogen burning has been reached. Due to mass loss of stellar wind, the primary and secondary now have masses of 34.4M$_\odot$ and 24.3M$_\odot$, respectively. Because of loss of mass and angular momentum, the orbital period of the binary system has increased to 20.45 days (corresponding to a distance of 122.24R$_\odot$ between the two stars). The first mass transfer takes only 6400 yrs, within which the primary has given up 17.42M$_\odot$ to the secondary. Therefore, at the end of the first mass transfer (point c), the mass of the primary is 16.46M$_\odot$ and that of the secondary is 41.97M$_\odot$, so the roles of the primary and the secondary have interchanged with each other. At this time, the orbital period of the binary system has increased to 35.17 days (corresponding to a distance of 175.21R$_\odot$ between the two components). It is obvious that case B's mass transfer is much shorter than case A's mass transfer. After completion of the first mass transfer, the abundance of hydrogen on the surface of the primary is only 0.36, and the evolutionary track of the primary moves quickly to the left in the H–R diagram and evolves quickly to a WR star. The secondary has become an O type supergiant. Therefore, the binary system has become a WR + O type system.

11.3.12 Evolution to a Low Mass X-ray Binary System

We employ the 15M$_\odot$ + 2M$_\odot$ binary system studied by Sutanyo (1992) as an example to illustrate the possibility for a not too massive binary

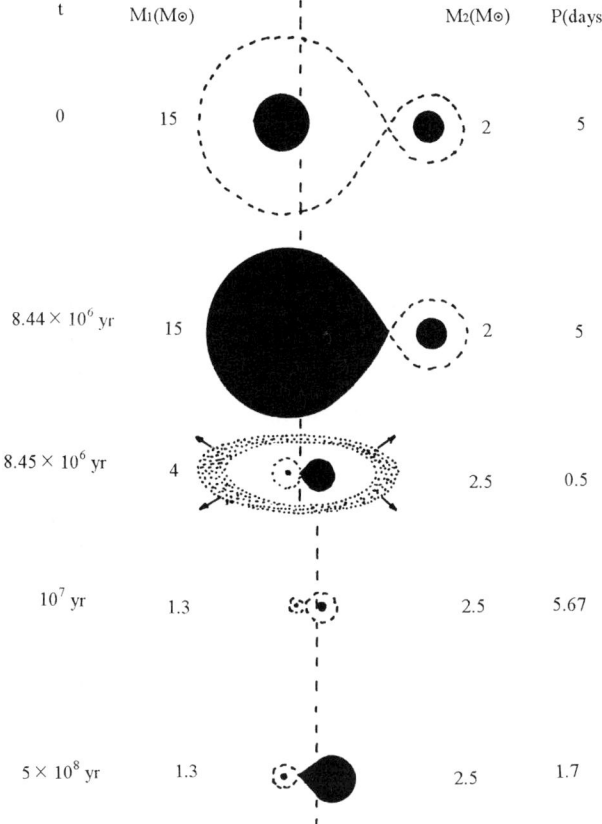

| t | M₁(M⊙) | | M₂(M⊙) | P(days) |

Fig. 11.23 Schematic diagram showing the evolution of a 15M⊙ + 2M⊙ binary system to a low mass X-ray binary system (after Sutantyo 1992).

system with a very small mass ratio to evolve into a low mass X-ray binary system. Figure 11.23 is the schematic diagram showing the evolution of the binary system. The initial masses of the primary and the secondary are $15M_\odot$ and $2M_\odot$, respectively, and the orbital period of the system is five days. After 8.44×10^6 yrs, the primary has evolved to a red giant and fills the Roche lobe. Since the mass of the secondary is too small, during the fast mass transfer, the secondary accretes too much matter from the primary and does not have enough time to carry out thermal adjustment and expands quickly, so the two stars spiral into a binary system with a common envelope. After the loss of the common envelope, the primary leaves a nucleus of $4M_\odot$ and becomes a helium star, and the secondary becomes a main sequence star with mass $2.5M_\odot$, and the orbital period has decreased to 0.5 days. After another 1.5×10^6 yrs, the helium star evolves to a supernova. During supernova explosion, the external envelope with a mass $2.7M_\odot$ has been ejected leaving behind a nucleus with a mass $1.3M_\odot$ and a neutron star is formed. The orbital period of the binary system has changed to 5.67 days. After supernova explosion, there are 5×10^8 peaceful years before the main sequence star fills up the Roche lobe and mass transfer starts. At this time, the neutron star accretes matter to form an X-ray source and the binary system becomes a low mass X-ray binary system. The orbital period is 1.7 days. The parameters of the binary system are consistent with those observed for the X-ray binary system HerX-1.

11.3.13 Evolution to a Massive X-ray Binary System

We use the evolution of a $20M_\odot + 8M_\odot$ binary system studied by De Loore *et al.* (1975) to illustrate the evolution of a massive binary system into a massive X-ray binary system.

Figure 11.24 gives the schematic diagram showing the evolution of this binary system. After 6.17×10^6 yrs, the primary with a mass of

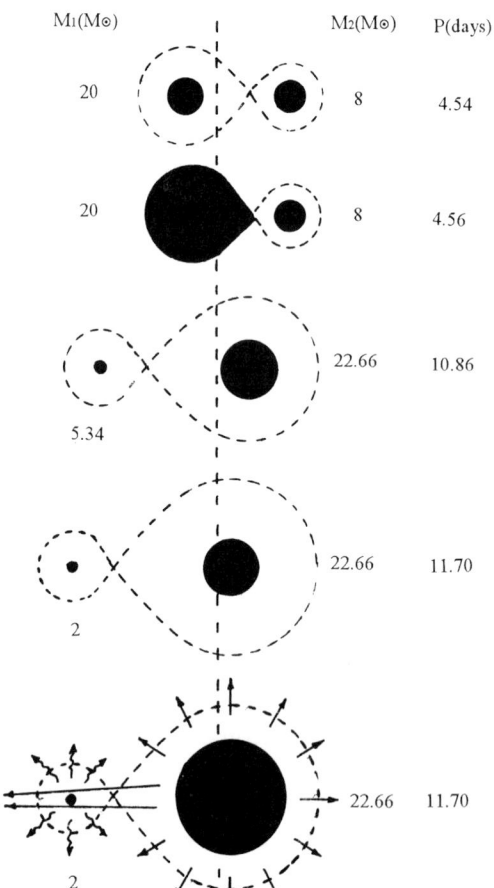

Fig. 11.24 Schematic diagram showing the evolution of a $20M_\odot + 8M_\odot$ binary system into a massive X-ray binary system. The primary with a mass of $20M_\odot$ becomes a helium star with a mass $5.34M_\odot$ after mass transfer, and continues to evolve to supernova explosion leaving a nucleus with a mass $2M_\odot$ and becomes a neutron star. The binary system has become a massive X-ray binary system (De Loore *et al.* 1975).

$20M_\odot$ fills the Roche lobe and mass transfer starts. After 3×10^4 yrs, the primary has lost a mass of $14.66M_\odot$ and becomes a helium star with a mass $5.34M_\odot$, while the mass of the secondary has changed from the original $8M_\odot$ to $22.66M_\odot$. The period of the binary system has changed from 4.54 days before mass transfer to 10.86 days. The helium star with a mass of $5.34M_\odot$ has evolved to a supernova after 5.6×10^5 yrs. The supernova explosion ejects the envelope with a mass of $3.34M_\odot$ and leaves a compact nucleus with a mass of $2M_\odot$, thus forming a neutron star. The period of the binary system has changed to 11.70 days at this time. After another 4.42×10^6 yrs, the main sequence star with a mass $22.66M_\odot$ has evolved as a super giant with great mass loss due to stellar wind. The stellar wind is accreted by the neutron star to produce X-ray emission. Therefore, the binary system becomes a massive X-ray binary system.

11.3.14 The Fate of Evolution of Massive X-ray Binary Systems

The massive X-ray binary system is a product in a relatively short evolutionary stage in the late evolutionary process of a massive binary system. Since the masses of the two components in a massive X-ray binary are significantly different, spiral-in must occur during mass transfer to form a system with a common envelope. After loss of the common envelope, it is possible to have a binary system consisting of two compact objects, or two solitary compact objects. Van den Heuvel (1992) discussed the following possible evolution fates:

(a) for a $16M_\odot + 1.4M_\odot$ (neutron star) massive X-ray binary system with a period less than one year: during mass transfer, the neutron star spirals into the core of the star M_2; after loss of the common envelope, one neutron star is left (Fig. 11.25(a)).

(b) for a $\leq 8M_\odot + 1.4M_\odot$ (neutron star) massive X-ray binary system with a period greater than one year : after the stage for a common envelope, the core of He, C and O is less than $2.0M_\odot$; after the second stage with a common envelope, a binary system consisting of a white dwarf ($<1.4M_\odot$) and a neutron star ($1.4M_\odot$), whose orbit is circular, is formed (Fig. 11.25(b)).

(c) for a $12M_\odot + 1.4M_\odot$ (neutron star) massive X-ray binary system with a period greater than one year: after the stage for a common envelope, the core of He, C and O has a mass of $3.0M_\odot$; then after supernova explosion, a binary system consisting of two neutron stars ($1.4M_\odot$) is formed (Fig. 11.25(c)).

Fig. 11.25 Several possible evolutionary fates of a massive X-ray binary system. (a) The period is less than one year. Through the stage of a common envelope, a neutron star is formed. (b) The period is greater than one year, and $M_2 \leq 8M_\odot$. After the stage for a common envelope, $M_2 \leq 2.0M_\odot$. After the second stage with a common envelope, a binary system which consists of a white dwarf $< 1.4M_\odot$ and a $1.4M_\odot$ neutron star is formed and a has a circular orbit. (c) The period is greater than one year, and $M_2 = 12M_\odot$. After the stage with a common envelope, $M_2 = 3.0M_\odot$. Then after supernova explosion, a binary system which consists of two $1.4M_\odot$ neutron stars is formed. (d) The period is greater than one year, and $M_2 \geq 16M_\odot$. After the stage for a common envelope, $M_2 \geq 4.2M_\odot$. Then after supernova explosion, the binary system is destroyed, leaving two single neutron stars.

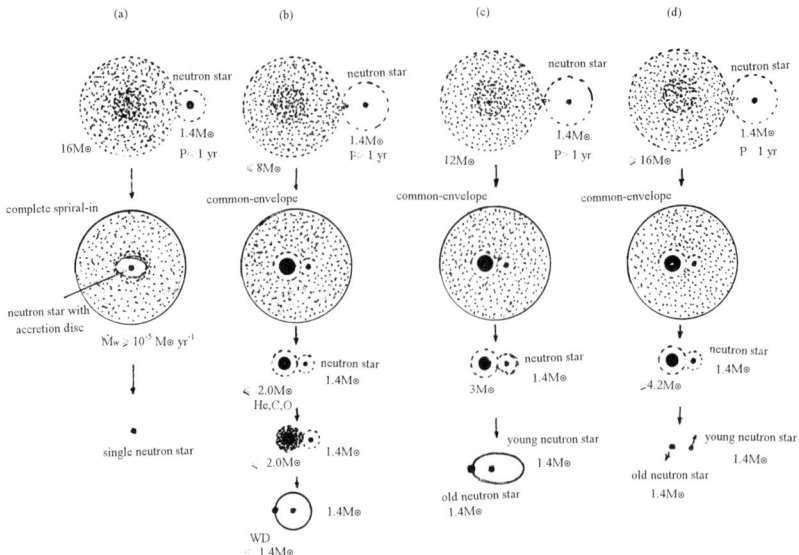

(d) for a $>16M_\odot + 1.4M_\odot$ (neutron star) massive X-ray binary system with a period greater than one year: after the stage for a common envelope, the core of He, C and O has a mass greater than $4.2M_\odot$ then after supernova explosion, the binary system may be destroyed leaving behind two single neutron stars (Fig. 11.25(d)).

11.3.15 Evolution of Massive Binaries with Shock Fronts of Colliding Winds

Observations have shown the existence of shock fronts of colliding winds in some massive and WR binaries. An important effect of the shock front of colliding winds is the production of X-rays, which makes massive and WR binaries with shock fronts of stellar winds become members of X-ray binaries. Another important effect of shock fronts of colliding winds is the hindrance of mass transfer between the two components. Supersonic gaseous outflows from the two components will collide on the shock front and then flow out of the binary system along the shock front. It is established that mass transfer between the two components will determine the evolution of the binary system, and drastically affect the mass and orbital period of the two components in a short time. Therefore, the existence of shock fronts of colliding winds and the resulting hindrance of mass transfer will have significant effects on the evolution of massive binaries.

In calculations of evolution of massive binaries with shock fronts of colliding winds, we have to consider in particular the loss of mass and angular momentum. Assume that the primary in the binary system has evolved to fill its Roche lobe and Roche lobe overflow starts. We denote ΔM_{1w} and ΔM_{2w} as the mass loss due to stellar winds from the primary and the secondary, ΔM_{1c} as the mass loss due to Roche lobe overflow. When there is a shock front of colliding winds, not only ΔM_{1w} and ΔM_{2w}, but also ΔM_{1c} will flow out of the binary system along the shock front, which takes away the angular momentum [see Eqs. (11.47) and (11.48)]

$$
\begin{aligned}
\Delta J &= \left(\frac{2}{3}R_1^2 + X_\omega^2\right)(\Delta M_{1w} + \Delta M_{1c})\omega + \left[\frac{2}{3}R_2^2 + (A - X_\omega)^2\right]\Delta M_{2w}\omega \\
&= \frac{G^{1/2}(M_1 + M_2)^{1/2}}{A^{3/2}}\left[\frac{2}{3}(\Delta M_{1w}R_1^2 + \Delta M_{1c}R_1^2 + \Delta M_{2w}R_2^2) + \frac{A^2 M_2^2}{(M_1 + M_2)^2}\right. \\
&\quad (\Delta M_{1w} + \Delta M_{1c} + \Delta M_{2w}) \\
&\quad \left.+ \Delta M_{2w}A^2\left(1 - \frac{2M_2}{M_1 + M_2}\right)\right]
\end{aligned}
$$

$$(11.104)$$

Since the mass loss rate caused by Roche lobe overflow is far greater than that of stellar winds, the angular momentum loss is far greater in the presence of shock front of colliding winds.

The orbital separation between the two components in the presence of shock front of colliding winds can still be revealed by Eq. (11.55). Huang (1997) calculated the evolution of a massive binary system with a $40M_\odot$ and a $30M_\odot$ star in the presence of a shock front of colliding winds, and proved that the period of the binary system is much shorter while the total mass of the system is much smaller in the presence of the shock front of colliding winds.

11.4 NON-CONSERVATIVE EVOLUTION OF LOW MASS BINARY SYSTEMS

11.4.1 Low Mass Binary Systems

It is now known that many cataclysmic variables, such as novae, dwarf novae, quasi-novae and many low mass X-ray binaries, are low mass binary systems. These low mass binary systems are composed of a more

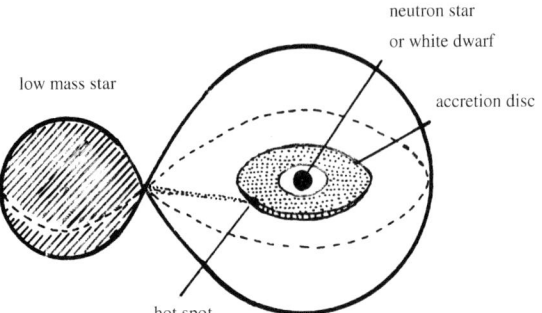

Fig. 11.26 The schematic diagram showing the structure of a low mass binary.

massive white dwarf (primary) and a less massive main sequence star (secondary). The less massive secondary has filled its Roche lobe and there is mass flowing out from the surface of the secondary to the white dwarf through the inner Lagrangian point. Since the mass flow has angular momentum, it cannot travel straight to the white dwarf, but forms an accretion disk around it. The position at which the mass flow hits the accretion disk will produce a very hot spot (see Fig. 11.26). Due to viscosity inside the accretion disk, the kinetic energy of the mass will be continuously dissipated as thermal energy which is radiated outwards, while the mass will continuously fall into the inner disk and finally onto the white dwarf. The mass continuously falling on the surface of the white dwarf is called the accretion mass, which is mainly composed of hydrogen. When the hydrogen layer accreted onto the surface of the white dwarf reaches a certain thickness, instantaneous hydrogen burning will take place which leads to nova explosion.

The orbital period of a low mass binary system is generally very short, ranging from 80 minutes to 12 hours, and is also shortened on a timescale of 10^4 to 10^9 y. Figure 11.27 shows the distribution of period of low mass binaries. It can be seen that there are two categories of low mass binaries. The first category has extremely short periods   ($80 \text{ min} \leq P \leq 2h$) while the other has longer periods ($3h \leq P \leq 12h$). However, there are extremely few low mass binaries in the interval $2h \leq P \leq 3h$, and we refer to this as the period gap. From observations, we further know that low mass binaries with $80 \text{ min} \leq P \leq 2h$ have mass transfer rates of $\dot{M}_c \sim 10^{-11}$ to 10^{-10} $M_\odot \text{yr}^{-1}$, while low mass binaries with $3h \leq P \leq 12h$ have mass transfer rates of $\dot{M}_c \sim 10^{-9}$ to 10^{-8} $M_\odot \text{yr}^{-1}$.

Gravitational energy will be released when the accreted mass falls onto the surface of the white dwarf, and the resulting accreted luminosity is

$$L_{acc} = \frac{GM_1 \dot{M}_{acc}}{R_1} \qquad (11.105)$$

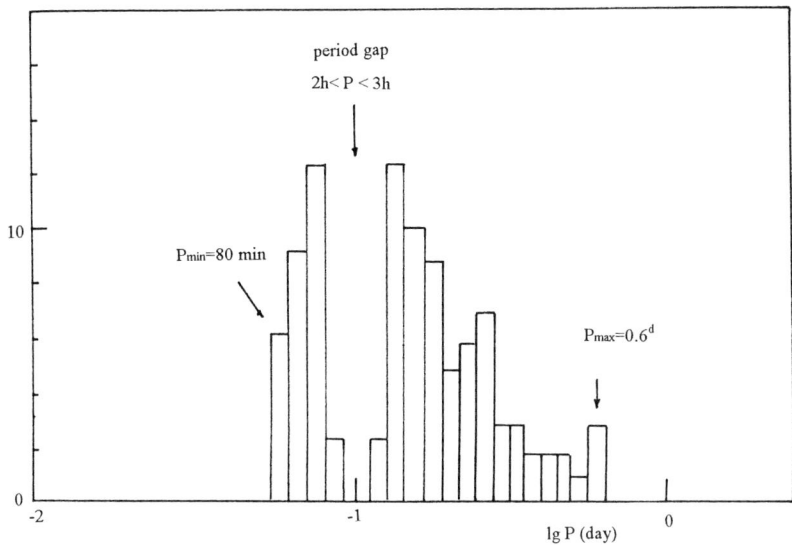

Fig. 11.27 The distribution of orbital periods of low mass binaries. There is a gap in the interval 2h $\leq$ P $\leq$ 3h (according to Patterson 1984).

where M_1 and R_1 are, respectively, the mass and radius of the white dwarf, and $\dot{M}_{acc}$ is the accretion rate. For the typical values of $M_1 = 1M_\odot$, $R_1 = 5 \times 10^8$cm, $\dot{M}_{acc} = 10^{-9}$ $M_\odot$yr^{-1}, we obtain $L_{acc} \cong 1.5 \times 10^{34}$ ergs$^{-1} \approx 4L_\odot$. Apparently, the accreted luminosity produced at the surface of the white dwarf is far larger than that of the secondary, so the luminosities observed for low mass binaries come mainly from the white dwarf and the accretion disk.

11.4.2 Non-conservative Low Mass Binaries

The observed low mass binaries are semi-detached. The less massive secondary has filled its Roche lobe and there is mass flowing out from the surface of the secondary through the inner Lagrangain point to the white dwarf. If a low mass binary system is conservative, the total mass and angular momentum are constants. From §11.2.3, for a conservative binary system, we know that when mass flows from the less massive star to a more massive star, the distance between the two stars and the volume of the Roche lobe will increase. In this way, the low mass binary system will develop from a semi-detached system to a detached system, which is contradictory to observations. Observations have shown that low mass binary systems maintain the mass transfer and remain as semi-detached systems for a long time, and the orbital period only decreases slowly. Therefore, it is certain that low mass binary systems are non-conservative, i.e., there is loss of mass and angular momentum.

11.4.3 Some Possible Mechanisms Driving the Non-conservative Evolution of Low Mass Binary Systems

For the secondary in the low mass binary system to keep filling its Roche lobe for a long time, and to keep some mass flowing to the accretion disk and finally onto the white dwarf, the binary system should be non-conservative, i.e., there is loss of mass and angular momentum from the binary system. Up to now, the following mechanisms have been proposed for the loss of mass or angular momentum.

(A) Gravitational radiation

Kraft *et al.* (1962) first pointed out from general relativity the significance of gravitational radiation in close binaries. After that, Paczynski (1967) and Faulkner (1971) suggested that gravitational radiation is a major mechanism driving the non-conservative evolution of low mass binary systems. From the Einstein quadrupole equation (see Landau and Lifschitz 1972), we can obtain the angular momentum loss due to gravitational radiation by

$$\frac{\dot{J}_{orb}}{J_{orb}} = -\frac{32G^3}{5c^5} \frac{M_1 M_2 (M_1 + M_2)}{A^4} \quad \text{s}^{-1} \tag{11.106}$$

where J_{orb} is the orbital angular momentum, M_1 and M_2 are, respectively, the mass of the primary and the secondary, A is the distance between the two stars and c is the speed of light.

According to the studies of Taam *et al.* (1980), Rapparport *et al.* (1982) and Taam (1983), the angular momentum lost through gravitational radiation can lead to a mass transfer rate of 10^{-11} to $10^{-10} M_\odot \text{yr}^{-1}$ in low mass binaries which are consistent with observed values for low mass binaries with periods smaller than two hours. This shows that gravitational radiation is the main mechanism for driving the non-conservative evolution of low mass binaries with periods smaller than two hours. However, the mass transfer rate ranges from 10^{-9} to $10^{-8} M_\odot \text{yr}^{-1}$ for low mass binaries with periods larger than three hours, so it is obvious that gravitational radiation is not the main mechanism for driving their non-conservative evolution.

(B) Magnetic braking

Skumanich (1972) pointed out that the slowing down of the rotational velocity of late type main sequence stars (single stars) is due to that the stellar wind can take away more angular momentum with magnetic coupling. This effect is called magnetic braking. Verbunt and Zwaan (1981) and Hameury *et al.* (1988) proposed that magnetic braking is the major mechanism for driving the non-conservative evolution of low mass binary systems with periods longer than three hours.

Yu and Huang (1995) studied the angular momentum loss due to magnetic braking and concluded that if the mass loss rate due to stellar wind from the secondary is only of the order 10^{-15} to $10^{-14} M_\odot yr^{-1}$, the angular momentum loss due to magnetic braking is extremely small and cannot be the major mechanism for driving the non-conservative evolution of low mass binaries with periods greater than three hours. For magnetic braking to become the major mechanism, the rate of mass loss from the secondary should reach the order 10^{-11} to $10^{-10} M_\odot yr^{-1}$.

11.4.4 The Evolutionary Fate of Low Mass Binary Systems

During the evolution of a low mass binary system, the secondary keeps filling its Roche lobe and mass keeps flowing to the accretion disk. Together with the continuous outflow of mass, the hydrogen burning region inside the secondary contracts inwards continuously and the outer convective layer extends inwards continuously, and at the same time, the boundary between the radiative equilibrium region and the convective region becomes vague. When the mass of the secondary becomes smaller than $0.3 M_\odot$, it will become an entirely convective star. When the mass of the secondary is smaller than a critical mass M_{cri} (the value of M_{cri} is closely related to the mass loss rate of the secondary), the mass loss will lead to no thermodynamical equilibrium of the secondary, so the radius and luminosity etc. of the secondary will fluctuate around their thermodynamical equilibrium values.

When the mass of the secondary is less than $0.095 M_\odot$, the internal hydrogen burning stops completely, and the secondary has evolved to a degenerate black dwarf, so the low mass binary system has evolved to one consisting of a white dwarf and a black dwarf.

CHAPTER TWELVE

STELLAR OSCILLATION

12.1 PULSATIONAL VARIABLES AND THEORY OF STELLAR OSCILLATION

There are a lot of variables among the multitude of stars. The luminosities, colors, radial velocities and the profiles of spectral lines of variables will all change with time. If these changes are regular and periodic, the variable is called a pulsational variable. There are many types of pulsational variables. Some of them are known to have radial oscillations around their equilibrium structures, but more of them have non-radial oscillations. The amplitudes and the periods are also different, and some even have many periods. Figure 1.3 gives the distribution of pulsational variables in the H-R diagram.

The theory of stellar oscillation studies why pulsational variables oscillate, the mode of oscillations, regions of propagation and period etc. It is based on the radiation hydrodynamical and thermodynamical equations which describe the motion of fluid elements, and employs the linear perturbation theory to establish a set of linear oscillation equations.

12.2 EQUATIONS OF RADIATION HYDRODYNAMICS

12.2.1 Introduction

Stars are composed of gas and radiation, so stellar oscillations are in fact motions of radiative fluids. In common stellar oscillation theories (e.g., Ledoux and Walraven 1958; Cox, 1980; Unno *et al.*, 1989; Huang and Li, 1990) the radiation field and the gas are treated as the same thermodynamical system, which has the same temperature. Therefore, the description of the motion of a radiative fluid element requires only three basic equations in hydrodynamics (i.e., the continuity equation, the equation of motion and the energy equation), the Poisson equation and thermodynamical equations. However, newest developments propose that the radiation field and the gas should be treated as two separate thermodynamical systems when the star oscillates, i.e., when the gas moves on a macroscopic scale, especially when this movement occurs inside the envelope of the star. In other words, the two systems have their own temperatures and other thermodynamical parameters. The changes in the states of the radiation field and the gas require different hydrodynamical and thermodynamical equations. Nevertheless, these different hydrodynamical and thermodynamical equations should also contain conjugate terms describing the interaction between the radiation field and the gas. In this way, a bisystem oscillation theory of stars (see Li, 1992b,c) has emerged which treats the radiation field and the gas as two thermodynamical systems. This theory has been successfully applied to studies of properties of some pulsational variables, and has given results well consistent with observational results (Li 1992a; Li and Stix 1994; Li and Gong 1994). Therefore, in this chapter, we will describe this new bisystem stellar oscillation theory.

When using the three hydrodynamical equations, viz., the continuity equation, the equation of motion and the energy equation, we have to note that two different methods can be used to describe the fluid motion, i.e., the Euler description and the Lagrangian description. The Euler method describes the change in the field with respect to time at each spatial point, while the Lagrangian method describes the motion of every element in the fluid. Since the same fluid motion can be described using two different methods, the expressions for the change of the same physical quantity with respect to time using the Euler method and the Lagrangian method should be interchangeable. For example, the change of an arbitrary physical quantity f (can be a vector or a scalar) with respect to time can be written as

$$\frac{Df}{Dt} = \frac{\partial f}{\partial t} + (\vec{v} \cdot \nabla)f \qquad (12.1)$$

The term $\frac{Df}{Dt}$ on the left-hand side of Eq. (12.1) represents the co-moving derivative and is a description of the Lagrangian method. The terms of local change $\frac{\partial f}{\partial t}$ and of translocation change $v\frac{\partial f}{\partial s}$ on the right-hand side of Eq. (12.1) arise because of the change in time and in the inhomogeneity of the field, and are descriptions of the Euler method. Therefore, Eq. (12.1) is the interconversion between the temporal change of an arbitrary physical quantity f expressed by the two methods. Given that f is arbitrary, the interconversion between the two expressions is in fact the conversion between the operators, i.e.,

$$\frac{D}{Dt} = \frac{\partial}{\partial t} + (\vec{v} \cdot \nabla) \qquad (12.2)$$

In a hydrodynamical equation, some physical quantities characterize properties of fluid elements, and others characterize properties of the field. The temporal change of physical quantities characterizing properties of fluid elements belongs to co-moving changes and should be expressed using the Lagrangian method while the change in physical quantities characterizing properties of the field should be expressed using the Euler method. However, for the sake of convenience, the entire equation can be converted to a single type of expression with the help of Eq. (12.2), i.e., all terms in the equation are expressed using either the Euler method or the Lagrangian method. The expressions of the Euler method can use all known formulae of field theory in mathematics, so the use of expressions of the Euler method has special advantages.

12.2.2 Continuity Equation

The continuity equation is a special expression of the law of conservation of mass in hydrodynamics. We take a finite volume τ in the star which is bounded by the surface s. The volume τ is made up of spatial points so it is fixed in space and does not change with time. We take the outward normal line to be the positive direction, and $\vec{n}$ as its unit vector. We now consider the change of fluid mass within the volume τ. There are two causes for the change. First, there is fluid flowing inwards or outwards through the surface s. The fluid flowing inwards or outwards per unit time is

$$\oint_s \rho v_n ds \tag{12.3}$$

where the positive sign represents the fluid flowing away from s. Second, there is a temporal change in the density within τ. The decrease in fluid per unit time due to change of density is

$$-\int_\tau \frac{\partial \rho}{\partial t} d\tau \tag{12.4}$$

where the negative sign represents the decrease in mass. According to the law of conservation of mass, Eqs. (12.3) and (12.4) should be equal, i.e.,

$$+\oint_s \rho v_n ds = -\int_\tau \frac{\partial \rho}{\partial t} d\tau \tag{12.5}$$

Using the Gauss theorem to change the surface integral to a volume integral, Eq. (12.5) can be written as

$$\int_\tau \left[\frac{\partial \rho}{\partial t} + \nabla \cdot (\rho \vec{v}) \right] d\tau = 0 \tag{12.6}$$

Since the volume τ is arbitrary, we have

$$\frac{\partial \rho}{\partial t} + \nabla \cdot (\rho \vec{v}) = 0 \tag{12.7}$$

which is the continuity equation in the Euler form.

12.2.3 Equation of Motion

When rotation, magnetic field and tidal effects are neglected in a star, the only forces acting on a fluid element are the gravitational force and the pressure. When these forces are not at equilibrium, there will be a motion for the fluid, and

$$\rho \frac{D\vec{v}}{Dt} = -\nabla(P_g + P_R) - \rho \nabla \Phi \tag{12.8}$$

where the first term on the right is the pressure term and the total pressure P has been written as the combination of gas pressure P_g and radiation pressure P_R; the second term on the right is the gravitational force term, Φ is the gravitational potential and the gravitational acceleration $\vec{g}$ is expressed as $-\nabla\Phi$.

Using Eq. (12.2), the equation of motion can be written in the Euler form:

$$\rho\left(\frac{\partial}{\partial t} + \vec{v} \cdot \nabla\right)\vec{v} = -\nabla(P_g + P_R) - \rho\nabla\Phi \qquad (12.9)$$

12.2.4 Energy Equation of Gas

The energy equation of gas is a special expression of the law of conservation of energy in hydrodynamics. The energy $\frac{Dq}{Dt}$ acquired by a fluid element in the star in a unit time can be treated as the combination of the energy generated in that element by thermonuclear reactions, the energy flowing into the element by convection and the absorption of radiation by gas in the element per unit time. Denoting ε as the rate of energy generation by thermonuclear reactions and $\vec{F}_{con}$ as the convective energy flux, the equation for the conservation of energy can be written as

$$\rho T_g\left(\frac{\partial}{\partial t} + \vec{v} \cdot \nabla\right)S_g = \rho\varepsilon - \nabla \cdot \vec{F}_{con} - ac\rho\kappa^T\left(T_g^4 - T_R^4\right) \qquad (12.10)$$

where T_g and S_g are the temperature and the entropy of the gas, T_R is the temperature of the radiation and κ^T is the true absorption coefficient. The third term on the right of Eq. (12.10) can be interpreted as follows. Inside the star, the radiation field is isotropic, and the relationship between the average radiation intensity J and the temperature of the radiation system T_R is

$$J = \frac{ac}{4\pi} T_R^4 \qquad (12.11)$$

where a is the radiation constant and c is the velocity of light in vacuum. On the other hand, a high temperature gas emits radiation according to the Planck's law for black body radiation, i.e.,

$$B = \frac{ac}{4\pi} T_g^4 \qquad (12.12)$$

The interaction between gas and radiation means that the gas absorbs or emits radiation. When $J > B$, the gas absorbs radiation; when $J < B$, the gas emits radiation. Therefore, the absorption of radiation in the fluid element is

$$4\pi\rho\kappa^T(J - B) = -ac\rho\kappa^T\left(T_g^4 - T_R^4\right).$$ (12.13)

Therefore, the third term on the right of Eq. (12.10) represents the absorption of radiation in the fluid element.

Equation (12.10) requires further clarification. Under most circumstances, the regions of oscillation and those of thermonuclear reactions in the star are separated, so we can ignore the contribution of energy generated by thermonuclear reactions. It is necessary to consider this contribution only when we study some special problems of stellar oscillations, where the regions of oscillation and those of thermonuclear reactions overlap and where the timescales of oscillation and of thermonuclear reactions are of the same order of magnitude.

The conductive energy flux has not been taken into account in Eq. (12.10) because the energy transported through conduction is far less than that transported through radiation and convection in normal cases. Conduction is significant when the star is degenerate.

Regarding effects of convection, the most difficult problem to date is the lack of a perfect time dependent convection theory to handle the coupling between convection and oscillation. Because of this, the usual method adopted is to neglect the direct effects of convection on oscillation. Convection is considered only when calculating the equilibrium model structure of the star.

12.2.5 Poisson Equation

The Poisson equation gives the relationship between gravitational potential Φ and gas density ρ:

$$\nabla^2\Phi = 4\pi G\rho$$ (12.14)

where G is the gravitational constant. Equation (12.14) is a second order differential equation. Usually, it is more convenient to write it as two first order differential equations, i.e.,

$$\vec{g} = -\nabla\Phi$$ (12.15)

and
$$\nabla \cdot \vec{g} = -4\pi G\rho \qquad (12.16)$$

where $\vec{g}$ is the gravitational acceleration.

12.2.6 Thermodynamical Equation of Gas

The thermodynamical equation of gas gives the relationship among entropy S_g, density ρ and gas pressure P_g as

$$d \ln P_g = \Gamma_1 d \ln \rho + \frac{\rho}{P_g}(\Gamma_3 - 1)T_g dS_g \qquad (12.17)$$

where
$$\Gamma_1 \equiv \left(\frac{\partial \ln P_g}{\partial \ln \rho}\right)_{S_g} \qquad (12.18)$$

$$\Gamma_3 - 1 \equiv \left(\frac{\partial \ln T_g}{\partial \ln \rho}\right)_{S_g} \qquad (12.19)$$

and
$$\frac{\Gamma_3 - 1}{\Gamma_1} \equiv \left(\frac{\partial \ln T_g}{\partial \ln P_g}\right)_{S_g} = \nabla_{ad} \qquad (12.20)$$

In order to get the relationship in Eq. (12.17), we have to prove the two relationships

$$\Gamma_3 - 1 = \frac{P_g \chi_T}{\rho C_V T_g} \qquad (12.21)$$

$$\Gamma_1 = \chi_\rho + \chi_T(\Gamma_3 - 1) \qquad (12.22)$$

where
$$\chi_T \equiv \left(\frac{\partial \ln P_g}{\partial \ln T_g}\right)_\rho \qquad (12.23)$$

$$\chi_\rho \equiv \left(\frac{\partial \ln P_g}{\partial \ln \rho}\right)_{T_g} \qquad (12.24)$$

Taking volume V and temperature T_g as independent variables of the system, the first law of thermodynamics can be written as

$$dq = du + P_g dV = \left[\left(\frac{\partial u}{\partial V}\right)_{T_g} + P_g\right]dV + \left(\frac{\partial u}{\partial T_g}\right)dT_g \qquad (12.25)$$

Dividing the whole equation by T_g, we have

$$dS_g = \frac{1}{T_g}\left[\left(\frac{\partial u}{\partial V}\right)_{T_g} + P_g\right]dV + \frac{1}{T_g}\left(\frac{\partial u}{\partial T_g}\right)_V dT_g \qquad (12.26)$$

As entropy is a function of state, we have

$$\frac{\partial}{\partial T_g}\left\{\frac{1}{T_g}\left[\left(\frac{\partial u}{\partial V}\right)_{T_g} + P_g\right]\right\} = \frac{\partial}{\partial V}\left[\frac{1}{T_g}\left(\frac{\partial u}{\partial T_g}\right)_V\right] \qquad (12.27)$$

The solution to this equation gives

$$\left(\frac{\partial u}{\partial V}\right)_{T_g} = T_g\left(\frac{\partial P_g}{\partial T_g}\right)_V - P_g \qquad (12.28)$$

or

$$-\rho u\left(\frac{\partial \ln u}{\partial \ln \rho}\right)_{T_g} = P_g\left(\frac{\partial \ln P_g}{\partial \ln T_g}\right)_\rho - P_g \qquad (12.29)$$

According to Eq. (12.23), the above equation can be written as

$$\left(\frac{\partial \ln u}{\partial \ln \rho}\right)_{T_g} = -\frac{P_g}{\rho u}(\chi_T - 1) \qquad (12.30)$$

which is an important relationship when deriving Eq. (12.21).

To derive Eq. (12.21), we choose ρ and T_g to be independent variables of the system, so the first law of thermodynamics becomes

$$\begin{aligned}
dq &= \left[\left(\frac{\partial u}{\partial \ln \rho}\right)_{T_g} - \frac{P_g}{\rho}\right]\frac{d\rho}{\rho} + \left(\frac{\partial u}{\partial \ln T_g}\right)_\rho \frac{dT_g}{T_g} \\
&= \left[u\left(\frac{\partial \ln u}{\partial \ln \rho}\right)_{T_g} - \frac{P_g}{\rho}\right]\frac{d\rho}{\rho} + C_V dT_g \qquad (12.31)
\end{aligned}$$

where we have used the relationship $C_V \equiv \left(\frac{\partial u}{\partial T_g}\right)_\rho$. For an adiabatic process, $dq = 0$, so

$$\Gamma_3 - 1 \equiv \left(\frac{\partial \ln T_g}{\partial \ln \rho}\right)_{S_g} = \frac{\left(\frac{P_g}{\rho}\right) - u\left(\frac{\partial \ln u}{\partial \ln \rho}\right)_{T_g}}{C_V T_g} \qquad (12.32)$$

Substituting Eq. (12.30) into (12.32), we obtain

$$\Gamma_3 - 1 = \frac{P_g \chi_T}{\rho C_V T_g} \qquad (12.21)$$

which is the required relationship in Eq. (12.21).

To derive Eq. (12.22), we treat P_g as a function of ρ and T_g, i.e., $P_g = P_g(\rho, T_g)$, so

$$d \ln P_g = \left(\frac{\partial \ln P_g}{\partial \ln T_g}\right)_\rho d \ln T_g + \left(\frac{\partial \ln P_g}{\partial \ln \rho}\right)_{T_g} d \ln \rho$$

$$= \chi_T d \ln T_g + \chi_\rho d \ln \rho \qquad (12.33)$$

Dividing the whole equation by $d \ln \rho$, we obtain

$$\frac{d \ln P_g}{d \ln \rho} = \chi_T \frac{d \ln T_g}{d \ln \rho} + \chi_\rho \qquad (12.34)$$

This relationship is valid in general. Using the equation for an adiabatic process, and considering Eqs. (12.18) and (12.19), we arrive at

$$\Gamma_1 = \chi_T(\Gamma_3 - 1) + \chi_\rho \qquad (12.22)$$

which is the required relationship in Eq. (12.22).

We now come to the derivation of Eq. (12.17). We take P_g and V to be independent variables of the system and obtain from the first law of thermodynamics that

$$dq = \left[\left(\frac{\partial u}{\partial V}\right)_{P_g} + P_g\right] dV + \left(\frac{\partial u}{\partial P_g}\right)_V dP_g \qquad (12.35)$$

Dividing the equation by T_g, we obtain

$$dS_g = \frac{dq}{T_g} = \frac{1}{T_g}\left[\left(\frac{\partial u}{\partial V}\right)_{P_g} + P_g\right] dV + \frac{1}{T_g}\left(\frac{\partial u}{\partial P_g}\right)_V dP_g \qquad (12.36)$$

Since dS_g is a total differential, we have

$$\frac{\partial}{\partial P_g}\left\{\frac{1}{T_g}\left[\left(\frac{\partial u}{\partial V}\right)_{P_g} + P_g\right]\right\} = \frac{\partial}{\partial V}\left[\frac{1}{T_g}\left(\frac{\partial u}{\partial P_g}\right)_V\right] \qquad (12.37)$$

Using $C_V = \left(\frac{\partial u}{\partial T_g}\right)_V$ to solve the above equation, we have

$$\left(\frac{\partial u}{\partial V}\right)_{P_g} + P_g = C_V \left(\frac{\partial T_g}{\partial V}\right)_{P_g} + T_g \left(\frac{\partial P_g}{\partial T_g}\right)_V \qquad (12.38)$$

Substituting Eq. (12.38) into (12.35), and using

$$\left(\frac{\partial u}{\partial P_g}\right)_V = C_V \left(\frac{\partial T_g}{\partial P_g}\right)_V, \qquad (12.39)$$

we arrive at

$$dq = \left[C_V \left(\frac{\partial T_g}{\partial V}\right)_{P_g} + T_g \left(\frac{\partial P_g}{\partial T_g}\right)_V\right] dV + C_V \left(\frac{\partial T_g}{\partial P_g}\right)_V dP_g \qquad (12.40)$$

which can be rewritten as

$$d \ln P_g = \left[\frac{P_g}{\rho C_V T_g} \left(\frac{\partial \ln P_g}{\partial \ln T_g}\right)_\rho^2 - \left(\frac{\partial \ln T_g}{\partial \ln \rho}\right)_{P_g} \left(\frac{\partial \ln P_g}{\partial \ln T_g}\right)_\rho\right] d \ln \rho$$
$$+ \frac{1}{C_V T_g} \left(\frac{\partial \ln P_g}{\partial \ln T_g}\right)_\rho dq \qquad (12.41)$$

Using the relationship

$$\left(\frac{\partial \ln P_g}{\partial \ln \rho}\right)_{T_g} = -\left(\frac{\partial \ln T_g}{\partial \ln \rho}\right)_{P_g} \left(\frac{\partial \ln P_g}{\partial \ln T_g}\right)_\rho \qquad (12.42)$$

and Eqs. (12.19), (12.23) and (12.24), and writing dq as $T_g dS_g$, Eq. (12.41) becomes

$$d \ln P_g = [(\Gamma_3 - 1)\chi_T + \chi_\rho] d \ln \rho + \frac{\chi_T}{C_V} dS_g \qquad (12.43)$$

Substituting Eq. (12.22), this becomes

$$d \ln P_g = \Gamma_1 d \ln \rho + \frac{\chi_T}{C_V} dS_g$$

which, when Eq. (12.21) is considered, finally becomes

$$d \ln P_g = \Gamma_1 d \ln \rho + \frac{\rho}{P_g} (\Gamma_3 - 1) T_g dS_g \qquad (12.17)$$

which is the required thermodynamical equation (12.17) for gas.

12.2.7 Radiative Momentum Equation

The radiative momentum equation shows that if a fluid element absorbs a certain amount of radiative energy, it will be affected by the radiation pressure which can be expressed as

$$\nabla P_R = -\frac{\rho \kappa}{c} \vec{F}_R \tag{12.44}$$

12.2.8 Radiative Energy Equation

The radiative energy equation can be derived using a method similar to that in the derivation of the gas energy equation (12.10). The net radiation travelling into the fluid element and the radiation emitted by the gas within the fluid element will heat up the radiation system inside the fluid element. The Euler expression can be written as

$$\rho T_R \left(\frac{\partial}{\partial t} + \vec{v}_R \cdot \nabla \right) S_R = -\nabla \cdot \vec{F}_R + ac\rho \kappa^T (T_g^4 - T_R^4) \tag{12.45}$$

where T_R and S_R are temperature and entropy of the radiation field, and $\vec{F}_R$ is the radiative energy flux.

12.2.9 Radiative Thermodynamical Equation

Assuming that the radiation field inside a star is isotropic, from Eqs. (2.54), (2.57) and (2.58), we can obtain the relationship among radiation pressure P_R, energy density u of the radiation field and radiation temperature T_R as $P_R = \frac{1}{3}u = \frac{a}{3}T_R^4$. If the density ρ and the radiation pressure P_R are taken as independent variables of the system, the first law of thermodynamics can be expressed as

$$dQ = d(V \cdot u) + P_R dV \tag{12.46}$$

or $$dq = \frac{1}{\rho} du - \frac{u}{\rho^2} d\rho - \frac{P_R}{\rho^2} d\rho = \frac{3}{\rho} dP_R - \frac{4P_R}{\rho^2} d\rho \tag{12.47}$$

which can give the radiative thermodynamical equation as

$$\rho T_R dS_R = 3dP_R - \frac{4P_R}{\rho} d\rho \tag{12.48}$$

12.2.10 Equations of Radiation Hydrodynamics

When the equations in §12.2.2 to §12.2.9 are appropriately rewritten, we can obtain a set of equations describing the motion of a radiative fluid, which are:

$$\frac{\partial \rho}{\partial t} + \nabla \cdot (\rho \vec{v}) = 0, \tag{12.49}$$

$$\rho \left(\frac{\partial}{\partial t} + \vec{v} \cdot \nabla \right) \vec{v} = -\nabla (P_g + P_R) + \rho \vec{g} \tag{12.50}$$

$$\nabla P_R = -\frac{\rho \kappa}{c} \vec{F}_R \tag{12.51}$$

$$\frac{1}{\Gamma_3 - 1} \left(\frac{\partial}{\partial t} + \vec{v} \cdot \nabla \right) P_g - \frac{\Gamma_1}{\Gamma_3 - 1} \frac{P_g}{\rho} \left(\frac{\partial}{\partial t} + \vec{v} \cdot \nabla \right) \rho = \rho \varepsilon - \nabla \cdot \vec{F}_{con} - \rho \kappa^T c (a T_g^4 - 3 P_R) \tag{12.52}$$

$$3 \left(\frac{\partial}{\partial t} + \vec{v} \cdot \nabla \right) P_R - \frac{4 P_R}{\rho} \left(\frac{\partial}{\partial t} + \vec{v} \cdot \nabla \right) \rho = -\nabla \cdot \vec{F}_R + \rho \kappa^T c (a T_g^4 - 3 P_R) \tag{12.53}$$

$$\vec{g} = -\nabla \Phi \tag{12.54}$$

and
$$\nabla \cdot \vec{g} = -4 \pi G \rho \tag{12.55}$$

Within this set of equations, the independent variables are time and spatial position, and the dependent variables are velocity $\vec{v}$, gas pressure P_g, radiation pressure P_R, convective energy flux $\vec{F}_{con}$, radiative energy flux $\vec{F}_R$, gravitational potential Φ and gravitational acceleration $\vec{g}$.

In studies of stellar oscillation problems, employment of the spherical coordinate system is more convenient. In order to change equations (12.49) to (12.55) to expressions in the spherical coordinate system, we have to note the following conversions:

$$\left.\begin{aligned}
\frac{D}{Dt} &= \frac{\partial}{\partial t} + v_r \frac{\partial}{\partial r} + \frac{v_\theta}{r} \frac{\partial}{\partial \theta} + \frac{v_\varphi}{r \sin \theta} \frac{\partial}{\partial \varphi} \\
\nabla \Phi &= \frac{\partial \Phi}{\partial r} \vec{e}_r + \frac{1}{r} \frac{\partial \Phi}{\partial \theta} \vec{e}_\theta + \frac{1}{r \sin \theta} \frac{\partial \Phi}{\partial \varphi} \vec{e}_\varphi \\
\nabla \cdot \vec{v} &= \frac{1}{r^2} \frac{\partial (r^2 v_r)}{\partial r} + \frac{1}{r \sin \theta} \frac{\partial (\sin \theta v_\theta)}{\partial \theta} + \frac{1}{r \sin \theta} \frac{\partial v_\varphi}{\partial \varphi} \\
\nabla^2 \Phi &= \frac{1}{r^2} \frac{\partial}{\partial r} \left(r^2 \frac{\partial \Phi}{\partial r} \right) + \frac{1}{r^2 \sin \theta} \frac{\partial}{\partial \theta} \left(\sin \theta \frac{\partial \Phi}{\partial \theta} \right) + \frac{1}{r^2 \sin^2 \theta} \frac{\partial^2 \Phi}{\partial \varphi^2}
\end{aligned}\right\} \tag{12.56}$$

12.3 LINEAR OSCILLATION EQUATIONS

12.3.1 Linear Oscillation Theory

The basic equations (12.49) to (12.55) describing the motion of a radiative fluid are non-linear equations. It is extremely difficult to directly solve these equations to obtain the various properties of stellar oscillation. The usual method adopted is to linearize the non-linear basic equations, which gives the linear oscillation theory. In fact, the linear oscillation theory has made use of the perturbation theory to study problems in stellar oscillation.

Suppose that there is a perturbation in the star at some instance such that the star has a small deviation from its equilibrium state. The inside of the star will change quickly. If the change can decrease the deviation mentioned earlier and causes the star to develop towards restoration of equilibrium, the star will not have self-excitative oscillation and the star is stable. On the contrary, if the change increases the deviation from the equilibrium state, the star will have self-excitative oscillations so the star is unstable.

The state of the star after perturbation is called the perturbed state. The various physical quantities of the perturbed state, such as density, pressure and velocity etc., can be treated as the sum of the equilibrium value and a perturbation value, e.g.,

$$\left.\begin{array}{l} P = P_0 + P_1 \\ T = T_0 + T_1 \\ \vec{v} = \vec{v}_0 + \vec{v}_1 \\ \vdots \end{array}\right\} \qquad (12.57)$$

where the subscript "0" refers to equilibrium values and the subscript "1" refers to perturbation values. Substituting Eq. (12.57) into the non-linear basic equation set (12.49) to (12.55), neglecting terms with higher orders of perturbation values, and noting that the equilibrium values fulfil equations (12.49) to (12.55), we can obtain a set of linearized equations called linear oscillation equations.

By employing the linear oscillation theory to study the stellar oscillation, we can get much important information such as the rather precise oscillation period, type and properties of oscillation waves, the region in the star in which the oscillation wave propagates, and the cause of stellar oscillation. However, the linear theory has its own limitations because we have eliminated all non-linear effects and information when linearizing the equations (12.49) to (12.55). There-fore, some observational phenomena belonging to the non-linear regime cannot be explained using the linear theory. For example, the observed light curve of variables and the radial velocity curves are usually non-sine curves, but those obtained by the linear theory are sine curves. The discrepancy is caused by non-linear effects. Another example is that there are a lot of oscillation modes predicted by the linear theory but in reality the star has only some and even one oscillation mode. The selection effect of oscillation modes also belong to non-linear effect.

Nevertheless, the linear theory is simpler, and is quicker and more convenient in giving a solution, and can give a lot of stellar oscillation properties, so it is widely employed.

12.3.2 Euler Perturbation and Lagrangian Perturbation

We have introduced in §12.2.1 the two methods to describe fluid motion, i.e., the Euler method and the Lagrangian method. As regards perturbations occurring inside the star, there are still two methods for the description, which are again the Euler method and the Lagrangian method. The former focuses on a spatial point while the latter focuses on a fluid element.

When linearizing the basic equations (12.49) to (12.55) for the radiative fluid, we have to note that each term in the equations represents a physical process, which is not only related to some physical quantities but also related to some differential operators. The perturbation acts on the whole physical process, so we have to consider the perturbation on the physical quantities as well as the perturbation on the differential operators to get accurate oscillation equations which can be interchange-able between the two types of perturbations.

We now proceed to discuss the perturbations defined by the Euler method and the Lagrangian method, and their inter-conversions. In the Euler method, a physical quantity f can be represented by

$$f = f(x_1, x_2, x_3, t) \tag{12.58}$$

where t is the time, and the combination of scalars x_1, x_2 and x_3 determines a spatial point. The defined region of vectors $\vec{r}\,(x_1, x_2, x_3)$ is a field, and $\vec{r}$ and t are independent variables.

In the Lagrangian method, the same physical quantity f can be represented by

$$f = f(a_1, a_2, a_3, t) \tag{12.59}$$

where the combination of scalars a_1, a_2 and a_3 characterizes a fluid element. When the values of a_1, a_2 and a_3 are used to define a vector $\vec{r}'$ (a_1, a_2, a_3), the defined region of $\vec{r}'$ (a_1, a_2, a_3) is not a field because $\vec{r}'$ is not a function of space but is a function of indices of fluid elements. Obviously, $\vec{r}'$ (a_1, a_2, a_3) is a function of time t and is not an independent variable. Nevertheless, the $f(\vec{r}, t)$ expressed by the Euler method and the $f(\vec{r}', t)$ expressed by the Lagrangian method are describing the same physical quantity so they must be interchangeable, i.e., the Lagrangian type $f(\vec{r}', t)$ should be able to change to the Euler type $f(\vec{r}, t)$ through certain conversion relationships. This conversion is equivalent to equalization between the physical quantity f of the fluid element defined by a_1, a_2 and a_3 at time t and the physical quantity f at a spatial point $\vec{r}$ at the same time t.

In the Euler method, the perturbation value f' of the physical quantity f is defined as

$$f' \equiv f(\vec{r}, t) - f_0(\vec{r}, t) \tag{12.60}$$

where $f(\vec{r}, t)$ is the pertubed physical quantity and $f_0(\vec{r}, t)$ is the physical quantity at equilibrium. In the Lagrangian method, the perturbation value δf of the physical quantity f is defined as

$$\delta f \equiv f(a_1, a_2, a_3, t) - f_0(a_1, a_2, a_3, t) \tag{12.61}$$

where $f(a_1, a_2, a_3, t)$ represents the perturbed physical quantity and $f_0(a_1, a_2, a_3, t)$ represents the physical quantity at equilibrium.

Through a conversion, δf can be written as

$$\delta f = f(\vec{r}, t) - f_0(\vec{r}_0, t) \tag{12.62}$$

where we have converted $f(a_1, a_2, a_3, t)$ to $f(\vec{r}, t)$, and the physical quantity at equilibrium to $f_0(\vec{r}_0, t)$, i.e., we have treated it as the physical quantity at a spatial point $\vec{r}_0$ at time t. Since there is only a very small deviation between the perturbed state and the equilibrium state, the spatial translation $\vec{\xi} = \vec{r} - \vec{r}_0$ is also a small quantity. By adding and subtracting $f_0(\vec{r}_0, t)$ from the right-hand side of Eq. (12.62), and using the definition of f' in Eq. (12.60), we obtain

$$\begin{aligned} \delta f &= f(\vec{r}, t) - f_0(\vec{r}, t) + [f_0(\vec{r}, t) - f_0(\vec{r}_0, t)] \\ &= f' + [f_0(\vec{r}, t) - f_0(\vec{r}_0, t)] \end{aligned} \tag{12.63}$$

Expanding $f_0(\vec{r}, t)$ around $\vec{r}_0$, and taking first-order terms, Eq. (12.63) becomes

$$\delta f = f' + \vec{\xi} \cdot \nabla f_0 \qquad (12.64)$$

where $\vec{\xi}$ is the translation of the fluid element and f_0 is the unperturbed value of the physical quantity f. Equation (12.64) is the conversion between the Euler perturbation value and the Lagrangian perturbation value.

We go on to discuss the change of differential operators under the two methods. From the definition of the Euler perturbation value f' in Eq. (12.60) and that both $\vec{r}$ and t are independent variables, it is easy to observe the following properties of differential operators of the Euler method under perturbation:

(1) $\left(\dfrac{\partial f}{\partial t}\right)' = \dfrac{\partial}{\partial t} f'$,   i.e., $'$ and $\dfrac{\partial}{\partial t}$ are interchangeable;

(2) $(\nabla f)' = \nabla f'$,   i.e., $'$ and ∇ are interchangeable;

(3) $(\nabla \cdot f)' = \nabla \cdot f'$, i.e., $'$ and $\nabla \cdot$ are interchangeable;

(4) $\left(\dfrac{Df}{Dt}\right)' \neq \dfrac{Df'}{Dt}$,   i.e., $'$ and the co-moving derivative $\dfrac{D}{Dt}$ are

not interchangeable.

$$(12.65)$$

From the derivation of the Lagrangian perturbation value δf in Eq. (12.61) and that $\vec{r}'\,(a_1, a_2, a_3, t)$ is not an independent variable, it is easy to obtain the following properties of differential operators of the Lagrangian method under perturbation:

(1) $\delta\left(\dfrac{\partial f}{\partial t}\right) \neq \dfrac{\partial}{\partial t} \delta f$,   i.e., δ and $\dfrac{\partial}{\partial t}$ are not interchangeable;

(2) $\delta(\nabla f) \neq \nabla \delta f$,   i.e., δ and ∇ are not interchangeable;

(3) $\delta(\nabla \cdot f) \neq \nabla \cdot \delta f$, i.e., δ and $\nabla \cdot$ are not interchangeable;

(4) $\delta\left(\dfrac{Df}{Dt}\right) = \dfrac{D\delta f}{Dt}$,   i.e., δ and the co-moving derivative

$\dfrac{D}{Dt}$ are interchangeable.

$$(12.66)$$

From Eq. (12.65), it can be seen that the Euler perturbation can be interchangeable with most differential operators while the Lagrangian perturbation is not. Therefore, it is simpler to write down the basic equations of the radiative fluid by the Euler method and to linearize the basic equations using Euler perturbation.

12.3.3 Linearized Oscillation Equations

For simplicity, we employ Euler perturbation to establish the linearized oscillation equations. Under the perturbed state, the corresponding physical quantities can be written as

$$
\left.
\begin{aligned}
\rho &= \rho_0 + \rho' \\
P &= P_0 + P' \\
T &= T_0 + T' \\
&\vdots
\end{aligned}
\right\}
\tag{12.67}
$$

where the subscript "0" refers to quantities at equilibrium, and primed quantities are the corresponding Euler perturbation values. Substituting Eq. (12.67) into Eqs. (12.49) to (12.55), neglecting terms with high orders of the perturbation value, considering that values at equilibrium fulfil Eqs. (12.49) to (12.55), and noting that $\vec{v}_0 = 0$ and $\vec{v} = \vec{v}' = \frac{\partial \vec{\xi}}{\partial t}$, we get the linearized oscillation equation set.

The linearized continuity equation (12.49) is $\frac{\partial \rho'}{\partial t} + \nabla \cdot (\rho_0 \frac{\partial \vec{\xi}}{\partial t}) = 0$.

Integrating the above equation with respect to time, and determining the integration constant from the boundary condition that the perturbation is zero at equilibrium positions, we can obtain

$$
\rho' + \nabla \cdot (\rho_0 \vec{\xi}) = 0
\tag{12.68}
$$

The linearized equation of motion (12.50) is

$$
\rho_0 \frac{\partial^2 \vec{\xi}}{\partial t^2} = -\nabla (P'_g + P'_R) + \rho' \vec{g}_0 + \rho_0 \vec{g}'
\tag{12.69}
$$

The linearized radiative momentum equation (12.51) is

$$
\nabla P'_R = -\frac{\rho_0 \kappa_0}{c} \vec{F}_{R0} \left(\frac{\rho'}{\rho_0} + \frac{\kappa'}{\kappa_0} \right) - \frac{\rho_0 \kappa_0}{c} \vec{F}'_R
\tag{12.70}
$$

The linearized gas energy equation (12.52) is

$$
\frac{1}{\Gamma_3 - 1} \frac{\partial}{\partial t} \left[P_g' - \Gamma_1 \frac{P_{go}}{\rho_o} \rho' + \vec{\xi}' \left(\nabla P_{go} - \Gamma_1 \frac{P_{go}}{\rho_o} \nabla \rho_o \right) \right]
$$
$$
= (\rho \varepsilon - \nabla \cdot \vec{F}_{con})' - ac\rho_o \kappa_o^T (T_{go}^4 - T_{Ro}^4) \left(\frac{\rho'}{\rho_o} + \frac{\kappa^{T'}}{\kappa_o^T} \right) - \rho_o \kappa_o^T c (4aT_{go}^3 T_g' - 3P_R')
$$

(12.71)

The linearized radiative thermodynamical equation (12.53) is

$$
\frac{\partial}{\partial t} \left[3P_R' - 4\frac{P_{Ro}}{\rho_o} \rho' + \vec{\xi} \cdot \left(3\nabla P_{Ro} - 4\frac{P_{Ro}}{\rho_o} \nabla \rho_o \right) \right]
$$
$$
= -\nabla \cdot \vec{F}_R' + ac\rho_o K_o^T (T_{go}^4 - T_{Ro}^4) \left(\frac{\rho'}{\rho_o} + \frac{\kappa^{T'}}{\kappa_o^T} \right) + \rho_o \kappa_o^T c (4aT_{go}^3 T_g' - 3P_R')
$$

(12.72)

The linearized form of equation (12.54) is

$$
\vec{g}' = -\nabla \Phi'
$$

(12.73)

The linearized Poisson equation (12.55) is

$$
\nabla \cdot \vec{g}' = -4\pi G \rho'
$$

(12.74)

12.3.4 Relationship Between Perturbation and Time and Angle

The linearized oscillation equations (12.68) to (12.74) have analytical solutions for time and angle so these will be determined first. Firstly, since the coefficients in equations (12.68) to (12.74) are time independent and only the operator $\frac{\partial^2}{\partial t^2}$ appears in the equations, all relationships between the perturbation values and time can be expressed by a time dependent term $e^{-i\sigma t}$ where σ is the oscillation frequency. Secondly, to obtain the relationship between the perturbation values and the angle, we express the linearized oscillation equations in the spherical coordinate system. In equations (12.68) to (12.74), only three equations, viz., Eqs. (12.69), (12.70) and (12.73), are vector equations with expressions in the radial direction and the horizontal direction. Only the component equations in the horizontal direction have an angle dependent relationship. These three component equations in the horizontal direction are:

$$\rho \frac{\partial^2 \vec{\xi}_h}{\partial t^2} = -\nabla_h(P'_g + P'_R + \rho\Phi') \tag{12.75}$$

$$\nabla_h P'_R = -\frac{\rho\kappa}{c} \vec{F}'_{Rh} \tag{12.76}$$

$$\vec{g}'_h = -\nabla_h \Phi' \tag{12.77}$$

where the subscript "h" refers to the component in the horizontal direction, and the operator ∇_h is the horizontal component of the gradient operator:

$$\nabla_h = \vec{e}_\theta \cdot \frac{1}{r}\frac{\partial}{\partial\theta} + \vec{e}_\varphi \frac{1}{r\sin\theta}\frac{\partial}{\partial\varphi} \tag{12.78}$$

where $\vec{e}_\theta$ and $\vec{e}_\varphi$ are unit vectors in the directions of θ and φ. In writing down equations (12.75) to (12.77), we have used the feature that the quantities ρ_0, P_0, T_0, ... at equilibrium are only dependent on radius r and are independent of direction, and we have ignored the subscript "o" for simplicity. From these equations, it can be observed that the horizontal component of an arbitrary vector perturbation can be obtained by applying the gradient operator on some scalar perturbations.

Using Eqs. (12.75) to (12.77), the horizontal components of vector perturbations in the linearized oscillation equations expressed in spherical coordinates can be eliminated; noting that $\frac{\partial}{\partial t} = -i\sigma$ and $\frac{\partial^2}{\partial t^2} = -\sigma^2$, and neglecting the terms for convection and energy generation, we have

$$\rho' + \frac{1}{r^2}\frac{d}{dr}(\rho r^2 \xi) - \frac{1}{\sigma^2}\nabla_h^2(P'_g + P'_R + \rho\Phi') = 0 \tag{12.79}$$

$$-\sigma^2\rho\xi = -\frac{dP'_g}{dr} - \frac{dP'_R}{dr} + \rho'g + \rho g' \tag{12.80}$$

$$\frac{dP'_R}{dr} = -\frac{\rho\kappa L_R}{4\pi r^2 c}\left(\frac{\rho'}{\rho} + \frac{\kappa'}{\kappa} + \frac{L'_R}{L_R}\right) \tag{12.81}$$

$$-\frac{i\sigma}{\Gamma_3 - 1}\left[P'_g - \Gamma_1\frac{P_g}{\rho}\rho' + \left(\frac{dP_g}{dr} - \Gamma_1\frac{P_g}{\rho}\frac{d\rho}{dr}\right)\xi\right]$$

$$= -ac\rho\kappa^T(T_g^4 - T_R^4)\left(\frac{\rho'}{\rho} + \frac{\kappa^{T'}}{\kappa^T}\right) - \rho\kappa^T c(4aT_g^3 T'_g - 3P'_R) \tag{12.82}$$

$$-3\sigma\left[P'_R - \frac{4}{3}\frac{P_R}{\rho}\rho' + \left(\frac{dP_R}{dr} - \frac{4}{3}\frac{P_R}{\rho}\frac{d\rho}{dr}\right)\xi\right] = -\frac{1}{4\pi r^2}\frac{dL'_R}{dr} + \frac{c}{\rho\kappa}\nabla_h^2 P'_R$$

$$+ac\rho\kappa^T(T_g^4 - T_R^4)\left(\frac{\rho'}{\rho} + \frac{\kappa^{T'}}{\kappa^T}\right) + \rho\kappa^T c(4aT_g^3 T'_g - 3P'_R) \qquad (12.83)$$

$$g' = -\frac{d\Phi'}{dr} \qquad (12.84)$$

$$\frac{1}{r^2}\frac{d}{dr}(r^2 g')\nabla_h^2\Phi' = -4\pi G\rho' \qquad (12.85)$$

where ξ, g' and F_R' are, respectively, radial components of the translation $\vec{\xi}$, perturbation $\vec{g}'$ of the gravitational acceleration and perturbation $\vec{F}'_R$ of the radiation flux. The perturbation of radiation luminosity L_R' is

$$L'_R = 4\pi r^2 F'_R \qquad (12.86)$$

The horizontal component of the Laplacian operator ∇_h^2 is

$$\nabla_h^2 = \frac{1}{r^2\sin^2\theta}\left[\sin\theta\frac{\partial}{\partial\theta}\left(\sin\theta\frac{\partial}{\partial\theta}\right) + \frac{\partial^2}{\partial\varphi^2}\right] \qquad (12.87)$$

In the equations (12.79) to (12.85), except that ∇_h^2 is related to angles θ and φ, all coefficients are functions of r only. Therefore, all scalar perturbations and vector perturbations in the equations can be written as functions of r and functions of the angles. A scalar perturbation f' is written as

$$f'(r, \theta, \varphi, t) = f'(r)Y_l^m(\theta, \varphi)e^{-i\sigma t} \qquad (12.88)$$

and a vector perturbation $\vec{F}'$ is written as

$$\vec{F}'(r, \theta, \varphi, t) = \left[F'(r), F'_h(r)\frac{1}{r}\frac{\partial}{\partial\theta}, F'_h(r)\frac{1}{r\sin\theta}\frac{\partial}{\partial\varphi}\right]Y_l^m(\theta, \varphi)e^{-i\sigma t} \quad (12.89)$$

Substituting Eqs. (12.88) and (12.89) into Eqs. (12.79) to (12.85), adding all equations together to form a single equation and rearranging terms, we will get an equation with functions of r only on the left and functions of angles θ, φ only on the right. After separation of variables, it can be proved that all functions $Y_l^m(\theta, \varphi)$ of perturbations related to angles fulfil the spherical harmonics equation

$$\left[\nabla_h^2 + \frac{l(l+1)}{r^2}\right] Y_l^m(\theta, \varphi) = 0 \qquad (12.90)$$

Therefore, $Y_l^m(\theta, \varphi)$ is the spherical harmonics

$$Y_l^m(\theta, \varphi) = (-1)^{(m+|m|)/2} \left[\frac{(l-|m|)!(2l+1)}{(l+|m|)!4\pi}\right]^{1/2} P_l^m(\cos\theta) e^{-im\varphi} \qquad (12.91)$$

Here, the spherical harmonics index l is an integer which can be 0, 1, 2, 3, ..., and the spherical index m is also an integer which can take values $-l, -l+1, -l+2, ..., 0, ..., l-2, l-1, l$, a total of $2l+1$ values. Some of the spherical harmonics with smaller l values are:

$$Y_0^0 = \frac{1}{\sqrt{4\pi}}, \quad Y_1^1 = \sqrt{\frac{3}{8\pi}} \sin\theta e^{i\varphi}, \quad Y_1^0 = \sqrt{\frac{3}{4\pi}} \cos\theta, \quad Y_1^{-1} = \sqrt{\frac{3}{8\pi}} \sin\theta e^{-i\varphi}$$

$$Y_2^2 = \sqrt{\frac{15}{32\pi}} \sin^2\theta e^{i2\varphi}, \quad Y_2^1 = \sqrt{\frac{15}{8\pi}} \sin\theta \cos\theta e^{i\varphi}, \quad Y_2^0 = \sqrt{\frac{5}{16\pi}}(3\cos^2\theta - 1)$$

$$Y_2^{-1} = \sqrt{\frac{15}{8\pi}} \sin\theta \cos\theta e^{-i\varphi}, \quad Y_2^{-2} = \sqrt{\frac{15}{32\pi}} \sin^2\theta e^{-i2\varphi}$$

From the above harmonics, we notice that when $l = 0$, Y_0^0 is a constant and is not related to θ or φ, so the perturbation is independent of angles. Therefore, the oscillation for $l = 0$ is a purely radial oscillation. From this, we also note that a radial oscillation is a special case of non-radial oscillations. Figure 12.1 gives schematic diagrams of some spherical harmonics. From Fig. 12.1, we can observe the procedures for constructing the diagrams based on values of l and m as follows: the index m represents how many great circles are cutting the spherical surface longitudinally, and the value $(l - |m|)$ represents how many great circles are cutting the spherical surface latitudinally.

After solving for the dependence of perturbations on angles and time, the change of perturbations with radius are expressed in the following equations:

$$\rho' + \frac{1}{r^2}\frac{d}{dr}(\rho r^2 \xi) - \frac{l(l+1)}{\sigma^2 r^2}(P_g' + P_R' + \rho\Phi') = 0 \qquad (12.92)$$

$$-\sigma^2 \rho\xi = -\frac{dP_g'}{dr} + \frac{\rho\kappa L_R}{4\pi r^2 c}\left(\frac{\rho'}{\rho} + \frac{\kappa'}{\kappa} + \frac{L_R'}{L_R}\right) + \rho' g + \rho g' \qquad (12.93)$$

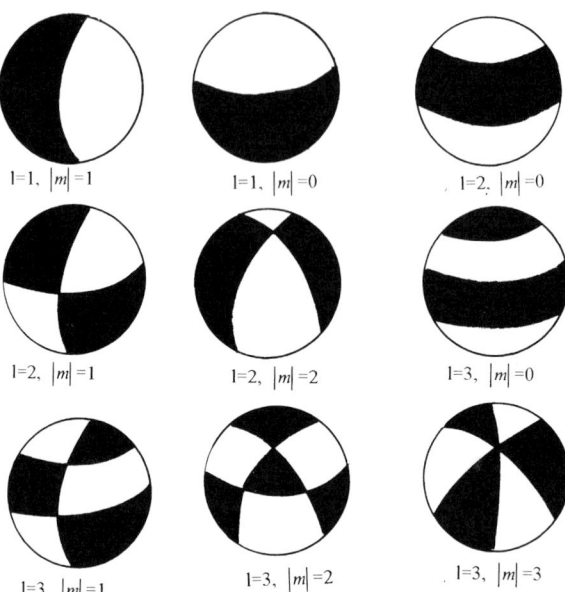

Fig. 12.1 Schematic diagrams of some spherical harmonics $Y_l^m (\theta, \varphi)$.

$$\frac{dP'_R}{dr} = -\frac{\rho \kappa L_R}{4\pi^2 c}\left(\frac{\rho'}{\rho} + \frac{\kappa'}{\kappa} + \frac{L'_R}{L_R}\right) \tag{12.94}$$

$$-\frac{i\sigma}{\Gamma_3 - 1}\left[P'_g - \Gamma_1 \frac{P_g}{\rho}\rho' + \left(\frac{dP_g}{dr} - \Gamma_1 \frac{P_g}{\rho}\frac{d\rho}{dr}\right)\xi\right]$$

$$= -ac\rho\kappa^T(T_g^4 - T_R^4)\left(\frac{\rho'}{\rho} + \frac{\kappa^{T'}}{\kappa^T}\right) - \rho\kappa^T c(4aT_g^3 T'_g - 3P'_R) \tag{12.95}$$

$$-i3\sigma\left[P'_R - \frac{4}{3}\frac{P_R}{\rho}\rho' + \left(\frac{dP_R}{dr} - \frac{4}{3}\frac{P_R}{\rho}\frac{d\rho}{dr}\right)\xi\right] = -\frac{1}{4\pi r^2}\frac{dL'_R}{dr} - \frac{l(l+1)c}{\rho\kappa r^2}P'_R$$

$$+ ac\rho\kappa^T(T_g^4 - T_R^4)\left(\frac{\rho'}{\rho} + \frac{\kappa^{T'}}{\kappa^T}\right) + \rho\kappa^T c(4aT_g^3 T'_g - 3P'_R) \tag{12.96}$$

$$g' = -\frac{d\Phi'}{dr} \tag{12.97}$$

$$\frac{1}{r^2}\frac{d}{dr}(r^2 g') + \frac{l(l+1)}{r^2}\Phi' = -4\pi G\rho' \tag{12.98}$$

12.3.5 Basic Equations for Radial Oscillation

When the spherical harmonics index $l = 0$, all perturbations variables are independent of angles and are functions of radius r only. In this way, the oscillation is spherically symmetric, and all spherical surfaces will move inwards and outwards periodically with time. Such oscillations are referred to as radial oscillations. Radial oscillation is a special case of non-radial oscillations.

By letting $l = 0$, the non-radial oscillation equations (12.92) to (12.98) will become radial oscillation equations. For example, when $l = 0$, Eq. (12.92) becomes

$$\rho' + \frac{1}{r^2}\frac{d}{dr}(\rho r^2 \xi) = 0 \qquad (12.99)$$

and Eq. (12.98) becomes

$$\frac{1}{r^2}\frac{d}{dr}(r^2 g') = -4\pi G\rho' \qquad (12.100)$$

After elimination of the density perturbation from these two equations, we have

$$\frac{1}{r^2}\frac{d}{dr}(r^2 g' - 4\pi G\rho r^2 \xi) = 0 \qquad (12.101)$$

Integrating the above equation with respect to r, and making use of the boundary condition that perturbation at the center of the star should be finite, we have

$$g' = 4\pi G\rho\xi \qquad (12.102)$$

In this way, the solution to the perturbation of gravitational acceleration is obtained.

By letting $l = 0$ in the non-radial oscillation equations (12.92) to (12.98), and using Eq. (12.102) to eliminate the perturbation of gravitational acceleration in the equation of motion, we get

$$\rho' + \frac{1}{r^2}\frac{d}{dr}(\rho r^2 \xi) = 0 \qquad (12.103)$$

$$-\sigma^2 \rho\xi = -\frac{dP'_g}{dr} + \frac{\rho\kappa L_R}{4\pi r^2 c}\left(\frac{\rho'}{\rho} + \frac{\kappa'}{\kappa} + \frac{L'_R}{L_R}\right) + \rho'g + 4\pi G\rho^2\xi \qquad (12.104)$$

$$\frac{dP'_R}{dr} = -\frac{\rho\kappa L_R}{4\pi r^2 c}\left(\frac{\rho'}{\rho} + \frac{\kappa'}{\kappa} + \frac{L'_R}{L_R}\right) \qquad (12.105)$$

$$-\frac{i\sigma}{\Gamma_3 - 1}\left[P'_g - \Gamma_1\frac{P_g}{\rho}\rho' + \left(\frac{dP_g}{dr} - \Gamma_1\frac{P_g}{\rho}\frac{d\rho}{dr}\right)\xi\right]$$

$$= -ac\rho\kappa^T(T_g^4 - T_R^4)\left(\frac{\rho'}{\rho} + \frac{\kappa^{T'}}{\kappa^T}\right) - \rho\kappa^T c(4aT_g^3 T'_g - 3P'_R) \qquad (12.106)$$

$$-i3\sigma\left[P'_R - \frac{4}{3}\frac{P_R}{\rho}\rho' + \left(\frac{dP_R}{dr} - \frac{4}{3}\frac{P_R}{\rho}\frac{d\rho}{dr}\right)\xi\right] = -\frac{1}{4\pi r^2}\frac{dL'_R}{dr}$$

$$+ ac\rho\kappa^T(T_g^4 - T_R^4)\left(\frac{\rho'}{\rho} + \frac{\kappa^{T'}}{\kappa^T}\right) + \rho\kappa^T c(4aT_g^3 T'_g - 3P'_R) \quad (12.107)$$

The equations (12.103) to (12.107) are radial oscillation equations.

12.3.6 The Dimensionless Non-Radial Oscillation Equations

Normally, the non-radial oscillation equations (12.92) to (12.98) can only be solved numerically. For convenience of integration, we first make them dimensionless, the key problem of which is to choose appropriate factors to simplify the equations. We adopt the dimensionless variables of Li (1990b):

$$\left.\begin{array}{l} y_1 = \dfrac{GM}{R^3}\dfrac{\xi}{g}, \quad y_2 = \dfrac{P'_g}{\rho g r}, \quad y_3 = \dfrac{P'_R}{\rho g r}, \quad y_4 = \dfrac{\kappa L'_R}{4\pi r^2 cg} \\[2ex] y_5 = \dfrac{\Phi'}{gr}, \quad y_6 = \dfrac{g'}{g}, \quad y_7 = \dfrac{\rho'}{\rho}, \quad \sigma^2 = \dfrac{GM}{R^3}\omega^2 \end{array}\right\} \quad (12.108)$$

Furthermore, T'_g, κ' and $\kappa^{T'}$ in the equations are not independent variables; they are functions of ρ' and P'_g and can be written as

$$\frac{T'_g}{T_g} = \theta_P\frac{P'_g}{P_g} + \theta_\rho\frac{\rho'}{\rho} \quad (12.109)$$

$$\frac{\kappa'}{\kappa} = \kappa_P\frac{P'_g}{P_g} + \kappa_\rho\frac{\rho'}{\rho} \quad (12.110)$$

and
$$\frac{\kappa^{T'}}{\kappa^T} = \kappa_P^T\frac{P'_g}{P_g} + \kappa_\rho^T\frac{\rho'}{\rho} \quad (12.111)$$

The coefficients of the above equations are

$$\left.\begin{array}{l} \theta_P = \left(\dfrac{\partial \ln T_g}{\partial \ln P_g}\right)_\rho, \quad \theta_\rho = \left(\dfrac{\partial \ln T_g}{\partial \ln \rho}\right)_{P_g} \\[2ex] \kappa_P = \left(\dfrac{\partial \ln \kappa}{\partial \ln T_g}\right)_\rho \theta_P, \quad \kappa_\rho = \left(\dfrac{\partial \ln \kappa}{\partial \ln \rho}\right)_{T_g} + \left(\dfrac{\partial \ln \kappa}{\partial \ln T_g}\right)_\rho \theta_\rho \\[2ex] \kappa_P^T = \left(\dfrac{\partial \ln \kappa^T}{\partial \ln T_g}\right)_\rho \theta_P, \quad \kappa_\rho^T = \left(\dfrac{\partial \ln \kappa^T}{\partial \ln \rho}\right)_{T_g} + \left(\dfrac{\partial \ln \kappa^T}{\partial \ln T_g}\right)_\rho \theta_\rho \end{array}\right\} \quad (12.112)$$

Using Eqs. (12.108) to (12.111), equations (12.92) to (12.98) can be written as dimensionless linear non-radial oscillation equations as below:

$$r\frac{dy_1}{dr} = (D - U)y_1 + \frac{l(l+1)}{\omega^2}(y_2 + y_3 + y_5) + Cy_7 \qquad (12.113)$$

$$r\frac{dy_2}{dr} = \omega^2 y_1 + (D - U + 1 - F_2)y_2 + y_4 + y_6 + (1 - F_7)y_7 \qquad (12.114)$$

$$r\frac{dy_3}{dr} = F_2 y_2 + (D - U + 1)y_3 - y_4 + F_7 y_7 \qquad (12.115)$$

$$r\frac{dy_4}{dr} = -\frac{i\omega}{C}\left(\frac{\Gamma_1}{\Gamma_3 - 1}C_g A_g + 4C_R A_R\right)y_1 + \frac{i\omega S}{\Gamma_3 - 1}y_2 + [i\omega 3S - l(l+1)]y_3$$

$$+ (W - U)y_4 + i\omega\left(\frac{\Gamma_1}{\Gamma_3 - 1}C_g + 4C_R\right)y_7 \qquad (12.116)$$

$$r\frac{dy_5}{dr} = (1 - U)y_5 - y_6, \qquad (12.117)$$

$$r\frac{dy_6}{dr} = -l(l+1)y_5 - U(y_6 - y_7) \qquad (12.118)$$

$$0 = -\frac{i\omega}{C}\frac{\Gamma_1}{\Gamma_3 - 1}C_g A_g y_1 + \left(B_2 + \frac{i\omega S}{\Gamma_3 - 1}\right)y_2 + 3B_3 y_3$$

$$+ \left(B_7 + i\omega\frac{\Gamma_1}{\Gamma_3 - 1}C_g\right)y_7 \qquad (12.119)$$

The dimensionless coefficients in the above equations set are defined as:

$$U = -\frac{4\pi G\rho r}{g}, \quad C = -\frac{GM}{R^3}\frac{r}{g}, \quad S = -\frac{\rho\kappa r^2}{c}\sqrt{\frac{GM}{R^3}}, \quad D = -\frac{d\ln\rho}{d\ln r}$$

$$W = -\frac{d\ln\kappa}{d\ln r}, \quad C_g = -\frac{P_g r\kappa}{cg}\sqrt{\frac{GM}{R^3}}, \quad A_g = \frac{d\ln\rho}{d\ln r} - \frac{1}{\Gamma_1}\frac{d\ln P_g}{d\ln r}$$

$$C_R = -\frac{P_R r\kappa}{cg}\sqrt{\frac{GM}{R^3}}, \quad A_R = \frac{d\ln\rho}{d\ln r} - \frac{3}{4}\frac{d\ln P_R}{d\ln r}, \quad F_2 = -\frac{\kappa L_R}{4\pi r^2 cg}\frac{\rho gr}{P_g}\kappa_P$$

$$F_7 = -\frac{\kappa L_R}{4\pi r^2 cg}(1 + \kappa_\rho), \quad B_2 = -ac\rho\kappa^T(T_g^4 - T_R^4)\frac{r\kappa}{cg}\frac{\rho gr}{P_g}\kappa_P - \frac{4aT_g^4 r\kappa\rho\kappa^T}{g}\frac{\rho gr}{P_g}\theta_P$$

$$B_3 = \rho^2 r^2 \kappa \kappa^T, \quad B_7 = -ac\rho\kappa^T(T_g^4 - T_R^4)\frac{r\kappa}{cg}(1+\rho) - \frac{4aT_g^4 r\kappa\rho\kappa^T}{g}\theta_\rho$$

$$(12.120)$$

12.4 PROPERTIES OF STELLAR OSCILLATIONS

The oscillation equations (12.113) to (12.119) contain the relationship among sophisticated interactions of various physical processes when the star oscillates non-radially, and determine the various properties of the stellar oscillation. However, it is difficult to understand the various properties of stellar oscillations by directly solving the equations. For example, it is difficult to directly know the restoring force of stellar oscillation, whether the oscillation wave belongs to a longitudinal wave, a transverse wave or another wave, i.e., the mode of oscillation, and the region for propagation of the oscillation wave in the star. However, if some approximations are applied to the oscillation equations (12.113) to (11.119), such as the adiabatic approximation assuming that the stellar oscillation is adiabatic, and the Cowling approximation assuming the perturbation of gravitational potential is minimal, we can analyze the above properties of stellar oscillation more directly. Before the discussion of numerical methods to solve the oscillation equations, the adiabatic approximation and the Cowling approximation will first be introduced in this section, and some properties of stellar oscillation will be analyzed.

12.4.1 Adiabatic Approximation

Equation (12.116) is obtained from combination of the radiative energy equation and the radiative thermodynamical equation, linearized and then made dimensionless. The equation contains the $i\omega$ term, which comes from the $T_R\frac{DS_R}{Dt}$ term in the original equation, so it describes how adiabatic the moving fluid element is. The other terms in Eq. (12.116) describe the radiative energy flowing into the fluid element and the radiation emitting from the gas within the fluid element so they represent the terms producing non-adiabatic effects. It can be proved for all stars that when we go deeper into the star, the relative value of the terms in Eq. (12.116) not containing $i\omega$, i.e. the non-adiabatic terms, become smaller. For example, when we take the total mass of the star to be $M \approx 10^{34}$ g, radius to be $R \approx 10^{13}$ cm, core gas pressure to be $P_g \approx 10^{17}$ dyn.cm^{-2}, core density to be $\rho \approx 10^2$ gcm^{-3} and opacity to be $\kappa \approx 1$ cm^{2}g^{-1}, we can first calculate the values of the coefficients C_g, W, ... using Eq. (12.120). Then, we calculate the values of non-adiabatic terms, i.e. terms without $i\omega$ in Eq. (12.116), around the center of the star, and find that they are smaller than adiabatic terms (terms with $i\omega$) by a factor of 10^9. This shows that it is highly adiabatic around the center of the star. On the contrary, when close to the surface of the star, the gas pressure is $P_g \approx 10^4$ dyn.cm^{-2} and the density is $\rho \approx 10^{-9}$ gcm^{-3}, so the values of non-adiabatic terms without $i\omega$ in Eq. (12.116) are only a factor of 10^{-2} of those of adiabatic terms. This shows that it is not highly adiabatic at the surface of the star.

Since oscillations are adiabatic in most regions inside the star, and are non-adiabatic only in extremely thin regions near the surface, the employment of the adiabatic approximation to study stellar oscillation should give very realistic oscillation properties. In other words, properties of non-adiabatic oscillation are small modifications of properties of adiabatic oscillation.

The employment of the adiabatic approximation in studies of stellar oscillations is equivalent to the introduction of the adiabatic condition $\frac{DS}{Dt} = 0$ into the stellar oscillation equations. The use of adiabatic approximation can give sufficiently precise oscillation periods, distribution of oscillation amplitudes inside the star, and special features and propagation regions of different kinds of oscillation. However, some phenomena arising from non-adiabatic properties, such as temporal change of oscillation amplitude, and cause for production of oscillation, etc., cannot be obtained from the adiabatic approximation.

12.4.2 Restoring Forces of Stellar Oscillation

After adoption of the adiabatic approximation, we can analyze more directly the restoring forces of stellar oscillation from the oscillation equations. Since the star is composed of gas and radiation which are themselves independent thermodynamical systems, there exist several possible adiabatic conditions, e.g., the entropy changes of individual systems of gas and radiation are zero (i.e., $\frac{DS_g}{Dt} = 0$ and $\frac{DS_R}{Dt} = 0$) when the star oscillates; or the entropy change of the combined system of gas and radiation is zero (i.e., $\frac{D(S_g+S_R)}{Dt} = 0$), etc. In the following, we will analyze the restoring forces of stellar oscillation based on different adiabatic oscillation cases. For simplicity, we will not use the dimensionless variables when discussing the restoring forces of stellar oscillation.

(A) Double adiabatic oscillation

Double adiabatic oscillation refers to the case when both the gaseous and radiative thermodynamical systems are adiabatic when oscillation occurs. The adiabatic conditions are $\frac{DS_g}{Dt} = 0$ and $\frac{DS_R}{Dt} = 0$.

From the gas thermodynamical equation (12.17) and the radiative thermodynamical equation (12.48), and considering adiabatic conditions, we have

$$\frac{1}{\Gamma_1}\frac{DP_g}{Dt} - \frac{P_g}{\rho}\frac{D\rho}{Dt} = 0 \tag{12.121}$$

and

$$\frac{3}{4}\frac{DP_R}{Dt} - \frac{P_R}{\rho}\frac{D\rho}{Dt} = 0 \tag{12.122}$$

When linear perturbations are applied to Eqs. (12.121) and (12.122), we have

$$P'_g - \Gamma_1\frac{P_g}{\rho}\rho' + \left(\frac{dP_g}{dr} - \Gamma_1\frac{P_g}{\rho}\frac{d\rho}{dr}\right)\xi = 0 \tag{12.123}$$

and

$$P'_R - \frac{4}{3}\frac{P_R}{\rho}\rho' + \left(\frac{dP_R}{dr} - \frac{4}{3}\frac{P_R}{\rho}\frac{d\rho}{dr}\right)\xi = 0 \tag{12.124}$$

Adding Eqs. (12.123) and (12.124), and solving for the perturbation of density, we have

$$\frac{\rho'}{\rho} = \frac{P'}{\Gamma_1 P_g + \frac{4}{3}P_R} + \left(\frac{1}{\Gamma_1 P_g + \frac{4}{3}P_R}\frac{dP}{dr} - \frac{1}{\rho}\frac{dP}{dr}\right)\xi \tag{12.125}$$

where the definitions of the total pressure P and its perturbation P' are

$$P = P_g + P_R \tag{12.126}$$

and

$$P' = P'_g + P'_R \tag{12.127}$$

We define an equivalent adiabatic compression coefficient

$$\Gamma_{1e} \equiv \frac{\Gamma_1 P_g + \frac{4}{3}P_R}{P} = \Gamma_1\beta + \frac{4}{3}(1 - \beta) \tag{12.128}$$

where β is the ratio of the gas pressure to the total pressure. Equation

(12.125) then becomes

$$\frac{\rho'}{\rho} = \frac{1}{\Gamma_{1e}} \cdot \frac{P'}{P} + \left(\frac{1}{\Gamma_{1e}} \frac{d \ln P}{dr} - \frac{d \ln \rho}{dr} \right) \xi. \qquad (12.129)$$

Note that the total pressure P has become an independent variable instead of the gas pressure P_g. The first term on the right of Eq. (12.129) represents a local pressure excess leading to compression of the fluid element, which is the pressure restoring force of the oscillation. The second term on the right-hand side of Eq. (12.129) represents the movement of the fluid element leading to a difference between its internal density and the ambient density, after subtraction of compression effects, and this is the condition for another restoring force of oscillation, namely the buoyancy. Therefore, Eq. (12.129) obtained from the adiabatic condition is in fact describing the properties of two main restoring forces – pressure and buoyancy. Furthermore, from Eq. (12.129), we know that pressure is always a restoring force, but buoyancy may not be one. Only when $\frac{d \ln \rho}{dr} < \frac{1}{\Gamma_{1e}} \frac{d \ln P}{dr}$ so that the buoyancy caused by the movement has a direction opposite to the movement itself does the force become a restoring force. On the contrary, when $\frac{d \ln \rho}{dr} > \frac{1}{\Gamma_{1e}} \frac{d \ln P}{dr}$, the directions of buoyancy and movement are the same, so the buoyancy will enhance the movement and dynamically unstable motion, the convective motion, occurs.

The equivalent adiabatic compression coefficient defined by Eq. (12.128) is in fact a weighted average of the adiabatic compression coefficients for gas and radiation, with individual pressures as weighting factors. In this way, the compression rate of the fluid element is not only related to the adiabatic compression coefficients of gas and radiation inside the fluid element, but is also related to individual pressures acting on the fluid element.

During a double adiabatic oscillation, the temperatures of gas and radiation inside the fluid element are different because their adiabatic compression coefficients are different. The double adiabatic case requires no heat exchange between the gas and the radiation, so the two systems will have their own adiabatic change.

(B) Equilibrium adiabatic oscillation

Common stellar oscillation theories treat gas and radiation as a single thermodynamical system which has a single temperature at all time and places. The adiabatic condition is that the entropy change of the combined system is zero, i.e. $\frac{D(S_g + S_R)}{Dt} = 0$, which leads to equilibrium adiabatic oscillation. During an equilibrium adiabatic oscillation, besides that it is adiabatic between the oscillating fluid element and the ambient environment, the gas and the radiation within the fluid element should adjust in time to keep their temperatures the same at all time.

Using a method similar to the derivation of Eq. (12.129), we can obtain an equation to be fulfilled by the equilibrium adiabatic oscillation as

$$\frac{\rho'}{\rho} = \frac{1}{\Gamma'_{1e}} \cdot \frac{P'}{P} + \left(\frac{1}{\Gamma'_{1e}} \frac{d \ln P}{dr} - \frac{d \ln \rho}{dr} \right) \xi \qquad (12.130)$$

Equations (12.129) and (12.130) are identical except for the adiabatic compression coefficients. The adiabatic coefficient Γ'_{1e} in Eq. (12.130) is for a mixture of a black body radiation and an ideal gas, and its expression is (see Cox and Giuli, 1968; Leoloux and Walraven 1958)

$$\Gamma'_{1e} = \beta + \frac{(4 - 3\beta)^2(\gamma - 1)}{\beta + 12(\gamma - 1)(1 - \beta)} \qquad (12.131)$$

The physical meanings expressed by Eqs. (12.129) and (12.130) are also the same; they are describing properties of two main restoring forces, pressure and buoyancy, of the oscillation.

Both the double adiabatic condition and the equilibrium adiabatic condition require the fluid element to be adiabatic with the surroundings during oscillation. However, the double adiabatic condition also requires the gas and radiation to be adiabatic, and the equilibrium adiabatic condition also requires thermodynamical equilibrium between the gas and the radiation.

In most regions inside the star, it is common that the gas and the radiation are at thermodynamical equilibrium, and the difference between temperatures of the gas and radiation is very small. However, in regions close to the surface of the star, the gas and the radiation will deviate from thermodynamical equilibrium and the corresponding temperatures will have significant differences. In order to explain these, we have the following analysis. Inside the star, if we do not consider effects of nuclear reaction and convection on the oscillation, the gas energy equation can be approximately derived from Eq. (12.10) and written as

$$\rho T_g \frac{DS_g}{Dt} \cong 4ac\rho\kappa^T T_g^4 \left(\frac{\Delta T}{T_g}\right) \tag{12.132}$$

where ΔT is the difference between temperatures of the gas and the radiation. Equation (12.132) should be valid when $\frac{\Delta T}{T_g} \ll 1$. On the other hand, from Eqs. (12.31) and (12.32), we can get

$$T_g DS_g = C_V DT_g - C_V T_g(\Gamma_3 - 1)\frac{D\rho}{\rho} \tag{12.133}$$

Neglecting the second term on the right-hand side of the above equation, and substituting this into Eq. (12.132), we can obtain

$$\frac{\Delta T}{T_g} \approx \frac{C_V}{4ac\kappa^T T_g^4}\frac{DT_g}{Dt} \tag{12.134}$$

It can be seen that the right-hand side of Eq. (12.134) is only related to the gas. Using this equation, we can estimate the difference between temperatures of the gas and the radiation in different regions inside the star. For a star, the specific heat capacity at constant volume C_V and the average absorption coefficient κ^T are two slowly varying functions, which can be respectively approximated by $C_V \approx 10^8$ ergK^{-1}mole^{-1} and $\kappa^T \approx 1$ cm^{2}g^{-1}, and the typical oscillation timescale is 10^5 s. Thus, Eq. (12.134) is mainly determined by the gas temperature. Near the core of the star, $T_g \approx 10^7$ K so $\frac{\Delta T}{T_g} \approx 10^{-13}$, which shows that the difference between temperatures of the gas and the radiation is extremely small here and that we can assume thermodynamical equilibrium between the gas and the radiation. However, in regions close to the surface of the star, $T_g \approx 10^4$ K so $\frac{\Delta T}{T_g} \approx 10^{-3}$, which shows that the difference between the temperatures is sufficiently large that non-equilibrium effects between the gas and the radiation become obvious.

From the above discussions, we see that the equilibrium adiabatic condition is a very good approximate description of the realistic stellar oscillation because it guarantees the same temperature for gas and radiation. This condition has discrepancies only in regions close to the surface of the star but is satisfactory in most regions inside the star. In regions close to the surface, there are strong non-adiabatic features and the stellar oscillation should be treated with a non-adiabatic theory. However, since the temperatures of the gas and the radiation differ here, the situation can also be roughly approximated by the double adiabatic condition.

12.4.3 Non-Adiabatic Effects

Adding Eqs. (12.95) and (12.96) together, and solving for the perturbation of density, we obtain

$$
\frac{\rho'}{\rho} = \frac{1}{\frac{\Gamma_1}{\Gamma_3 - 1} P_g + 4P_R} \left(\frac{P'_g}{\Gamma_3 - 1} + 3P'_R \right) + \left[\frac{1}{\frac{\Gamma_1}{\Gamma_3 - 1} P_g + 4P_R} \left(\frac{1}{\Gamma_3 - 1} \frac{dP_g}{dr} \right) \right.
$$

$$
\left. + 3\frac{dP_R}{dr} \right) - \frac{1}{\rho}\frac{d\rho}{dr} \right] \xi + i \frac{1}{\left(\frac{\Gamma_1}{\Gamma_3 - 1} P_g + 4P_R \right)\sigma} \nabla \cdot \vec{F}'_R \qquad (12.135)
$$

Defining some adiabatic coefficients:

$$
\left(\frac{\Gamma_1}{\Gamma_3 - 1} \right)_e \equiv \frac{\frac{\Gamma_1}{\Gamma_3 - 1} P_g + 4P_R}{P}
$$

$$
\Gamma^P_{1e} \equiv \left(\frac{\Gamma_1}{\Gamma_3 - 1} \right)_e \left(\frac{P'}{\frac{1}{\Gamma_3 - 1} P'_g + 3P'_R} \right) \qquad (12.136)
$$

$$
\Gamma^B_{1e} \equiv \left(\frac{\Gamma_1}{\Gamma_3 - 1} \right)_e \frac{\frac{dP}{dr}}{\frac{1}{\Gamma_3 - 1} \frac{dP_g}{dr} + 3\frac{dP_R}{dr}}
$$

Eq. (12.135) can then be written as

$$
\frac{\rho'}{\rho} = \frac{1}{\Gamma^P_{1e}} \frac{P'}{P} + \left(\frac{1}{\Gamma^B_{1e}} \frac{d\ln P}{dr} - \frac{d\ln \rho}{dr} \right)\xi + i \frac{1}{\left(\frac{\Gamma_1}{\Gamma_3 - 1} \right)_e} \frac{1}{P\sigma} \nabla \cdot \vec{F}'_R \qquad (12.137)
$$

The third term on the right-hand side of Eq. (12.137), i.e., the divergence term, represents the net energy flux flowing out from the fluid element caused by the perturbation of radiation field during oscillation of the fluid element. In general, this term is much smaller than other terms in the equation. It can be observed that on the right-hand side of Eq. (12.137), besides the two terms describing restoring forces from pressure and buoyancy, there is also a third term caused by non-adiabatic features which has a phase difference with the other two terms. Therefore, the restoring force of oscillation given here is no longer strictly out of phase with the translation, but has a phase lead or phase lag. This phase difference is caused by non-adiabatic features, i.e., the third term on the right of Eq. (12.137). Since non-adiabatic features can cause the restoring force not to be strictly opposite to the translation, these can enhance or suppress the oscillation. Therefore, it is obvious that the non-adiabatic theory should be employed to study the cause of stellar oscillation since the increase or decrease of oscillation amplitudes are caused by non-adiabatic features. On the contrary, under the adiabatic condition, the third term on the right-hand side of Eq. (12.137) does not exist, so there is no increase or decrease in the oscillation amplitude and we cannot analyze the cause for oscillation.

12.4.4 Propagation Properties of Oscillation Waves

The characteristics of propagation of oscillation waves are mainly determined by properties of the restoring force of the oscillations, and is not very much dependent on the increase or decrease in oscillation energy. Therefore, we can use the adiabatic theory to study propagation properties of oscillation waves. However, using the adiabatic condition alone is not enough. In order to study propagation properties of oscillation waves, we have to add an approximation, i.e., the Cowling approximation (Cowling 1941) assuming that the perturbation of gravitational potential Φ' is minimal and negligible. In fact, when there are a lot of nodes on the oscillation wave, the perturbation of gravitational potential Φ' will be very small and the Cowling approximation will be close to the realistic situation.

Using the Cowling approximation $\Phi' = 0$ and the density perturbation in Eq. (12.125) obtained under the adiabatic condition, and considering definitions of the total pressure P and the adiabatic compression coefficients in Eq. (12.126) to (12.128), we obtain from Eqs. (12.92) to (12.94):

$$\frac{d}{dr}(\rho r^2 \xi) = \left[\frac{l(l+1)}{\sigma^2} + \frac{1}{\Gamma_{1e}}\frac{\rho r^2}{P}\right]P' - \left(\frac{1}{\Gamma_{1e}}\frac{d\ln P}{dr} - \frac{d\ln\rho}{dr}\right)\rho r^2 \xi \quad (12.138)$$

and

$$\frac{dP'}{dr} = \rho\left[\sigma^2 + \left(\frac{1}{\Gamma_{1e}}\frac{d\ln P}{dr} - \frac{d\ln\rho}{dr}\right)g\right]\xi + \frac{1}{\Gamma_{1e}}\frac{g\rho}{P}P' \quad (12.139)$$

Using new variables u and w to replace ξ and P', where

$$u = r^2 \xi P^{1/\Gamma_{1e}} \quad (12.140)$$

and

$$w = P'/P^{1/\Gamma_{1e}} \quad (12.141)$$

and the adiabatic compression coefficient Γ_{1e} is treated as a constant here, Eqs. (12.138) and (12.139) can be written as

$$\frac{du}{dr} = \left[\frac{l(l+1)}{\sigma^2} - \frac{1}{\Gamma_{1e}}\frac{\rho r^2}{P}\right]\frac{P^{2/\Gamma_{1e}}}{\rho}w \tag{12.142}$$

and

$$\frac{dw}{dr} = \frac{1}{r^2}\left[\sigma^2 + \left(\frac{1}{\Gamma_{1e}}\frac{d\ln P}{dr} - \frac{d\ln\rho}{dr}\right)g\right]\frac{\rho u}{P^{2/\Gamma_{1e}}} \tag{12.143}$$

We adopt the local approximation to qualitatively discuss propagation properties of oscillation waves. Under the local approximation, the relationships between the coefficients in Eqs. (12.142) and (12.143) and r are neglected, i.e., the coefficients are treated as constants. The local approximation is valid when there are a lot of oscillation nodes, and the physical quantities within a wave node can be treated as constants. It is easy to observe that the solutions to Eqs. (12.142) and (12.143) have the form

$$u(r), w(r) \sim e^{iK_r r}. \tag{12.144}$$

Substituting Eq. (12.144) into (12.142) and (12.143), we can obtain the dispersion relationship as

$$K_r^2 = \left[\sigma^2 + \left(\frac{1}{\Gamma_{1e}}\frac{d\ln P}{dr} - \frac{d\ln\rho}{dr}\right)g\right]\left[\frac{1}{\Gamma_{1e}}\frac{\rho}{P} - \frac{l(l+1)}{\sigma^2 r^2}\right] \tag{12.145}$$

where K_r is the radial wave number. In regions where K_r is a real number, the oscillation propagates in form of waves; in regions where K_r is imaginary, the oscillation wave cannot propagate and decays exponentially. We define critical frequencies by setting $K_r = 0$, i.e.,

$$L_l^2 = \frac{l(l+1)}{\sigma^2}\frac{\Gamma_{1e}P}{\rho} \tag{12.146}$$

and

$$N^2 = -\left(\frac{1}{\Gamma_{1e}}\frac{d\ln P}{dr} - \frac{d\ln\rho}{dr}\right)g \tag{12.147}$$

where L_l is the Lamb frequency and N is the Brunt–Väisälä frequency. Furthermore, the local velocity of sound C_s is defined as

$$C_s^2 = \frac{\Gamma_{1e}P}{\rho} \tag{12.148}$$

Using Eqs. (12.146) to (12.148), the dispersion relationship in Eq. (12.145) can also be written as

$$K_r^2 = \frac{(\sigma^2 - N^2)(\sigma^2 - L_l^2)}{\sigma^2 C_s^2} \qquad (12.149)$$

It can be seen that when $\sigma > L_l$ and $\sigma > N$, or when $\sigma < L_l$ and $\sigma < N$, the radial wave number K_r is real and the oscillation wave can propagate radially; when $L_l > \sigma > N$, or $N > \sigma > L_l$, K_r is imaginary and the wave is dissipative.

We further introduce the horizontal wave number K_h as

$$K_h^2 = \frac{l(l+1)}{r^2} \qquad (12.150)$$

The dispersion relationship in Eq. (12.149) can then be written as

$$\sigma^4 - (N^2 + K^2 C_s^2)\sigma^2 + N^2 K_h^2 C_s^2 = 0 \qquad (12.151)$$

where the total wave number is

$$K^2 = K_r^2 + K_h^2 \qquad (12.152)$$

Employing the equation for solving for roots of a quadratic equation, the approximations to the two roots of Eq. (12.151) are

$$\sigma_+^2 \cong L_l^2 + K_r^2 C_s^2 + N^2 = K^2 C_s^2 + N^2 \qquad (12.153)$$

and $$\sigma_-^2 \cong \frac{L_l^2 N^2}{L_l^2 + K_r^2 C_s^2 + N^2} = \frac{K_h^2}{K^2 + N^2/C_s^2} N^2 \qquad (12.154)$$

It is easy to observe from the left-hand equations of Eqs. (12.153) and (12.154) that when K_r is real, $\sigma_+ > L_l$ and $\sigma_+ > N$, and $\sigma_- < L_l$ and $\sigma_- < N$. It can also be observed from the right-hand side of Eqs. (12.153) and (12.154) that when the total wave number K is very large, the second term on the right-hand side of Eq. (12.153) can be neglected and the equation degenerates to the dispersion relationship of sound in fluids ($\sigma^2 = K^2 C_s^2$); and the second term in the denominator on the right-hand side of Eq. (12.154) can also be neglected, and this equation represents the internal wave of an incompressible fluid in a gravitational field. In stellar oscillation theories, oscillations represented by σ_+ are called p-mode oscillations and those represented by σ_- are called g-mode oscillations. The information on the acoustic wave are employed for p-mode oscillations while the information on the gravitational wave are employed for g-mode oscillations. Acoustic waves and gravitational waves are common phenomena on earth, and their properties have already been well understood. For example, an acoustic wave is the propagation of pressure perturbation in a fluid. The magnitudes and the directions of its phase velocity and group velocity are the same: the direction is along the propagation of the wave and the magnitude is the local sound speed C_s. A gravitational wave is caused by inhomogeneities of the gravitational field. Its frequency is only dependent on the direction of the wave vector and is independent of its magnitude. The wave with a wave vector parallel to the gravitational field does not exist. The frequency for the wave vector perpendicular to the gravitational field is the highest, and is the Brunt-Väisälä frequency N. The phase velocity of the gravitational wave is parallel to the wave vector and the group velocity is perpendicular to the wave vector.

We have to note that the acoustic wave and the gravitational wave are approximate behaviors when the wave number K of the stellar oscillation wave is very large. When K is not very large, the stellar oscillation is something between the acoustic wave and the gravitational wave. The restoring forces of stellar oscillations including pressure and buoyancy will produce oscillations of acoustic wave and gravitational wave, and a mixture of the two under certain conditions.

For radial oscillations, since $l = 0$, the horizontal wave number is zero, so the oscillation with frequency σ_- does not exist. When the radial wave number K_r is very large, we know from Eq. (12.153) that the radial oscillation is a family of radially propagating acoustic waves.

In the above discussions, we have explained the properties of stellar oscillation waves. We have also studied the conditions for an oscillation wave to propagate in a region. When $K_r^2 > 0$, the wave can propagate; when $K_r^2 < 0$, it cannot. When $\sigma > L_l$ and $\sigma > N$, it can only propagate in the p-mode; when $\sigma < L_l$ and $\sigma < N$, it can only propagate in the g-mode. When $L_l > \sigma > N$ or $N > \sigma > L_l$, the wave cannot propagate. Another interesting topic is about the propagation region of waves with different properties in the star. In order to illustrate this, we have to know the relationship among L_l^2, N^2 and r.

From Eqs. (12.146) and (12.148), we have

$$L_l^2 = \frac{l(l+1)C_s^2}{r^2} \tag{12.155}$$

from which we know that when $r \to 0$, $L_l^2 \to \infty$, and L_l^2 decreases as r increases. Furthermore, the curve of L_l^2 versus r is closely related to the l value of the oscillation mode. From Eq. (12.147), we have

$$
\begin{aligned}
N^2 &= \frac{-g}{r}\left(\frac{d\ln\rho}{d\ln r} - \frac{1}{\Gamma_{1e}}\frac{d\ln P}{d\ln r}\right) \\
&= -\frac{GM_r}{r^3}\frac{d\ln P}{d\ln r}\left(\frac{d\ln\rho}{d\ln P} - \frac{1}{\Gamma_{1e}}\right)
\end{aligned}
\tag{12.156}
$$

For the equilibrium state of the star, we obtain from the condition for hydrostatic equilibrium in Eq. (7.7) that

$$
\frac{d\ln P}{d\ln r} = -\frac{r}{P}g\rho = -\frac{GM_r\rho}{rP}
\tag{12.157}
$$

In addition, the total pressure P can be written as

$$
P = P_g + P_R = \frac{\mathfrak{R}}{\mu}\rho T + \frac{1}{3}aT^4
\tag{12.158}
$$

where $\mathfrak{R}$ is the gas constant and μ is the mean molecular weight.

Differentiating Eq. (12.158), we have

$$
Pd\ln P = \frac{\mathfrak{R}}{\mu}\rho T(d\ln\rho + d\ln T - d\ln\mu) + \frac{4}{3}aT^4 d\ln T
\tag{12.159}
$$

Defining

$$
\left.
\begin{aligned}
\nabla &\equiv \frac{d\ln T}{d\ln P} \\
\nabla_\mu &\equiv \frac{d\ln\mu}{d\ln P} \\
\beta &\equiv P_g/P
\end{aligned}
\right\}
\tag{12.160}
$$

we get from Eqs. (12.159) and (12.160)

$$
\frac{d\ln\rho}{d\ln P} = -\frac{4-3\beta}{\beta}\nabla + \nabla_\mu + \frac{1}{\beta}
\tag{12.161}
$$

Applying Eq. (12.161) to the adiabatic case, we obtain

$$
\frac{1}{\Gamma_{1e}} = -\frac{4-3\beta}{\beta}\nabla_{ad} + \nabla_{\mu ad} + \frac{1}{\beta}
\tag{12.162}
$$

where $\Gamma_{1e} = \left(\dfrac{\partial \ln P}{\partial \ln \rho}\right)_{ad}$, $\nabla_{ad} = \left(\dfrac{\partial \ln T}{\partial \ln P}\right)_{ad}$ and $\nabla_{\mu ad} = \left(\dfrac{\partial \ln \mu}{\partial \ln P}\right)_{ad}$ (12.163)

Substituting Eqs. (12.157), (12.161) and (12.162) into (12.156), we have

$$N^2 = \left(\frac{GM_r}{r^2}\right)^2 \frac{\rho}{P}\left[\nabla_\mu - \nabla_{\mu ad} + \frac{4 - 3\beta}{\beta}(\nabla_{ad} - \nabla)\right] \qquad (12.164)$$

When the matter in a star is completely ionized, $\nabla_{\mu ad}$ can be neglected so we have

$$N^2 = \left(\frac{GM_r}{r^2}\right)^2 \frac{\rho}{P}\left[\nabla_\mu + \frac{4 - 3\beta}{\beta}(\nabla_{ad} - \nabla)\right] \qquad (12.165)$$

From the above equation, we see that the variation between N^2 and r is mainly affected by the values of ∇_μ and $(\nabla_{ad} - \nabla)$.

After knowing the relationship among L_l^2, N^2 and r, we can plot a graph to show the different regions inside a star which allow the propagation of oscillation waves with different properties. Figure 12.2 shows the propagation regions of the internal oscillation wave for a main sequence star with a mass of $10M_\odot$ when $l = 2$ (Unno et al. 1979, p. 96). In the figure, the vertical axis is σ^2 and the horizontal axis is r/R. Both the cross hatched P region ($\sigma > L_l$ and $\sigma > N$) and the hatched G region ($\sigma < L_l$ and $\sigma < N$) fulfil $K_r^2 > 0$ and allow the propagation of oscillation waves; the P region allows propagation of the p-mode oscillation wave while the G region allows propagation of the g-mode oscillation wave. The regions outside the P region and the G region are dissipative regions of the oscillation wave with $K_r^2 > 0$. The L_l^2 curve in Fig. 12.2 is obtained for $l = 2$. If l increases, the L_l^2 curve will move towards the upper right direction. The core of a main sequence star with a mass of $10M_\odot$ has a hydrogen rich convective core which has a homogeneous chemical composition because of the convection, so there is no gradient for the abundance, i.e., $\nabla_\mu = 0$. At the same time, in the convective core, $(\nabla_{ad} - \nabla) \cong 10^{-6}$ to 10^{-7} which is an extremely small value. Therefore, N^2 is also extremely small in the convective core and can be neglected. From Fig. 12.2, it can be seen that N^2 is zero in the region $r < 0.24R$.

When this star evolves with time, the hydrogen in the convective core has gradually been burnt to form helium and at the same time the convective core contracts inwards continuously. Thus, a region with varying hydrogen abundance emerges at the outer border of the convective core, which is called the μ gradient region. The size of the

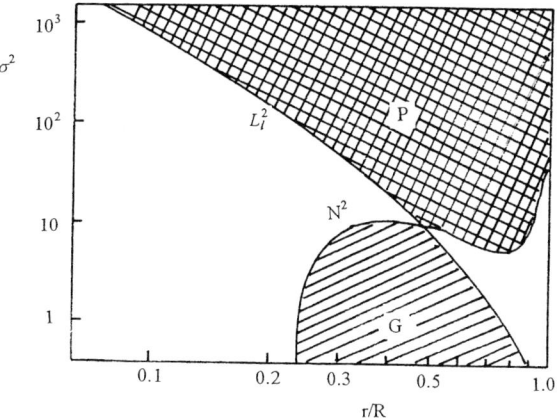

Fig. 12.2 The propagation regions of the internal oscillation wave for a main sequence star with a mass $10M_\odot$ when $l = 2$ (Unno *et al.* 1979).

μ gradient region changes with time. Inside the μ gradient region, ∇_μ has a larger value so N^2 has also a larger value. Figure 12.3 shows the propagation regions of the oscillation wave for a star with a mass of $10M_\odot$ at two different evolutionary stages. When comparing Figs. 12.3 and 12.2, we see that the N^2 curve has a significant change, so the G region has also a significant change. From the figure, we also learn that the μ gradient region inside the star is the main propagation region for g-mode oscillation, and p-mode oscillation propagates mainly in the envelope of the star. When l is very small, the L_l^2 and N^2 curves touch each other so the P region and the G region are connected. However, when l increases, the L_l^2 curve moves towards the upper right direction. When l is very large, the P region and the G region are separated. Therefore, for a star with its mass concentrated in the core and with a large value of l, the p-mode oscillation propagates mainly in the envelope of the star and the g-mode oscillation propagates mainly in the internal μ gradient region.

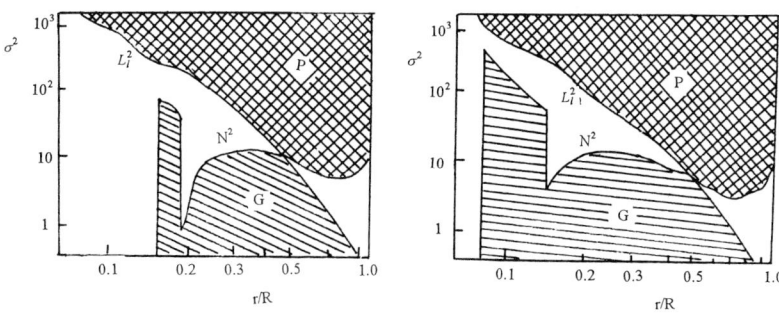

Fig. 12.3 The propagation regions of the oscillation wave for a star with a mass of $10M_\odot$ at two different evolutionary stages ($l = 2$) (Unno *et al.* 1979).

12.5 BOUNDARY CONDITIONS AND NUMERICAL METHODS

12.5.1 Boundary Conditions

To solve the non-adiabatic non-radial oscillation equations (12.113) to (12.119), it requires appropriate boundary conditions, i.e., conditions fulfilled by the equations at the center and at the surface of the star. These conditions are obtained from some physical requirements.

(A) Boundary conditions at the center

Near the center of the star, the adiabatic condition is satisfactorily fulfilled, and the gas temperature is almost exactly the same as the radiation temperature, so the oscillation should be an equilibrium adiabatic oscillation. In addition, all physical quantities at the center of the star should be finite and should not tend to infinity.

The procedures for choosing boundary conditions at the center are as follows. The oscillation equations are solved near the center of the star to get a general solution. A reasonable solution is then chosen according to the above conditions, which is then used for determination of the boundary conditions at the center.

Since the oscillation is under adiabatic equilibrium at the center, we can use the dimensionless variables to express the adiabatic relationship in Eq. (12.130) as

$$y_7 = \frac{A}{C}y_1 - V(y_2 + y_3) \qquad (12.166)$$

where
$$A = \frac{d \ln \rho}{d \ln r} - \frac{1}{\Gamma_{1e}} \frac{d \ln P}{d \ln r} \qquad (12.167)$$

$$V = -\frac{1}{\Gamma_{1e}} \frac{\rho g r}{P}, \qquad (12.168)$$

Γ_{1e} is the equivalent adiabatic compression coefficient and the dimensionless coefficient C has been defined in Eq. (12.120). On the right-hand side of Eq. (12.166), the first term represents the buoyancy and the second term represents the pressure restoring force.

When $r \to 0$, $\rho g r/P \to 0$ and $\frac{d \ln \rho}{d \ln r} \to 0$, so $A \to 0$ and $V \to 0$. Therefore, at the center of the star, $y_7 \to 0$, i.e., the density perturbation can be ignored. Besides, $U = -\frac{4\pi G \rho r}{g} = 3\rho/\bar{\rho}_r$ where $\bar{\rho}_r$

is the mean density of stellar matter with radius r. Since the local density ρ is similar to $\bar{\rho}_r$ around the center, $U \to 3$. Under these conditions, the equations (12.113) to (12.119) becomes:

$$r\frac{dy_1}{dr} = -3y_1 + \frac{l(l+1)}{\omega^2}(y_2 + y_3 + y_5) \qquad (12.169)$$

$$r\frac{dy_2}{dr} = \omega^2 y_1 - 2y_2 + y_4 + y_6 \qquad (12.170)$$

$$r\frac{dy_3}{dr} = -2y_3 - y_4 \qquad (12.171)$$

$$r\frac{dy_4}{dr} = -l(l+1)y_3 - 3y_4 \qquad (12.172)$$

$$r\frac{dy_5}{dr} = -2y_5 - y_6 \qquad (12.173)$$

and $$r\frac{dy_6}{dr} = -l(l+1)y_5 - 3y_6 \qquad (12.174)$$

It can be seen that Eqs. (12.173) and (12.174) are a set of independent constant-coefficient linear differential equations which can be solved directly. By setting $y_5 = Kr^n$, we get $y_6 = -(n+2)Kr^n$ from Eq. (12.173). Substituting y_5 and y_6 into Eq. (12.174), we get the characteristic equation

$$(n+2)(n+3) - l(l+1) = 0 \qquad (12.175)$$

which has two roots as

$$n_1 = l - 2, \quad n_2 = -l - 3 \qquad (12.176)$$

Since the physical quantities should have finite values at the center of the star, n_2 should be rejected; otherwise y_6 will become infinite at the center. Therefore, we have

and $$\left. \begin{array}{c} y_5 = Kr^{l-2} \\ y_6 = -lKr^{l-2} \end{array} \right\} \qquad (12.177)$$

Since K can take any arbitrary value, we get from Eq. (12.177) the relationship

$$ly_5 + y_6 = 0 \qquad (12.178)$$

which is one of the boundary conditions required at the center. It describes the change in the gravitational field so it is also called the gravitational boundary condition.

The coefficients in Eqs. (12.171) and (12.172) are exactly the same as those in Eqs. (12.173) and (12.174). Therefore, another boundary condition at the center can be obtained from Eqs. (12.171) and (12.172) as

$$ly_3 + y_4 = 0 \tag{12.179}$$

which describes the change in the radiation field and is thus called the radiative boundary condition.

Using Eqs. (12.172) and (12.174) to eliminate y_3 and y_5 in Eq. (12.169), and changing variables by

$$z_1 = -y_2, \; z_2 = \omega^2 y_1 + y_4 + y_6 \tag{12.180}$$

Eqs. (12.169) and (12.170) become

$$r\frac{dz_1}{dr} = -2z_1 - z_2 \tag{12.181}$$

and

$$r\frac{dz_2}{dr} = -l(l+1)z_1 - 3z_2 \tag{12.182}$$

This set of equations have coefficients exactly the same as those in the previous two solved sets of equations, so we obtain another relationship:

$$lz_1 + z_2 = 0 \tag{12.183}$$

Substituting Eq. (12.180) into the above equation, we have

$$\omega^2 y_1 - l(y_2 + y_3 + y_5) = 0 \tag{12.184}$$

which is a boundary condition describing the change in the velocity field of the fluid and is thus called the mechanical boundary condition.

It can be observed that the relationships between all the six variables and r near the center of the star have the form r^{l-2}, which makes it rather easy to obtain the boundary conditions at the center. This is one of the objectives we want to achieve when we introduced the dimensionless variables in §12.3.6.

(B) Boundary conditions at the surface

The outer boundary condition of the gravitational field is first considered. At the surface of the star, $\rho \to 0$ so $U \to 0$, and Eqs. (12.117) and (12.118) become

$$r\frac{dy_5}{dr} = y_5 - y_6 \tag{12.185}$$

and
$$r\frac{dy_6}{dr} = -l(l+1)y_5 \tag{12.186}$$

This is a set of constant-coefficient linear differential equations, which have solutions with the form r^j. We can also get the characteristic equation related to j as

$$j(j-1) - l(l+1) = 0 \tag{12.187}$$

which has two roots as $j_1 = -l$ and $j_2 = l + 1$. At an infinite distance ($r \to \infty$), the gravitational field should be zero, so j_2 should be rejected. The gravitational outer boundary condition thus obtained is

$$l(l+1)y_5 - y_6 = 0 \tag{12.188}$$

We proceed to consider the outer boundary condition of the radiation field. We choose a point at a sufficient depth from the surface of the star as the outer boundary point of the radiation field. As regards Eq. (12.116), since the non-adiabatic term is far greater than the adiabatic term near the surface of the star, the adiabatic term, i.e., the term with $i\omega$, in the equation can be neglected. Besides, the value of opacity κ is very small near the surface of the star so $F_2 \to 0$ and $F_7 \to 0$, and Eqs. (12.115) and (12.116) become

$$r\frac{dy_3}{dr} = (D+1)y_3 - y_4 \tag{12.189}$$

and
$$r\frac{dy_4}{dr} = -l(l+1)y_3 + Wy_4 \tag{12.190}$$

This is also a set of constant-coefficient linear differential equations which have solutions of the form r^m. The characteristic equation of m is

$$m^2 - (D+W+1)m - l(l+1) + W(D+1) = 0 \tag{12.191}$$

which has two solutions as

$$m_1 = \frac{1}{2}\left[W + D + 1 + \sqrt{(W - D - 1)^2 + 4l(l+1)}\right] \qquad (12.192)$$

and $$m_2 = \frac{1}{2}\left[W + D + 1 - \sqrt{(W - D - 1)^2 + 4l(l+1)}\right] \qquad (12.193)$$

Since there is no radiation source in the envelope of the star, the amplitude of the luminosity should not increase with the radius so m_1 should be rejected. Substituting m_2 into equations (12.189) and (12.190), we obtain the relationship

$$\left[W - D - 1 - \sqrt{(W - D - 1)^2 + 4l(l+1)}\right]y_3 + 2y_4 = 0 \qquad (12.194)$$

which is the radiative boundary condition at the outer boundary.

When the stellar atmosphere is homogeneous, the density ρ is a constant, so $W \to 0$ and $D \to 0$, $m_2 \to -l$ and Eq. (12.194) becomes

$$(l+1)y_3 - y_4 = 0 \qquad (12.195)$$

In such cases, it can be seen that the dependence of the radiation field on radius is exactly the same as that of the gravitational field.

We finally come to the mechanical boundary condition at the outer boundary. Since we are studying properties of the oscillation wave at the outer boundary, we can employ the adiabatic approximation and the Cowling approximation, and choose the outer boundary point at a sufficient depth below the surface of the star. Under the adiabatic approximation and the Cowling approximation, the oscillation wave is described by Eqs. (12.142) and (12.143). Using Eq. (12.144), Eq. (12.142) can be written as

$$iK_r\xi = \left[\frac{l(l+1)}{\sigma^2 r^2} - \frac{1}{\Gamma_{1e}}\frac{\rho}{P}\right]\frac{P'}{\rho} \qquad (12.196)$$

or the dimensionless form

$$iK_r r y_1 = \left[\frac{l(l+1)}{\omega^2} - VC\right](y_2 + y_3) \qquad (12.197)$$

Based on the different properties of the oscillation wave at the outer boundary, we can obtain the corresponding K_r value and substitute it into Eq. (12.197) to obtain the corresponding mechanical boundary condition at the surface.

Under the adiabatic approximation and the Cowling approximation, the dispersion relation of the oscillation wave can be obtained from Eq. (12.149) as

$$K_r^2 = \frac{(\sigma^2 - N^2)(\sigma^2 - L_l^2)}{\sigma^2 C_s^2} \tag{12.198}$$

When the oscillation frequency σ is higher than both the critical frequencies N and L_l, the oscillation wave is a p-mode acoustic wave which can transmit through the outer boundary and propagate to outer regions. When the oscillation frequency σ is lower than both the critical frequencies, the oscillation wave is a g-mode gravitational wave which can transmit through the outer boundary and propagate to outer regions. When the frequency is in between the two critical frequencies, the oscillation wave is dissipative at the outer boundary so the oscillation energy will not leak to regions outside the outer boundary or, in other words, the wave is reflected here.

We write the dispersion relation in Eq. (12.198) in another form:

$$K_r^2 r^2 = \left(\omega^2 + \frac{A}{C} \right) \left(VC - \frac{l(l+1)}{\omega^2} \right) \tag{12.199}$$

When the frequency σ is very high, the terms $\frac{A}{C}$ and $\frac{l(l+1)}{\omega^2}$ in Eq. (12.204) can be neglected, so we have

$$K_r r = \pm \omega \sqrt{VC} \tag{12.200}$$

At this time, the wave front is determined by the equation

$$\pm \omega \sqrt{VC} = \sigma t + \phi_0 \tag{12.201}$$

where ϕ_0 is the initial phase. Obviously, the positive sign refers to the wave with the wave front propagating parallel to the direction of the radius, and the negative sign refers to the wave with the wave front propagating opposite to the direction of the radius. When the frequency σ is very high, the oscillation is a p-mode acoustic wave. It is known that the directions of group velocity and phase velocity for an acoustic wave are the same, which is also the same as the direction of movement of the wave front. Therefore, the equation taking the positive sign in Eq. (12.201) represents an oscillation wave which can transmit from the inside of the star through the outer boundary to outer regions, so this equation should be accepted. The equation taking the negative sign represents an oscillation energy that can flow from the outside into the oscillation region, which is unphysical, so the equation should be rejected. We take the solution with a positive sign which represents the transfer of energy outwards and substitute it into Eq. (12.197) to obtain the mechanical boundary condition at the outer boundary for the oscillation wave with properties of the acoustic wave, which is

$$i\omega y_1 + \sqrt{VC}(y_2 + y_3) = 0 \qquad (12.202)$$

When the frequency is very low, we obtain from Eq. (12.199) that

$$K_r r = \pm \frac{1}{\omega} \sqrt{-\frac{A}{C} l(l+1)} \qquad (12.203)$$

At this time, the properties of the oscillation wave are similar to those of the g-mode gravitational wave. The projections of the group velocity and the phase velocity of the gravitational wave on the radial direction are opposite, so the equation adopting the positive sign in Eq. (12.203) represents the case where the wave energy transmits from regions outside the oscillation region, through the outer boundary and into the oscillation region, which is unphysical, so the equation should be rejected. On the other hand, the equation adopting the negative sign represents the case where the wave energy transmits through the outer boundary and leaks to the outside of the oscillation region so the solution should be accepted. When it is substituted into Eq. (12.197), the mechanical boundary condition at the outer boundary for an oscillation wave which has the properties of a gravitational wave can be obtained as

$$i\sqrt{-\frac{A}{C}} y_1 + \frac{\sqrt{l(l+1)}}{\omega}(y_2 + y_3) = 0 \qquad (12.204)$$

When the frequency σ is in-between the two critical frequencies, since $V >> 0$ and $A << 0$, we obtain from Eq. (12.199) that

$$K_r r = \pm i \sqrt{-AV} \qquad (12.205)$$

Obviously, the negative sign solution in Eq. (12.205) is unphysical because it makes the amplitude of the oscillation wave increase exponentially with radius, so it should be rejected. The positive sign solution should be accepted and is substituted into Eq. (12.197) to obtain the mechanical boundary condition as

$$\sqrt{-AV}y_1 - VC(y_2 + y_3) = 0 \qquad (12.206)$$

From the above discussions, we notice that the boundary conditions obtained at the center of the star are exact, while those obtained at the surface are not and might involve some approximations. In many cases, the dependence of stellar oscillation on boundary conditions is rather weak, and different boundary conditions almost give the same oscillation. Nevertheless, there are some special cases in which the effects of boundary conditions on the oscillation are not insignificant. In these cases, we have to choose the conditions very carefully; on one hand we have to choose the approximation to simplify the problem, and on the other hand we should not oversimplify the problem and overlook the factors we should have considered. In practice, we can compare the effects of different boundary conditions on the oscillation, and choose the most appropriate one.

The boundary conditions for radial oscillation problems can be obtained from the boundary conditions for the corresponding non-radial oscillation problems by setting $l = 0$. For example, Eq. (12.198) can be written as

$$K_r^2 = \frac{1}{C_s^2}(\sigma^2 - N^2) \qquad (12.207)$$

From this equation, we can see that when $\sigma > N$, the oscillation wave can propagate to the outside of the outer boundary; when $\sigma < N$, the oscillation wave cannot propagate near the outer boundary. It is known that the properties of the radial oscillation are similar to those of the radially propagating acoustic waves. Therefore, when $\sigma > N$, the solution for energy travelling outwards is

$$K_r = \frac{1}{C_s}\sqrt{\sigma^2 - N^2} \qquad (12.208)$$

and the corresponding mechanical boundary condition at the outer boundary is

$$i\sqrt{\omega^2 + \frac{A}{C}} \cdot y_1 + \sqrt{VC}(y_2 + y_3) = 0 \qquad (12.209)$$

and when $\sigma < N$, since the oscillation wave is dissipative at the outer boundary, its amplitude should not increase outwards, so we still choose the positive sign solution

$$K_r = i\frac{1}{C_s}\sqrt{N^2 - \sigma^2} \qquad (12.210)$$

and the corresponding mechanical boundary condition at the outer boundary is

$$\sqrt{-\left(\omega^2 + \frac{A}{C}\right)y_1 - \sqrt{VC}(y_2 + y_3)} = 0 \qquad (12.211)$$

12.5.2 Numerical Methods

The stellar oscillation equations (12.113) to (12.119) are linear harmonic ordinary differential equations, with the oscillation frequency σ as the parameter. For adiabatic problems, these equations are defined in the region of real numbers; for non-adiabatic problems, these equations are defined in the region of complex numbers. The derivative of these equations with respect to space is of order six. For these equations, there are three boundary conditions at the center, i.e., Eqs. (12.178), (12.179) and (12.184), and two boundary conditions near the surface, i.e., Eqs. (12.188), (12.195) and one equation chosen from Eqs. (12.206), (12.209) and (12.211) according to different σ values. It is well known that when the number of boundary conditions matches the order, the equation set becomes an eigenvalue problem. Therefore, the problem of stellar oscillation is an eigenvalue problem. Only when the oscillation frequency σ takes some special discrete values will the equations have a non-zero solution. From the physics point of view, the star is also equivalent to a big resonant cavity. Only standing waves can be expressed as oscillation waves, while non-standing waves will cancel each other due to superposition of reflected waves.

Since the coefficients of stellar oscillation equations are functions of radius, and these functions are also closely related to the initial mass of the star, the chemical composition and the stellar structure at different evolutionary stages, it is very difficult to get analytical solutions for the stellar oscillation equations and normally we can only use numerical methods.

The numerical method for solving stellar oscillation equations is very similar to the Henyey method for the model stellar structure (see section 7.4). The first step of this method is to divide the region for calculation into N discrete mesh points and to number these points. The corresponding variable on the nth mesh point is represented by y_i^n. We should note that the relative change of all variables on two neighboring mesh points should be small during the division of mesh points, so that the differential equations and the boundary conditions can be rewritten as difference equations. For example, the oscillation equations and the boundary conditions can be written in a gauge form:

$$\frac{dy_i}{dx} = H_{ij}y_j \qquad (12.212)$$

where the subscript takes the same value as the number of unknown variables. When the same subscript appears twice in the same expression, the expression represents summation of possible values of that subscript.

Between the $(n-1)$th and the nth mesh points, the oscillation equations can be written as difference equations

$$y_i^n - y_i^{n-1} = \frac{1}{2}(x^n - x^{n-1})(H_{ij}^n y_j^n + H_{ij}^{n-1} y_j^{n-1}) \qquad (12.213)$$

From the difference equations (12.213), a set of null functions can be defined by

$$F_i^n(y_j^n, y_j^{n-1}, \sigma) \equiv y_i^n - y_i^{n-1} - \frac{1}{2}(x^n - x^{n-1})(H_{ij}^n y_j^n + H_{ij}^{n-1} y_j^{n-1}) = 0$$
$$i, j = 1, 2, ..., 6; \quad n = 1, 2, ..., N$$

$$(12.214)$$

When solving the equations (12.213) using the Henyey method, we have to provide an initial solution, i.e., a set of y_{jo}^n, σ_0 ($j = 1, 2, ..., 6$; $n = 1, 2, ..., N$). Since the initial solution is not very precise, y_{jo}^n, σ_0 cannot fulfil Eq. (12.213). Therefore we have to find corrections δy_j^n, $\delta\sigma$ to the initial solution to make the modified solution fulfil Eq. (12.213) more satisfactorily, i.e., when we substitute $y_j^n = y_{jo}^n + \delta y_j^n$ and $\sigma = \sigma_0 + \delta\sigma$ into the function F_i^n, it will get closer to zero. In order to get equations for calculation of corrections, we can expand Eq. (12.214) by Taylor's expansion and take only the linear terms, so we have

$$\left. \begin{array}{c} F_{io}^n + \sum_j \frac{\partial F_i^n}{\partial y_j^n} \delta y_j^n + \sum_j \frac{\partial F_i^n}{\partial y_j^{n-1}} \delta y_j^{n-1} + \frac{\partial F_i^n}{\partial \sigma} \delta\sigma = 0 \\[2mm] i, j = 1, 2, ..., 6 \\[1mm] n = 1, 2, ..., N \end{array} \right\} \qquad (12.215)$$

where the values of the coefficients $\frac{\partial F_i^n}{\partial y_j^n}$, $\frac{\partial F_i^n}{\partial y_j^{n-1}}$ and $\frac{\partial F_i^n}{\partial \sigma}$ can be obtained by substituting the values of the initial solution.

It has to be pointed out that the above process of finding corrections to the initial solution to get the final solution is by iteration. In other words, the solution after each correction will be treated as the initial solution for the next iteration. The corrections are calculated repeatedly until the absolute value of the corrections tend to zero.

The employment of the above Henyey method for a numerical solution requires the provision of an initial solution, and the speed of convergence for the solution will be greatly determined by how well we have chosen the initial solution. Castor (1971) introduced a method for choosing the initial solution which is described as follows. If one boundary condition is canceled, it is no longer an eigenvalue problem. Therefore the oscillation equations can be solved under an arbitrarily chosen σ value. However, the solution y_i^n obtained in this way may not satisfy the canceled boundary condition at the same time. Only when the canceled boundary condition is also satisfied will the σ value and the corresponding y_i^n be the required initial solution.

Li (1990b) further simplified the Castor method. Since the canceled boundary condition should be fulfilled by the eigenfrequency, the former can be used to determine the latter. The procedures are as follows. Suppose the canceled boundary condition is

$$K_{ij}y_j = 0, \tag{12.216}$$

so we can define a function

$$f = f(\omega) = K_{ij}y_j \tag{12.217}$$

which is called the discrimination function and is a function of the frequency σ. The σ value which makes f zero is the eigenfrequency. The zero point of f is determined using Newton's iteration method. For an initial value of σ_K, we have the iteration equation

$$\sigma_{K+1} = \sigma_K - \frac{f(\sigma_K)}{\dot{f}(\sigma_K)} \tag{12.218}$$

At the Kth step, the oscillation frequency σ_K and the discrimination function $f(\sigma_K)$ can be known through solving the oscillation equations. However, we do not know the derivative of the discrimination function $\dot{f}(\sigma_K)$ with respect to frequency, which can be approximated using values of the Kth and the $(K+1)$th steps as

$$\dot{f}(\sigma_K) \approx \frac{f(\sigma_K) - f(\sigma_{K-1})}{\sigma_K - \sigma_{K-1}} \tag{12.219}$$

Therefore, after taking two random and close oscillation frequencies and calculating the corresponding values of the discrimination function, we can use Eqs. (12.218) and (12.219) for iteration until the difference in the oscillation frequencies obtained by two consecutive iterations is less than a required precision and at the same time the values of the discrimination function is less than the required value. In this way, we have obtained the solution of this oscillation mode.

12.6 EXCITATIVE MECHANISMS OF STELLAR OSCILLATIONS

Excitative mechanisms of stellar oscillations are due to some physical processes within the star which can convert energies of other forms in the star into oscillation energy. Inside the star, there exist factors in exciting the oscillation as well as factors in dissipating the energy of oscillation. If the effect of excitative factors is greater than that of dissipative factors, small perturbations inside the star can develop gradually to stellar oscillations with considerably large amplitudes. On the contrary, if the effect of dissipative factors is greater than that of excitative factors, the star will be stable.

In general, there are two problems in the study of excitative mechanisms of stellar oscillations. The first one is to know whether there are oscillation instabilities inside the star, i.e., how a small oscillation in a fluid element in the star will develop to a stellar oscillation with a large amplitude. The second one is to know the physical cause of oscillation instabilities, i.e., the excitative mechanism of the oscillation. These two problems will be discussed in the following sections.

12.6.1 Criteria for Oscillation Instabilities

If an oscillation mode is unstable, the amplitude will grow with time so the oscillation energy also grows with time. Therefore, we can base on the variation rate of oscillation energy to decide whether an oscillation mode is unstable. If the oscillation energy increases with time, the oscillation mode is unstable; if not, the oscillation mode is stable.

The variation of oscillation energy can be derived from the continuity equation (12.68) and the equation of motion (12.69). The dot product of Eq. (12.69) with the complex conjugate $\vec{v}^*$ of the velocity gives

$$\rho\vec{v}^* \cdot \frac{\partial \vec{v}}{\partial t} = -\vec{v}^* \cdot \nabla P' - \rho\vec{v}^* \cdot \nabla\Phi' + \frac{\rho'}{\rho}\vec{v}^* \cdot \nabla P$$

$$= -\nabla \cdot [(P' + \rho\Phi')\vec{v}^*] + P'\nabla \cdot \vec{v}^* + \Phi'\nabla \cdot (\rho\vec{v}^*) + \frac{\rho'}{\rho}\vec{v}^* \cdot \nabla P \quad (12.220)$$

Using the complex conjugate of the continuity equation (12.68), Eq. (12.220) can be further written as

$$\rho\vec{v}^* \cdot \frac{\partial \vec{v}}{\partial t} = -\nabla \cdot [(P' + \rho\Phi')\vec{v}^*] - P'\frac{\partial}{\partial t}\left(\frac{\delta\rho}{\rho}\right)^* - \Phi'\frac{\partial \rho'^*}{\partial t} + \frac{\rho'}{\rho}\vec{v}^* \cdot \nabla P$$

$$= -\nabla \cdot [(P' + \rho\Phi')\vec{v}^*] - (\delta P - \vec{\xi} \cdot \nabla P)\frac{\partial}{\partial t}\left(\frac{\delta\rho}{\rho}\right)^* - \Phi'\frac{\partial \rho'^*}{\partial t}$$

$$+ \left(\frac{\delta\rho}{\rho} - \frac{1}{\rho}\vec{\xi} \cdot \nabla\rho\right)\vec{v}^* \cdot \nabla P$$

$$= -\nabla \cdot [(P' + \rho\Phi')\vec{v}^*] - \delta P\frac{\partial}{\partial t}\left(\frac{\delta\rho}{\rho}\right)^* + i\sigma^*\left[(\vec{\xi} \cdot \nabla P)\left(\frac{\delta\rho}{\rho}\right)^* + (\vec{\xi}^* \cdot \nabla P)\left(\frac{\delta\rho}{\rho}\right)\right.$$

$$\left. - \frac{1}{\rho}(\vec{\xi} \cdot \nabla\rho)(\vec{\xi}^* \cdot \nabla P) - \Phi'\rho'^*\right] \quad (12.221)$$

The real part of Eq. (12.221) is

$$\sigma_I\rho\vec{v} \cdot \vec{v}^* = -\mathcal{R}e\left[\nabla \cdot (P' + \rho\phi')\vec{v}^* + \delta P\frac{\partial}{\partial t}\left(\frac{\delta\rho}{\rho}\right)^*\right] + \sigma_I\left[(\vec{\xi} \cdot \nabla P)\left(\frac{\delta\rho}{\rho}\right)^*\right.$$

$$\left. + (\vec{\xi}^* \cdot \nabla P)\left(\frac{\delta\rho}{\rho}\right) - \frac{1}{\rho}(\vec{\xi}^* \cdot \nabla\rho)(\vec{\xi}^* \cdot \nabla P) - \Phi'\rho'^*\right] \quad (12.222)$$

and the imaginary part of Eq. (12.221) is

$$-\sigma_R\rho\vec{v} \cdot \vec{v}^* = -\mathrm{Im}\left[\nabla \cdot (P' + \rho\phi')\vec{v}^* + \delta P\frac{\partial}{\partial t}\left(\frac{\delta\rho}{\rho}\right)^*\right] + \sigma_R\left[(\vec{\xi} \cdot \nabla P)\left(\frac{\delta\rho}{\rho}\right)^*\right.$$

$$\left. + (\vec{\xi}^* \cdot \nabla P)\left(\frac{\delta\rho}{\rho}\right) - \frac{1}{\rho}(\vec{\xi} \cdot \nabla\rho)(\vec{\xi}^* \cdot \nabla P) - \Phi'\rho'^*\right] \quad (12.223)$$

where $\mathcal{R}e$ and Im represent the real part and the imaginary part of a complex number, and σ_R and σ_I represent the real and imaginary parts of σ.

Multiplying Eq. (12.222) with σ_R, and Eq. (12.223) with σ_I, and then adding them together, we can obtain

$$
2\sigma_R\sigma_I\left[(\vec{\xi}\cdot\nabla P)\left(\frac{\delta\rho}{\rho}\right)^* + (\vec{\xi}^*\cdot\nabla P)\left(\frac{\delta\rho}{\rho}\right) - \frac{1}{\rho}(\vec{\xi}\cdot\nabla\rho)(\vec{\xi}^*\cdot\nabla P) - \Phi'\rho'^*\right]
$$

$$
= \sigma_R\mathcal{R}e\left[\nabla\cdot(P'+\rho\Phi')\vec{v}^* + \delta P\frac{\partial}{\partial t}\left(\frac{\delta\rho}{\rho}\right)^*\right] + \sigma_I\mathrm{Im}\left[\nabla\cdot(P'+\rho\Phi')\vec{v}^* + \delta P\frac{\partial}{\partial t}\left(\frac{\delta\rho}{\rho}\right)^*\right]
$$

$$
= 2\sigma_R\sigma_I\mathcal{R}e\left[\nabla\cdot(P'+\rho\Phi')\vec{\xi}^* + \delta P\left(\frac{\delta\rho}{\rho}\right)^*\right]
$$

$$
+ (\sigma_I^2 - \sigma_R^2)\mathrm{Im}\left[\nabla\cdot(P'+\rho\Phi')\vec{\xi}^* + \delta P\left(\frac{\delta\rho}{\rho}\right)^*\right] \tag{12.224}
$$

The complex conjugate equation of Eq. (12.221) is

$$
\rho\vec{v}\cdot\frac{\partial\vec{v}^*}{\partial t} = -\nabla\cdot(P'^*+\rho\Phi'^*)\vec{v} - \delta P^*\frac{\partial}{\partial t}\left(\frac{\delta\rho}{\rho}\right) - i\sigma\left[(\vec{\xi}\cdot\nabla P)\left(\frac{\delta\rho}{\rho}\right)^*\right.
$$

$$
\left. + (\vec{\xi}^*\cdot\nabla P)\frac{\delta\rho}{\rho} - \frac{1}{\rho}(\vec{\xi}\cdot\nabla\rho)(\vec{\xi}^*\cdot\nabla P) - \Phi'\rho'^*\right] \tag{12.225}
$$

Adding Eqs. (12.221) and (12.225) together, and then multiplying by σ_R, we have

$$
\sigma_R\frac{\partial}{\partial t}(\rho\vec{v}\cdot\vec{v}^*) = -\sigma_R\left[\nabla\cdot(P'+\rho\Phi')\vec{v}^* + \delta P\frac{\partial}{\partial t}\left(\frac{\delta\rho}{\rho}\right)^*\right] - \sigma_R[\nabla\cdot(P'^*+\rho\Phi')\vec{v}
$$

$$
+ \delta P^*\frac{\partial}{\partial t}(\frac{\delta\rho}{\rho})] + i\sigma_R(\sigma^*-\sigma)\left[(\vec{\xi}\cdot\nabla P)\left(\frac{\delta\rho}{\rho}\right)^* + (\vec{\xi}^*\cdot\nabla P)\frac{\delta\rho}{\rho}\right.
$$

$$
\left. -\frac{1}{\rho}(\vec{\xi}\cdot\nabla\rho)(\vec{\xi}^*\cdot\nabla P) - \Phi'\rho'^*\right]
$$

$$
= -2\sigma_R\mathcal{R}e\left[\nabla\cdot(P'+\rho\Phi')\vec{v}^* + \delta P\frac{\partial}{\partial t}\left(\frac{\delta\rho}{\rho}\right)^*\right] + 2\sigma_R\sigma_I\left[(\vec{\xi}\cdot\nabla P)\left(\frac{\delta\rho}{\rho}\right)^*\right.
$$

$$
\left. + (\vec{\xi}^*\cdot\nabla P)\frac{\delta\rho}{\rho} - \frac{1}{\rho}(\vec{\xi}\cdot\nabla\rho)(\vec{\xi}^*\cdot\nabla P) - \Phi'\rho'^*\right]
$$

Substituting Eq. (12.224) into the earlier equation, we have

$$
\sigma_R\frac{\partial}{\partial t}(\rho\vec{v}\cdot\vec{v}^*) = \sigma\sigma^*\mathrm{Im}\left[\nabla\cdot(P'+\rho\Phi')\vec{\xi}^* + \delta P\left(\frac{\delta\rho}{\rho}\right)^*\right] \tag{12.226}
$$

which is the equation describing the variation rate of oscillation energy. The left-hand side of the equation gives the variation rate of oscillation energy with time, while the right-hand side describes the cause of such a variation. The divergence term on the right-hand side of Eq. (12.226) describes the mechanical energy flowing into or out of a fluid element. If no oscillation energy flows from the outside through the surface of the star into the oscillation region, this term only describes the oscillation energy flow between different regions inside the star and is independent of the increase or decrease of the total oscillation energy, and will vanish when integration is performed over the entire volume of the star. The last term on the right-hand side of Eq. (12.226) represents the work done on the fluid element by the environment which increases the oscillation energy of the fluid element, so this is the term that represents the effects of different excitative mechanisms. When we eliminate the divergence term, we obtain an equation which describes the factors for excitation or dissipation of the oscillation as

$$\frac{\partial}{\partial t}\left(\frac{1}{2}\rho\vec{v}^2\right) = \frac{|\sigma|^2}{2\sigma_R}\,\mathrm{Im}\left[\delta P\left(\frac{\delta\rho}{\rho}\right)^*\right] \tag{12.227}$$

where the bracketed quantity on the left is the kinetic energy of oscillation within a unit volume of the fluid element and the right-hand side is the work done on the fluid element from the environment. From Eq. (12.227), we see that when the pressure perturbation and the density perturbation are in phase or out of phase, the work done by the environment on the fluid element is zero so the oscillation kinetic energy cannot change with time. When the density perturbation $\delta\rho$ leads the pressure perturbation δP by a phase from 0 to π, the environment has positive work done on the fluid element to increase its oscillation kinetic energy; when the density perturbation $\delta\rho$ lags behind the pressure perturbation δP by a phase from 0 to π, the environment has negative work done on the fluid element to dissipate its oscillation energy.

From the gas thermodynamics equation (12.17) and the radiative thermodynamics equation (12.48), we obtain

$$\delta P_g = (\Gamma_3 - 1)\rho T_g\delta S_g + \Gamma_1 P_g\frac{\delta\rho}{\rho} \tag{12.228}$$

and
$$\delta P_R = \frac{1}{3}\rho T_R\delta S_R + \frac{4}{3}P_R\frac{\delta\rho}{\rho} \tag{12.229}$$

Substituting Eqs. (12.228) and (12.229) into (12.227), we have

$$\frac{\partial}{\partial t}\left(\frac{1}{2}\rho v^2\right) = \frac{|\sigma|^2}{2\sigma_R}\mathrm{Im}\left[(\Gamma_3 - 1)\rho T_g \delta S_g \left(\frac{\delta\rho}{\rho}\right)^* + \frac{1}{3}\rho T_R \delta S_R \left(\frac{\delta\rho}{\rho}\right)^*\right]$$

$$= \frac{|\sigma|^2}{2\sigma_R}\mathrm{Im}\left[(\Gamma_3 - 1)\delta Q_g \left(\frac{\delta\rho}{\rho}\right)^* + \frac{1}{3}\delta Q_R \left(\frac{\delta\rho}{\rho}\right)^*\right] \qquad (12.230)$$

where δQ_g is the heat absorbed by a unit volume of gas, and δQ_R is the heat absorbed by a unit volume of radiation. This equation describes the effects on the oscillation when the fluid element absorbs or releases heat. We define an equivalent adiabatic expansion coefficient $(\Gamma_3 - 1)_e$ as

$$(\Gamma_3 - 1)_e = \frac{(\Gamma_3 - 1)\delta Q_g + \frac{1}{3}\delta Q_R}{\delta Q} \qquad (12.231)$$

where $\delta Q = \delta Q_g + \delta Q_R$ is the total heat energy absorbed by a unit volume of the fluid element. In this way, Eq. (12.230) can also be written as

$$\frac{\partial}{\partial t}\left(\frac{1}{2}\rho v^2\right) = \frac{|\sigma|^2}{2\sigma_R}\mathrm{Im}\left[(\Gamma_3 - 1)_e \delta Q \left(\frac{\delta\rho}{\rho}\right)^*\right] \qquad (12.232)$$

From Eq. (12.232), we see that if we use the heat absorption δQ as a reference phase, when the density perturbation $\delta\rho$ leads the reference phase by a phase of 0 to π, part of the absorbed heat will be changed into oscillation kinetic energy; and when the density perturbation $\delta\rho$ lags behind the reference phase by a phase of 0 to π, the oscillation kinetic energy will be dissipated as heat energy. The above phenomena can be illustrated by the S–ρ plot shown in Fig. 12.4. Here, the fluid element is treated as a heat engine and the oscillation can be treated as a cyclic process of the heat engine. On the S–ρ plot, if the fluid element absorbs

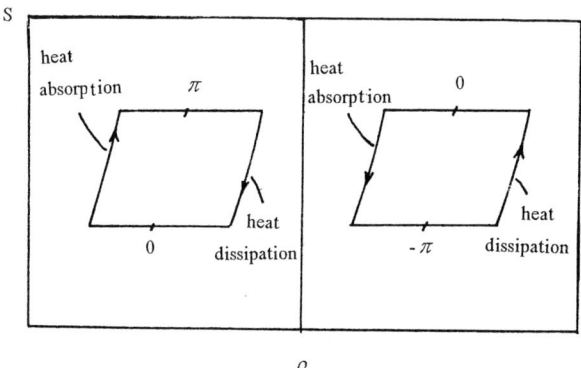

Fig. 12.4 The variation on the S–ρ plot of the state of the fluid element during oscillation. The left plot represents the case with net excitative factors while the right plot represents the case with net dissipative factors.

heat energy when it contracts and its density increases (i.e., the case for the density perturbation $\delta\rho$ to lead the reference phase by a phase o to π), and if the fluid element releases heat energy when it expands and its density decreases, the cyclic process is clockwise, i.e., equivalent to the one shown on the left of Fig. 12.4, and part of the heat energy will be converted to oscillation kinetic energy. Under such circumstances, the oscillation is unstable or there are excitative factors for the oscillations. On the contrary, if the fluid element absorbs energy when it expands and its density decreases (i.e., the case for the density perturbation to lag behind the reference phase by a phase o to π), and if the fluid element releases heat energy when it contracts and its density increases, the cyclic process is anti-clockwise, i.e., equivalent to the one shown on the right of Fig. 12.4, and the oscillation energy will be dissipated as heat energy. From Eq. (12.231), we see that the ratio of the part converted to oscillation kinetic energy to the total heat energy absorbed by the fluid element is $(\Gamma_3 - 1)_e$.

It should be noted that Eqs. (12.227) and (12.232) are strictly valid only when the integration is performed over the entire stellar volume, since the effects from the redistribution of oscillation energy inside the star will then cancel each other. We define a work integral W_k as

$$W_k = \frac{\partial}{\partial t} \int \frac{1}{2} \rho \vec{v} \cdot \vec{v}^* dV = \frac{|\sigma|^2}{2\sigma_R} \operatorname{Im} \int \delta P \left(\frac{\delta\rho}{\rho}\right)^* dV \qquad (12.233)$$

From the above discussions, we know that when $W_k > 0$, the oscillation mode is unstable and there are factors for excitative oscillations; when $W_k < 0$, the oscillation mode is stable and the oscillation energy will be dissipated. The integration in Eq. (12.233) is performed over the entire volume of the star.

The total kinetic energy of the oscillation K_E is

$$K_E = \int \frac{1}{2} \rho \vec{v} \cdot \vec{v}^* dV \qquad (12.234)$$

and the relationship between K_E and W_k is

$$W_k = 2\sigma_I K_E \qquad (12.235)$$

We further define an oscillation instability coefficient η as

$$\eta = \frac{\sigma_I}{\sigma_R} = \frac{W_k}{2\sigma_R K_E} \qquad (12.236)$$

Through solving the oscillation equations (i.e., solving the eigenvalue problem), we can directly calculate σ_R and σ_I to get η. On the other hand, we can also calculate η through determining the work integral W_k and the total oscillation kinetic energy K_E. The comparison between these two values of η can be used as a test for reliability and accuracy for the calculations. On the whole, we can calculate the work integral W_k and the instability coefficient η to determine whether an oscillation mode is unstable.

12.6.2 Some Excitative Mechanisms of Stellar Oscillations

The study on excitative mechanisms of stellar oscillation is equivalent to the determination of physical processes which can convert energies of other forms inside the star into oscillation kinetic energy. This needs further analysis of Eq. (12.232). The terms on the left of Eqs. (12.71) and (12.72) of the linear oscillation equations represent the heat change δQ_g and δQ_R of the gas and the radiation inside the fluid element. Substituting these equations into Eq. (12.232), we can separate excitative and dissipative factors of the oscillation to reveal the various excitative and dissipative mechanisms. However, the equations are relatively complicated, so it is difficult to observe their physical meanings quickly. For simplicity, we employ the quasi-adiabatic approximation and use the quasi-adiabatic solutions in these equations for an easier understanding of various excitative mechanisms.

Substituting Eqs. (12.71) and (12.72) into (12.232), and using the quasi-adiabatic approximation, we have

$$
\frac{\partial}{\partial t}\left(\frac{1}{2}\rho v^2\right) = \frac{\Gamma_3 - 1}{2}[(\rho\varepsilon - \nabla \cdot \vec{F}_{con})' - ac\rho\kappa^T(T_g^4 - T_R^4)\left(\frac{\rho'}{\rho} + \frac{\kappa^{T'}}{\kappa^T}\right)
$$
$$
- 4ac\rho\kappa^T(T_g^3 T_g' - T_R^3 T_R')]\left(\frac{\delta\rho}{\rho}\right)^* + \frac{1}{6}[-\nabla \cdot \vec{F}_R'
$$
$$
+ ac\rho\kappa^T(T_g^4 - T_R^4)\left(\frac{\rho'}{\rho} + \frac{\kappa^{T'}}{\kappa^T}\right) + 4ac\rho\kappa^T(T_g^3 T_g' - T_R^3 T_R')]\left(\frac{\delta\rho}{\rho}\right)^*
$$

$$(12.237)$$

It can be seen that the right-hand side of Eq. (12.237) consists of two parts, i.e.

$$
W_g = \frac{\Gamma_3 - 1}{2}[(\rho\varepsilon - \nabla \cdot \vec{F}_{con})' - ac\rho\kappa^T(T_g^4 - T_R^4)\left(\frac{\rho'}{\rho} + \frac{\kappa^{T'}}{\kappa^T}\right)
$$
$$
- 4ac\rho\kappa^T(T_g^3 T_g' - T_R^3 T_R')]\left(\frac{\delta\rho}{\rho}\right)^*
$$

$$(12.238)$$

and

$$W_R = \frac{1}{6}[-\nabla \cdot \vec{F}_R' + ac\rho\kappa^T(T_g^4 - T_R^4)\left(\frac{\rho'}{\rho} + \frac{\kappa^{T'}}{\kappa^T}\right)$$
$$+ 4ac\rho\kappa^T(T_g^3 T_g' - T_R^3 T_R')]\left(\frac{\delta\rho}{\rho}\right)^* \qquad (12.239)$$

The following is the analysis of the possibility of various excitative mechanisms inside the star:

(A) ε mechanism

We first study the excitative mechanism represented by the first term on the right-hand side of Eq. (12.237), which is

$$W_\varepsilon = \frac{\Gamma_3 - 1}{2}(\rho\varepsilon)'\left(\frac{\delta\rho}{\rho}\right)^* \qquad (12.240)$$

From this equation, we study the case when an oscillation occurs within a region having thermonuclear reactions. When a fluid element contracts, its temperature and density will increase so the nuclear generation rate ε will also increase to release more heat, which is equivalent to that the fluid element absorbs heat when it contracts. On the contrary, when the fluid element expands, its temperature and density will decrease so the nuclear generation rate ε will also decrease, which is equivalent to that the fluid element releases heat when it expands. According to the above discussions of Eq. (12.232), we know that if the fluid element absorbs heat when it contracts and its density increases, or if the fluid element releases heat when it expands and its density decreases, these are the factors for excitative oscillations, which are called the ε mechanism in the present case.

Since thermonuclear reactions normally occur in deep regions with very high temperatures inside the star, the ε mechanism is normally effective in deep regions in the star which have oscillation modes with relatively large oscillation amplitudes, e.g., the g-mode non-radial oscillations. Another condition for the mechanism to be effective is that the timescale of thermonuclear reactions and the period of stellar oscillation should have the same order of magnitude. If the nuclear reaction timescale of an element is much longer than the oscillation period, the abundance of that element can be treated as a constant within an oscillation period. If the nuclear reaction timescale of the element is much shorter than the oscillation period, the abundance of the element remains at the equilibrium abundance within an oscillation period. Here, we have used the variation of the abundance of elements to represent the variation of the nuclear energy generation rate.

(B) Temperature freezing mechanism

The excitative mechanism represented by the fourth term on the right-hand side of Eq. (12.237) can be written as

$$W_t = \frac{\Gamma_3 - 1}{2} 4ac\rho\kappa^T (T_R^3 T_R' - T_g^3 T_g') \left(\frac{\delta\rho}{\rho}\right)^* \tag{12.241}$$

Obviously, this term illustrates that when the oscillation causes a difference between the temperature of the gas and the radiation, the gas can absorb or emit radiation, and can excite or dissipate the oscillation. In order to show clearly its physical meaning, we can assume that the temperatures of the gas and radiation of the unperturbed state are the same. Furthermore, the double adiabatic approximation is introduced, (i.e., assuming that both the gas and the radiation are adiabatic). In this way, we have

$$\frac{\delta T_g}{T_g} = (\Gamma_3 - 1)\frac{\delta\rho}{\rho} \tag{12.242}$$

and

$$\frac{\delta T_R}{T_R} = \frac{1}{3}\frac{\delta\rho}{\rho} \tag{12.243}$$

Therefore, Eq. (12.241) can be written as

$$\begin{aligned} W_t &= \frac{\Gamma_3 - 1}{2} 4ac\rho\kappa^T T^4 \left(\frac{\delta T_R}{T_R} - \frac{\delta T_g}{T_g}\right) \left(\frac{\delta\rho}{\rho}\right)^* \\ &= \frac{\Gamma_3 - 1}{2} 4ac\rho\kappa^T T^4 \left[\frac{1}{3} - (\Gamma_3 - 1)\right] \left|\frac{\delta\rho}{\rho}\right|^2 \end{aligned} \tag{12.244}$$

From Eq. (12.244), it is easy to observe that the adiabatic expansion coefficient of the radiation is the constant 1/3, and that the adiabatic expansion coefficient of the gas, $(\Gamma_3 - 1)$, is a function of the temperature and density of the gas. When $(\Gamma_3 - 1) > \frac{1}{3}$, for an increasing density $(\delta\rho > 0)$, the gas temperature increases more quickly than the radiation temperature so the former will be higher than the latter, and the fluid element releases heat when it contracts; on the contrary, for a decreasing density ($\delta\rho < 0$), the gas temperature decreases more quickly than the radiation temperature so the latter will be higher than the former, and the fluid element absorbs heat when it expands. To sum up, this mechanism has a dissipative effect on the oscillation. On the other hand, when ($\Gamma_3 - 1$) $< \frac{1}{3}$, the above conclusions will be reversed. At this time, since the adiabatic expansion coefficient of the gas is relatively small, the change in the gas temperature caused by the change in the density is suppressed. When the fluid element contracts, the radiation temperature is higher than the gas temperature and the gas absorbs heat; when the fluid element expands, the gas temperature is higher than the radiation temperature and the gas releases heat. Therefore, this will excite the oscillation. It is the small adiabatic expansion coefficient of the gas for this mechanism that "freezes" the temperature so that the gas temperature does not rise when the fluid element contracts and leads to an excitation. That is why this mechanism is called the temperature freezing mechanism (see Li 1992c).

The ionization of gaseous matter inside the star is the main reason for the small gas adiabatic expansion coefficient, because ionization increases the degree of freedom of matter thus increasing the heat capacity of matter. When there is external work done, the fluid element does not convert all the work done to internal energy of thermal motion, but converts part of it as ionization energy to be stored within the fluid element. If the ionization of an element is strong enough in a certain region of the star to fulfil the criterion ($\Gamma_3 - 1$) $< \frac{1}{3}$, the temperature freezing mechanism will excite oscillations in this region and will dissipate oscillation in other regions.

It should be pointed out that most of the known excitative mechanisms for variable stars are the temperature freezing mechanism. For example, the oscillation of Cepheid variables (see Baker and Kippenhahn 1962, 1965; Cox 1963), the RR Lyrae variables (see Baker 1966; Iben 1971) and the δ-Scuti variables (see Chevalier 1971; Stellingwerf 1979) are the results of combined ionization effects of hydrogen and helium in the envelope of the stars. The oscillations of DAV type stars (or called the ZZ Ceti variables) of white dwarf variables (see Dziembowski and Koester 1981; Starrfield et al. 1983) are excited by ionization effects of hydrogen; DBV type stars by ionization effects of helium; and DOV type stars by ionization effects of carbon and oxygen. The excitative mechanisms in these stars are often called the κ-mechanism, because when the fluid element contracts and its density increases, the ionization of some elements will be enhanced and the opacity κ of matter will increase so that the gas absorbs radiation energy. If the opacity of matter is represented by the Kramers equation

$$\kappa = \kappa_0 \rho T^{-3.5} \tag{12.245}$$

it is easy to notice that when the fluid element contracts so that the density increases, the temperature will also increase and the κ value will decrease. Only when the temperature is "frozen" by ionization effects will the κ value increase with the density. Therefore, the main physical process of the κ-mechanism is the temperature "freezing" caused by ionization.

(C) Modulation mechanism

The excitative mechanism represented by the third term on the right-hand side of Eq. (12.237) can be written as

$$W_m = -\frac{\Gamma_3 - 1}{2} ac\rho\kappa^T (T_g^4 - T_R^4) \left(\frac{\rho'}{\rho} + \frac{\kappa^{T'}}{\kappa^T} \right) \left(\frac{\delta\rho}{\rho} \right)^* \tag{12.246}$$

The factor $ac\rho\kappa^T (T_g^4 - T_R^4)$ on the right-hand side of this equation shows the difference between the gas temperature and the radiation temperature when there is no oscillation, which causes the gas to absorb or emit radiation. If the gas temperature is higher than the radiation temperature, the gas will emit radiation with a quantity proportional to the absorption coefficient κ. Therefore, if the absorption coefficient κ varies with time when the star oscillates, the process of radiation emission from the gas of the unperturbed state can be modulated, and that is why the mechanism is called the modulation mechanism. Whether the modulation mechanism has excitative or dissipative effects depends on two factors: the first is whether the gas temperature is higher than or lower than the radiation temperature; and the second is whether the variation in the absorption coefficient κ and the variation in the density $\delta\rho$ are in phase or out of phase during the oscillation (i.e., whether there is an ionization process). If the gas temperature is higher than the radiation temperature in the unperturbed state, and at the same time $\kappa^{T'}$ and $\delta\rho$ are out of phase during the oscillation, the density will reach a maximum when the absorption coefficient reaches a minimum, which means that the fluid element will emit less energy when it contracts. On the contrary, the absorption coefficient will reach a maximum when the density reaches a minimum, which means that the fluid element will emit excess energy when it expands, or that the fluid element releases energy when it expands. To sum up, the modulation here will have excitative effects on the oscillation. A similar analysis will show that if the gas temperature is higher than the radiation temperature in the unperturbed state, and $\kappa^{T\prime}$ and $\delta\rho$ are in phase during the oscillation, the modulation will have dissipative effects on the oscillation. Similarly, if the gas temperature is lower than the radiation temperature in the unperturbed state, the condition that $\kappa^\prime$ and $\delta\rho$ are out of phase during the oscillation will have dissipative effects, and that the condition that $\kappa^\prime$ and $\delta\rho$ are in phase will have excitative effects.

It is necessary to point out that the emergence of a difference between the gas temperature and the radiation temperature in a stable star is very rare when there is no oscillation, so the effect of the modulation mechanism is minimal. However, when there is a fast changing process in the star, it can cause a difference between the gas temperature and the radiation temperature in local regions. If the timescale of this fast process is greater than the timescale of the oscillation, the modulation mechanism will be effective. For example, if an unstable process such as a thermal pulse occurs in a region when there are thermonuclear reactions, i.e., thermonuclear reactions of an element are extraordinarily violent within a relatively short time to release a large amount of energy which cannot be transported away within the short time, the gas temperature can be higher than the radiation temperature in the unperturbed state. To cite another example, a difference in the two temperatures can also be caused by a fast expansion or a fast contraction of a region in the star; a region with a fast expansion will lead to a gas temperature lower than the radiation temperature while a region with a fast contraction will lead to a gas temperature higher than the radiation temperature. From these, we see that when there are violent processes inside the star at some evolutionary stages, the modulation mechanism can have excitative or dissipative effects on the oscillation.

(D) Convective coupling mechanism

The excitative mechanism represented by the second term on the right-hand side of Eq. (12.237) can be written as

$$W_c = -\frac{\Gamma_3 - 1}{2} \nabla \cdot \vec{F}'_{con} \left(\frac{\delta\rho}{\rho}\right)^* \tag{12.247}$$

which represents the mutual interaction of the convective motion and the oscillation. A deeper understanding of this mechanism involves the time dependent convection theory and the knowledge of mutual interaction between convective motion and oscillation, which are rather lacking at present.

Finally, we have to point out that we have only discussed the possibility of the first four terms on the right-hand side of Eq. (12.237) in forming excitative or dissipative mechanisms, i.e., we have only discussed the terms on the right-hand side of Eq. (12.238), and have not discussed the effects of the 5th term to the 8th term on the right-hand side of Eq. (12.237), i.e., the terms on the right-hand side of Eq. (12.239). The reason is as follows. From Eq. (12.237), an order-of-magnitude estimation shows that

$$\frac{W_R}{W_g} \approx \frac{P_R}{P_g} \tag{12.248}$$

where W_g represents the 1st to the 4th terms on the right-hand side of Eq. (12.237) and W_R represents the 5th to the 8th terms on the right-hand side of Eq. (12.237) (see Eqs. (12.238) and (12.239)). In general, the radiation pressure is much smaller than the gas pressure inside a star, so the effects of W_R are much smaller than those of W_g and we only analyze the first four terms on the right-hand side of Eq. (12.237).

12.7 CEPHEID VARIABLES

Cepheid variables refer to two types of variables. The first type is called classical Cepheid variables which belong to star population I, while the second type is called W Virginis variables which belong to star population II. They are distributed in a narrow strip on the H-R diagram, which is called the Cepheid pulsation strip (Fig. 1.3). Classical Cepheid variables are luminous periodically pulsating variables, and a typical one is the δ Cephei. Figure 12.5 gives the changes in the visual magnitude, effective temperature, radius and radial velocity of δ Cephei in a period. From the figure, we see that the change in the visual magnitude is 0.86 within a period, which means that the maximum brightness is twice the minimum brightness. The effective temperature corresponding to the maximum brightness is 6600 K while that corresponding to the minimum brightness is 5800 K, which means that the spectral type can change from F5Ib to G1Ib. Within a period, the change in the radial velocity is about 40 km/s, and the difference between the maximum and minimum values of the radius is about 10%

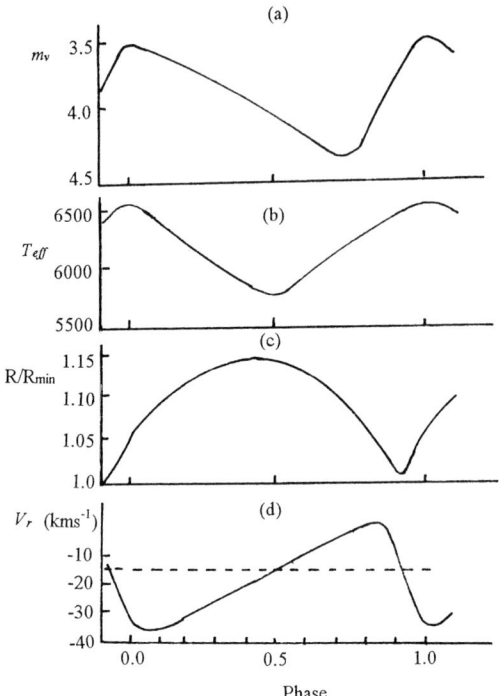

Fig. 12.5 The change in the visual magnitude, effective temperature, radius and radial velocity of δ Cephei in a period.

(i.e., from $26R_\odot$ to $23R_\odot$). It is worth noting that there is a phase difference between the changes in radius and brightness, i.e., the largest brightness will not occur at the smallest radius, but at a time when the radius has left the minimum value and is increasing. Furthermore, the variation curves in Fig. 12.5 are not sine curves, but have precise periods. Table 12.1 displays some characteristics of typical Cepheid variables in the galaxy given by Cox (1974).

An important property of Cepheid variables is the close relationship between the period and the luminosity, which is known as the period-

Table 12.1 Characteristics of typical Cepheid variables in the galaxy.

Properties	Range of Variation
Period P (days)	1–50
Mean luminosity L ($L_\odot$)	300–26000
Mean spectral type S_p	F5–G5
Mean radius R ($R_\odot$)	14–200
Mass M ($M_\odot$)	3.7–14

luminosity relationship. Another important property is the close relationship between the period and the radius, which is known as the period-radius relationship. We can use these relationships to determine the luminosities of variables from observed oscillation periods, and then determine the distance of variables from us. Since the observed period-luminosity relationships are the same for Cepheid variables in different star clusters and galaxies, the relationship is also used as a tool to determine the cosmological distance. Once a Cepheid variable is found in a stellar cluster or a galaxy, we can use the observed period and the period-luminosity relation to find out the distance of that stellar cluster or galaxy.

The period-radius relation observed by Gieren *et al.* (1989) can be written as

$$\log(R/R_\odot) = 1.108 + 0.743 \log P \tag{12.249}$$

where the period P is in units of days. Based on observational data of 98 Cepheid variables, Fernie (1990 a,b) obtained the relation between the color index $(B-V)$ and the period as

$$(B - V) = 0.311 + 0.438 \log P \tag{12.250}$$

According to the relation between the color index and the effective temperature (Flower 1977; Iben and Tuggle 1975) of

$$\log T_{eff} = 3.869 - 0.175(B - V) \tag{12.251}$$

and the relations among the luminosity, radius and the effective temperature

$$L = 4\pi R^2 \sigma T_{eff}^4, \tag{12.252}$$

the period-luminosity relation is established as

$$\log(L/L_\odot) = 2.430 + 1.179 \log P \tag{12.253}$$

Backer and Kippenhahn (1965) were the first to apply the linear non-adiabatic radial oscillation theory to classical Cepheid variables. They have studied a model for a star with a mass $5M_\odot$. All evolutionary models of this star starting from the main sequence are treated as equilibrium structure models, and the non-adiabatic radial oscillation equations are solved. At the same time, the work integrals are evaluated, and whether the evolution model will have radial oscillation depends on whether the value of the work integral is greater than zero. According to the calculations, this star does not have self-excitative radial oscillation in most regions on the H–R diagram. Only when the star evolves into a specific strip on the H–R diagram will the radial oscillation mode with fundamental frequency and first harmonic become unstable to cause oscillations in the star which then becomes a pulsating variable.

Figure 12.6 shows the relationship between the dimensionless work integral $w = W/W_P$ (W_P is the work integral in the envelope) and $\log P_0$ (P_0 is the pressure of the equilibrium model) for the two self-excitative oscillations for the $5M_\odot$ star model given by them. From the figure, we can see that the region in the star with the largest self-excitative radial oscillation is the HeII ionization region close to the surface, but the HI and HeI ionization regions also have significant contributions to the self-excitative oscillation. They referred to the mechanism for excitation of oscillations as the κ mechanism. Based on the method of Baker and Kippenhahn (1965), Hofmeister (1967) performed similar calculations for

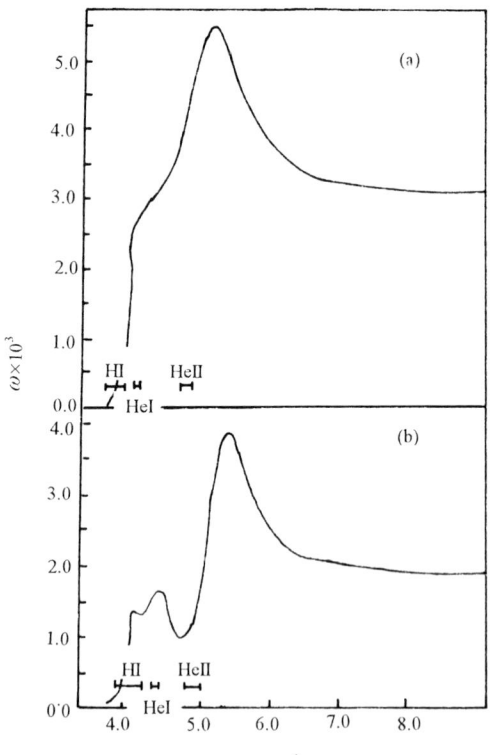

Fig. 12.6 The work integral curves of the two self-excitative models. Model a: $5M_\odot$, $X = 0.602$, $Z = 0.044$, $\alpha = 1.5$; model b: $5M_\odot$, $X = 0.602$, $Z = 0.044$, $\alpha = 2.3$. $w = W/W_p$ (Baker and Kippenhahn 1965).

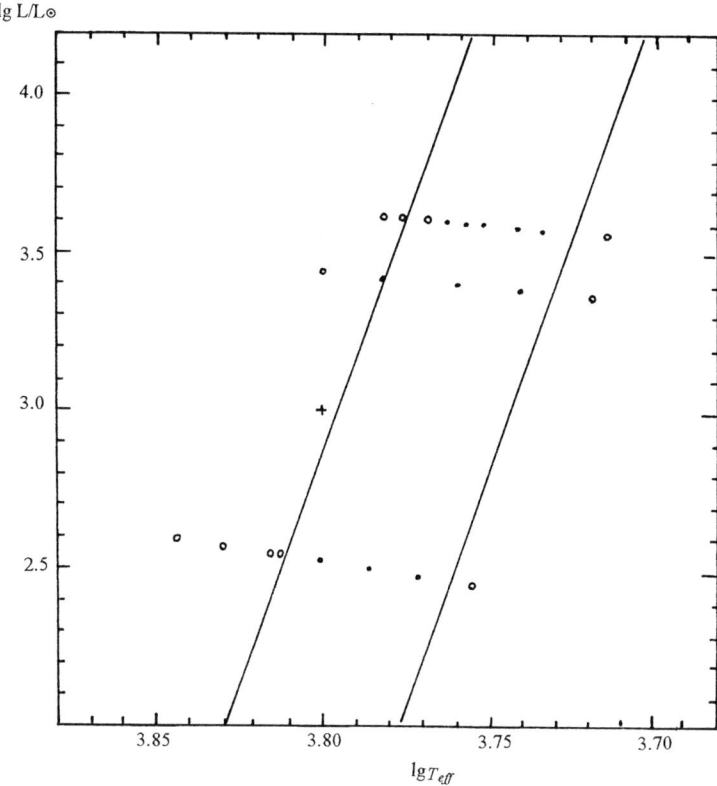

Fig. 12.7 The position of the Cepheid pulsation strip on the H–R diagram. The outlined circles represent data where the oscillation mode is stable; the filled circles represent the data where the oscillation mode is unstable (Li 1992a).

5–$9M_\odot$ stellar models, and found that all of them had oscillation instabilities in the same strip on the H–R diagram, which coincided with the position of the Cepheid pulsation strip. In this way, it became certain that pulsations of Cepheid variables could be explained using the linear non-adiabatic radial oscillation theory. However, their calculations could only give the blue edge of the Cepheid pulsation strip, but not the red edge. Li (1992a) used the bi-system oscillation theory to calculate the model oscillation of $9M_\odot$ and $5M_\odot$ stars, and the results could give both the blue and red edges of the Cepheid pulsation strip.

Figure 12.7 shows the position of the Cepheid pulsation strip on the H–R diagram obtained by the bi-system oscillation theory. In the figure, the outlined circles represent data where the oscillation mode is stable while the filled circles represent data where the oscillation mode is unstable. In this way, the red and the blue edges of the Cepheid pulsation strip are clearly identified, which agree very well with the observed data.

Using the bi-system oscillation theory, the period-radius relation for the oscillation mode with fundamental frequency is

$$\log(R/R_\odot) = 1.200 + 0.720 \log P \qquad (12.254)$$

and the period-luminosity relation is

$$\log(L/L_\odot) = 2.540 + 1.182 \log P \qquad (12.255)$$

Comparing Eqs. (12.254) and (12.255) with Eqs. (12.249) and (12.253) obtained from observations, it is found that the observed curves are always systematically lower than the theoretical curves. This may be due to that the helium abundance adopted for the theoretical model is too low. If the helium abundance is taken as 0.28, the discrepancy between the two will be greatly reduced.

12.8 δ-SCUTI VARIABLES

The δ-Scuti variables are also called dwarf Cepheid variables, and are situated near the main sequence in the Cepheid strip on the H–R diagram (Fig. 1.3). They are pulsational variables with ultra-short periods and small amplitudes. They have spectral types A–F, and belong to young stars of star population I. A typical example is the δ-Scuti. Up to now, more than 100 stars in this category have been discovered. There are great changes in the shape of their light curves, but the changes in the amplitude are usually less than 0.3 magnitude. The periods vary approximately from 0.05 to 0.15 days. Figure 12.8 gives

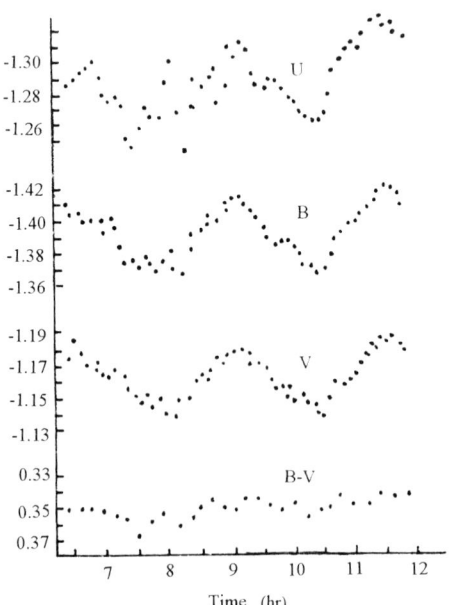

Fig. 12.8 The *U*, *B*, *V* and (*B–V*) light curves of β Cephei (according to Millis 1966).

Table 12.2 Some properties of δ-Scuti variables (from Petit 1987).

Star	m_V	Δm_V	period (d)	period (h, m)	spectral type
ε Cephei	4.17	0.05	0.042	1 h 00 m	F0IV
τ Pegasi	4.60	0.02	0.0543	1 h 19 m	A5IV
F_Z Vela	5.18	0.02	0.065	1 h 34 m	F0III
χ_2 Bootis	6.09	0.04	0.0678	1 h 38 m	F3V
τ Cygni	3.73	0.03	0.083	2 h 00 m	F2IV
β Cassiopeiae	2.78	0.04	0.1043	2 h 30 m	F2III–IV
Monoceros V474	6.29	0.07	0.1352	3 h 14 m	F2IV
δ-Scuti	4.71	0.15	0.1937	4 h 40 m	F3IIIp

the light curves of the β Cephei (a δ-Scuti variable). Table 12.2 gives the properties of eight δ-Scuti variables.

In the light curves of most δ-Scuti variables, there are two oscillations with different periods, which differ by a factor of 20 to 50. For example, the two periods of δ-Scuti are $P_1 = 0.1937$ day and $P_2 = 5.2477$ days. A lot of workers (Chevalier 1971; Cox *et al.* 1973; Dziebhowski and Kozlowski 1974; King and Cox 1987) have investigated the oscillation properties of δ-Scuti variables. They have found that when population I stars with masses of about $2M_\odot$ are situated in or have just left the main sequence to enter the Cepheid strip, there will be oscillation instabilities caused by the κ mechanism. There will also be many oscillation modes simultaneously, which include radial and non-radial oscillation modes. The most unstable oscillation mode is not the fundamental radial frequency, but is a higher radial harmonic. The blue edge of the pulsation strip can be precisely determined from theories, and their positions can be proved to be a function of the helium abundance, which is consistent with observational results. However, as in the cases for Cepheid variables, the red edge for the pulsation strip cannot be obtained. Another famous problem in the research of δ-Scuti variables is the mass paradox. The ratio between the periods of the fundamental frequency radial oscillation mode and the first harmonic oscillation mode obtained theoretically exceeds the same ratio obtained in observations by an amount far greater than that caused by statistical uncertainties. In fact, this problem applies to all pulsating variables inside the Cepheid strip. An immediate solution is to propose that the masses of variables obtained using different methods are inconsistent. This problem has been solved recently. The

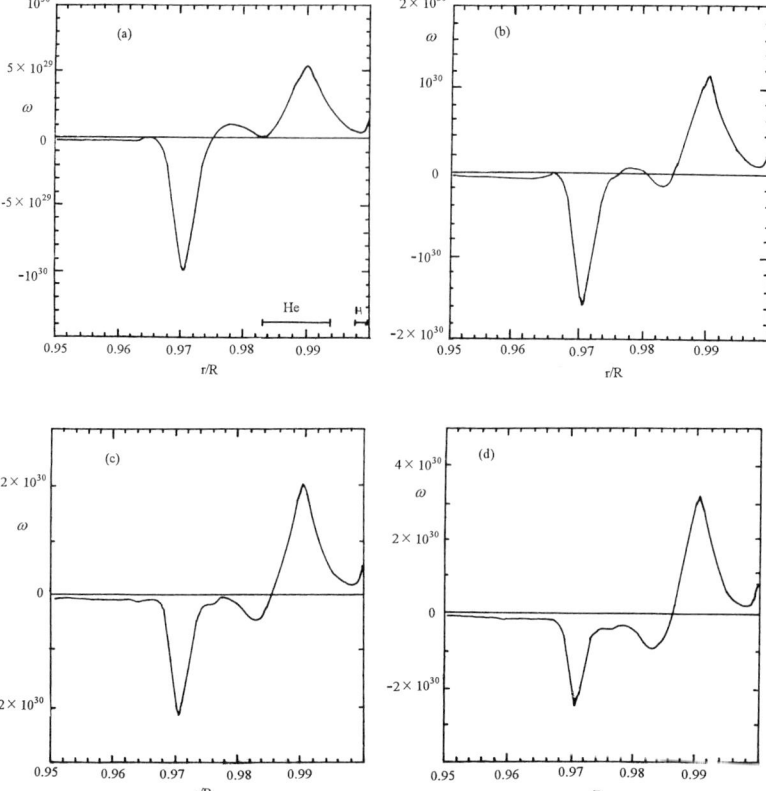

Fig. 12.9 The distribution curves of the work integral for the oscillation model of δ-Scuti variables with 2M$_\odot$. (a) Fundamental frequency radial oscillation mode (b) first harmonic radial oscillation mode (c) second harmonic radial oscillation mode (d) third harmonic radial oscillation mode (Li 1992a).

underlying reason is that the ionization of iron in the temperature range $\log T = 5.2$–5.9 can increase the opacity by a factor of 2.5 (Andreasen 1988; Andreasen and Peterson 1988). Li (1992a) used the bisystem oscillation theory to study δ-Scuti variables with 2M$_\odot$.

Figure 12.9 shows the distribution of the work integral when the 2M$_\odot$ star evolves to enter the Cepheid strip given by Li (1992a). From this figure, we see that many oscillation modes are unstable. There is a strong dissipative peak at $r/R \cong 0.97$ inside the star. There are a total of two excitative peaks; the broad and high one closer to the inside of the star is caused by secondary ionization effects of helium while the more-peaked one closer to the surface of the star is caused by combined effects of first ionization of helium and ionization of hydrogen. The excitative mechanism of the oscillation is mainly the temperature freezing mechanism caused by ionization.

Figure 12.10 gives the positions of pulsation strip on the H–R diagram calculated for a 2M$_\odot$ star model using the bi-system oscillation theory. In the figure, the outlined circles represent the data where the oscillation mode is stable while the filled circles represent the

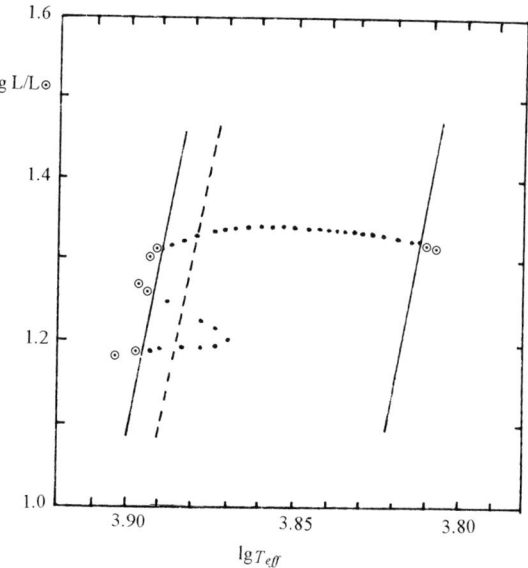

Fig. 12.10 The position of 2M$_\odot$ population I δ-Scuti variables on the H–R diagram. The outlined circles represent the data where the oscillation mode is stable and the filled circles represent the data where the oscillation mode is unstable. These are obtained according to the fundamental frequency radial oscillation mode.

data where the oscillation mode is unstable. It is noticed that both the blue and red edges can be precisely determined using the bi-system oscillation theory, the positions of which are consistent with observational data.

12.9 WHITE DWARF VARIABLES

It has been discovered that some white dwarfs have periodic light curves and the periods are longer than those calculated using the radial oscillation theory by one to two orders of magnitude. For example, for the white dwarf HLTau-76 first discovered by Landolt in 1964, the observed light curve period is 12.4 min while the period calculated using the radial oscillation theory is only 10 seconds. A large amount of white dwarfs of this type have been discovered up to now, and are called DAV type white dwarfs or ZZ Ceti variables. Robinson (1979) pointed out that these stars have the following common features:

(a) All of them are white dwarfs. The spectral type is DA type, i.e., their spectra have hydrogen absorption lines only. Due to the very strong gravitational acceleration g on the surface (log g = 7.8–8.1), the Balmer absorption lines show very large pressure broadening.

(b) The range of their color changes is $0.16 \leq B-V \leq 0.29$, so the range of temperature change is 10500–13500 K.

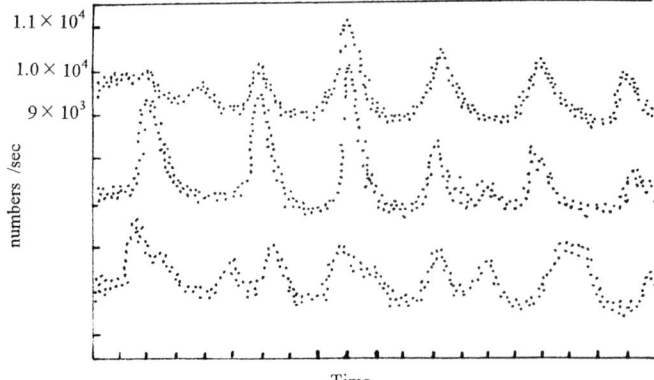

Fig. 12.11 The light curves of the white dwarf HLTau-76 with the largest amplitude (according to Warner and Nather 1972).

(c) These stars have multiple oscillation periods, and the range of the periods is 100–1200 seconds.

(d) They are distributed in the extended part of the Cepheid pulsation strip (Fig. 1.3). Figure 12.11 gives the light curves of HLTau-76.

Besides DAV type white dwarf variables with a hydrogen surface layer, Winget *et al.* (1982) theoretically predicted the existence of white dwarf variables with a helium surface layer which are called the DBV type white dwarf variables. Their prediction was soon proved to be correct by observations. The first discovered DBV type white dwarf variable was GD358. Its spectrum consists of helium absorption lines only. Its surface temperature is 20000 K and its magnitude is +0.3. A total of 26 different periods can be identified from the light curves of this star, and the range of periods is 140–950 seconds. Some time later, another DBV type white dwarf variable, PG1654+160, was discovered. Subsequently, another type of white dwarf variables with a surface layer consisting of helium, carbon, oxygen and metals was discovered, called DOV type white dwarf variables, or GW Vir white dwarf variables, or PG1159-035 type variables. The stars within some planetary nebulae belong to this type of variables.

Cox *et al.* (1987) studied oscillation properties of DAV type white dwarf variables. They have studied a $0.6M_\odot$ white dwarf model in which the mass of the envelope is 1.0×10^{-4} $M_\odot$ consisting of pure hydrogen and the inside is a $0.6M_\odot$ core of carbon and oxygen. The structure of the model is given in Table 12.3, where T_e is the effective temperature, α is the ratio between the convective mixing length and the pressure scale height, R is the radius of the star, L is the luminosity, T_c and ρ_c are the core temperature and the core density of the star.

The calculations of the linear, non-adiabatic, non-radial oscillation of the model shows that when the effective temperature is in the range

Table 12.3 The structure of a 0.6M$_\odot$ white dwarf model studied by Cox *et al.* (1987) in which the mass of the envelope of 1.0×10^{-4} M$_\odot$ consisting of pure hydrogen and the inside is a 0.6 M$_\odot$ core of carbon and oxygen, where T_e is the effective temperature, α is the ratio between the convective mixing length and the pressure scale height, R is the radius of the star, L is the luminosity, T_c and ρ_c are the core temperature and the core density of the star.

Model	T_e (10^3 K)	α	R (10^6 m)	L (10^{23} Js^{-1})	T_c (10^6 K)	ρ_c (10^9 kgm^{-3})
1	11.0	2.5	9.40	9.22	10.4	2.8
2	11.5	2.5	9.40	11.0	11.2	2.8
3	12.0	3.0	9.45	13.2	12.2	2.8

9000 to 11500 K, the model is unstable for oscillation modes of $l = 1, 2, 3, 4$ for periods in the range of 100 to 1000 seconds, and the oscillation modes belong to the non-radial *g*-mode. Figure 12.12 gives the relationship between the increase rate η of the oscillation amplitude and the period P for a model with $T_e = 11500$ K and $\alpha = 2.5$. It can be seen that the instability increases when the index l increases. Calculations show that when $T_e = 12000$ K, the model is always stable. Therefore, the blue edge of the pulsation strip obtained theoretically is between 11500 and 12000 K. For the low temperature side, oscillation

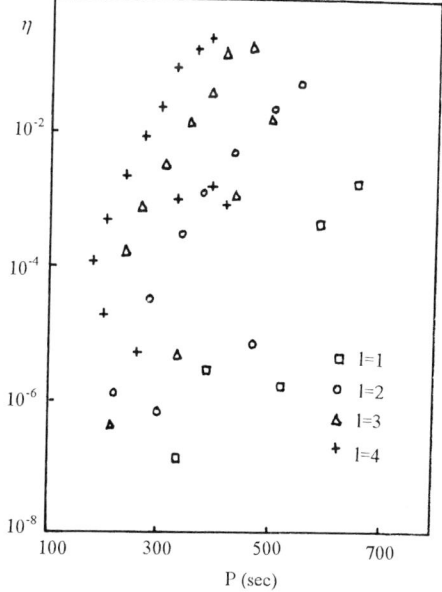

Fig. 12.12 The relationship between the increase rate of η of the oscillation amplitude and the period P for a 0.6M$_\odot$ white dwarf with $T_e = 11500$ K and $\alpha = 2.5$.

instability has been found at $T_e = 9000$ K. However, the red edge of the pulsation strip is not precisely known. Pesnell (1987) proposed that the excitative κ mechanism for ZZ Ceti variables is the "convective blockage effect". For a model with $T_e = 11500$ K and $\alpha = 2.5$, they found that the largest excitative region has a temperature reaching about 100000 K which is higher than the ionization temperature of hydrogen and helium so the mechanism cannot produce excitative effects. Calculations have shown that this region belongs to the bottom of the convective region where convective motions develop very quickly and take away a large amount of the outward energy flux. Assuming that there are no nuclear reactions at the bottom of the convective layer, the energy equilibrium equation requires

$$\nabla \cdot (\vec{F}_R + \vec{F}_{con}) = 0 \qquad (12.256)$$

Since $\nabla \cdot \vec{F}_{con} > 0$, we have $\nabla \cdot \vec{F}_R < 0$ or $\frac{dL_R}{dr} < 0$, i.e., the radiation luminosity is higher for inner regions. When a fluid element moves inwards, its volume contracts and its temperatures rises. However, the luminosity increase of the fluid element caused by its temperature increase is slower than the increase in the ambient luminosity. Therefore, there will be net radiative energy entering the fluid element, i.e., it will absorb energy under the greatest compression, which is the requirement for producing oscillation instabilities. Cox *et al.* (1987) called this effect the "convective blockage effect".

Cox *et al.* (1987) have also studied the oscillation of DBV type white dwarf variables with a helium envelope and $T_e = 25000$ to 27000 K. For the model with $T_e = 25000$ K, oscillation instability is obtained for $\alpha = 2.5$; for the model with $T_e = 27000$ K, oscillation instability is obtained for $\alpha = 3.0$. The cause for excitation of oscillations is still the convective blockage effect, and the theoretical period is consistent with the observed period.

Starrfield *et al.* (1985) have studied oscillation properties of DOV type white dwarf variables. They calculated the linear, non-adiabatic, non-radial oscillation of a late type high-luminosity white dwarf model with a mass of $0.6 M_\odot$ of which half is carbon and half is oxygen. When $l = 1, 2, 3, 4$, and T_e is in the range of 1.5×10^5 to 1.65×10^5 K, oscillation instability is obtained. The period is between 1000 and 4000 s. All the oscillations belong to high harmonic modes. The number of radial nodes is 40 to 50. The causes for excitation of oscillation are the ionization effects of carbon and oxygen in the temperature range 2×10^6 K to 5×10^6 K, which are consistent with the observed position of the pulsation strip and the observed periods of DOV type white dwarf variables.

CHAPTER THIRTEEN

STELLAR ATMOSPHERE

The stellar atmosphere refers to the layer which transfers the radiation produced from the stellar surface outwards. Its thickness can reach several hundreds of kilometers to several thousands of kilometers for most stars. Nonetheless, they are very thin layers when compared to the stellar radius.

Stellar spectra are generated in the stellar atmosphere. The properties of the continuous part and the absorption lines in stellar spectra have embedded information for the physical and chemical structure of the stellar atmosphere, e.g., the distribution of density, temperature and pressure and the chemical composition in the atmosphere. The objective of the theory of stellar atmosphere is to reveal the information embedded in stellar spectra for understanding of physical and chemical structures of the stellar atmosphere.

After making some simplified assumptions for the geometrical and physical conditions of the stellar atmosphere, a basic set of equations can be established which reflects the physical processes inside the stellar atmosphere, and the material function related to this set of basic equations (such as density and opacity etc.) and the corresponding boundary conditions, i.e., a model atmosphere can be established. On solving the model atmosphere, not only the properties of the stellar spectra (mainly the continuous part) can be obtained, but also the physical and chemical structure inside the stellar atmosphere can be obtained. If the properties of a spectrum obtained from model calculations and from observations agree with each other, the physical and chemical structures given by the model atmosphere can be said to reflect the true structures. Based on the obtained atmospheric structure, we can further compute properties of the spectral lines, such as profile and equivalent width etc., which are known as line formation calculations. The comparison between the properties of spectral lines obtained theoretically and those from observations is also an important step in determining the atmospheric structure.

In this chapter, we will separately introduce the construction and calculations of the model atmosphere, and construction and calculations of the line model, the problem of line formation in a moving atmosphere and the methodology in comparing theoretical and observational results to determine relevant physical parameters in the stellar atmosphere.

13.1 MODEL ATMOSPHERE

13.1.1 Basic Considerations, Basic Equations and Different Model Atmospheres

Before we construct a model atmosphere, we should fix three basic parameters relevant to the stellar atmosphere, which are the effective temperature T_{eff}, surface gravitational acceleration g and the chemical composition X_i of the stellar atmosphere. After fixing these parameters, in order to simplify the problem, we have to make some simplified assumptions on the geometrical structure and physical condition of the stellar atmosphere which are described as follows.

(1) We assume a semi-infinite plane-parallel atmosphere, and the physical and chemical properties are uniform in each layer. This assumption has enabled us to treat all physical parameters in the stellar atmosphere, such as density, pressure and temperature etc. to be simply one-dimensional variables depending only on the depth of the stellar atmosphere. This will greatly simplify the solution of equations of the model atmosphere.

The thickness of the stellar atmosphere is very small compared to the stellar radius for most stars, so the assumption of plane-parallel structure inside the stellar atmosphere is justified.

However, for supergiant stars, especially those with outward expansion of the atmosphere, the thickness of the stellar atmosphere is not very small compared to the stellar radius, and the plane parallel structure is not justified, and it is more appropriate to use the spherical symmetric structure here.

There is no convection inside the stellar atmosphere of early type stars, so it is reasonable to assume that the physical quantities of the stellar atmosphere are uniform in the horizontal direction. For late type stars, there is strong convection inside the stellar atmosphere. For example, the observed solar atmosphere has two-dimensional structures. However, two-dimensional atmospheric structures are very complicated, which are still unsolved until now. After all, the assumption of uniformity in the horizontal direction can give mean values of physical quantities for each layer.

(2) We assume the stellar atmosphere is at hydrostatic equilibrium. We also assume that the only forces acting on an arbitrary fluid element in the stellar atmosphere are the gravitational force and the pressure, so we have neglected effects of magnetic force, rotation and tidal force. This assumption will simplify the integration of the hydrostatic equilibrium equation, and enables us to use mass instead of depth as the basic variable, which is beneficial for model atmosphere calculations.

(3) We assume that the stellar atmosphere is composed of hydrogen and helium. For early type stars, the ionization of hydrogen is the main source of free electrons in the stellar atmosphere, and the contribution from other metals is negligible. For late type cool stars, we have to consider other elements.

(4) We assume the stellar atmosphere is stable and is time independent. This assumption neglects all time dependent phenomena in the stellar atmosphere, and requires the occupancies of atomic energy levels to be time independent. Based on this assumption, occupancies of atomic energy levels can still be calculated using the statistical equilibrium equation when the atmosphere deviates from local thermodynamic equilibrium.

(5) We assume the atmosphere is under radiative equilibrium, and there is no energy source in the atmosphere. Based on this assumption, the atmosphere should obey the energy theorem.

Under the previous simplified assumptions, basic equations can be set up for determination of the structure of the stellar atmosphere, which are described in the following.

(A) Radiative transfer equation

Inside an atmosphere with a plane-parallel structure, the energy transfer should fulfil the radiative transfer equation given by Eq. (2.59):

$$\mu \frac{dI_\nu}{dZ} = -\kappa_\nu I_\nu + \eta_\nu \tag{13.1}$$

or written as

$$\mu \frac{dI_\nu}{d\tau_\nu} = I_\nu - S_\nu \tag{13.2}$$

where Z is the depth of the atmosphere (direction from bottom to top defined as positive), τ_ν is the optical depth, κ_ν and η_ν are the absorption coefficient and the emission coefficient of matter in the atmosphere. From discussions in Chapter 5, their expressions are

$$
\kappa_\nu = \sum_\alpha \sum_{\beta=0}^{\beta_\alpha} \left[\sum_{i=1}^{l_{\alpha,\beta-1}} \sum_{j>i}^{l_{\alpha,\beta}} a_{ij}^{\alpha\beta}(\nu) \left(n_i^{\alpha\beta} - \frac{g_i^{\alpha\beta}}{g_j^{\alpha\beta}} n_j^{\alpha\beta} \right) \right.
$$
$$
+ \sum_{i=1}^{l_{\alpha,\beta}} \sum_{K=1}^{l_{\alpha,\beta+1}} a_{iK}^{\alpha\beta}(\nu)(n_i^{\alpha\beta} - n_i^{*\alpha\beta} e^{-h\nu/kT}) \tag{13.3}
$$
$$
\left. + n_e a_{KK}^{\alpha\beta} \left(\sum_{i=1}^{l_{\alpha,\beta+1}} n_i^{\alpha,\beta+1} \right)(1 - e^{-h\nu/kT}) \right] + n_e a_e
$$

$$
\eta_\nu = \frac{2h\nu^3}{c^2} \sum_\alpha \sum_{\beta=0}^{\beta_\alpha} \left[\sum_{i=1}^{l_{\alpha,\beta-1}} \sum_{j>i}^{l_{\alpha,\beta}} n_j^{\alpha\beta} \frac{g_i^{\alpha\beta}}{g_j^{\alpha\beta}} a_{ij}^{\alpha\beta}(\nu) + e^{-h\nu/kT} \left(\sum_{i=1}^{l_{\alpha,\beta}} \sum_{K=1}^{l_{\alpha,\beta+1}} n_i^{*\alpha\beta} a_{iK}^{\alpha\beta}(\nu) \right. \right.
$$
$$
\left. \left. + n_e a_{KK}^{\alpha\beta}(\nu) \sum_{i=1}^{l_{\alpha,\beta}} n_i^{\alpha,\beta+1} \right) \right] + n_e a_e J_\nu
$$

$$\tag{13.4}$$

where

α:	represents the species of atoms
β:	represents the degree of ionization ($\beta = 0$ means neutral)
i, j:	represents the energy levels of atoms
$n_i^{\alpha\beta}$:	represents the occupancy number on the energy level i of α-type atoms with the degree of ionization β
$a_{ij}^{\alpha\beta}$:	represents the atomic absorption cross section of the above atom for the transition from a bound state i to the bound state j

$a_{iK}^{\alpha\beta}$: represents the atomic absorption cross section of the above atom for bound-free transition

$a_{KK}^{\alpha\beta}$: represents the atomic absorption cross section of the above atom for free-free transition

a_e: represents the scattering cross section for scattering on a free electron

β_α: represents the electron number possessed by the α-type atom

$\ell_{\alpha\beta}$: represents the total number of bound states of the α-type atoms with the degree of ionization β

$n_i^{*\alpha\beta}$: represents the occupancy number under local thermodynamic equilibrium

The boundary condition for the radiative transfer equation is:
At the surface of the atmosphere, there is no incoming radiation, so

$$\tau_\nu = 0: \quad I_\nu(\tau_\nu = 0, \mu) = 0 , \quad \text{when } 0 > \mu \geq -1 \qquad (13.5)$$

At the bottom of the atmosphere, we have

$$\tau_\nu = \tau_{\nu n}: \quad I_\nu(\tau_{\nu n}, \mu) = I_\nu^+ \qquad (13.6)$$

where I_ν^+ represents the outward radiation intensity from the bottom of the atmosphere.

(B) Hydrostatic equilibrium equation

In an atmosphere with a plane-parallel structure, we consider a fluid element with surface area dF and thickness dZ. When not considering the magnetic field, rotation and tidal force, the only forces acting on the fluid element are the gravitational force $g\rho dFdZ$ pointing at the center of the star, and the pressure $-dPdF$ pointing in the opposite direction. At hydrostatic equilibrium, these two forces are equal, so

$$dP = -\rho g dZ \qquad (13.7)$$

where P is the total pressure, which is constituted by the gas pressure P_g and the radiation pressure P_R. Using Eq. (4.23) for the gas pressure and Eq. (2.16) for the radiation pressure, the hydrostatic equilibrium equation becomes

$$\frac{d}{dZ}(n_g kT) + \frac{d}{dZ}\frac{1}{c}\int_\nu\int_\omega I_\nu\mu^2 d\omega d\nu = -g\rho \qquad (\mu = \cos\theta) \qquad (13.8)$$

When Z in Eq. (13.8) is replaced by the optical depth τ, we obtain another form of the expression as

$$\frac{dP_g}{d\tau} = \frac{g}{\kappa} - \frac{dP_R}{d\tau} \tag{13.9}$$

The boundary condition for the hydrostatic equilibrium equation is

$$\text{on the stellar surface where } \tau = 0: \ P_g = 0 \tag{13.10}$$

(C) Energy equilibrium equation and the energy theorem

When the stellar atmosphere is under radiative equilibrium and there is no energy source inside the atmosphere, all the energy is transferred in form of radiation. For an arbitrary volume element in the atmosphere, the energy absorbed per second should be equal to the energy radiated away, i.e., the energy equilibrium equation

$$\int_\nu \int_\omega [\kappa_\nu I_\nu(\tau_\nu, \mu) - \eta_\nu(\tau_\nu)] \frac{d\omega}{4\pi} d\nu = 0 \tag{13.11}$$

should be satisfied. Integrating the radiative transfer equation (13.1) with respect to the frequency ν and the spatial angle ω, and using the energy equilibrium equation (13.11) and the radiation flux in Eq. (2.9), we have

$$\int_\nu \int_\omega \mu \frac{dI_\nu}{dZ} d\nu d\omega = \frac{d}{dZ} \int_\nu \int_\omega I_\nu \mu d\omega d\nu$$

$$= \frac{d}{dZ} \int_\nu \pi F_\nu d\nu = 0 \tag{13.12}$$

from which we can get

$$\frac{d}{dZ}(\pi F) = 0 \tag{13.13}$$

which is the energy theorem and is only valid for the total radiation flux. Equation (13.13) is also equivalent to the expression

$$\pi F = \frac{L}{4\pi R^2} = \sigma T_{eff}^4 = const \tag{13.14}$$

where R is the stellar radius, L is the stellar luminosity, T_{eff} is the effective temperature and σ is the Stefan–Boltzmann constant.

(D) Boltzmann equation and Saha equation

The Boltzmann equation gives the relationship among the occupancy numbers of different excited states for the same degree of ionization [see Eq. (4.232)]:

$$\frac{n_s^{\alpha\beta}}{n_i^{\alpha\beta}} = \frac{g_s^{\alpha\beta}}{g_i^{\alpha\beta}} e^{-\Delta\varepsilon^{\alpha\beta}/kT} \tag{13.15}$$

where $\Delta\varepsilon^{\alpha\beta} = \varepsilon_s^{\alpha\beta} - \varepsilon_i^{\alpha\beta}$ is the excitation energy.

The Saha equation gives the relationship between the ground state occupancy numbers of two neighboring degrees of ionization [see Eq. (4.233)]:

$$\frac{n_o^{\alpha\beta+1}}{n_o^{\alpha\beta}} n_e = \frac{g_o^{\alpha\beta+1}}{g_o^{\alpha\beta}} \cdot 2 \cdot \frac{(2\pi m_e kT)^{3/2}}{h^3} e^{-\chi_\beta^\alpha/kT} \tag{13.16}$$

where χ_β^α is the ionization energy.

(E) Statistical equilibrium equation

From §4.4.2, we know that the statistical equilibrium equation can be written as

$$n_i^{\alpha\beta}\left[\sum_{j\neq i}^{l_{\alpha,\beta}} P_{ij}^{\alpha\beta} + P_{iK}^{\alpha\beta}\right] - \sum_{j\neq i}^{l_{\alpha,\beta}} n_j^{\alpha\beta} P_{ji}^{\alpha\beta} - n_K^{*\alpha\beta} P_{Ki}^{\alpha\beta} = 0 \tag{13.17}$$

where $P_{ij}^{\alpha\beta}$ and $P_{ji}^{\alpha\beta}$ are the rates of transition of atoms from state i to j and from state j to i respectively, and $P_{iK}^{\alpha\beta}$ is the rate of bound-free transition. $P_{ij}^{\alpha\beta}$, $P_{ji}^{\alpha\beta}$, $P_{iK}^{\alpha\beta}$ and $P_{Ki}^{\alpha\beta}$ are composed of the rates of radiative transition and collisional transition:

$$\begin{aligned} P_{ij}^{\alpha\beta} &= R_{ij}^{\alpha\beta} + C_{ij}^{\alpha\beta}, & P_{ji}^{\alpha\beta} &= R_{ji}^{\alpha\beta} + C_{ji}^{\alpha\beta} \\ P_{iK}^{\alpha\beta} &= R_{iK}^{\alpha\beta} + C_{iK}^{\alpha\beta}, & P_{Ki}^{\alpha\beta} &= R_{Ki}^{\alpha\beta} + C_{Ki}^{\alpha\beta} \end{aligned} \tag{13.18}$$

where

$$
\left. \begin{aligned}
R_{ij}^{\alpha\beta} &= 4\pi \int_0^\infty \frac{\alpha_{ij}^{\alpha\beta}(v)}{hv} J_v dv, \ i < j \\[2mm]
R_{ji}^{\alpha\beta} &= 4\pi \left(\frac{n_i^{\alpha\beta}}{n_j^{\alpha\beta}}\right)^* \int_0^\infty \frac{\alpha_{ij}^{\alpha\beta}(v)}{hv}\left(J_v + \frac{2hv^3}{c^2}\right)e^{-hv/kT} dv, \ i < j \\[2mm]
R_{iK}^{\alpha\beta} &= 4\pi \int_0^\infty \frac{\alpha_{iK}^{\alpha\beta}(v)}{hv} J_v dv \\[2mm]
R_{Ki}^{\alpha\beta} &= 4\pi \frac{n_i^{*\alpha\beta}}{n_e} \int_0^\infty \frac{\alpha_{iK}^{\alpha\beta}}{hv}\left(J_v + \frac{2hv^3}{c^2}\right)e^{-hv/kT} dv
\end{aligned} \right\}
\tag{13.19}
$$

$$
\left. \begin{aligned}
C_{ij}^{\alpha\beta} &= n_e \Omega_{ij}(T) \\[2mm]
C_{ij}^{\alpha\beta} &= n_e \left(\frac{n_i^{\alpha\beta}}{n_j^{\alpha\beta}}\right)^* \Omega_{ij}(T) \\[2mm]
\Omega_{ij}(T) &= \int_0^\infty \sigma_{ij}(v)f(v)v\,dv
\end{aligned} \right\}
\tag{13.20}
$$

where $f(v)$ is the Maxwellian distribution of velocities, $\sigma_{ij}(v)$ is the atomic collision cross section

$$
\left. \begin{aligned}
C_{iK}^{\alpha\beta} &= n_e \Omega_{iK} \\[2mm]
C_{Ki}^{\alpha\beta} &= \left(\frac{n_i^{\alpha\beta}}{n_K^{\alpha\beta}}\right)^* C_{iK}^{\alpha\beta}
\end{aligned} \right\}
\tag{13.21}
$$

(F) Equation for conservation of total particles [see Eq. (4.262)]

$$
n_g = n_e + \sum_\alpha \sum_\beta \sum_i n_i^{\alpha\beta}
\tag{13.22}
$$

(G) Equation for conservation of the particles for elements [see Eq. (4.263)]

$$
n_g X_\alpha = \sum_{\beta=0}^{\beta_\alpha} \sum_{i=1}^{l_{\alpha\beta}} n_i^{\alpha\beta}
\tag{13.23}
$$

where X_α is the abundance of the element α.

(H) Equation for conservation of electric charges [see Eq. (4.264)]

$$n_e = \sum_{\alpha} \sum_{\beta} \beta \sum_i n_i^{\alpha\beta} \qquad (13.24)$$

All the above equations are coupled together. For example, to solve the radiative transfer equation, we have to know the absorption coefficient κ_v and the emission coefficient η_v for the material. To obtain κ_v and η_v, we have to know the occupancy numbers on each energy state for the atoms inside all the layers in the atmosphere, which require the hydrostatic equilibrium equation, statistical equilibrium equation and the equation for conservation of particles.

The earliest and the simplest model atmosphere is the gray atmosphere. In a gray atmosphere, the absorption coefficient of the stellar matter is assumed to be independent of the frequency, i.e., $\kappa_v = \kappa$. On assuming a gray atmosphere, the radiative transfer equation is decoupled from the rest of the equation set. If we further assume that the atmosphere is under local thermodynamic equilibrium, we can integrate the radiative transfer equation by itself and combine the results from the integration with the energy theorem to get the temperature distribution in the atmosphere and the crudest atmospheric structure.

Although the assumption for a gray atmosphere might be different from the realistic situation so that the obtained temperature distribution differs from the realistic one, the distribution obtained is still meaningful in that it can be used as part of the approximate initial solution when solving for the model atmosphere.

Current model atmospheres can be divided into two categories, one being the local thermodynamic equilibrium (LTE) model atmosphere and the other being the non-local thermodynamic equilibrium (non-LTE) model atmosphere. The LTE model atmosphere assumes the atmosphere to be at local thermodynamic equilibrium so the occupancy number of atomic energy states can be calculated with the Boltzmann equation and the Saha equation. Since the non-LTE model atmosphere has deviated from local thermodynamic equilibrium, the Boltzmann equation and the Saha equation are no longer valid, and the occupancy number of the atomic energy states should be calculated using the statistical equilibrium equation. Since different stars have different effective temperatures and gravitational accelerations, some stars can be calculated using the LTE model atmosphere while others should be calculated using the non-LTE model atmosphere.

It is relatively easy to give the self-consistent basic equation set and boundary conditions for the model atmosphere. However, it will be relatively complicated to solve the basic equation set since it is a high order non-linear set. It is common to solve the equations numerically. There are different types of numerical methods. One of them was introduced by Auer and Mihalas (1969) which first linearized the radiative transfer equation and other equations (the other basic equations can be collectively called the constraint equations) and then solving these linearized equations simultaneously. This method is called the complete linearized method. The biggest advantages of this method are the stability of calculations and the fast convergence. Nonetheless, since all basic equations are solved at the same time, the number of equations and variables are large. Due to restrictions in computer storage and speed, the model calculations cannot accommodate a lot of line blanketing effects. Another numerical method separately solves the radiative transfer equation and the constraint equation set, without performing the linearization. There are many methods using this idea, e.g., the temperature correction method (see Kurucz 1979) and the Λ operator method (see Werner 1986) etc., and can be called the separate solution method. The advantage of this method is the ability to consider more line blanketing effects in the model atmosphere calculations. Nevertheless, the convergence of the method is rather poor. Recently, there is a method which combines the advantages of the above two methods. It first linearizes the radiative transfer equation and the constraint equation set, and then solves the linearized radiative transfer equation and the linearized constraint equation set separately through mutual iterations. This method has good stability and convergence, and can include a lot of line blanketing effects.

It should be pointed out that the above numerical methods can be employed for either LTE or non-LTE model atmospheres or line formation calculations. To let the readers get more familiar with these numerical methods, they will be introduced in the sections on LTE or non-LTE model atmosphere and line formation calculations.

In general, when the density of the stellar atmosphere is lower and the temperature is higher, it is more deviated from the LTE state and the non-LTE model atmosphere should be employed. For example, for early type main sequence stars with effective temperature higher than 25000 to 30000 K, hot super giant stars with temperatures higher than 25000 K, and cool super giant stars with temperatures exceeding 10000 K, the non-LTE model atmosphere should be employed; and for main sequence stars with temperatures lower than 25000 K, the LTE model atmosphere can be used.

13.1.2 Temperature Distribution in the Gray Atmosphere

If the opacity of matter inside the atmosphere is independent of frequency, i.e., $\kappa_\nu = \kappa$, the atmosphere is called a gray atmosphere. Under the assumption of a gray atmosphere, the coupling between the radiative transfer equation and other equations in the basic equation set can be neglected, as these can be solved individually. If we further assume that the atmosphere is under local thermodynamic equilibrium, the solution of the radiative transfer equation can be combined with the energy theorem to give the temperature distribution of the atmosphere which is known as the temperature distribution of the gray atmosphere.

(A) Solution of the radiative transfer equation for the gray atmosphere

Integrating the radiative transfer equation (13.1) over all frequencies, we have

$$\mu \frac{d}{dZ} \int_0^\infty I_\nu d\nu = - \int_0^\infty \kappa_\nu I_\nu d\nu + \int_0^\infty \kappa_\nu S_\nu d\nu \qquad (13.25)$$

Using the assumption $\kappa_\nu = \kappa$ for the gray atmosphere, the above equation becomes

$$\mu \frac{dI}{d\tau} = I - S \qquad (13.26)$$

where

$$I = \int_0^\infty I_\nu d\nu \qquad S = \int_0^\infty S_\nu d\nu \qquad (13.27)$$

and the direction of τ is opposite to that of Z.

Equation (13.26) shows that, under the gray atmosphere condition, the total radiation intensity is only determined by the radiative transfer equation. This is different from the cases of non-gray atmospheres, in which the radiation intensity at each frequency is determined by one radiative transfer equation, and the total radiation intensity is determined by the radiative transfer equations for all frequencies.

Substituting the assumption for the gray atmosphere into the energy equilibrium equation (13.11), we can obtain

$$\int_0^\infty \kappa J_\nu d\nu = \int_0^\infty \kappa S_\nu d\nu \qquad (13.28)$$

from which we get

$$S = J \qquad (13.29)$$

where

$$J = \int_0^\infty J_\nu d\nu \qquad (13.30)$$

Substituting Eq. (13.29) into Eq. (13.26), we have

$$\mu \frac{dI}{d\tau} = I - J = I - \frac{1}{2} \int_{-1}^{+1} I d\mu \qquad (13.31)$$

which is the integral-differential equation related to the total radiation intensity $I(\tau, \mu)$ for the gray atmosphere. Since there is no inward radiation at the stellar surface where $\tau = 0$, the surface boundary condition can be written as:

$$I(0, \mu) = 0, \quad -1 \le \mu \le 0 \qquad (13.32)$$

We can also derive the equation for the mean radiation intensity. From Eq. (2.85), and considering $S = J$, we have

$$J(\tau) = \frac{1}{2} \int_0^\infty J(t) E_1 \mid t - \tau \mid dt \qquad (13.33)$$

which is the required linear integral equation for the mean radiation intensity J (or S) for the gray atmosphere. The problem now is to get the expression for J from Eqs. (13.31) or (13.33). Integrating the integral-differential equation (13.31) over all angles, and considering Eqs. (2.24) and (2.25), we have

$$\frac{d(F/4)}{d\tau} = J - J = 0 \qquad (13.34)$$

from which we have

$$F/4 = H = const \tag{13.35}$$

Multiplying Eq. (13.31) by μ, then integrating over all angles, and using Eqs. (2.25) and (2.26), we have

$$\frac{dK}{d\tau} = \frac{F}{4} - J \int_{-1}^{+1} \mu d\mu = \frac{F}{4} \tag{13.36}$$

Substituting Eq. (13.35) into Eq. (13.36), we have

$$\frac{dK}{d\tau} = \frac{F}{4} = const \tag{13.37}$$

from which we get the solution

$$K(\tau) = \frac{1}{4}F \cdot \tau + const \tag{13.38}$$

From Eq. (2.104), we know that $K = (1/3)J$ for $\tau \gg 1$, i.e., at a large depth of the atmosphere. Substituting Eq. (2.104) into Eq. (13.38), we obtain for $\tau \gg 1$ the equation

$$J(\tau) = \frac{3}{4}F \cdot \tau \tag{13.39}$$

which shows that $J(\tau)$ is linearly related to τ at a large depth of the atmosphere with F as the proportionality constant. Nonetheless, the above approximate solution deviates from the true solution near the surface of the atmosphere. We can adopt the following general expression

$$J(\tau) = \frac{3}{4}F\left[\tau + q(\tau)\right], \tag{13.40}$$

where $q(\tau)$ represents the deviation of the approximate solution from the true solution, and is called the Hopf function. If the Hopf function is known, $J(\tau)$ for the stellar atmosphere will be known. To obtain the Hopf function, Eddington made the approximation

$$q(\tau) = C = const \tag{13.41}$$

It is easy to see that this approximation favors the approximate solution for $\tau \gg 1$, since $(3/4)F\tau$ will be far greater than the constant C for large τ so $J(\tau) \to (3/4)F\tau$ which is the approximate solution. The constant C can be determined using the boundary condition. According to $S = J$, we have from Eq. (13.40)

$$S = \frac{3}{4}F(\tau + C) \tag{13.42}$$

Substituting Eq. (13.42) into the general solution in Eq. (2.78) for the outward radiation ($\mu \geq 0$), we have

$$
\begin{aligned}
I(\tau, \mu) &= \int_{\tau}^{\infty} S(t)e^{-(t-\tau)/u}\frac{dt}{\mu} \\
&= \frac{3}{4}F\int_{\tau}^{\infty}(t + C)e^{-(t-\tau)/\mu}\frac{dt}{\mu} \\
&= \frac{3}{4}F\int_{\tau}^{\infty}(t - \tau + C + \tau)e^{-(t-\tau)/\mu}\frac{dt}{\mu}
\end{aligned}
\tag{13.43}
$$

Denoting $y = (t - \tau)/\mu$, and since

$$\mu\int_{0}^{\infty} ye^{-y}dy = \mu\left[-e^{-y}(y + 1)\right]_{0}^{\infty} = \mu$$

and

$$(C + \tau)\int_{0}^{\infty} e^{-y}dy = -(C + \tau)\left[e^{-y}\right]_{0}^{\infty} = C + \tau$$

(13.44)

we have from Eq. (13.43)

$$I(\tau, \mu) = \frac{3}{4}F(\mu + \tau + C) \tag{13.45}$$

At the surface where $\tau = 0$, we should have

$$I(0, \mu) = \frac{3}{4}F(\mu + C) \tag{13.46}$$

$$F^{+}(0) = 2\int_{0}^{1} I(0, \mu)\mu d\mu$$

$$= \frac{3}{2}F \int\limits_{0}^{1} (\mu^2 + C\mu)d\mu$$

$$= \frac{3}{2}F\left(\frac{1}{3} + \frac{C}{2}\right) \tag{13.47}$$

Since the inward radiation flux at the stellar surface is zero, i.e., $F^-(0) = 0$, so $F^+(0) = F$. Substituting this relation into Eq. (13.47), we obtain $C = 2/3$, so from Eqs. (13.40) and (13.41), we can get

$$J(\tau) = \frac{3}{4}F\left(\tau + \frac{2}{3}\right) \tag{13.48}$$

which is the Eddington approximate solution for the gray atmosphere.

(B) Temperature in the gray atmosphere

If we further assume that the stellar atmosphere is under local thermodynamic equilibrium, according to the Kirchoff law and the Planck radiation law, we have $S = J = B = [\sigma T^4(\tau)/\pi]$. Further considering $\pi F = \sigma T_{eff}^4$ and Eq. (13.48), we get

$$T^4(\tau) = \frac{3}{4}T_{eff}^4\left(\tau + \frac{2}{3}\right) \tag{13.49}$$

which is the temperature distribution in the gray atmosphere. From Eq. (13.49), we know

when $\tau = 2/3$, $T(2/3) = T_{eff}$ $\qquad\qquad\qquad\qquad$ (13.50)

when $\tau = 0$, $T(0) = (1/2^{1/4})T_{eff} = 0.841\,T_{eff}$ $\qquad\qquad$ (13.51)

13.1.3 LTE Model Atmosphere

(A) Introduction

In this section, we will introduce the LTE model atmosphere calculations developed by Kurucz (1979) which separately solves the radiative transfer equation and the constraint equations and has temperature corrections. In §13.1.1, we have pointed out that the LTE model atmosphere can be calculated using either the complete linearized method and separate solution method. The method developed by Kurucz is described in the following. For fixed effective temperature T_{eff}, surface gravitational acceleration g and atmospheric chemical composition X_i (elemental abundance), the following steps are taken to calculate the LTE model atmosphere.

(1) We assume a temperature distribution $T(\tau)$, e.g., the temperature distribution for a gray atmosphere. Using this temperature distribution, we then solve the hydrostatic equilibrium equation, the Boltzmann equation, Saha equation and conservative equations for different particles to obtain the pressure distribution, density distribution, the absorption coefficient of matter and the optical depths in the atmosphere, i.e., we obtain $P(\bar{\tau})$, $T(\bar{\tau})$, $\kappa_\nu(\bar{\tau})$, $\tau_\nu(\bar{\tau})$, ...

(2) The radiative transfer equation is solved using the above results, and the total radiation flux $\pi F(\tau)$ is calculated.

(3) In general, the total radiation flux obtained above does not fulfil the energy theorem $\pi F = \sigma T_{eff}^4$, because the temperature distribution for the assumed gray atmosphere is only an approximate distribution, so we have to correct the distribution, i.e., to find $\Delta T(\bar{\tau})$.

Using the corrected temperature distribution $T(\bar{\tau}) + \Delta T(\bar{\tau})$, we repeat the above calculations from step (1) to step (3). The iterations continue until $\pi F(\bar{\tau})$ fulfils the energy theorem.

The $\pi F(\nu, Z = 0)$ obtained above should also agree with the flux distribution in the observed stellar spectrum. If they do not agree, the original chosen values of T_{eff}, g and X_i should be corrected and the model atmosphere should then be calculated again. These procedures are repeated until they agree. At this time, $P(\bar{\tau})$, $T(\bar{\tau})$, $\rho(\bar{\tau})$, $\kappa_\nu(\bar{\tau})$, ..., T_{eff}, g and X_i are the physical and chemical structures of the stellar atmosphere.

The above has outlined the procedures for calculating the LTE model atmosphere. The method in comparing the theoretical and observational results to determine the major parameters T_{eff}, g and X_i in the atmosphere will be discussed in §13.3.

The continuous spectrum of a star is superimposed by many absorption lines, which will significantly affect the energy flux distribution of the continuous spectrum as well as the temperature distribution in the atmosphere, which is referred to as the line blanketing effect. The line blanketing effect should be considered in calculations of the model atmosphere to give results close to realistic properties of the spectrum. The effect will be discussed in this section too.

(B) Integration of the hydrostatic equilibrium equation

We write the hydrostatic equilibrium equation in Eq. (13.9) in the form

$$\frac{dP_g}{d\bar{\tau}} = \frac{g}{\bar{\kappa}_\nu} - \frac{dP_R}{d\bar{\tau}} \qquad (13.52)$$

where $\bar{\kappa}_\nu$ is the Rosseland mean value for the opacity, $\bar{\tau}$ is the optical depth calculated using the Rosseland mean value for the opacity.

Before integrating Eq. (13.52), we first study the last term in that equation, i.e., the expression $(dP_R/d\bar{\tau})$. When a radiation beam with intensity I_ν passes through a layer of the atmosphere with a thickness dx subtending an angle θ with the line normal to the layer (see Fig. 13.1), the radiation energy absorbed by the gas is

$$dE_\nu = \kappa_\nu I_\nu d\nu d\omega dt dA ds \cos\theta \qquad (13.53)$$

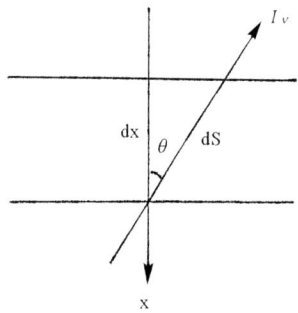

Fig. 13.1 A radiative beam with intensity I_ν passes through a layer of the atmosphere with a thickness dx.

Multiplying the above equation by $(1/c)$, we obtain the momentum transferred from the radiation to the gaseous material. If this is further multiplied by $\cos\theta$, the component along the normal line is obtained. Therefore, the momentum flux per unit time per unit area along the normal line is

$$\frac{1}{c}\kappa_\nu I_\nu d\nu d\omega \cos^2\theta ds = \frac{1}{c}\kappa_\nu I_\nu \mu d\nu d\omega dx \qquad (13.54)$$

Integrating the above equation with respect to frequency and solid angle, we get the total momentum flux transferred from the radiation to the layer, or the radiation pressure difference for that gas layer:

$$dP_R = \frac{1}{c}dx \int_0^\infty \kappa_\nu d\nu \oint_\omega I_\nu \mu d\omega \qquad (13.55)$$

According to Eq. (2.9), the above equation can be written as

$$\frac{dP_R}{dx} = \frac{\pi}{c} \int_0^\infty \kappa_\nu F_\nu d\nu \qquad (13.56)$$

In terms of the optical depth $\bar{\tau}$, the above equation becomes

$$\frac{dP_R}{d\bar{\tau}} = \frac{\pi}{c} \int_0^\infty \kappa_\nu F_\nu d\nu / \bar{\kappa}_\nu \qquad (13.57)$$

However, we do not know F_ν in Eq. (13.57) at the beginning when we integrate the hydrostatic equilibrium equation (13.52), so we have to use the gray atmosphere approximation, and we have

$$\frac{dP_R}{d\bar{\tau}} = \frac{\pi F}{c} = \frac{\sigma T_{eff}^4}{c} \qquad (13.58)$$

Now that the expression for $(dP_R/d\bar{\tau})$ is known, we will discuss the integration of Eq. (13.52). Since the change in the atmospheric pressure can be very drastic which can reach several orders of magnitude, for the sake of convenience, we write Eq. (13.52) in the logarithmic form:

$$\frac{d\log P_g}{d\bar{\tau}} = \left(\frac{g}{\bar{\kappa}_\nu} - \frac{dP_R}{d\bar{\tau}} \right) / (2.3025851 P_g) \qquad (13.59)$$

To integrate Eq. (13.59), we assume a temperature distribution for a gray atmosphere

$$T^4(\bar{\tau}) = \frac{3}{4} T_{eff}^4 \left[\bar{\tau} + q(\bar{\tau}) \right] \qquad (13.60)$$

where $q(\bar{\tau})$ can be written in the form

$$q(\bar{\tau}) = 0.710 - 0.1333 \exp(-3.4488\bar{\tau}) \qquad (13.61)$$

The integration of Eq. (13.59) can be performed through the following steps.

Step 1: Integration from $\bar{\tau} = 0$ to the first $\bar{\tau}_1$ value. The $\bar{\tau}_1$ value should be very small, e.g., $\bar{\tau}_1 = 10^{-4}$, and we assume $\bar{\kappa}_\nu$ has very little variation between $\bar{\tau} = 0$ to $\bar{\tau}_1$ and can be treated as a constant. We have to first estimate a value for $\bar{\kappa}_\nu$, which is written as $\bar{\kappa}_\nu^{(1)}$. Since $\bar{\tau}_1$ is very small, Eq. (13.52) can be written as

$$(P_g + P_R) - 0 = (\bar{\tau}_1 - 0) \frac{g}{\bar{\kappa}_\nu^{(1)}} \qquad (13.62)$$

in which P_R can be obtained from Eq. (13.58). In Eq. (13.62), P_R, $\bar{\tau}_1$, g and $\bar{\kappa}_\nu^{(1)}$ are known already, so $P_g^{(1)}$ can be calculated. Since $T(\bar{\tau})$ has been given beforehand, we can get $T^{(1)}$ at $\bar{\tau}_1$. After evaluating $P_g^{(1)}$ and $T^{(1)}$, a new value of $\bar{\kappa}_\nu$ can be calculated which is written as $\bar{\kappa}_\nu^{(2)}$. There are two methods in calculating $\bar{\kappa}_\nu^{(2)}$, namely, through interpolation from the known $\bar{\kappa}_\nu(T, P_g)$ table and the other through direct calculations of the occupancy numbers on all energy states for all types of atoms, and then multiplied by the corresponding atomic absorption cross sections.

After getting $\bar{\kappa}_\nu^{(2)}$, they can be put back into Eq. (13.62) to calculate a new $P_g^{(2)}$. Iterations are repeated until

$$\left| \log P_g^{(2)} - \log P_g^{(1)} \right| \leq \varepsilon \tag{13.63}$$

where ε is a small precision term such as $\varepsilon = 10^{-4}$. In this way, we will obtain a set of initial values: $\bar{\tau}_1$, T_1, P_{g_1}, and $\bar{\kappa}_{\nu 1}$.

Step 2: Starting from step 2, we will integrate Eq. (13.59). The integration starts from the point τ_1. Since T_1, P_{g_1}, and $\bar{\kappa}_{\nu 1}$ are already known, $(dP_R/d\bar{\tau})$ can be given by Eq. (13.58) [The first integration employs Eq. (13.58), and in the repeated calculations after the temperature correction, $(dP_R/d\bar{\tau})$ should be calculated using Eq. (13.57)]. Therefore, we can use the Runge–Kutta method to obtain P_{g_2} at $\bar{\tau}_2$. Since $T(\bar{\tau})$ has been given beforehand, the values of T_2 at $\bar{\tau}_2$ can be obtained. When P_{g_2} and T_2 are known, $\bar{\kappa}_{\nu 2}$ can also be calculated. On using the same method three times, the values for P_g, T, $\bar{\kappa}_\nu$ and $\bar{\tau}$ can be evaluated.

Step 3: After getting the four sets of initial values in the above, we use the Adams–Mouthon prediction-correction algorithm to integrate Eq. (13.59). In this method, the prediction term is

$$y_{n+1}^{(P)} = y_n + \frac{\Delta x}{24} \left(55 f_n - 59 f_{n-1} + 37 f_{n-2} - 9 f_{n-3} \right) \tag{13.64}$$

and the correction term is

$$y_{n+1}^{(C)} = y_n + \frac{\Delta x}{24} \left[9 f \left(x_n, y_{n+1}^{(P)} \right) + 19 f_n - 5 f_{n-1} + f_{n-2} \right] \tag{13.65}$$

For each step of integration, the difference between the prediction term and the correction term should be less than 10^{-5}. Furthermore, the step length is required to be shorter than and is about one-fourth that required in the Runge–Kutta method. The integration is continued until the expected value of $\bar{\tau}$ is reached.

(C) Calculations of the number density

During the discussion for the integration of the hydrostatic equilibrium equation, we have pointed out that there are two methods for determining $\bar{\kappa}_\nu$ for known values of T and P. The first one uses the known $\bar{\kappa}_\nu(T, P_g)$ table for interpolation, and the second calculates the occupancy numbers in each energy state for all types of atoms, which are then multiplied by the corresponding atomic absorption cross sections.

We now discuss the number density of atoms and ions of different elements under the LTE condition. We suppose the atmosphere is composed of hydrogen and helium, the abundance of hydrogen and helium are X_H and X_{He} respectively, and the total number density of atoms and ions is n_A, so the equations for conservation of particles for different elements are

$$n_A X_H = n(HI) + n(HII) + 2n(H_2) + 2n(H_2^+) + n(H^-) \tag{13.66}$$

$$n_A X_{He} = n(HeI) + n(HeII) + n(HeIII) \tag{13.67}$$

the equation for conservation of electric charge is

$$n_e = n(HII) + n(H_2^+) + n(HeII) + 2n(HeIII) - n(H^-) \tag{13.68}$$

and the equation for conservation of the total number of particles is

$$n_g = n_e + n_A = P_g/kT \tag{13.69}$$

All the terms on the right hand side of Eqs. (13.66) to (13.68) can be written as functions of the number density of neutral atoms and electrons through the Saha equation. We use Eq. (13.66) as our example, in which

$$n(H_2) = n(HI) \cdot n(HI) \cdot \frac{U(H_2)\left[2\pi m(H_2)kT/h^2\right]^{3/2}}{\left\{U(HI)\left[2\pi m(HI)kT/h^2\right]^{3/2}\right\}^2} e^{-\left[E(H_2)-E(HI)\right]/kT}$$

$$\tag{13.70}$$

$$n(HII) = \frac{n(HI)}{n_e} \frac{U_e\left(2\pi m_e kT/h^2\right)^{3/2} U(HII)\left[2\pi m(HII)kT/h^2\right]^{3/2}}{U(HI)\left[2\pi m(HI)kT/h^2\right]^{3/2}}$$

$$e^{-[E(HII)-E(HI)]/kT} \tag{13.71}$$

$$n(H_2^+) = \frac{n(HI)n(HI)}{n_e} \frac{U_e\left(2\pi m_e kT/h^2\right)^{3/2} U(HII)\left[2\pi m(HII)kT/h^2\right]^{3/2}}{\left\{U(HI)\left[2\pi m(HI)kT/h^2\right]^{3/2}\right\}^2}$$

$$e^{-\left[E(H_2^+)-E(HI)\right]/kT} \tag{13.72}$$

$$n(H^-) = n(HI)n_e \frac{U(H^-)\left[2\pi m(H^-)kT/h^2\right]^{3/2}}{U(HI)\left[2\pi m(HI)kT/h^2\right]^{3/2}U_e\left(2\pi m_e kT/h^2\right)^{3/2}}$$

$$e^{-\left[E(H^-)-E(HI)\right]/kT} \tag{13.73}$$

where $U(HII)=1$, $U_e=2$, $m(HI) \approx m(HII) \approx m(H^-)$, and $E(He) < E(H^-)$ $< E(HI) < E(H_2^+) < E(HII)$.

After converting the equations for conservation for different types of particles into functions of the number density of neutral atoms and electons, i.e., n_A and n_e, the equations have the form

$$\left. \begin{array}{l} f_g(n_A, n_e, n_H, n_{He}) = 0 \\ f_H(n_A, n_e, n_H, n_{He}) = 0 \\ f_{He}(n_A, n_e, n_H, n_{He}) = 0 \\ f_e(n_A, n_e, n_H, n_{He}) = 0 \end{array} \right\} \tag{13.74}$$

The equation set (13.74) is not a set of linear equations, and can be solved using a method similar to the Henyey method described in §7.4. For this, we should first give a set of initial approximate solutions for the number densities n_A, n_e, n_H and n_{He}. Due to the imprecision of the initial solutions, the equation set (13.74) cannot be fulfilled, and we need to find the correction terms δn_A, δn_e, δn_H and δn_{He}, which can be obtained from the linear equation set

$$f_i + \sum_j \frac{\partial f_i}{\partial n_j}\delta n_j = 0 \tag{13.75}$$

The correction terms are determined through repeated iterations. If the correction terms δn_A, δn_e, δn_H and δn_{He} for any step are not less than a certain precision ε, the values after correction, i.e., $n_A + \delta n_A$, $n_e + \delta n_e$, $n_H + \delta n_H$ and $n_{He} + \delta n_{He}$ are used as new initial approximate solutions to calculate the new correction terms. The steps are repeated until the correction terms are all less than a given precision ε.

When the number density for neutral atoms is obtained for hydrogen and helium, it is not difficult to calculate the atomic occupancy number at each energy state for each degree of ionization for hydrogen and helium through the Boltzmann equation [see Eq. (4.4)] and the Saha equation [see Eq. (4.10)].

If the stellar temperature is relatively high, there will be no cross terms for different elements in the equations for conservation of different particles, so a simpler method can be employed for solving the number densities of the atoms and ions. At this time, the equation for conservation of particles for each element can be written as

$$n_A \cdot X_Q = n(Q) + n(Q^+) + n(Q^{++}) + \dots \tag{13.76}$$

and

$$n_A = n_g - n_e \tag{13.77}$$

We first estimate a value for the electron number density n_e so every term in Eq. (13.76) can be expressed as a function of the number density $n(Q)$ for neutral atoms and the number density n_e for electrons for that element through the Saha equation, e.g.,

$$n(Q^+) = \frac{n(Q)}{n_e} K(Q) \tag{13.78}$$

$$n(Q^{++}) = \frac{n(Q^+)}{n_e} K(Q^+) = \frac{n(Q)}{n_e} K(Q) \frac{K(Q^+)}{n_e} \tag{13.79}$$

Substituting Eqs. (13.78) and (13.79) into Eq. (13.76), we have

$$n_A \cdot X_Q = n(Q) + \frac{n(Q)}{n_e} K(Q) + \frac{n(Q)}{n_e} K(Q) \frac{K(Q^+)}{n_e} + \dots \tag{13.80}$$

Since n_e is given, n_A can be solved from Eq. (13.77) [n_g in Eq. (13.77) can be calculated from P_g], so $n(Q)$ can be solved from Eq. (13.80). We also solve for $n(Q^+)$, $n(Q^{++})$, ... from equations (13.78), (13,79), ... In this way, the electric charges provided by the element Q can be solved as

$$n(Q^+) + 2n(Q^{++}) + 3n(Q^{+++}) + \dots \tag{13.81}$$

Adding the electric charge provided by each element, we get the electron number density which is written as $n_e^{(1)}$. If this calculated $n_e^{(1)}$ does not agree with the estimated n_e (i.e., the difference is not smaller than the required precision ε), we can use $n_e^{(1)}$ as the new estimated value and then repeat the above calculations. The steps are repeated until the calculated n_e is consistent with the input value. At this time, the number densities for the atoms and ions obtained in Eqs. (13.78), (13.79), ... are the ones we require.

(D) Calculations of the optical depth

On integrating the hydrostatic equilibrium equation, we get $P_g(\bar{\tau})$, $T(\bar{\tau})$, $\bar{\kappa}_\nu(\bar{\tau})$, ... Here, $\bar{\tau}$ is the average optical depth. The next step is to integrate the radiative transfer equation. In the radiative transfer equation, the optical depth at a certain frequency τ_ν is chosen as an independent variable. We have to first determine the relation between τ_ν and $\bar{\tau}$, and then integrate the radiative transfer equation using the results from the integration of the hydrostatic equilibrium equation. The relation between τ_ν and $\bar{\tau}$ can be obtained through the formula

$$\tau_\nu(\bar{\tau}) = \int_0^{\bar{\tau}} (\kappa_\nu^T(t) + \sigma_\nu(t))dt \tag{13.82}$$

where the opacity κ_ν has been expressed as the combination of the true absorption coefficient κ_ν^T and the scattering absorption coefficient σ_ν. To integrate Eq. (13.82), we can first expand the integrand as a hyperbolic function, and then use the summation to approximate the integration, such as

$$\int_0^{X_n} f(x)dx \cong \sum_j \int_{X_j}^{X_{j+1}} f(x)dx = \sum_j \int_{X_j}^{X_{j+1}} (a_j + b_j x + c_j x^2)dx$$

$$= \sum_j \left[a_j(x_{j+1} - x_j) + b_j \frac{(x_{j+1}^2 - x_j^2)}{2} + c_j \frac{x_{j+1}^3 - x_j^3}{3} \right]$$

$$= \sum_j \left[a_j + b_j \frac{x_{j+1} + x_j}{2} + c_j \frac{(x_{j+1}^2 + x_{j+1}x_j + x_j^2)}{3} \right] (x_{j+1} - x_j) \tag{13.83}$$

where a_j, b_j and c_j are

$$\left. \begin{array}{l} a_j = f_{j-1} - x_{j-1} \dfrac{f_j - f_{j-1}}{x_j - x_{j-1}} + x_j x_{j-1} c_j \\[3mm] b_j = \dfrac{f_j - f_{j-1}}{x_j - x_{j-1}} - (x_j + x_{j-1}) c_j \\[3mm] c_j = \dfrac{f_{j+1}}{(x_{j+1} - x_j)(x_{j+1} - x_{j-1})} - \dfrac{f_j}{(x_j - x_{j-1})(x_{j+1} - x_j)} + \dfrac{f_{j-1}}{(x_j - x_{j-1})(x_{j+1} - x_{j-1})} \end{array} \right\} \tag{13.84}$$

In realistic calculations, it is common not to use the above coefficients a_j, b_j and c_j directly. To be more precise, we can define a weighting function w_j and use it to calculate the mean values $\bar{a}_j$, $\bar{b}_j$ and $\bar{c}_j$, which are then substituted into Eq. (13.83). The mean coefficients are evaluated by

$$w_j = \frac{|c_{j+1}|}{|c_{j+1}| + |c_j|} \tag{13.85}$$

$$\left.\begin{aligned}
\bar{a}_j &= w_j a_j + (1 - w_j)a_{j+1} \\
\bar{b}_j &= w_j b_j + (1 - w_j)b_{j+1} \\
\bar{c}_j &= w_j c_j + (1 - w_j)c_{j+1}
\end{aligned}\right\} \tag{13.86}$$

(E) Integration of the radiative transfer equation

The integration of the radiative transfer equation can be divided into several parts.

(1) Calculation of the mean radiation intensity and the radiation flux
From the general solution of the radiative transfer equation, we can obtain the expressions for the mean radiation intensity $J_\nu(\tau_\nu)$, the radiation flux $F_\nu(\tau_\nu)$ and the K-integral $K_\nu(\tau_\nu)$ [see Eqs. (2.85), (2.87) and (2.89)]:

$$J_\nu(\tau_\nu) = \frac{1}{2} \int_0^\infty S_\nu(t_\nu)E_1|t_\nu - \tau_\nu|dt \tag{13.87}$$

$$F_\nu(\tau_\nu) = 2 \int_{\tau_\nu}^\infty S_\nu(t_\nu)E_2|t_\nu - \tau_\nu|dt_\nu - 2 \int_0^{\tau_\nu} S_\nu(t_\nu)E_2|t_\nu - \tau_\nu|dt_\nu \tag{13.88}$$

$$K_\nu(\tau_\nu) = \frac{1}{2} \int_0^\infty S_\nu(t_\nu)E_3|t_\nu - \tau_\nu|dt_\nu \tag{13.89}$$

where $S_\nu(t_\nu)$ is the source function, $E_n|t_\nu - \tau_\nu|$ is the nth order exponential integral. If the above expressions are written as a unified expression with the omission of subscripts, we have

$$M_n(\tau) = \frac{1}{2} \int_0^\infty sign(t - \tau)^{n-1}S(t)E_n|t - \tau|dt \tag{13.90}$$

Comparing Eq. (13.90) with Eqs. (13.87), (13.88) and (13.89), we have

$$\left. \begin{array}{l} J(\tau) = M_1(\tau) \\[6pt] H(\tau) = \dfrac{1}{4}F(\tau) = M_2(\tau) \\[6pt] K(\tau) = M_3(\tau) \end{array} \right\} \tag{13.91}$$

Our objective is to integrate the radiative transfer equation, and finally obtain the expression for $H(\tau)$. Therefore, we have to compute the function expressed by Eq. (13.90) and obtain $M_2(\tau)$. Dividing the integration region into N small intervals, and writing the integration in Eq. (13.90) in the form of summation, we have

$$M_{nl} \equiv M_n(\tau_l) = \frac{1}{2} \sum_{j=1}^{N} sign(\tau_j - \tau_l)^{n-1} \int_{\tau_j}^{\tau_{j+1}} S(t)E_n(t - \tau_l)dt \tag{13.92}$$

The integral on the right of the above equation can be evaluated using the method in Eq. (13.83), i.e., by assuming that the source function S can be expanded as a hyperbolic function in the region (τ_j, τ_{j+1}):

$$S(t) = \sum_{K=1}^{3} t^{K-1} \sum_{i=1}^{N} C_{jKi}S_i \tag{13.93}$$

where C_{jKi} is the interpolation coefficient and is a function of τ_i. Substituting Eq. (13.93) into Eq. (13.92), we have

$$M_{nl} = \sum_{j=1}^{N} R_{nlj}S_j \tag{13.94}$$

where

$$R_{nlj} = \sum_{K=1}^{3} \sum_{i=1}^{N} \left[\frac{1}{2} sign(\tau_j - \tau_l)^{n-1} \int_{\tau_j}^{\tau_{j+1}} t^{K-1} E_n |\tau_l - t| dt \right] C_{iKj} \tag{13.95}$$

is the matrix element for the operator matrix R_n. If the operator matrix R_n is known, Eq. (13.92) can be written as

$$M_n = R_n S \tag{13.96}$$

(2) Calculation of the source function

When the source function S is known, we know from the above that we can multiply the operator matrix R_2 by the source function S to get the radiation flux H, i.e., we can achieve the objective in solving the radiative transfer equation.

When the atmosphere is under LTE, and the radiation comes from isotropic coherent scattering, from Eqs. (2.35), (2.40), (2.53) and (2.58), we have

$$S_v = \frac{\eta_v}{\kappa_v} = \frac{\eta_v^T}{\kappa_v^T + \sigma_v} + \frac{\eta_v^s}{\kappa_v^T + \sigma_v}$$

$$= \frac{\kappa_v^T}{\kappa_v^T + \sigma_v} B_v + \frac{\sigma_v}{\kappa_v^T + \sigma_v} J_v \qquad (13.97)$$

Denoting

$$\rho_v \equiv \frac{\sigma_v}{\kappa_v^T + \sigma_v} \qquad (13.98)$$

Equation (13.97) can be written as

$$S_v = (1 - \rho_v)B_v + \rho_v J_v \qquad (13.99)$$

which is the integral equation for S_v. According to Eq. (13.87), J_v is also a function of S_v. Equation (13.99) can also be written in the form of a vector equation. Considering Eq. (13.96), we have

$$\vec{S}_v = (I - \rho_v)\vec{B}_v + \rho_v R_{1v}\vec{S}_v, \qquad (13.100)$$

where $\vec{B}_v$ and $\vec{S}_v$ are N-dimensional vectors:

$$\vec{B}_v = (B_{1v}, B_{2v}, ..., B_{Nv}) \qquad (13.101)$$

$$\vec{S}_v = (S_{1v}, S_{2v}, ...S_{Nv}) \qquad (13.102)$$

I is the identity matrix, ρ_v is a diagonal matrix with $(\rho_v)_{ii} = \rho_{iv}$, and R_{1v} is the operator matrix determined by Eq. (13.95). Equation (13.100) can also be written as

$$(I - \rho_v R_{1v})\vec{S}_v = (I - \rho_v)\vec{B}_v \qquad (13.103)$$

which can be solved by iterations. At the beginning, we assume one $\vec{S}_v{}^0$ such as $\vec{S}_v^0 = \vec{B}_v$, which is multiplied with the matrix $(I - \rho_v R_{1v})$ to get a vector. This vector is not equal to $(I - \rho_v)\vec{B}_v$ and the difference is Δ_v, i.e.,

$$\Delta_v = (I - \rho_v R_{1v})\vec{S}_v^o - (I - \rho_v)\vec{B}_v \tag{13.104}$$

Our objective is to find the correction $\Delta\vec{S}_v$ for $\Delta\vec{S}_v^o$ in such a way that

$$(I - \rho_v R_{1v})(\vec{S}_v^o + \Delta\vec{S}_v) = (I - \rho_v)\vec{B}_v \tag{13.105}$$

for which the solution is

$$\begin{aligned}
\Delta\vec{S}_v &= (I - \rho_v R_{1v})^{-1}\big[(I - \rho_v)\vec{B}_v - (I - \rho_v R_{1v})\vec{S}_v^o\big] \\
&= -(I - \rho_v R_{1v})^{-1}\Delta_v
\end{aligned} \tag{13.106}$$

Since $(I - \rho_v R_{1v})$ is a diagonal matrix, the inverse matrix is

$$(I - \rho_v R_{1v})^{-1} = \frac{1}{I - \rho_{iv}(R_{1v})_{ii}} \tag{13.107}$$

from which we get

$$\Delta\vec{S}_v = \frac{-\Delta_v}{I - \rho_{iv}(R_{1v})_{ii}} \tag{13.108}$$

We write $\vec{S}_v^{(1)} = \vec{S}_v^o + \Delta\vec{S}_v$, and repeat the calculations for $\Delta\vec{S}_v$. The steps are repeated until $(\Delta\vec{S}_v/\vec{S}_v^{(1)})$ is less than a chosen precision.

We now summarize the calculation steps in the above.

Step 1: Calculation for ρ_v and $\vec{B}_v$: ρ_v can be obtained from Eq. (13.98) and $\vec{B}_v$ can be obtained from the definition of Planck's function.

Step 2: Take $\vec{S}_v^o = \vec{B}_v$ as the initial value and then calculate $\Delta\vec{S}_v$ using Eq. (13.108). To calculate $\Delta\vec{S}_v$, we have to first calculate ρ_v, R_{1v} and Δ_v in the equation, where
ρ_v: calculated in Step 1
R_{1v}: the matrix elements can be calculated from Eq. (13.95)
$\Delta_v = (I - \rho_v R_{1v})\vec{S}_v^o - (I - \rho_v)\vec{B}_v$. Substituting $\vec{S}_v^o = \vec{B}_v$, we can get Δ_v.

Step 3: Taking $\vec{S}_v^{(1)} = \vec{S}_v^o + \Delta\vec{S}_v$, the calculations for $\Delta\vec{S}_v$ are repeated. During repeated calculations, we only need to replace $\vec{S}_v^o$ in Eq. (13.104) by $\vec{S}_v^{(1)}$. Such iterations are carried out until $\Delta\vec{S}_v/\vec{S}_v^{(1)} < 10^{-5}$. Then we obtain the required $\vec{S}_v$.

(3) Table for the operator matrix

In the above calculations, the most complicated and time-consuming calculations are those for the operator matrix. We have to calculate R_{1lj} to solve for the source function, and R_{2lj} to calculate the radiation flux. In the ATLAS LTE model atmosphere, Kurucz divided the optical depth τ of the atmosphere into 43 mesh points and calculated the operator matrix corresponding to these mesh points, i.e., they gave the table relating the operator matrix and the optical depth. In this way, one needs only to interpolate the corresponding values of the operator matrices from the table according to a τ_l value with realistic calculations, and the amount of calculations is greatly reduced.

(4) Total radiation flux H

For a fixed frequency ν, we can first determine its source function $\vec{S}_\nu$, and then find H_ν from the matrix equation (13.96). Integrating H_ν with respect to the frequency, the total radiation flux H can be obtained:

$$H = \int_0^\infty H_\nu d\nu \tag{13.109}$$

(F) Correction for the temperature distribution

When integrating the hydrostatic equilibrium equation, we first assume a temperature distribution in the atmosphere, such as the temperature distribution of a gray atmosphere. If the assumed distribution is correct, and the atmosphere has no convection at the same time, the total radiation flux H (denoted as H°) obtained from integrating the radiative transfer equation should fulfil the energy theorem, i.e.,

$$\left. \begin{array}{c} H^\circ = \dfrac{\sigma T^4_{eff}}{4\pi} \\[2ex] \text{and for the entire atmosphere, we have } \dfrac{dH^\circ}{d\tau} = const \end{array} \right\} \tag{13.110}$$

However, since the assumed temperature distribution might not be accurate enough, and since there might be convection inside the atmosphere layer, the H° obtained from integrating the radiative transfer equation might not fulfil the energy theorem in Eq. (13.110). If $H^{(1)}$ represents the true total radiation flux, i.e.,

$$H^{(1)} = \frac{\sigma T_{eff}^4}{4\pi} \qquad (13.111)$$

the deviation of the total radiation flux is

$$\Delta H = H^{(1)} - H^{\circ} \qquad (13.112)$$

Our objective is to correct the temperature distribution so that the obtained H value approaches the true value, or ΔH approaches zero.

If the depth of the atmosphere is measured through the variable m, which represents the mass inside the cylinder with a base area of 1 cm² extending from the stellar surface into the star, i.e.,

$$m = \int_0^x \rho(x)dx$$

or written in the differential form

$$dm = \rho dx \qquad (13.113)$$

the correction to the temperature distribution can be calculated through

$$\Delta T(m) = -\frac{dT}{dm} \int_0^m \frac{H^{\circ}(y) - H^{(1)}(y)}{H^{\circ}(y)} dy \qquad (13.114)$$

The corrections for the temperature are calculated through iterations. If the chosen $\Delta T(m)$ leads to a smaller ΔH from new calculations and does not change the sign, we can multiply the chosen $\Delta T(m)$ by 1.25 and repeat the above calculations. If the chosen $\Delta T(m)$ leads to a larger ΔH from new calculations or changes the sign, we can multiply the chosen $\Delta T(m)$ by 0.5 and repeat the above calculations. The above steps are repeated until ΔH is smaller than a certain precision.

(G) Line-blanketing effect

A lot of absorption lines are embedded in the continuous spectra of stars. The presence of absorption lines will significantly affect the flux distribution in the continuous spectrum and the temperature distribution inside the stellar atmosphere. Since the absorption line regions are much darker than the surrounding continuous regions, a large amount of absorption lines will decrease the energy flux, which is called the "blockage effect". However, according to the energy theorem, the total energy flux should be a constant. If the energy flux is "blocked", this should lead to a redistribution of the temperature inside the atmosphere. Only when the temperature is raised in some layers in the atmosphere will the total energy flux be maintained as a constant. The redistribution of temperature will in turn change the flux distribution of the continuous spectrum. On the other hand, for a volume element close to the surface of the atmosphere, the presence of absorption lines will decrease the average radiation intensity J_ν. This will decrease the energy absorbed by the volume element. However, the energy emitted by the volume element can be directly radiated away from the stellar surface, so the energy absorbed by the element is less than the energy emitted, and the temperature of the volume element will be automatically decreased to attain a new equilibrium between the absorbed and the emitted energy. In this way, the presence of absorption lines can lower the temperature of the surface of the atmosphere. From this, we can see that a large amount of absorption lines embedded in the continuous spectrum can change the flux distribution of the continuous spectrum and the temperature inside the atmosphere, which lowers the temperature at the surface of the atmosphere. All these effects are collectively known as the "blanketing effects".

The theoretically calculated model atmosphere should take into consideration the blanketing effects; otherwise, the results obtained will not agree with the realistic stellar atmosphere.

The consideration of blanketing effects will include the corresponding absorption coefficients for the absorption lines when calculating the absorption coefficient. In general, the methods include the following.

(1) Direct calculations
Direct calculations include the absorption coefficients for absorption lines appearing in the spectrum when calculating the absorption coefficient, and take as many mesh points as possible inside the frequency range of every spectral line.

(2) Statistical method
The statistical method was introduced by Kurucz (1979). It divides the entire stellar spectrum into many frequency intervals $\Delta\nu$, and each frequency interval $\Delta\nu$ is sufficiently small so that the continuous absorption coefficient κ_ν^c and the continuous source function S_ν^c can be

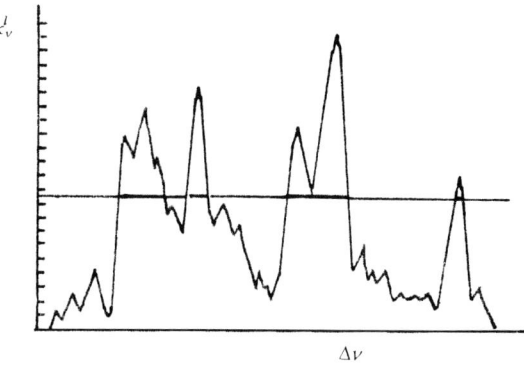

Fig. 13.2 The distribution of absorption coefficient inside a frequency interval $\Delta\nu$.

treated as constants inside the intervals. Figure 13.2 shows the distribution of the absorption coefficient inside a frequency interval $\Delta\nu$. We then express the absorption coefficient in this frequency interval in terms of a smooth curve determined by a single parameter, which is called the distribution curve of absorption coefficient and is shown in Fig. 13.3. The detailed derivation of the distribution curve of absorption coefficient is as follows. We first divide the vertical axis in Fig. 13.2 into several small intervals, which represent different values of κ_ν^ℓ. For each value of κ_ν^ℓ, we construct a horizontal line which crosses the curves of absorption coefficients at certain points separated by different lengths as shown by the bold lines in Fig. 13.2. The ratio of the total length of these bold lines to the total length in this interval is denoted as f. When we plot κ_ν^ℓ against f, we get the distribution curve of absorption coefficient as shown in Fig. 13.3. When the distribution curve of absorption coefficient is divided into n intervals (see Fig. 13.4). Each interval has a weighting factor w_i and corresponds to an

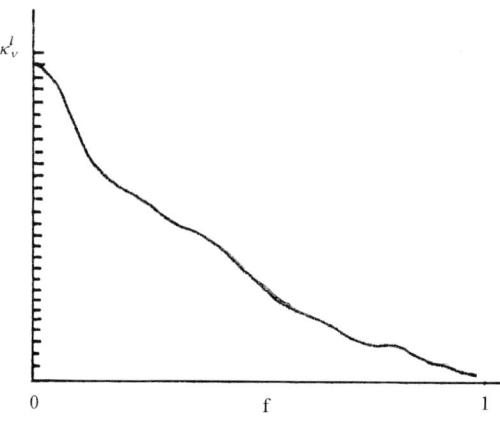

Fig. 13.3 The distribution curve of absorption coefficient.

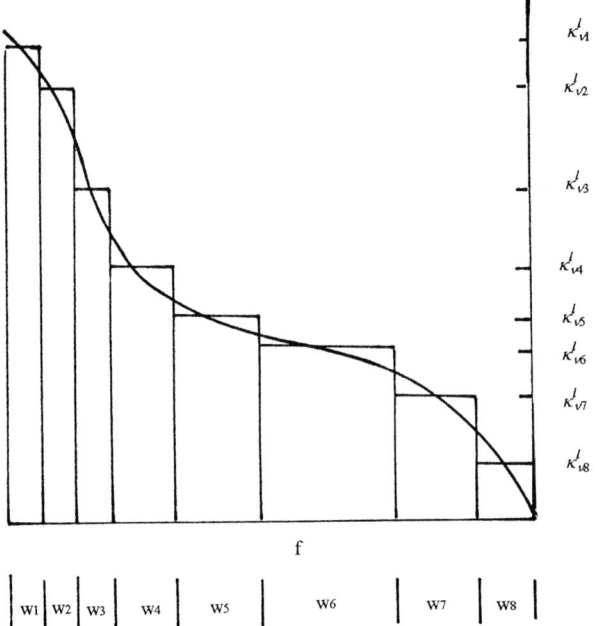

Fig. 13.4 The division of the distribution curve of the absorption coefficient into n levels.

absorption coefficient κ_{vi}^{ℓ}. The magnitude of the weighting factor w_i is inversely proportional to the width of that interval. Therefore, the absorption coefficient in this frequency interval Δv is

$$\kappa_v^l = \sum_{i=1}^{n} w_i \kappa_{vi}^l \tag{13.115}$$

To prevent two different spectra from having the same distribution curves of absorption coefficient in the same frequency interval, we can require the same shape of distribution of the line absorption coefficient for the same frequency interval in the entire atmosphere, such as that shown in Fig. 13.5.

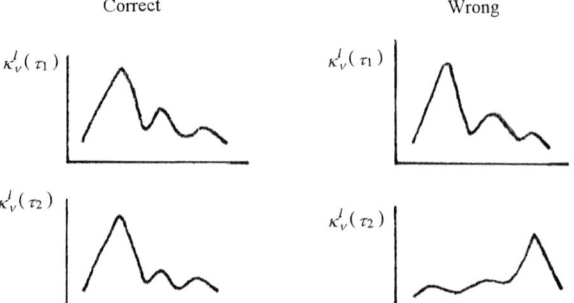

Fig. 13.5 The requirement of same shape of the distribution of the line absorption coefficient for the same frequency interval.

For every given frequency interval, all the absorption coefficient distribution curves are calculated according to different temperatures, pressures, electron number densities and elemental abundance, which are then stored in a magnetic tape. In calculations of the model atmosphere, the required line absorption coefficient is then obtained according to the temperature and pressure for each atmosphere layer.

13.1.4 Non-LTE Model Atmosphere

Some stellar atmospheres might have very small densities or very high effective temperatures, so they deviate considerably from the local thermodynamic equilibrium state. In calculating the model atmospheres for these stars, we have to adopt non-local thermodynamic equilibrium (non-LTE) model atmospheres.

The basic difference between the non-LTE model atmosphere and the LTE model atmosphere is the adoption of the statistical equilibrium equation instead of the Boltzmann equation and the Saha equation in the calculation of occupancy number of atomic energy states in the non-LTE model atmosphere. In this section, we will introduce the complete linearized method introduced by Auer and Mihalas (1969), and the linearized separate solution method introduced by Wu (1992) and Zhou (1995) to calculate the non-LTE model atmosphere.

The complete linearized method can be divided into the following steps.

(A) Rewriting the basic equation set

The basic equation set for the non-LTE model atmosphere has been given in §13.1.1. We now need to rewrite this set of equations, which will be divided into three steps. First, we introduce a new variable and express the equation set in terms of the new variable. At the same time, the first order radiative transfer equation is rewritten as a second order differential equation. Second, the above rewritten basic equations are converted into difference equations. Third, the Eddington factor is introduced to simplify the equation set.

We now introduce the new variable m, which is the mass of matter inside a cylinder with a base area of 1 cm² extending from the surface of the atmosphere into the star:

$$dm = \rho dx \qquad (13.116)$$

where x is the geometric depth of the atmosphere and $x = 0$ is equivalent to the surface of the atmosphere, and ρ is the density.

In terms of the variable m, the basic equations are rewritten as follows:

(1) Radiative transfer equation

When m is the variable, the radiative transfer equation (13.1) can be written as

$$\mu \frac{dI_\nu(m, \mu)}{dm} = w_\nu \left[I_\nu(m, \mu) - \frac{\alpha_\nu}{\kappa_\nu} - \frac{n_e a_e}{\kappa_\nu} J_\nu) \right] \qquad (13.117)$$

where $w_\nu = \kappa_\nu/\rho$ and according to Eq. (13.4), we have written η_ν as

$$\eta_\nu = \alpha_\nu + n_e a_e J_\nu \qquad (13.118)$$

and

$$\alpha_\nu = \frac{2h\nu^3}{c^2} \sum_\alpha \sum_{\beta=0}^{\beta_\alpha} \left[\sum_i^{l_{\alpha,\beta-1}} \sum_{j>i}^{l_{\alpha,\beta}} n_j^{\alpha\beta} \frac{g_i^{\alpha\beta}}{g_j^{\alpha\beta}} a_{ij}^{\alpha\beta}(\nu) + e^{-h\nu/kT} \right. $$
$$\left. \times \left(\sum_{i=1}^{l_{\alpha,\beta}} \sum_{K=1}^{l_{\alpha,\beta+1}} n_i^{*\alpha\beta} a_{iK}^{\alpha\beta}(\nu) + n_e a_{KK}^{\alpha\beta}(\nu) \sum_{i=1}^{l_{\alpha,\beta}} n_i^{\alpha,\beta+1} \right) \right] \qquad (13.119)$$

Equation (13.117) is a first order differential equation. We have to rewrite it as a second order differential equation since the boundary value for a second order differential equation is more stable than that for a first order differential equation. The conversion of a first order differential equation into a second order differential equation was first proposed by Feautrier (1964). To achieve this, he introduced two quantities $u_\nu(m, \mu)$ and $v_\nu(m, \mu)$ to replace I_ν, which are defined by

$$u_\nu(m, \mu) \equiv \frac{1}{2} \left[I_\nu(m, \mu) + I_\nu(m, -\mu) \right]$$
$$v_\nu(m, \mu) \equiv \frac{1}{2} \left[I_\nu(m, \mu) - I_\nu(m, -\mu) \right] \qquad (13.120)$$

Equation (13.117) is valid for μ in the range $(-1, +1)$. If μ is restricted to vary in $(0, 1)$, Eq. (13.117) should be changed to two equations:

$$\mu \frac{dI_\nu(m, \mu)}{dm} = w_\nu(m) \left[I_\nu(m, \mu) - \frac{\alpha_\nu}{\kappa_\nu} - \frac{n_e a_e}{\kappa_\nu} J_\nu \right] \qquad (13.121)$$

$$-\mu \frac{dI_\nu(m, -\mu)}{dm} = w_\nu(m)\left[I_\nu(m, -\mu) - \frac{\alpha_\nu}{\kappa_\nu} - \frac{n_e a_e}{\kappa_\nu}J_\nu\right] \qquad (13.122)$$

Subtracting the above two equations from each other, we get

$$\mu \frac{du_\nu(m, \mu)}{dm} = w_\nu(m) \cdot v_\nu(m, \mu) \qquad (13.123)$$

while adding the two equations will give

$$\mu \frac{dv_\nu(m, \mu)}{dm} = w_\nu(m)\left[u_\nu(m, \mu) - \frac{\alpha_\nu}{\kappa_\nu} - \frac{n_e a_e}{\kappa_\nu}J_\nu\right] \qquad (13.124)$$

Substituting Eq. (13.123) into Eq. (13.124), we obtain

$$\mu^2 \frac{d}{dm}\frac{1}{w_\nu}\frac{du_\nu(m, \mu)}{dm} = w_\nu(m)\left[u_\nu(m, \mu) - \frac{\alpha_\nu}{\kappa_\nu} - \frac{n_e a_e}{\kappa_\nu}J_\nu\right] \qquad (13.125)$$

In this way, we have changed the first order differential equation in I_ν into a second order differential equation in u_ν. Here μ only varies within (0, 1). We proceed to discuss the change in the boundary conditions. Adding the two equations in Eq. (13.120), we obtain

$$v_\nu(m, \mu) = I_\nu(m, \mu) - u_\nu(m, \mu) \qquad (13.126)$$

The boundary condition in Eq. (13.5) at the stellar surface can be written as

$$I(0, -\mu) = 0 \qquad (13.127)$$

The boundary condition at the stellar surface can be obtained from Eqs. (13.127), (13.126) and (13.123) as

$$m = 0: \quad \mu \frac{du_\nu(0, \mu)}{dm} = w_\nu u_\nu(0, \mu) \qquad (13.128)$$

From the boundary condition at the bottom of the atmosphere in Eq. (13.6)

$$I_\nu(m_N, \mu) = I_\nu^+ \qquad (13.129)$$

and Eqs. (13.126) and (13.123), we obtain the boundary condition at the bottom of the atmosphere as

$$m = m_N \; : \; \mu \frac{du_\nu(m_N, \mu)}{dm} w_\nu(m_N)\left[I_\nu^+ - u_\nu(m_N, \mu)\right] \qquad (13.130)$$

In addition, after the introduction of u_ν, the expressions for the average radiation intensity J_ν, the Eddington radiation flux H_ν and the K_ν integral should become

$$J_\nu = \frac{1}{2}\int_{-1}^{+1} I_\nu(\mu)d\mu = \frac{1}{2}\left(\int_{-1}^{0} I_\nu(\mu)d\mu + \int_{0}^{1} I_\nu(\mu)d\mu\right)$$

$$= \frac{1}{2}\left(\int_{0}^{1} I_\nu(-\mu)d\mu + \int_{0}^{1} I_\nu(\mu)d\mu\right) = \int_{0}^{1} u_\nu(\mu)d\mu$$

$$(13.131)$$

$$H_\nu = \int_{0}^{1} v_\nu(\mu)\mu d\mu = \int_{0}^{1}\left[I_\nu(\mu) - u_\nu(\mu)\right]\mu d\mu \qquad (13.132)$$

$$K_\nu = \int_{0}^{1} u_\nu(\mu)\mu^2 d\mu \qquad (13.133)$$

Using Eq. (13.131) the radiative transfer equation (13.125) can be written as

$$\mu^2 \frac{d}{dm}\frac{1}{w_\nu}\frac{du_\nu(m, \mu)}{dm} = w_\nu\left[u_\nu(m, \mu) - \frac{\alpha_\nu}{\kappa_\nu} - \frac{n_e a_e}{\kappa_\nu}\int_{0}^{1} u_\nu(\mu)d\mu\right] \qquad (13.134)$$

(2) Hydrostatic equilibrium equation
When m is used as the variable, the hydrostatic equilibrium equation (13.8) can be changed through Eq. (13.131) to

$$\frac{d}{dm}(n_g kT) + \frac{d}{dm}\frac{4\pi}{c}\int_{\nu}\int_{0}^{1} u_\nu(\mu)\mu^2 d\mu d\nu = g \qquad (13.135)$$

(3) Energy equilibrium equation

Using Eq. (13.131), the energy equilibrium equation (13.11) can be rewritten as

$$\int_\nu \int_0^1 \left[(\kappa_\nu - n_e a_e) u_\nu(\mu) - \alpha_\nu \right] d\mu \, d\nu = 0 \qquad (13.136)$$

(4) Other equations

It is not necessary to change the equations for conservation of total particles, electric charge and particles for different elements. The statistical equilibrium equation (13.17) has included the radiative transition rate R_{ij} and R_{iK}. We need only to replace J_ν by Eq. (13.131) according to their expressions in Eq. (13.19).

We have now rewritten the basic equations in terms of the variable m. For the convenience of numerical calculations, we still need to convert the above basic equations to difference equations. We first change the continuous regions for the variables m, ν, μ to a series of discrete values, i.e.,

$$\{ m_d \}, \quad d = 1, 2..., N$$
$$\{ \nu_k \}, \quad k = 1, 2..., K$$
$$\{ \mu_j \}, \quad j = 1, 2..., M$$

(i) Rewriting the radiative transfer equation as a difference equation

Dividing Eq. (13.134) by $w(m)$, replacing the integral by the following equations

$$\int_0^1 f(\mu) d\mu \rightarrow \sum_{j=1}^M a_j f(\mu_j), \qquad \int_0^\infty f(\nu) d\nu \rightarrow \sum_{k=1}^K b_k f(\nu_k) \qquad (13.137)$$

and replacing differentiation by differences

$$\left(\frac{dx}{d\tau} \right)_{d+\frac{1}{2}} \approx \left(\frac{\Delta x_{d+\frac{1}{2}}}{\Delta \tau_{d+\frac{1}{2}}} \right) = (x_{d+1} - x_d)/(\tau_{d+1} - \tau_d) \qquad (13.138)$$

$$\left(\frac{d^2 x}{d\tau^2} \right)_d \approx \left[\left(\frac{dx}{d\tau} \right)_{d+\frac{1}{2}} - \left(\frac{dx}{d\tau} \right)_{d-\frac{1}{2}} \right] / \left[\frac{1}{2} \left(\Delta \tau_{d+\frac{1}{2}} + \Delta \tau_{d-\frac{1}{2}} \right) \right], \qquad (13.139)$$

where

$$\Delta \tau_{d\pm\frac{1}{2}} \equiv \frac{1}{2} \left[w(m_{d\pm1}) + w(m_d) \right] \cdot \left| m_{d\pm1} - m_d \right| \tag{13.140}$$

$$\Delta \tau_d \equiv \frac{1}{2} \left[\Delta \tau_{d+\frac{1}{2}} + \Delta \tau_{d-\frac{1}{2}} \right] \tag{13.141}$$

Eq. (13.134) becomes the difference equation

$$\frac{\mu_j^2}{\Delta \tau_d(v_k) \Delta \tau_{d-\frac{1}{2}}(v_k)} u(m_{d-1}, v_k, \mu_j) - \frac{\mu_j^2}{\Delta \tau_d(v_k)} \left(\frac{1}{\Delta \tau_{d+\frac{1}{2}}(v_k)} + \frac{1}{\Delta \tau_{d-\frac{1}{2}}(v_k)} \right)$$

$$u(m_d, v_k, \mu_j)$$

$$+ \frac{\mu_j^2}{\Delta \tau_d(v_k) \Delta \tau_{d+\frac{1}{2}}(v_k)} u(m_{d+1}, v_k, \mu_j) = u(m_d, v_k, \mu_j) - \frac{\alpha(m_d, v_k)}{\kappa(m_d, v_k)}$$

$$- \frac{n_e(m_d) a_e}{\kappa(m_d, v_k)} \left[\sum_{j'=1}^{M} a_{j'} u(m_d, v_k, \mu_{j'}) \right] \tag{13.142}$$

To obtain the surface boundary condition for the radiative transfer equation, we expand u_v at m_1 as

$$u_v(m_2, \mu) = u_v(m_1, \mu) + \left. \frac{du_v}{d\tau_v} \right|_{m_1} \cdot \Delta \tau_{3/2} + \left. \frac{d^2 u_v}{d\tau_v^2} \right|_{m_1} \cdot \frac{1}{2} \Delta \tau_{3/2}^2 \tag{13.143}$$

and replace the first order and second order differentials of Eq. (13.143) by Eqs. (13.128) and (13.125), and get

$$\frac{\mu_j}{\Delta \tau_{3/2}} \left[\frac{\mu_j}{\Delta \tau_{3/2}} \left(u_2(v_k, \mu_j) - u_1(v_k, \mu_j) \right) - u_1(v_k, \mu_j) \right]$$

$$= \frac{1}{2} \left[u_1(v_k, \mu_j) - \frac{\alpha(1, v_k)}{\kappa(1, v_k)} - \frac{n_e(1) a_e}{\kappa(1, v_k)} \sum_{j=1}^{M} a_j u_v \right] \tag{13.144}$$

From Eq. (13.130), the boundary condition at the bottom of the atmosphere is

$$\frac{\mu_j \left[u(m_N, v_k, \mu_j) - u(m_{N-1}, v_k, \mu_j) \right]}{\Delta \tau_{N-\frac{1}{2}}} = I^+(v_k, \mu_j) - u(m_N, v_k, \mu_j) \tag{13.145}$$

In the differential equation, we can require $I^+ = B_\nu$ but in the difference equation, I^+ is the value between τ_N and τ_{N-1}. Therefore, we expand B_ν at τ_N and take the linear term, so we get

$$I^+(\nu_k, \mu_j) = B(\nu_k, T_N)$$

$$+ \frac{\partial B(\nu_k, T_N)}{\partial T_N} \cdot \frac{3}{\kappa_N(\nu_k)} \cdot \frac{\frac{\sigma T_{eff}^4}{4\pi} - \frac{1}{2}\sum\limits_{j'=1}^{M}\sum\limits_{k'=1}^{k} a_{j'}, b_{k'}, \mu_{j'}\left[B(\nu_{k'}, T_N) - u(\nu_{k'}, \mu_{j'}, m_N)\right]}{\sum\limits_{k'=1}^{k} b_{k'} \frac{1}{\kappa_N(\nu_{k'})} \frac{\partial B(\nu_{k'}, T_N)}{\partial T_N}}$$

$$(13.146)$$

(ii) Rewriting the hydrostatic equilibrium equation as a difference equation

$$\frac{k\left[n_g(m_d)T(m_d) - n_g(m_{d-1})T(m_{d-1})\right]}{m_d - m_{d-1}} + \frac{4\pi}{c}\sum_{j=1}^{M}\sum_{k=1}^{K} a_j b_k \mu_j^2 \qquad (13.147)$$

$$\left[u(m_d, \nu_k, \mu_j) - u(m_{d-1}, \nu_k, \mu_j)\right]/(m_d - m_{d-1}) = g$$

At the surface of the atmosphere, since $P_g(0) = 0$, we change the first term of the hydrostatic equilibrium equation to $kn_g(m_1)T(m_1)/m_1$, and replace the differential term in the second term by Eq. (13.128), and obtain the boundary condition for the hydrostatic equilibrium equation at the surface of the atmosphere as

$$\frac{kn_g(m_1)T(m_1)}{m_1} + \frac{4\pi}{c}\sum_{j=1}^{M}\sum_{k=1}^{K} a_j b_k \mu_j w(m_1, \nu_k) u(m_1, \nu_k, \mu_j) = g \quad (13.148)$$

(iii) Rewriting the energy equilibrium equation as a difference equation

$$\sum_{j=1}^{M}\sum_{k=1}^{K} a_j b_k \left[\left(\kappa(m_d, \nu_k) - n_e(m_d)a_e\right)u(m_d, \nu_k, \mu_j) - \alpha(m_d, \nu_k)\right] = 0$$

$$(13.149)$$

(iv) Rewriting the statistical equilibrium equation as a difference equation

Normally, we will not treat all energy states under the non-LTE case, and only treat δ out of S bound states of the hydrogen atom under the non-LTE case, and treat the occupancy number on the remaining $(S - \delta)$ bound states under the LTE case. The ground state of the helium atom is treated using non-LTE. Such a treatment is due to the fact that the deviation of the atomic occupancy number of high energy states is extremely small from that obtained from the Boltzmann distribution. In addition, the treatment of only several main discrete bound states by non-LTE can be handled by realistic volume and speed of computers.

Considering Eqs. (13.18) to (13.21), Eq. (13.17) can be rewritten as

$$
n_i^{\alpha,\beta}\left(\sum_{j \neq i}^{\delta} P_{ij}^{\alpha\beta} + \sum_{j=\delta+1}^{S} n_e\Omega_{ij}^{\alpha\beta}\right) + n_i^{\alpha\beta}(R_{iK}^{\alpha\beta} + n_e\Omega_{iK}^{\alpha\beta}) - \sum_{j \neq i}^{\delta} n_j^{\alpha\beta} P_{ji}^{\alpha\beta}
$$

$$
- \sum_{j=\delta+1}^{S} n_j^{*\alpha\beta} n_e\Omega_{ji}^{\alpha\beta} = n_i^{*\alpha\beta}(R_{Ki}^{\alpha\beta} + n_e\Omega_{iK}^{\alpha\beta})
$$

$$(13.150)$$

Defining

$$
\hat{\Omega}_{iK}^{\alpha\beta} = \Omega_{iK}^{\alpha\beta} + \sum_{j=\delta+1}^{S} \Omega_{ij}^{\alpha\beta}, \quad i = 1, ..., \delta \tag{13.151}
$$

using Eq. (13.151) and $C_{ij}^{\alpha\beta} = \left(\frac{n_j^{\alpha\beta}}{n_i^{\alpha\beta}}\right)^* C_{ji}^{\alpha\beta}$, Eq. (13.150) can be written as

$$
n_i^{\alpha\beta}\sum_{j \neq i}^{\delta} P_{ij}^{\alpha\beta} + n_i^{\alpha\beta}(R_{iK}^{\alpha\beta} + n_e\hat{\Omega}_{iK}^{\alpha\beta}) - \sum_{j \neq i}^{\delta} n_j^{\alpha\beta} P_{ji}^{\alpha\beta} = n_i^{*\alpha\beta}(R_{Ki}^{\alpha\beta} + n_e\hat{\Omega}_{iK}^{\alpha\beta})
$$

$$
i = 1, ..., \delta
$$

$$(13.152)$$

In the following, we write the symbol $\hat{\Omega}_{iK}$ as Ω_{iK}.

(v) Rewriting the equation of conservation of particles into a difference equation

For an atmosphere composed of hydrogen and helium, the equation of conservation of particles [Eq. (13.23)] for hydrogen and helium can be replaced by the two equations

$$
Y = n_{He}/n_H \tag{13.153}
$$

$$\tilde{n}_H = (n_g - n_e) \frac{1 + 4Y}{1 + Y} \tag{13.154}$$

where n_{He} and n_H represent the number of helium and hydrogen particles per unit volume, $\tilde{n}_H$ is the number of equivalent hydrogen atoms of a mixed gas in a unit volume, n_g and n_e are the total number of particles and number of electrons in a unit volume. Equation (13.154) can be obtained from the equation set

$$\rho = \sum_i n_i m_i = n_H \cdot m_H + n_{He} \cdot 4m_H = \tilde{n}_H \cdot m_H \tag{13.155}$$

$$n_\alpha = n_H + n_{He} = (1 + Y)n_H \tag{13.156}$$

$$n_g = n_\alpha + n_e \tag{13.157}$$

$$n_{He} = \frac{n_{He}}{n_\alpha} \cdot n_\alpha = \frac{n_{He}}{(1 + Y)n_H} \cdot n_\alpha = \frac{Y \cdot n_\alpha}{1 + Y} \tag{13.158}$$

We combine the statistical equilibrium equation (13.152) with equations (13.153) and (13.154) for conservation of particles, and express in the form of a vector equation

$$A \cdot \vec{n} = \vec{b} \tag{13.159}$$

which is called the rate equation and where the vector $\vec{n}$ is

$$\vec{n} \equiv (n_1(H), n_2(H), ... n_\delta(H), n_H, n_{HeI}, n_{HeII}, n_{HeIII}) \tag{13.160}$$

and the vector $\vec{b}$ is

$$\begin{cases} b_i = n_i^*(R_{Ki} + n_e \Omega_{iK}) & i = 1, ..., \delta \\ b_{\delta+1} = (n_g - n_e) \dfrac{1 + 4Y}{1 + Y} \\ b_{\delta+2} = 0 \\ b_{\delta+3} = 0 \\ b_{\delta+4} = Y \cdot n_P(1 + MM \cdot SUM1) \end{cases} \tag{13.161}$$

and the matrix element for the matrix A are

$$
\begin{cases}
A_{ij} = 0, \quad i = 1, ..., \delta, \quad j > \delta \\[2mm]
A_{ii} = \sum_{j=1}^{i-1}(R_{ij} + \frac{n_j^*}{n_i^*} n_e \Omega_{ji}) + \sum_{j=i+1}^{\delta}(R_{ij} + n_e \Omega_{ij}) + (R_{iK} + n_e \Omega_{iK}), \quad i = 1, ..., \delta \\[2mm]
A_{ij} = -(R_{ji} + n_e \Omega_{ji}), \quad i = 1, ..., \delta, \quad j < i \\[2mm]
A_{ij} = -(R_{ji} + \frac{n_i^*}{n_j^*} n_e \Omega_{ij}), \quad i = 1, ..., \delta, \quad j > i \\[2mm]
A_{ii} = 1, \quad i = \delta + 1 \\[2mm]
A_{ij} = 0, \quad i = \delta + 1, j \neq i \\[2mm]
A_{ii} = R_{1K} + n_e \Omega_{1K}, \quad i = \delta + 2 \\[2mm]
A_{ij} = -\frac{MM}{4} e^{\chi(1)/kT}(R_{K1} + n_e \Omega_{1K}), \quad i = \delta + 2, j = i + 1 \\[2mm]
A_{ij} = 0, \quad i = \delta + 2, j \neq i, i + 1 \\[2mm]
A_{ii} = R_{1K} + n_e \Omega_{1K}, \quad i = \delta + 3 \\[2mm]
A_{ij} = -MM \cdot e^{\chi(2)/kT}(R_{K1} + n_e \Omega_{1K}), \quad i = \delta + 3, j = \delta + 4 \\[2mm]
A_{ij} = -Y, \quad i = \delta + 4, j = 1, ..., \delta \\[2mm]
A_{ij} = 1, \quad i = \delta + 4, j = \delta + 2 \\[2mm]
A_{ij} = (1 + MM \cdot SUM3), \quad i = \delta + 4, j = \delta + 3 \\[2mm]
A_{ij} = (1 + MM \cdot SUM5), \quad i = \delta + 4, j = \delta + 4
\end{cases}
$$

$$(13.162)$$

where

$MM = \frac{n_e h^3}{(2\pi m_e k)^{3/2} T^{2/3}}$, SUM1, SUM3, SUM5 are the sum of states for hydrogen atoms, helium atoms, the first and second order helium ions, and $U_i = \sum_i g_i e^{-\chi_i/kT}$.

We have now rewritten the basic equation set as a difference equation set. However, we see that there are a large number of equations since the variables include the depth m, the frequency ν and the direction μ. To simplify the calculations and to reduce the number of equations, we introduce the Eddington factors to eliminate the variable μ.

(1) Rewriting the radiative transfer equation
We introduce the Eddington factors,

$$
M_{d,k} \equiv \frac{\int \mu u_d(\nu_k) d\mu}{\int u_d(\nu_k) d\mu}, \quad F_{d,k} \equiv \frac{\int \mu^2 u_d(\nu_k) d\mu}{\int u_d(\nu_k) d\mu}
$$

$$(13.163)$$

Integrating the radiative transfer equation (13.142) with respect to μ, we obtain

$$
\begin{aligned}
&\frac{F_{d-1,k}}{\Delta\tau_d(v_k)\Delta\tau_{d-1/2}(v_k)}J_{d-1}(v_k) - J_d(v_k)\frac{F_{d,k}}{\Delta\tau_d(v_k)}\left(\frac{1}{\Delta\tau_{d+1/2}(v_k)} + \frac{1}{\Delta\tau_{d-1/2}(v_k)}\right) \\
&+ \frac{F_{d+1,k}}{\Delta\tau_d(v_k)\Delta\tau_{d+1/2}(v_k)}J_{d+1}(v_k) = J_d(v_k) - \frac{\alpha_d(v_k)}{\kappa_d(v_k)} - \frac{n_e(m_d)a_e}{\kappa_d(v_k)}J_d(v_k)
\end{aligned}
$$

$$(13.164)$$

Integrating the boundary condition in Eq. (13.144), and considering Eq. (13.163), we have

$$
\begin{aligned}
\left(\frac{F_{1,k}}{\Delta\tau_{3/2}} + M_{1,k}\right)J_1(v_k) - \frac{F_{2,k}}{\Delta\tau_{3/2}}J_2(v_k) =& \frac{\Delta\tau_{3/2}}{2}\frac{\alpha_1(v_k)}{\kappa_1(v_k)} + \frac{\Delta\tau_{3/2}}{2}\frac{n_e(1)a_e}{\kappa_1(v_k)}J_1(v_k) \\
&- \frac{\Delta\tau_{3/2}}{2}J_1(v_k)
\end{aligned}
$$

$$(13.165)$$

Multiplying the boundary condition in Eq. (13.145) at the bottom of the atmosphere with μ and then integrating with respect to μ, we obtain the boundary condition at the bottom of the atmosphere as

$$
\frac{F_{N,k}J_N(v_k) - F_{N-1,k}J_{N-1}(v_k)}{\Delta\tau_{N-1/2}} = \frac{1}{2}(B(v_k, T_N) - J_N(v_k))
$$

$$
+ \frac{\partial B(v_k, T_N)}{\partial T_N}\frac{1}{\kappa_N(v_k)} \cdot \frac{\frac{\sigma T_{eff}}{4\pi} - \frac{1}{2}\sum_{k'=1}^{k} b_{k'}(B(v_{k'}, T_N) - J_N(v_{k'}))}{\sum_{k'=1}^{k} b_{k'}\frac{1}{\kappa_N(v_{k'})}\frac{\partial B(v_{k'}, T_N)}{\partial T_N}} \qquad (13.166)
$$

(2) Rewriting the hydrostatic equilibrium equation
Integrating Eq. (13.147) with respect to μ, and considering Eq. (13.163), we have

$$
\frac{k[n_g(m_d)T(m_d) - n_g(m_{d-1})T(m_{d-1})]}{m_d - m_{d-1}} + \frac{4\pi}{c}\sum_{k=1}^{K} b_k \qquad (13.167)
$$

$$
[F_{d,k}J_d(v_k) - F_{d-1,k}J_{d-1}(v_k)]/(m_d - m_{d-1}) = g
$$

Integrating Eq. (13.148) with respect to μ, and considering Eq. (13.163), we obtain the surface boundary condition for the hydrostatic equilibrium equation as

$$\frac{kn_g(m_1)T(m_1)}{m_1} + \frac{4\pi}{c}\sum_{k=1}^{K} b_k w(m_1) M_{1,k} J_1(\nu_k) = g \tag{13.168}$$

(3) Rewriting the energy equilibrium equation
We need not introduce Eddington factors into the energy equilibrium equation, and need only rewrite Eq. (13.149) in the form

$$\sum_{k=1}^{K} b_k \left[\left(\kappa(m_d, \nu_k) - n_e(m_d)a_e \right) J_d(\nu_k) - \alpha(m_d, \nu_k) \right] = 0 \tag{13.169}$$

The other equations, such as the statistical equilibrium equation and the equation for conservation of the particles, do not contain Eddington factors.

(B) Complete linearized method

It is very complicated and difficult to simultaneously solve the earlier rewritten equations at all depth and frequency points, since they are all non-linear equations and the variables are inter-related at all depth points. Auer and Mihalas (1969) proposed a method to solve the equations simultaneously, which is called the complete linearized method and will be described in the following.

If δ bound states for the hydrogen atom are treated by non-LTE, there will be $K + 8 + \delta$ equations for each depth point m_d, which are

Name of the equation	Number
Radiative transfer equations	1 to K
Hydrostatic equilibrium equation	$K + 1$
Energy equilibrium equation	$K + 2$
Equation for the conservation of the total particles	$K + 3$
Equation for the conservation of the total electric charge	$K + 4$
Rate equations	$K + 5$ to $K + 8 + \delta = KL$

where the variables are

Name of the variable	Number
$J_d(v_1), ..., J_d(v_k)$	1 to K
$n_g(m_d)$	$K+1$
$T(m_d)$	$K+2$
$n_e(m_d)$	$K+3$
$\tilde{n}_H(m_d)$	$K+4$
$n_P(m_d)(= n_H(m_d))$	$K+5$
$n_l(m_d)\ l = 1, ..., \delta$	$K+5+1$ to $K+5+\delta$
$n_{HeI}(m_d)$	$K+6+\delta$
$n_{HeII}(m_d)$	$K+7+\delta$
$n_{HeIII}(m_d)$	$K+8+\delta = KL$

We define a vector $\vec{Z}^d$ as

$$\vec{Z}^d \equiv (Z_i^d) \qquad (13.170)$$

which represents all the above variables, i.e.,

$$
\left.
\begin{aligned}
Z_i^d &\equiv J_d(v_i), \quad i = 1, ..., K \\
Z_i^d &\equiv n_g(m_d), \quad i = K+1 \\
Z_i^d &\equiv T(m_d), \quad i = K+2 \\
Z_i^d &\equiv n_e(m_d), \quad i = K+3 \\
Z_i^d &\equiv \tilde{n}_H(m_d), \quad i = K+4 \\
Z_i^d &\equiv n_P(m_d), \quad i = K+5 \\
Z_i^d &\equiv n_l(m_d), \quad l = 1, ..., \delta, i = K+5+l \\
Z_i^d &\equiv n_{HeI}(m_d), \quad i = K+6+\delta \\
Z_i^d &\equiv n_{HeII}(m_d), \quad i = K+7+\delta \\
Z_i^d &\equiv n_{HeIII}(m_d), \quad i = K+8+\delta
\end{aligned}
\right\}
$$

At the depth point m_d ($1 < d < N$), there are KL equations which are constituted by Eqs. (13.164), (13.167), (13.169), the rate equation (13.159), the equation (13.22) for the conservation of total particles, and the equation (13.24) for the conservation of total electric charge. These equations can be written in the form

$$f_i^d(\vec{Z}^{d-1}, \vec{Z}^d, \vec{Z}^{d+1}) = 0 \qquad i = 1, ..., KL \qquad d = 2, ..., N-1 \quad (13.171)$$

However, at the depth point m_1, since Eq. (13.164) is replaced by Eq. (13.165) and Eq. (13.167) is replaced by Eq. (13.168), the KL equations have the form

$$f_i^1(\vec{Z}^1, \vec{Z}^2) = 0 \qquad i = 1, ..., KL \qquad\qquad (13.172)$$

On the other hand, at the bottom of the atmosphere m_N, since Eq. (13.164) is replaced by Eq. (13.166), the KL equations have the form

$$f_i^N(\vec{Z}^{N-1}, \vec{Z}^N) = 0 \qquad i = 1, ..., KL \qquad\qquad (13.173)$$

There are $N \times KL$ equations in the earlier section. To solve these equations, we should first give an approximate solution $\{\vec{Z}_0^d\}_{d=1,...,N}$ which is the initial solution. Since the initial solution does not fulfil the above $N \times KL$ equations, we have to find the correction terms $\delta\vec{Z}^d$ for the initial solution in such a way that $\vec{Z}_0^d + \delta\vec{Z}^d$ fulfil the above equations, i.e.,

$$f_i^d(\vec{Z}_0^{d-1} + \delta\vec{Z}^{d-1}, \delta\vec{Z}_0^d + \vec{Z}^d, \vec{Z}_0^{d-1} + \delta\vec{Z}^{d+1}) = 0 \; i = 1, ..., KL \; d = 1, ..., N$$
$$(13.174)$$

To determine the correction terms $\delta\vec{Z}^d$, we can expand Eqs. (13.171) to (13.173) by Taylor expansion, and take only the first order differential terms, so the equations for $\delta\vec{Z}^d$ are

$$\sum_{l=1}^{KL} \frac{\partial f_i^d}{\partial Z_l^{d-1}} \delta Z_l^{d-1} + \sum_{l=1}^{KL} \frac{\partial f_i^d}{\partial Z_l^d} \delta Z_l^d + \sum_{l=1}^{KL} \frac{\partial f_i^d}{\partial Z_l^{d+1}} \delta Z_l^{d+1} = -f_i^d(\vec{Z}_0^{d-1}, \vec{Z}_0^d, \vec{Z}_0^{d+1}),$$
$$i = 1, ..., KL, \quad 1 < d < N$$

$$\sum_{l=1}^{KL} \frac{\partial f_i^1}{\partial Z_l^1} \delta Z_l^1 + \sum_{l=1}^{KL} \frac{\partial f_i^1}{\partial Z_l^2} \delta Z_l^2 = -f_i^1(\vec{Z}_0^1, \vec{Z}_0^2), \quad i = 1, ..., KL$$

$$\sum_{l=1}^{KL} \frac{\partial f_i^N}{\partial Z_l^{N-1}} \delta Z_l^{N-1} + \sum_{l=1}^{KL} \frac{\partial f_i^N}{\partial Z_l^N} \delta Z_l^N = -f_i^N(\vec{Z}_0^{N-1}, \vec{Z}_0^N), \quad i = 1, ..., KL$$

$$(13.175)$$

There are $N \times KL$ equations in the equation set (13.175), which can be used to determine $N \times KL$ correction terms $\delta\vec{Z}_l^d$. Since they are all linear equations, this method is referred to as the complete linearized method.

To solve the equation set (13.175), we first write them in form of vectors, and define the following matrix elements:

$$
\left.
\begin{aligned}
-(G_d)_{i,l} &\equiv \frac{\partial f_i^d}{\partial Z_l^{d-1}} \\
(B_d)_{i,l} &\equiv \frac{\partial f_i^d}{\partial Z_l^d} \\
-(C_d)_{i,l} &\equiv \frac{\partial f_i^d}{\partial Z_l^{d+1}} \qquad i,l = 1, ..., KL , \ d = 1, ..., N
\end{aligned}
\right\}
\tag{13.176}
$$

Therefore Eq. (13.175) can be written in the form of vectors as

$$
-G_d \delta \vec{Z}^{d-1} + B_d \delta \vec{Z}^d - C_d \delta \vec{Z}^{d+1} = \vec{Q}^d
\tag{13.177}
$$

where the vector $\vec{Q}^d$ is defined as

$$
(Q^d)_i \equiv -f_i^d(\vec{Z}_0^{d-1}, \vec{Z}_0^d, \vec{Z}_0^{d+1})
\tag{13.178}
$$

It should be noted that all matrix elements of G_1 and C_N in Eq. (13.177) are zero.

We first consider the vector equation (13.177) at the depth point $d = 1$. Since all matrix elements of G_1 are zero, we can solve $\delta \vec{Z}^1$ as

$$
\delta \vec{Z}^1 = B_1^{-1}(C_1 \delta \vec{Z}^2 + \vec{Q}^1)
\tag{13.179}
$$

We write Eq. (13.179) in the form

$$
\delta \vec{Z}^1 = D_1 \delta \vec{Z}^2 + \vec{E}^1
\tag{13.180}
$$

where

$$
\left.
\begin{aligned}
D_1 &\equiv B_1^{-1} C_1 \\
\vec{E}^1 &\equiv B_1^{-1} \vec{Q}^1
\end{aligned}
\right\}
\tag{13.181}
$$

We then consider the vector equation (13.177) at the depth point $d = 2$

$$
-G_2 \delta \vec{Z}^1 + B_2 \delta \vec{Z}^2 - C_2 \delta \vec{Z}^3 = \vec{Q}^2
\tag{13.182}
$$

Substituting $\delta \vec{Z}^1$ in Eq. (13.180) into Eq. (13.182), and solving for $\delta \vec{Z}^2$, we get

$$\delta \vec{Z}^2 = [-G_2 D_1 + B_2]^{-1} C_2 \delta \vec{Z}^3 + [-G_2 D_1 + B_2]^{-1}(G_2 \vec{E} + \vec{Q}^2)$$

which can also be written as

$$\delta \vec{Z}^2 = D_2 \delta \vec{Z}^3 + \vec{E}^2 \qquad (13.183)$$

where

$$\left.\begin{aligned} D_2 &\equiv \left[-G_2 D_1 + B_2 \right]^{-1} C_2 \\ \vec{E}^2 &\equiv \left[-G_2 D_1 + B_2 \right]^{-1}(G_2 \vec{E}^1 + \vec{Q}^2) \end{aligned}\right\} \qquad (13.184)$$

Using the methods described above in turn for each depth point, for depth points $d(3 \leq d \leq N - 1)$, we obtain

$$\delta \vec{Z}^d = D_d \delta \vec{Z}^{d+1} + \vec{E}^d \qquad (13.185)$$

where

$$\left.\begin{aligned} D_d &\equiv \left[-G_d D_{d-1} + B_d \right]^{-1} C_d \\ \vec{E}^d &\equiv \left[-G_d D_{d-1} + B_d \right]^{-1}(G_d \vec{E}^{d-1} + \vec{Q}^d) \end{aligned}\right\} \qquad (13.186)$$

At the boundary point at the bottom where $d = N$, the vector equation (13.177) becomes

$$-G_N \delta \vec{Z}^{N-1} + B_N \delta \vec{Z}^N = \vec{Q}^N \qquad (13.187)$$

Substituting Eq. (13.185) into Eq. (13.187), we have

$$\delta \vec{Z}^N = (-G_N D_{N-1} + B_N)^{-1}(G_N \vec{E}^{N-1} + \vec{Q}^N) \equiv \vec{E}^N \qquad (13.188)$$

We can now summarize our steps for calculating the correction terms $\delta \vec{Z}^d$. We first start from the depth point $d = 1$, i.e., from the surface of the atmosphere, and then calculate the matrix element D_d and the vector $\vec{E}_d$ depth point by depth point inwards from the surface. Then, the correction terms $\delta \vec{Z}^d$ are calculated depth point by depth point starting from the bottom layer outwards according to the equations

$$\left.\begin{aligned} \delta \vec{Z}^N &= \vec{E}^N \\ \delta \vec{Z}^d &= D_d \delta \vec{Z}^{d+1} + \vec{E}^d , \quad d = 1, ..., N - 1 \end{aligned}\right\} \qquad (13.189)$$

After the correction terms $\delta \vec{Z}^d$ have been obtained for all depth points, the solutions for corrections are obtained as

$$\vec{Z}^d = \vec{Z}_o^d + \delta \vec{Z}^d \qquad (13.190)$$

The earlier corrected solution $\vec{Z}^d$ is then treated as a new initial solution. The earlier calculations are repeated to get the new corrections and the new solutions. These steps will continue until $\vec{Z}^d$ fulfils the equation sets (13.171) to (13.173) or, in other words, the absolute values of $\delta \vec{Z}^d$ tend to zero or less than the precision ε, i.e.,

$$\left| \delta \vec{Z}^d \right| \leq \varepsilon \qquad (13.191)$$

In this way, the solution $\vec{Z}^d$ can be obtained for all depth points.

(C) Some problems in the complete linearized method

(1) The choice of discrete values for the independent variables
The independent variables m and ν in the numerical calculations have adopted a series of discrete values $\{m_d\}$ and $\{\nu_k\}$. The choice of discrete values are very important. We first discuss the choice of discrete values of depth points m_d. Since the values of m vary significantly in the atmosphere, sometimes reaching several orders of magnitude, we adopt $\log m$ as the measurement scale for depth. Normally, we can take about 100 depth points in the region from $\log m = -7$ to $\log m = 3$.

When choosing discrete values for the frequency, we have to consider the following. To achieve a higher precision in calculations, we choose as many frequency points as possible. However, the computation time is proportional to $(N \times K)^3$, so too many frequency points will lead to a too long computation time. The typical frequency discrete points for O and B type stars used in the programs of Auer and Mihalas (1969) are listed in Tables 13.1 and 13.2.

(2) Choice of the initial model
When we use the complete linearized method to solve the basic equations, we need an approximate initial solution $\{Z_o^d\}_{d=1,\ldots,N}$, which is the initial model. The choice of the initial model is very important; the calculations may not converge for an incorrect choice of the initial model.

Usually, the calculations for a non-LTE model atmosphere should be separated into several steps. First, an initial model is calculated using

Table 13.1 The typical discrete frequency points used for O type stars (adapted from the program of Auer and Mihalas 1969).

k	ν_k (Hz)	Absorption Edges and Lines
1	$2.6320\ 0000 \times 10^{16}$	
2	$2.2510\ 0000 \times 10^{16}$	
3	$1.8700\ 0001 \times 10^{16}$	
		CNO II*
4	$1.8699\ 9999 \times 10^{16}$	
5	$1.5930\ 0000 \times 10^{16}$	
6	$1.3160\ 0001 \times 10^{16}$	
		HeII ($n = 1$)
7	$1.3159\ 9999 \times 10^{16}$	
8	$1.1600\ 0001 \times 10^{16}$	
		CNO III*
9	$1.1599\ 9999 \times 10^{16}$	
10	$9.5250\ 0000 \times 10^{15}$	
11	$7.4500\ 0001 \times 10^{15}$	
		CNO II*
12	$7.4499\ 9999 \times 10^{15}$	
13	$6.6993\ 1000 \times 10^{15}$	
14	$5.9486\ 2001 \times 10^{15}$	
		HeI ($n = 1$)
15	$5.9486\ 1999 \times 10^{15}$	
16	$5.1890\ 1428 \times 10^{15}$	
17	$4.4294\ 0856 \times 10^{15}$	
18	$4.0496\ 0572 \times 10^{15}$	
19	$3.6698\ 02858 \times 10^{15}$	
20	$3.4799\ 0143 \times 10^{15}$	
21	$3.2900\ 0001 \times 10^{15}$	
		H ($n = 1$)
22	$3.2899\ 9999 \times 10^{15}$	
23	$3.0843\ 7500 \times 10^{15}$	
		L_γ
24	$2.9244\ 4444 \times 10^{15}$	
		L_β
25	$2.4675\ 0000 \times 10^{15}$	
		L_α
26	$1.9648\ 6111 \times 10^{15}$	
27	$1.4622\ 2233 \times 10^{15}$	
		He ($n = 3$)
28	$1.4622\ 2221 \times 10^{15}$	
29	$1.1423\ 6111 \times 10^{15}$	
30	$8.2250\ 0001 \times 10^{14}$	
		H ($n = 2$)
31	$8.2249\ 9999 \times 10^{14}$	
32	$7.1968\ 7500 \times 10^{14}$	
33	$6.1687\ 5000 \times 10^{14}$	
		H_β
34	$5.2640\ 0001 \times 10^{14}$	
		HeII ($n = 5$)
35	$5.2639\ 9999 \times 10^{14}$	
36	$4.5694\ 4444 \times 10^{14}$	
		H_α
37	$3.6555\ 5556 \times 10^{14}$	
		H ($n = 3$)
38	$3.6555\ 5555 \times 10^{14}$	
39	$2.8559\ 0278 \times 10^{14}$	
40	$2.0562\ 5001 \times 10^{14}$	
		H ($n = 4$)
41	$2.0562\ 4999 \times 10^{14}$	
42	$1.5993\ 0555 \times 10^{14}$	
		P_α
43	$1.3160\ 0001 \times 10^{14}$	
44	$1.3159\ 9999 \times 10^{14}$	
		H ($n = 5$)
45	$6.6300\ 0000 \times 10^{13}$	
46	$1.0000\ 0000 \times 10^{12}$	

* These absorption lines need not be considered for an atmosphere composed of hydrogen and helium.

Table 13.2 The typical discrete frequency points used for B type stars (adapted from the program of Auer and Mihalas 1969).

k	ν_k	Edges and Lines
1	$1.3160\ 0001 \times 10^{16}$	$HeII\ (n = 1)$
2	$1.3659\ 9999 \times 10^{16}$	
3	$1.1600\ 0001 \times 10^{16}$	
		$CNO\ III^*$
4	$1.1599\ 9999 \times 10^{16}$	
5	$9.5250\ 0000 \times 10^{15}$	
6	$7.4500\ 0001 \times 10^{15}$	
		$CNO\ II^*$
7	$7.4499\ 9999 \times 10^{15}$	
8	$6.6993\ 1000 \times 10^{15}$	
9	$5.9486\ 2001 \times 10^{15}$	
		$HeI\ (n = 1)$
10	$5.9486\ 1999 \times 10^{15}$	
11	$5.1890\ 1428 \times 10^{15}$	
12	$4.4294\ 0856 \times 10^{15}$	
13	$4.0496\ 0572 \times 10^{15}$	
14	$3.6698\ 0286 \times 10^{15}$	
15	$3.4799\ 0143 \times 10^{15}$	
16	$3.2900\ 0001 \times 10^{15}$	
		$H\ (n = 1)$
17	$3.2899\ 9999 \times 10^{15}$	
18	$2.6307\ 4074 \times 10^{15}$	
19	$2.0714\ 8148 \times 10^{15}$	
20	$1.7668\ 5158 \times 10^{15}$	
21	$1.4622\ 2223 \times 10^{15}$	
		$HeII\ (n = 3)$
22	$1.4622\ 2221 \times 10^{15}$	
23	$1.1423\ 6111 \times 10^{15}$	
24	$8.2250\ 0001 \times 10^{14}$	
		$H\ (n = 2)$
25	$8.2249\ 9999 \times 10^{14}$	
26	$6.9090\ 0000 \times 10^{14}$	H_γ
27	$6.1687\ 5000 \times 10^{14}$	H_β
28	$5.2640\ 0001 \times 10^{14}$	
		$He_{II}\ (n = 5)$
29	$5.2639\ 9999 \times 10^{14}$	
30	$4.5694\ 4444 \times 10^{14}$	H_α

* These absorption lines need not be considered for an atmosphere composed of hydrogen and helium.

the assumption of a gray atmosphere, and a LTE-continuum model is calculated through this initial model. Second, we use this LTE-continuum model as the initial model to calculate the non-LTE continuum model. Third, we use this non-LTE continuum model as the initial model to calculate the non-LTE line model.

In the following, we will use the gray atmosphere assumption to build our initial model. We need to perform the following calculations.

(1) Integration of the hydrostatic equilibrium equation
We start from the depth point m_1 at the stellar surface. According to the temperature distribution of the gray atmosphere, the temperature T_1 at m_1 can be obtained as

$$T_1 = T_{eff} \left[\frac{3}{4} (\bar{\tau}_1 + q) \right]^{1/4} \tag{13.192}$$

From the hydrostatic equilibrium equations (13.9) and (13.57), the gas pressure P_{g1} at point m_1 is

$$P_{g1} = n_{g1} k T_1 = \left(g - \frac{4\pi H}{c} \frac{\bar{\kappa}_1}{\rho_1} \right) m_1 \tag{13.193}$$

where H is the Eddington radiation flux, which can be taken as $H = (1/4)\sigma T_{eff}^4$ at m_1, but should be calculated from Eq. (13.57) at other depth points m_d.

The average optical depth $\bar{\tau}$ in Eq. (13.192) is

$$\bar{\tau}_1 = \bar{\kappa}_1 m_1 / \rho_1 \tag{13.194}$$

and the average opacity $\bar{\kappa}_1$ at m_1 can simply be considered as the contribution from electron scattering, so

$$\bar{\kappa}_1 = \frac{\rho_1}{m_P} \frac{(1 + 2Y)}{(1 + 4Y)} a_e \tag{13.195}$$

where $Y = n_{He}/n_H$ and a_e is the Thompson scattering cross section.

The values at other depth points m_d can be solved using the prediction-correction algorithm, with the prediction terms

$$\bar{\tau}_d^o = \bar{\tau}_{d-1} + \left(\frac{d\bar{\tau}}{dm} \right)_{d-1/2} (m_d - m_{d-1}) = \bar{\tau}_{d-1} + \left(\frac{\bar{\kappa}_{d-1}}{\rho_{d-1}} \right) (m_d - m_{d-1}) \tag{13.196}$$

$$T_d^o = T_{eff}\left[\frac{3}{4}(\bar{\tau}_d^o + q)\right]^{1/4} \tag{13.197}$$

$$P_{gd}^o = n_{gd-1}kT_{d-1} + \left(\frac{dP_g}{dm}\right)_{d-1/2}(m_d - m_{d-1})$$

$$= n_{gd-1}kT_{d-1} + \left(g - \frac{4\pi H}{c}\frac{\bar{\kappa}_{d-1}}{\rho_{d-1}}\right)(m_d - m_{d-1}) \tag{13.198}$$

and the correction terms

$$\bar{\tau}_d = \bar{\tau}_{d-1} + \left(\frac{d\bar{\tau}}{dm}\right)_{d-1/2}(m_d - m_{d-1}) = \bar{\tau}_{d-1} + \frac{1}{2}\left(\frac{\bar{\kappa}_{d-1}}{\rho_{d-1}} + \frac{\bar{\kappa}_d^o}{\rho_d^o}\right)(m_d - m_{d-1}) \tag{13.199}$$

$$T_d = T_{eff}\left[\frac{3}{4}(\bar{\tau}_d^o + q)\right]^{1/4} \tag{13.200}$$

$$P_{gd} = n_{gd-1}kT_{d-1} + \left[g - \frac{2\pi H}{c}\left(\frac{\bar{\kappa}_{d-1}}{\rho_{d-1}} + \frac{\bar{\kappa}_d^o}{\rho_d^o}\right)(m_d - m_{d-1})\right] \tag{13.201}$$

In the correction terms, $\bar{\kappa}_d^o$ is calculated from known values of T_d^o and P_{gd}^o. The reader can refer to the calculations in Eqs. (13.66) to (13.81) in §13.1.3.

(2) Solution for the radiative transfer equation
The radiative transfer equation can be solved by the Feautrier method. Writing Eq. (13.164) in the form

$$f_i^d(J_{d-1i}, J_{di}, J_{d+1i}) = o \quad d = 2, ..., N-1, i = 1, ..., K \tag{13.202}$$

the surface boundary condition in Eq. (13.165) can be written as

$$f_i^1(J_{1i}, J_{2i}) = o, \quad i = 1, ..., K \tag{13.203}$$

and the boundary condition at the bottom of the atmosphere in Eq. (13.166) can be written as

$$f_i^N(J_{N-1i}, J_{Ni}) = o, \quad i = 1, ..., K \tag{13.204}$$

We first consider Eq. (13.203) at m_1. From this equation, we can solve J_{1i} as the function of J_{2i}:

$$J_{1i} = \alpha_{1i} J_{2i} + \beta_{1i} \tag{13.205}$$

We proceed to consider Eq. (13.202) at m_2:

$$f_i^2(J_{1i}, J_{2i}, J_{3i}) = 0 \tag{13.206}$$

Substituting Eq. (13.205) into Eq. (13.206), we obtain an equation with only two unknowns J_{2i} and J_{3i}, and can solve J_{2i} as a function of J_{3i}:

$$J_{2i} = \alpha_{2i} J_{3i} + \beta_{2i} \tag{13.207}$$

Using the same method point by point, we obtain for depth points $m_d (3 \leq d \leq N - 1)$

$$J_{di} = \alpha_{di} J_{d+1i} + \beta_{di} \tag{13.208}$$

At the boundary point m_N at the bottom, we have

$$f_i^N(J_{N-1i}, J_{Ni}) = 0 \tag{13.209}$$

Substituting Eq. (13.208), Eq. (13.209) will become an equation with only one unknown J_{Ni} which can be solved from this equation.

We now summarize the above steps. Starting from m_1 at the surface inwards depth point by depth point, the coefficients α_{di} and β_{di} at each depth point are calculated. After solving J_{Ni} from Eq. (13.209), the values J_{di} at each depth point are calculated using Eq. (13.208) outwards from the inside.

(3) Linearization of the radiative transfer equation
We simplify the second order radiative transfer equation (13.164) of the frequency ν_k in the form

$$\frac{F_{d-1} J_{d-1}}{\Delta \tau_d \Delta \tau_{d-1/2}} - \frac{F_d J_d}{\Delta \tau_d} \left(\frac{1}{\Delta \tau_{d-1/2}} + \frac{1}{\Delta \tau_{d+1/2}} \right) + \frac{F_{d+1} J_{d+1}}{\Delta \tau_d \Delta \tau_{d+1/2}} = J_d - \frac{\alpha_d}{\kappa_d} - \frac{n_{e,d} a_e J_d}{\kappa_d} \tag{13.210}$$

and define

$$\lambda_d \equiv \frac{F_d J_d - F_{d+1} J_{d+1}}{\Delta \tau_{d+1/2} \Delta \tau_d} \tag{13.211}$$

$$\gamma_d \equiv \frac{F_d J_d - F_{d-1} J_{d-1}}{\Delta \tau_{d-1/2} \Delta \tau_d} \tag{13.212}$$

$$\beta_d \equiv \lambda_d + \gamma_d \tag{13.213}$$

$$a_d \equiv \frac{1}{(w_d + w_{d-1})}\left(\gamma_d + \frac{1}{2}\beta_d \frac{\Delta\tau_{d-1/2}}{\Delta\tau_d}\right) \tag{13.214}$$

$$c_d \equiv \frac{1}{(w_d + w_{d+1})}\left(\lambda_d + \frac{1}{2}\beta_d \frac{\Delta\tau_{d+1/2}}{\Delta\tau_d}\right) \tag{13.215}$$

$$b_d \equiv a_d + c_d \tag{13.216}$$

where

$$w_d = \kappa_d/\rho_d \tag{13.217}$$

Further using the equations

$$\delta(\Delta\tau_{d+1/2}) = \frac{\delta w_d + \delta w_{d+1}}{w_d + w_{d+1}} \cdot \Delta\tau_{d+1/2} \tag{13.218}$$

$$\delta\left(\frac{1}{\Delta\tau_{d+1/2}}\right) = -\frac{1}{\Delta\tau_{d+1/2}} \cdot \frac{\delta w_d + \delta w_{d+1}}{w_d + w_{d+1}} \tag{13.219}$$

$$\delta\left(\frac{1}{\Delta\tau_d}\right) = -\frac{\delta(\Delta\tau_{d-1/2}) + \delta(\Delta\tau_{d+1/2})}{2(\Delta\tau_d)^2} \tag{13.220}$$

so, after linearization, Eq. (13.210) becomes

$$\frac{F_{d-1}}{\Delta\tau_d\Delta\tau_{d-1/2}}\delta J_{d-1} + \frac{a_d}{\rho_{d-1}}\left(\frac{\partial\kappa_{d-1}}{\partial T_{d-1}}\delta T_{d-1} + \frac{\partial\kappa_{d-1}}{\partial n_{e,d-1}}\delta n_{e,d-1} + \sum_{l=1}^{\delta}\frac{\partial\kappa_{d-1}}{\partial n_{l,d-1}}\delta n_{l,d-1}\right.$$

$$\left. -\frac{\kappa_{d-1}}{\tilde{n}_{H,d-1}}\delta\tilde{n}_{H,d-1}\right) - \left[\frac{F_d}{\Delta\tau_d}\left(\frac{1}{\Delta\tau_{d+1/2}} + \frac{1}{\Delta\tau_{d-1/2}}\right) + 1 - \frac{n_{e,d}a_e}{\kappa_d}\right]\delta J_d$$

$$+\left(d_d\frac{\partial\kappa_d}{\partial T_d} + \frac{1}{\kappa_d}\frac{\partial\alpha_d}{\partial T_d}\right)\delta T_d + \left(d_d\frac{\partial\kappa_d}{\partial n_{e,d}} + \frac{1}{\kappa_d}\frac{\partial\alpha_d}{\partial n_{e,d}} + \frac{a_e J_d}{\kappa_d}\right)\delta n_{e,d}$$

$$+\sum_{l=1}^{\delta}\left(d_d\frac{\partial\kappa_d}{\partial n_{l,d}} + \frac{1}{\kappa_d}\frac{\partial\alpha_d}{\partial n_{l,d}}\right)\delta n_{l,d} - \frac{b_d\kappa_d}{\rho_d\tilde{n}_{H,d}}\delta\tilde{n}_{H,d} + \frac{F_{d+1}}{\Delta\tau_d\Delta\tau_{d+1/2}}\delta J_{d+1}$$

$$+\frac{c_d}{\rho_{d+1}}\left(\frac{\partial\kappa_{d+1}}{\partial T_{d+1}}\delta T_{d+1} + \frac{\partial\kappa_{d+1}}{\partial n_{e,d+1}}\delta n_{e,d+1} + \sum_{l=1}^{\delta}\frac{\partial\kappa_{d+1}}{\partial n_{l,d+1}}\delta n_{l,d+1}\right.$$

$$\left. -\frac{\kappa_{d+1}}{\tilde{n}_{H,d+1}}\delta\tilde{n}_{H,d+1}\right) = J_d + \beta_d - \frac{n_{ed}a_e}{\kappa_d}J_d - \frac{\alpha_d}{\kappa_d} \tag{13.221}$$

where

$$d_d \equiv \frac{b_d}{\rho_d} - \frac{1}{\kappa_d^2}(\alpha_d + n_{e,d}a_eJ_d) \qquad (13.222)$$

The surface boundary condition in Eq. (13.165) for the frequency ν_k can be simplified as

$$\frac{F_1J_1 - F_2J_2}{\Delta\tau_{3/2}} = \frac{1}{2}(S_1 - J_1)\Delta\tau_{3/2} - M_1J_1 \qquad (13.223)$$

where

$$S_1 = (\alpha_1 + n_{e,1}a_eJ_1)/\kappa_1 \qquad (13.224)$$

We define

$$\lambda_1 \equiv \frac{F_2J_2 - F_1J_1}{\Delta\tau_{3/2}} \qquad (13.225)$$

After linearization, Eq. (13.223) becomes

$$\left[\frac{F_1}{\Delta\tau_{3/2}} + M_1 + \frac{1}{2}\Delta\tau_{3/2}\left(1 - \frac{n_{e,1}a_e}{\kappa_1}\right)\right]\delta J_1 + \left(C_1\frac{\partial\kappa_1}{\partial T_1} - d_1\frac{\partial\alpha_1}{\partial T_1}\right)\delta T_1$$

$$+\left(C_1\frac{\partial\kappa_1}{\partial n_{e,1}} - d_1\frac{\partial\alpha_1}{\partial n_{e,1}} - d_1a_eJ_1\right)\delta n_{e,1} + \sum_{l=1}^{\delta}\left(C_1\frac{\partial\kappa_1}{\partial n_{l,1}} - d_1\frac{\partial\alpha_1}{\partial n_{l,1}}\right)\delta n_{l,1}$$

$$-\frac{b_1}{\rho_1}\frac{\kappa_1}{\tilde{n}_{H,1}}\delta\tilde{n}_{H,1} - \frac{F_2}{\Delta\tau_{3/2}}\delta J_2 + \frac{b_1}{\rho_2}\left[\frac{\partial\kappa_2}{\partial T_2}\delta T_2 + \frac{\partial\kappa_2}{\partial n_{e,2}}\delta n_{e,2} + \sum_{l=1}^{\delta}\frac{\partial\kappa_2}{\partial n_{l,2}}\delta n_{l,2}\right.$$

$$\left.-\kappa_2\frac{\delta\tilde{n}_{H,2}}{\tilde{n}_{H,2}}\right] = \frac{1}{2}(S_1 - J_1)\Delta\tau_{3/2} + \lambda_1 - M_1J_1 \qquad (13.226)$$

where

$$c_1 \equiv \frac{b_1}{\rho_1} + \frac{1}{2}\frac{\Delta\tau_{3/2}S_1}{\kappa_1} \qquad (13.227)$$

$$d_1 \equiv \frac{1}{2}\frac{\Delta\tau_{3/2}}{\kappa_1} \qquad (13.228)$$

$$b_1 \equiv \frac{1}{(w_1 + w_2)}\left[\lambda_1 + \frac{1}{2}\Delta\tau_{3/2}(J_1 - S_1)\right] \qquad (13.229)$$

The boundary condition at the bottom of the atmosphere in Eq. (13.166) for the frequency ν_k can be simplified as

$$\frac{F_N J_N - F_{N-1} J_{N-1}}{\Delta \tau_{N-1/2}} = \frac{1}{2}(B_N - J_N) + \frac{1}{\kappa_N} \frac{dB_N}{dT} \left[\frac{\frac{\sigma T_{eff}^4}{4\pi} - \frac{1}{2} \sum\limits_{k'=1}^{k} b_{k'}(B_{N,k'} - J_{N,k'})}{\sum\limits_{k'=1}^{k} \frac{b_{k'}}{\kappa_{Nk'}} \frac{dB_{N,k'}}{dT}} \right]$$

$$(13.230)$$

We denote

$$\lambda_N = \frac{F_N J_N - F_{N-1} J_{N-1}}{\Delta \tau_{N-1/2}} \tag{13.231}$$

$$a_N = \frac{\lambda_N}{w_N + w_{N-1}} \tag{13.232}$$

$$\Phi = \left(\frac{1}{\kappa} \frac{dB_k}{dT} \right)_N \tag{13.233}$$

After linearization, Eq. (13.230) becomes

$$-\frac{F_{N-1}}{\Delta \tau_{N-1/2}} \delta J_{N-1} - \left(\frac{a_N}{\rho_{N-1}} \frac{\partial \kappa_{N-1}}{\partial T_{N-1}} \right) \delta T_{N-1} - \left(\frac{a_N}{\rho_{N-1}} \frac{\partial \kappa_{N-1}}{\partial n_{e,N-1}} \right) \delta n_{e,N-1}$$

$$-\sum_{l=1}^{\delta} \left(\frac{a_N}{\rho_{N-1}} \frac{\partial \kappa_{N-1}}{\partial n_{l,N-1}} \right) \delta n_{l,N-1} + \frac{a_N}{\rho_{N-1}} \frac{\kappa_{N-1}}{\tilde{n}_{H,N-1}} \delta \tilde{n}_{H,N-1} + \frac{F_N}{\Delta \tau_{N-1/2}} \delta J_N$$

$$+\frac{1}{2} \delta J_N - \frac{\frac{1}{2}\Phi}{\sum\limits_{k'=1}^{k} b_{k'} \Phi_{k'}} \left(\sum_{k'=1}^{k} b_{k'} \delta J_{N,k'} \right) + \frac{a_N \kappa_N}{\rho_N} \frac{1}{\tilde{n}_{H,N}} \delta \tilde{n}_{H,N}$$

$$-\delta T_N \left\{ \frac{a_N}{\rho_N} \frac{\partial \kappa_N}{\partial T_N} + \frac{1}{2} \frac{\partial B}{\partial T_N} - \frac{\frac{1}{2}\Phi}{\sum\limits_{k'=1}^{k} b_{k'} \Phi_{k'}} \left(\sum_{k'=1}^{k} b_{k'} \frac{dB_{k'}}{dT} \right) + Z\Phi \left[\left(\frac{d \ln \frac{dB}{dT}}{dT} \right. \right. \right.$$

$$\left. \left. \left. -\frac{1}{\kappa} \frac{\partial \kappa_N}{\partial T} \right) - \frac{1}{\sum\limits_{k'=1}^{k} b_{k'} \Phi_{k'}} \sum_{k'=1}^{k} b_{k'} \Phi_{k'} \left(\frac{d \ln \frac{dB_{k'}}{dT}}{dT} - \frac{1}{\kappa_{N,k'}} \frac{\partial \kappa_{N,k'}}{\partial T} \right) \right] \right\}$$

$$
+ \delta n_{e,N} \left[-\frac{a_n}{\rho_N} \frac{\partial \kappa_N}{\partial n_{e,N}} + Z\Phi \left(\frac{1}{\kappa_N} \frac{\partial \kappa_N}{\partial n_{e,N}} - \frac{1}{\sum_{k'=1}^{k} b_{k'} \Phi_{k'}} \sum_{k'=1}^{k} \frac{b_{k'} \Phi_{k'}}{\kappa_{N,k'}} \frac{\partial \kappa_{N,k'}}{\partial n_{e,N}} \right) \right]
$$

$$
+ \sum_{l=1}^{\delta} \delta n_{l,N} \left[-\frac{a_N}{\rho_N} \frac{\partial \kappa_N}{\partial n_{l,N}} + Z\Phi \left(\frac{1}{\kappa_N} \frac{\partial \kappa_N}{\partial n_{l,N}} - \frac{1}{\sum_{k'=1}^{k} b_{k'} \Phi_{k'}} \sum_{k'=1}^{k} \frac{b_{k'} \Phi_{k'}}{\kappa_{k'}} \frac{\partial \kappa_{N,k'}}{\partial n_{l,N}} \right) \right]
$$

$$
= \frac{1}{2}(B - J_N) + Z\Phi - \lambda_N \tag{13.234}
$$

where

$$
Z = \frac{\frac{\sigma T_{eff}^4}{4\pi} - \frac{1}{2} \sum_{k'=1}^{k} b_{k'} (B_{k'} - J_{N,k'})}{\sum_{k'=1}^{k} \frac{b_{k'}}{\kappa_{k'}} \frac{dB_{k'}}{dT}} \tag{13.235}
$$

After linearization, the hydrostatic equilibrium equation in (13.167) becomes

$$
-\frac{4\pi}{c} \sum_{k'=1}^{k} b_{k'} F_{d-1,k'} \delta J_{d-1,k'} - T_{d-1} \delta n_{g,d-1} - n_{g,d-1} \delta T_{d-1} + \frac{4\pi}{kc} \sum_{k'=1}^{k} b_{k'} F_{d,k'} \delta J_{d,k'}
$$

$$
+ T_d \delta n_{g,d} + n_{g,d} \delta T_d = \frac{g}{k}(m_d - m_{d-1}) - n_{g,d} T_d + n_{g,d-1} T_{d-1}
$$

$$
-\frac{4\pi}{kc} \sum_{k'=1}^{k} b_{k'} (F_{d,k'} J_{d,k'} - F_{d-1,k'} J_{d-1,k'}) \tag{13.236}
$$

After linearization, the surface boundary condition in Eq. (13.168) for the hydrostatic equilibrium equation becomes

$$
\frac{T_1}{m_1} \delta n_{g,1} + \left[\frac{n_{g,1}}{m_1} + \frac{4\pi}{kc\rho_1} \left(\sum_{k'=1}^{k} b_{k'} M_1 J_{1,k'} \frac{\partial \kappa_{1,k'}}{\partial T_1} \right) \right] \delta T_1 + \frac{4\pi}{kc\rho_1} \left[\sum_{k'=1}^{k} b_{k'} \kappa_{1,k'} M_1 \delta J_{1,k'} \right.
$$

$$
+ \left(\sum_{k'=1}^{k} b_{k'} M_1 J_{1,k'} \frac{\partial \kappa_{1,k'}}{\partial n_{e,1}} \right) \delta n_{e,1} + \sum_{l=1}^{\delta} \left(\sum_{k'=1}^{k} b_{k'} \frac{\partial \kappa_{1,k'}}{\partial n_{l,1}} M_1 J_{1,k'} \right) \delta n_{l,1}
$$

$$-\left(\sum_{k'=1}^{k} b_{k'} M_1 J_{1,k'} \kappa_{1,k'}\right) \frac{\delta \tilde{n}_{H,1}}{\tilde{n}_{H,1}}\right] = \frac{g}{k} - \frac{n_{g,1} T_1}{m_1} - \frac{4\pi}{kc\rho_1} \sum_{k'=1}^{k} b_{k'} \kappa_{1,k'} M_1 J_{1,k'}$$

$$(13.237)$$

After linearization, the energy equilibrium equation in (13.167) becomes

$$\sum_{k'=1}^{k} b_{k'}(\kappa_{d,k'} - n_{e,d}a_e)\delta J_{d,k'} + \delta T_d \cdot \sum_{k'=1}^{k} b_{k'}\left(J_{d,k'}\frac{\partial \kappa_{d,k'}}{\partial T_d} - \frac{\partial \alpha_{d,k'}}{\partial T_d}\right)$$

$$+ \delta n_{e,d} \cdot \sum_{k'=1}^{k} b_{k'}\left(J_{d,k'}\frac{\partial \kappa_{d,k'}}{\partial n_{e,d}} - \frac{\partial \alpha_{d,k'}}{\partial n_{e,d}} - a_e J_{d,k'}\right) + \sum_{l=1}^{\delta} \delta n_{l,d}\left[\sum_{k'=1}^{k} b_{k'}\right.$$

$$\left.\left(J_{d,k'}\frac{\partial \kappa_{d,k'}}{\partial n_{l,d}} - \frac{\partial \alpha_{d,k'}}{\partial n_{l,d}}\right)\right] = \sum_{k'=1}^{k} b_{k'}[\alpha_{d,k'} - (\kappa_{d,k'} - n_{e,d}a_e)J_{d,k'}]$$

$$(13.238)$$

and the equation of conservation of total particles in Eq. (13.22) can be written in the form

$$n_{g,d} = n_{e,d} + (1 + Y)\left[\sum_{l=1}^{\delta} n_{e,d} + \widetilde{\sum}_{H,d} + n_{P,d}\right],$$

$$(13.239)$$

where

$$\widetilde{\sum}_{H,d} = \sum_{l=\delta+1}^{s} n_{e,d} n_{P,d}\left(\frac{h^2}{2\pi m_e kT}\right)^{3/2} g_l e^{-\Delta\chi_1/kT}$$

$$(13.240)$$

After linearization, Eq. (13.239) becomes

$$-\delta n_{g,d} + \delta n_{e,d}\left[1 + (1 + Y)\widetilde{\sum}_{H,d}/n_{e,d}\right] + \delta T_d(1 + Y)\frac{d\widetilde{\sum}_{H,d}}{dT}$$

$$+ (1 + Y)\left[1 + \widetilde{\sum}_{H,d}/n_{P,d}\right]\delta n_{P,d} + (1 + Y)\sum_{l=1}^{\delta} \delta n_{l,d} = n_{g,d} - n_{e,d}$$

$$-(1 + Y)\left[\sum_{l=1}^{\delta} n_{l,d} + \widetilde{\sum}_{H,d} + n_{P,d}\right]$$

$$(13.241)$$

The linearization of the rate equation in Eq. (13.159) gives

$$\frac{\partial \vec{n}}{\partial T}\delta T + \frac{\partial \vec{n}}{\partial n_e}\delta n_e + \sum_i \frac{\partial \vec{n}}{\partial J_i}\delta J_i - \delta \vec{n} = 0 \qquad (13.242)$$

where

$$\frac{\partial \vec{n}}{\partial x} = A^{-1}\left(\frac{\partial \vec{b}}{\partial x} - \frac{\partial A}{\partial x}\vec{n}\right)$$

The equation of conservation of electric charge in Eq. (13.24) can be written as

$$n_{e,d} = n_{P,d} + n_{HeII,d} + 2n_{HeIII,d} + n_{e,d}\cdot n_{HeIII,d}\cdot MM\cdot SUM5 \qquad (13.243)$$

where

$$MM = \left(\frac{h^2}{2\pi m_e kT}\right)^{3/2} \qquad (13.244)$$

$$SUM5 = \sum_{i=2}^{32} g_i e^{\Delta\varepsilon_i/kT_d} \qquad (13.245)$$

The linearization of Eq. (13.243) gives

$$\delta n_{P,d} + \delta n_{HeII,d} + (2 + n_{e,d}\cdot MM\cdot SUM5)\delta n_{HeIII,d} + (MM\cdot SUM5\cdot n_{HeIII,d} - 1)\delta n_{e,d}$$

$$- n_{e,d}\cdot MM\cdot n_{HeIII,d}\cdot \frac{1}{T_d}\left(\frac{3}{2}SUM5 + SUM6\right)\delta T_d = n_{e,d} - n_{P,d} - n_{HeII,d}$$

$$- 2n_{HeIII,d} - n_{HeIII,d}\cdot n_{e,d}\cdot MM\cdot SUM5 \qquad (13.246)$$

where

$$SUM6 = \sum_{i=2}^{32} g_i e^{-\Delta\varepsilon_i/kT}\frac{\Delta\varepsilon_i}{T_d} \qquad (13.247)$$

(D) Linearized separate solution method

In the earlier section, we have described the complete linearized method for calculation of the non-LTE model atmosphere. We now introduce the linearized separate solution method developed by Wu

(1992) and Zhou (1995). This method has the same stability and convergence for the calculations as those in the complete linearized method, and can also take into account more spectal line blanketing effects. The basic features of this method are as follows.

We write the radiative transfer equation (13.164) in the form:

$$f^d(\vec{J}^{d-1}, \vec{J}^d, \vec{J}^{d+1}, \vec{n}^{d-1}, \vec{n}^d, \vec{n}^{d+1}) = 0 \ , \ d = 2, ..., N - 1 \qquad (13.248)$$

where the vector

$$\vec{J} = \{J_1, J_2, ..., J_K\} \qquad (13.249)$$

$$\vec{n} = \{n_g, T, n_e, \tilde{n}_H, n_P, n_l(H), n_{HeI}, n_{HeII}, n_{HeIII}\} \qquad (13.250)$$

The surface boundary condition in Eq. (13.165) for the radiative transfer equation can be written as

$$f^1(\vec{J}^1, \vec{J}^2, \vec{n}^1, \vec{n}^2) = 0 \qquad (13.251)$$

and the boundary condition at the bottom in Eq. (13.166) for the radiative transfer equation can be written as

$$f^N(\vec{J}^{N-1}, \vec{J}^N, \vec{n}^{N-1}, \vec{n}^N) = 0 \qquad (13.252)$$

The linearization of Eq. (13.248) gives

$$\frac{\partial f^d}{\partial \vec{J}^{d-1}} \delta\vec{J}^{d-1} + \frac{\partial f^d}{\partial \vec{J}^d} \delta\vec{J}^d + \frac{\partial f^d}{\partial \vec{J}^{d+1}} \delta\vec{J}^{d+1}$$

$$= -f^d(\vec{J}_0^{d-1}, \vec{J}_0^d, \vec{J}_0^{d+1}, \vec{n}_0^{d-1}, \vec{n}_0^d, \vec{n}_0^{d+1}) - \frac{\partial f^d}{\partial \vec{n}^{d-1}} \delta\vec{n}^{d-1} - \frac{\partial f^d}{\partial \vec{n}^d} \delta\vec{n}^d - \frac{\partial f^d}{\partial \vec{n}^{d+1}} \delta\vec{n}^{d+1}$$

$$= -f^d(\vec{J}_0^{d-1}, \vec{J}_0^d, \vec{J}_0^{d+1}, \vec{n}_0^{d-1} + \delta\vec{n}^{d-1}, \vec{n}_0^d + \delta\vec{n}^d, \vec{n}_0^{d+1} + \delta\vec{n}^{d+1}) \qquad (13.253)$$

Similarly, the linearization of Eqs. (13.251) and (13.252) give

$$\frac{\partial f^1}{\partial \vec{J}^1} \delta\vec{J}^1 + \frac{\partial f^1}{\partial \vec{J}^2} \delta\vec{J}^2 = -f^1(\vec{J}_0^1, \vec{J}_0^2, \vec{n}_0^1 + \delta\vec{n}^1, \vec{n}_0^2 + \delta\vec{n}^2) \qquad (13.254)$$

$$\frac{\partial f^N}{\partial \vec{J}^{N-1}} \delta\vec{J}^{N-1} + \frac{\partial f^N}{\partial \vec{J}^N} \delta\vec{J}^N = -f^N(\vec{J}_0^{N-1}, \vec{J}_0^N, \vec{n}_0^{N-1} + \delta\vec{n}^{N-1}, \vec{n}_0^N + \delta\vec{n}^N)$$

$$(13.255)$$

We combine all equations in the basic equation set except the radiative transfer equation and the hydrostatic equilibrium equation to form an equation called the constraint equation, which has the general form

$$\Psi^d(\vec{J}^d, \vec{n}^d) = 0 \qquad d = 2, ..., N-1 \tag{13.256}$$

The linearization of Eq. (13.256) gives

$$\frac{\partial \Psi^d}{\partial \vec{n}^d} \delta \vec{n}^d = -\Psi^d(\vec{J}_0^d, \vec{n}_0^d) - \frac{\partial \Psi^d}{\partial \vec{J}^d} \delta \vec{J}^d$$

$$= -\Psi^d(\vec{J}_0^d + \delta \vec{J}^d, \vec{n}_0^d) \tag{13.257}$$

The hydrostatic equilibrium equation (13.167) can be written as

$$g^d(\vec{J}^{d-1}, \vec{J}^d, \vec{n}^{d-1}, \vec{n}^d) = 0 \qquad d = 2, ..., N-1 \tag{13.258}$$

The linearization of Eq. (13.258) gives

$$\frac{\partial g^d}{\partial \vec{n}^{d-1}} \delta \vec{n}^{d-1} + \frac{\partial g^d}{\partial \vec{n}^d} \delta \vec{n}^d = -g^d(\vec{J}_0^{d-1} + \delta \vec{J}^{d-1}, \vec{J}_0^d + \delta \vec{J}^d, \vec{n}_0^{d-1}, \vec{n}_0^d) \tag{13.259}$$

The surface boundary condition in Eq. (13.168) for the hydrostatic equilibrium equation can be written as

$$g^1(\vec{J}^1, \vec{n}^1) = 0 \tag{13.260}$$

The linearization of Eq. (13.260) gives

$$\frac{\partial g^1}{\partial \vec{n}^1} \delta \vec{n}^1 = -g^1(\vec{J}_0^1 + \delta \vec{J}^1, \vec{n}_0^1) \tag{13.261}$$

The steps for the linearized separate solution method are as follows. We first give an initial solution $\vec{J}_0^d$ and $\vec{n}_0^d$ and assume $\delta \vec{n}^d = 0$, which is then used for solving the radiative transfer equations (13.253), (13.254) and (13.255) to get $\delta \vec{J}^d$. We then use $\vec{J}^d = \vec{J}_0^d + \delta \vec{J}^d$ and $\vec{n}_0^d$ to solve the hydrostatic equations (13.261) and (13.259) and the constraint equation (13.257) to get a new $\delta \vec{n}^d$. We then use $\vec{n}^d = \vec{n}_0^d + \delta \vec{n}^d$ and $\vec{J}^d$ to solve the radiative transfer equation. The above procedures are repeated until the correction terms $\delta \vec{J}^d$ and $\delta \vec{n}^d$ become less than an acceptable precision.

13.1.5 Comparisons Between Results from LTE and Non-LTE Model Atmospheres

To better understand the results calculated from the non-LTE model atmosphere, we compare the results from LTE and non-LTE model atmospheres in terms of occupancy numbers in energy states, temperature distribution in the atmosphere and spectral features.

(1) Occupancy numbers in energy states
Figure 13.6 gives the variation of atomic occupancy number of the three lowest energy states for the hydrogen atom with the depth calculated by Kudritzki (1978). In the figure, all models correspond to an atmosphere composed of hydrogen and helium, $T_{eff} = 15000$ K and $\log g = 4$. From the figure, we notice the following features. Starting from a certain depth, the occupancy number of atomic energy states obtained from the non-LTE model atmosphere deviates from that obtained from the LTE model atmosphere. For the groundstate, the atomic occupancy number obtained from the non-LTE model atmosphere is larger than that obtained from the LTE model atmosphere, i.e., $n_1 > n_1^*$ (in Fig. 13.6, for n_1, the curve a is below curves b and c). The deviation for the non-LTE continuum model is the largest, and the deviation for the non-LTE line model with the three spectral lines for H_α, H_β and P_α is smaller. For excited states, the atomic occupancy number obtained from the non-LTE model atmosphere is smaller than that obtained from the LTE model atmosphere, i.e., $n_i < n_i^*$ ($i = 2, 3$). In Fig. 13.6, for n_2 and n_3, the curve a is above curves b and c. For the excited state $n = 2$, the case resembles that for the groundstate, i.e., the deviation of atomic occupancy numbers for the non-LTE line model with the three spectral lines for H_α, H_β and P_α is smaller. However, for the excited state $n = 3$, the case is completely different; the deviation of atomic occupancy numbers obtained for the non-LTE spectral line model is even larger than that obtained from the LTE model.

We will discuss the phenomenon that the atomic occupancy number for groundstate obtained from the non-LTE continuum model is larger than that obtained from the LTE continuum model, and that the atomic occupancy number for the excited state is smaller than that obtained from the LTE continuum model. We first inspect from Table 13.3 the relation between the radiative transition rate R_{iK} and the total transition rate P_{iK} with depth under LTE. In outer layers of the atmosphere, the radiative transition rate is equal to the total transition rate. This shows that there are very few particle collisions in outer

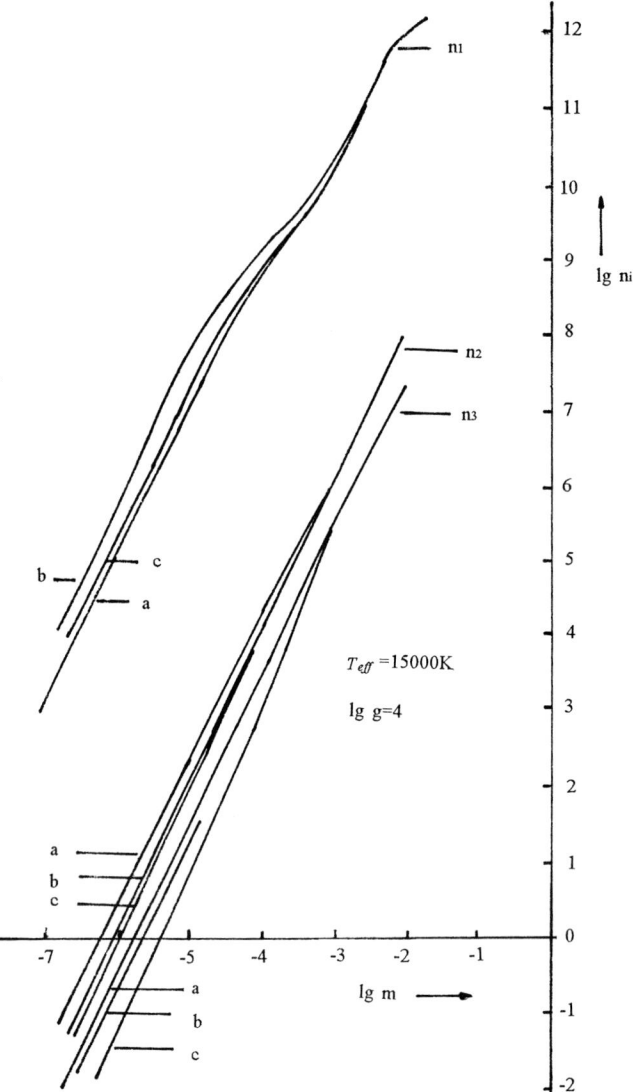

Fig. 13.6 The variation of atomic occupancy number with depth. The atmosphere is composed of hydrogen and helium, and T_{eff} = 15000 K and log g = 4. Curve a: LTE-continuum model; curve b: non-LTE continuum model; curve c: non-LTE line model with spectral lines for H_α, H_β and P_α (from Kudritzki 1978).

layers of the atmosphere, so the collisional transition rate is zero. Here, the atomic occupancy numbers in the energy states cannot be calculated using the Saha equation and the Boltzmann equation; instead, they should be calculated using the statistical equilibrium equation. For the non-LTE continuum model, due to the zero collision term, the statistical equilibrium equation gives

$$n_i R_{iK} = n_i^* R_{Ki} \qquad (13.262)$$

Table 13.3 The relation between the radiative transition rate R_{iK} and the total transition rate P_{iK} with depth under LTE for $T_{eff} = 15000$ K and log $g = 4$ (from Kudritzki 1978).

logm	R_{1K}	R_{K1}	P_{1K}	P_{K1}	R_{2K}	R_{K2}	P_{2K}	P_{K2}	R_{3K}	R_{K3}	P_{3K}	P_{K3}
-7.0	1.86×10	2.14×10	1.86×10	2.14×10	1.19×10^6	6.99×10^5	1.19×10^6	6.99×10^5	2.12×10^6	2.01×10^6	2.12×10^6	2.01×10^6
-5.0	1.88×10	2.16×10	1.88×10	2.16×10	1.19×10^6	7.01×10^5	1.19×10^6	7.01×10^5	2.12×10^6	2.02×10^6	2.13×10^6	2.02×10^6
-4.0	2.85×10	3.15×10	2.85×10	3.15×10	1.19×10^6	7.8×10^5	1.19×10^6	7.8×10^5	2.12×10^6	2.13×10^6	2.16×10^6	2.17×10^6
-3.0	1.03×10^2	1.03×10^2	1.03×10^2	1.03×10^2	1.19×10^6	1.1×10^6	1.20×10^6	1.10×10^6	2.13×10^6	2.55×10^6	2.50×10^6	2.92×10^6
-2.0	1.73×10^2	1.73×10^2	1.73×10^2	1.73×10^2	1.33×10^6	1.28×10^6	1.43×10^6	1.37×10^6	2.28×10^6	2.77×10^6	6.69×10^6	7.18×10^6
-1.0	7.16×10^3	7.16×10^3	7.16×10^3	7.16×10^3	3.93×10^6	3.90×10^6	5.89×10^6	5.86×10^6	5.22×10^6	5.47×10^6	5.82×10^7	5.85×10^7

or

$$n_i = b_i n_i^*$$ (13.263)

where

$$b_i = R_{Ki}/R_{iK}$$ (13.264)

Since

$$
\left.
\begin{aligned}
R_{iK} &= 4\pi \int_{\nu_i}^{\infty} \frac{a_{iK}}{h\nu} J_\nu d\nu \\
R_{Ki} &= 4\pi \int_{\nu_i}^{\infty} \frac{a_{iK}}{h\nu} J_\nu \left(1 + \frac{2h\nu^3}{c^2 J_\nu}\right) e^{-h\nu/kT} d\nu \\
B_\nu &= \frac{2h\nu^3}{c^2} \frac{1}{e^{h\nu/kT} - 1}
\end{aligned}
\right\}
$$ (13.265)

we know that when $J_\nu = B_\nu$,

$$\left(1 + \frac{2h\nu^3}{c^2 J_\nu}\right) e^{-h\nu/kT} = 1$$ (13.266)

so

$$b_i = 1$$ (13.267)

If $J_\nu \neq B_\nu$,

$$\left(1 + \frac{2h\nu^3}{c^2 J_\nu}\right) e^{-h\nu/kT} \begin{matrix} > 1, & J_\nu < B_\nu \\ < 1, & J_\nu > B_\nu \end{matrix}$$ (13.268)

so

$$b_i \begin{matrix} > 1, & J_\nu < B_\nu \\ < 1, & J_\nu > B_\nu \end{matrix}$$ (13.269)

From this, we obtain the following: in outer layers of the atmosphere, if the average radiation intensity in a certain region is larger than Planck's function ($J_\nu > B_\nu$), the enhancement of ionization process will decrease the atomic occupancy number in that energy state, so $n_i < n_i^*$, i.e., the atomic occupancy number in that energy state is smaller than that obtained under LTE. Conversely, if $J_\nu < B_\nu$ in some region, $n_i > n_i^*$. The above shows that the variation of occupancy numbers of atomic energy states are caused by the discrepancy between J_ν and B_ν. Under LTE, $J_\nu = B_\nu$.

Using values for R_{iK} and R_{Ki} in Table 13.3 for our calculations, we obtain $b_1 \approx 1.1$, $b_2 \approx 0.6$, $b_3 \approx 0.9$. This shows that at the ground state the atomic occupancy number n_1 obtained for the non-LTE continuum model is larger than n_1^* obtained for the LTE case, and that at the excited state atomic occupancy number n_i is smaller than n_i^* ($i = 2, 3$) obtained for the LTE case. It has to be pointed out that the above values b_1, b_2 and b_3 are obtained from the LTE model, so they only qualitatively illustrate properties of the atomic occupancy number for the non-LTE continuum model.

To illustrate features of the atomic occupancy number obtained for the non-LTE line model which has considered spectral lines, we can consult the radiative transition rate R_{ij} and the total transition rates P_{ij} given by the non-LTE line model (see Table 13.4). For the non-LTE line model, the radiative transition rate is also equal to the total transition rate in outer layers of the atmosphere, which means that there are extremely few collisional transitions in these regions. On comparing R_{ij} and R_{ji}, we find that the rates for transitions from excited states to lower energy states are larger than those for transitions to higher energy states, i.e., $R_{ji} > R_{ij}$. We know that $J_\nu = B_\nu$ so $n_i^* R_{ij}^* = n_j^* R_{ji}^*$ under LTE. However, under non-LTE, and when there are spectral lines, $J_\nu < B_\nu$, so

$$\frac{R_{ij}}{R_{ji}} = \frac{n_j^*}{n_i^*} \frac{\int\limits_0^\infty \frac{a_{ij}}{h\nu} J_\nu d\nu}{\int\limits_0^\infty \frac{a_{ij}}{h\nu} J_\nu \left(1 + \frac{2h\nu^3}{c^2 J_\nu}\right) d\nu} < \frac{n_j^*}{n_i^*} = \frac{R_{ij}^*}{R_{ji}^*} < 1, \quad j > i \qquad (13.270)$$

Since $(R_{ij}/R_{ji}) < (R_{ij}^*/R_{ji}^*)$, $n_2 < n_2^*$ and $n_3 < n_3^*$. This means that the atomic occupancy number for excited states under non-LTE is smaller than those for LTE. On the other hand, since $(R_{ij}/R_{ji}) < 1$, the excited states become optically transparent, so the transition rate and thus the absorption coefficient of excited atoms to higher energy states decreases which implies even greater optical transparency. However, the above situation is not valid for transitions between the two lowest energy states, which are in fact contrary to the above situation.

Figure 13.7 shows the relation between the depth $\log m$ and the deviation coefficient b_i of energy state occupancy numbers of hydrogen

Table 13.4 The transition rates for bound-free and the bound-bound processes given by the non-LTE line model for $T_{eff} = 15000$ K and log$g = 4$ (from Kudritzki 1978).

log$m$	P_{1K}	P_{K1}	R_{1K}	R_{K1}	P_{2K}	P_{K2}	R_{2K}	R_{K2}	P_{3K}	P_{K3}	R_{3K}	R_{K3}
-7.0	7.01	1.36×10^3	7.01	1.36×10^3	1.19×10^3	2.33×10^6	1.19×10^6	2.33×10^6	2.11×10^6	3.78×10^6	2.11×10^6	3.78×10^6
-5.0	7.12	1.0×10^3	7.12	1.0×10^3	1.19×10^6	2.13×10^6	1.19×10^6	2.13×10^6	2.12×10^6	3.61×10^6	2.11×10^6	3.60×10^6
-4.0	1.47×10	6.0×10^2	1.47×10	6.0×10^2	1.19×10^6	1.83×10^6	1.19×10^6	1.83×10^6	2.16×10^6	3.37×10^6	2.11×10^6	3.32×10^6
-3.0	8.46×10	9.14×10	8.46×10	9.14×10^3	1.21×10^6	1.07×10^6	1.19×10^6	1.06×10^6	2.52×10^6	2.9×10^6	2.12×10^6	2.5×10^6
-2.0	1.83×10^2	1.79×10^2	1.83×10^2	1.79×10^2	1.47×10^6	1.4×10^6	1.37×10^6	1.29×10^6	7.14×10^6	8.74×10^6	2.3×10^6	2.88×10^6
-1.0	6.17×10^3	6.18×10^3	6.17×10^3	6.16×10^3	5.54×10^6	5.50×10^6	3.76×10^6	3.72×10^6	5.42×10^7	6.92×10^7	5.01×10^6	5.31×10^6

log$m$	P_{23}	P_{32}	R_{23}	R_{32}	P_{24}	P_{42}	R_{24}	R_{42}	P_{34}	P_{43}	R_{34}	R_{43}
-7.0	2.63×10^6	4.45×10^6	2.63×10^6	4.54×10^7	6.26×10^5	8.61×10^5	6.26×10^5	8.61×10^6	6.11×10^6	1.25×10^7	6.11×10^6	1.25×10^7
-5.0	2.64×10^6	4.54×10^6	2.64×10^6	4.54×10^7	6.27×10^5	8.61×10^5	6.27×10^5	8.61×10^6	6.16×10^6	1.25×10^7	6.12×10^6	1.25×10^7
-4.0	3.89×10^6	4.6×10^6	3.88×10^6	4.6×10^7	6.76×10^5	8.62×10^5	8.62×10^5	6.75×10^6	7.11×10^6	1.32×10^7	6.71×10^6	1.28×10^7
-3.0	1.47×10^7	5.09×10^7	1.46×10^7	5.08×10^7	2.06×10^6	8.98×10^6	2.03×10^6	8.96×10^6	1.75×10^7	2.13×10^7	1.37×10^7	1.67×10^7
-2.0	1.68×10^7	5.52×10^7	1.56×10^7	5.12×10^7	2.45×10^6	9.34×10^6	2.39×10^6	9.05×10^6	5.97×10^7	2.85×10^7	1.53×10^7	1.76×10^7
-1.0	3.95×10^7	8.48×10^7	2.6×10^7	5.58×10^7	5.38×10^6	1.13×10^7	4.58×10^6	9.59×10^6	3.91×10^8	3.82×10^9	2.18×10^7	2.13×10^7

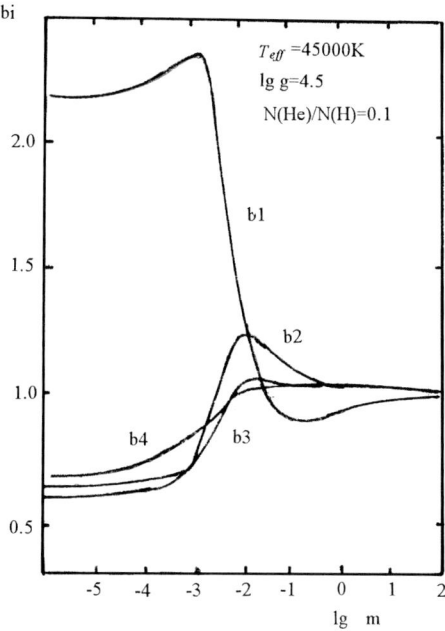

Fig. 13.7 The relation between the depth log m and the deviation coefficient b_i of energy state occupancy numbers of hydrogen atoms given in the non-LTE continuum model for an atmosphere with $T_{eff} = 45000$ K, $\log g = 4.5$ and $N_{He}/N_H = 0.1$ (equivalent to the atmosphere of an O type main sequence star) (from Kudritzki 1981).

atoms given in the non-LTE continuum model for an atmosphere composed of hydrogen and helium with $T_{eff} = 45000$ K, $\log g = 4.5$ and $N_{He}/N_H = 0.1$ obtained by Kudritzki (1981). From the figure, we see that $b_1 > 1$ for the ground state which means that the atomic occupancy number under non-LTE is larger than that under LTE, and $b_i < 1$ ($i = 2, 3, 4$) for excited states which means that the atomic occupancy number under non-LTE is smaller than that under LTE.

(2) Temperature distribution in the atmosphere

Figure 13.8 shows the relation between the depth $\log m$ and the temperature given by different models for an atmosphere composed entirely of hydrogen with $T_{eff} = 15000$ K and $\log g = 4$ calculated by Kudritzki (1978). From the figure, we see that there are two plateaus in the relation between the temperature and the depth for the LTE continuum (curve a). The first plateau occurs between $\log m = -2$ to -4. When deeper than $\log m = -2$ in the atmosphere, the Balmer and the Paschen continua have formed without the Lyman continuum. The Lyman continuum is formed in the region from $\log m = -2$ to -4 (see Table 13.5). The second plateau occurs after the formation of the Lyman continuum, which refers to the region from $\log m = -4$ to -7.

The temperature increases after formation of the Lyman continuum for the non-LTE continuum model (curve b), and increases even more for non-LTE line models (curves c and d). This phenomenon is

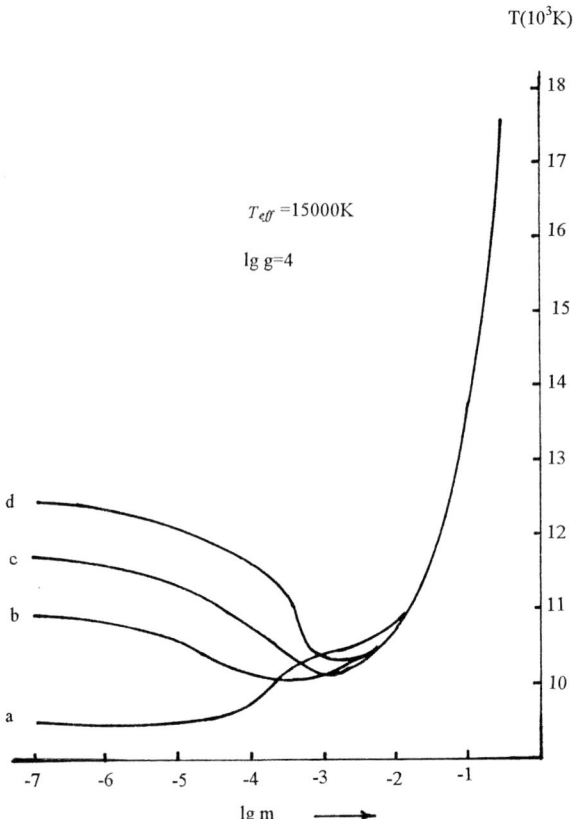

Fig. 13.8 The relation between the depth log m and the temperature given by different models for an atmosphere composed entirely of hydrogen with T_{eff} = 15000 K and $\log g = 4$ calculated by Kudritzki (1981). Curve a: LTE continuum model; curve b: non-LTE continuum model; curve c: non-LTE + H_α line model; curve d: non-LTE + H_α, H_β, P_α lines model (from Kudritzki 1978).

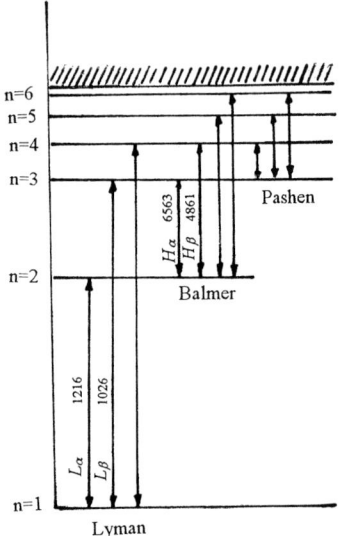

Fig. 13.9 The energy states and spectral lines for the hydrogen atom.

contrary to the LTE model atmosphere with the consideration of spectral line blanketing effects, which decrease the surface temperature.

To study the features of temperature distribution in different model atmospheres, we can use the energy state diagram for the hydrogen atom (Fig. 13.9) to show the formation region of different continua. At the outermost layer of the atmosphere, the temperature is lowest and most atoms are at the ground state so the Lyman series of spectral lines

Table 13.5 The regions for $\tau_\nu = 1$ for different transition processes in an atmosphere composed of entirely hydrogen and $T_{eff} = 15\,000$K and $\log g = 4$.

Process	1,K	2,K	3,K	2,3	2,4	3,4
$\log m$	−4.1	−1.9	−1.3	−4.0	−3.6	−3.8

are formed. When going inwards from the stellar surface, the temperature increases and the hydrogen atoms are at excited states so the Balmer series and the Paschen series of spectral lines are formed. Table 13.5 shows the depths corresponding to different spectral lines.

If there are a lot of atomic ionizations in the regions for formation of Lyman and Blamer continua, and if the average kinetic energy of free electrons produced from ionizations is higher than that of atmospheric particles in the region, the temperature of the region will increase. To prove that the above phenomenon will occur under non-LTE, we first prove that an atomic ionization in the region for Lyman and Balmer continua will produce free electrons with an average kinetic energy higher than that for atmospheric particles in that region.

It is known that the temperature of regions for formation of the Lyman continuum ($1 \to K$ transition) and the Balmer continuum ($2 \to K$ transition) is about 12000 K. For this temperature, the average kinetic energy of an atmospheric particle is $(3/2)kT \approx 2.5 \times 10^{-12}$ erg, which corresponds to $\nu = 3.77 \times 10^{14}$ s^{-1}. In the atmosphere, the probability for a hydrogen atom to absorb a photon with a frequency ν and to undergo the transition $i \to K$ is

$$\tilde{P}_{iK}(\nu) = \frac{\kappa_{iK}(\nu)J_\nu}{\int\limits_{\nu_i}^{\infty} \kappa_{iK}(\nu)J_\nu d\nu} \qquad (13.271)$$

Considering the absorption coefficient in Eq. (5.78) for bound-free transitions of hydrogen atoms, the above equation becomes

$$\tilde{P}_{iK}(\nu) = \frac{J_\nu/\nu^3}{\int\limits_{0}^{\infty} \frac{J_\nu d\nu}{\nu^3}} \qquad (13.272)$$

from which we obtain the average frequency for the absorbed photon to be

$$\bar{\nu}_{iK} = \frac{\int\limits_{\nu_i}^{\infty} \frac{J_\nu}{\nu^3} \nu d\nu}{\int\limits_{\nu_i}^{\infty} \frac{J_\nu d\nu}{\nu^3}} \qquad (13.273)$$

Using the average radiation intensities given in Fig. 13.10 and from

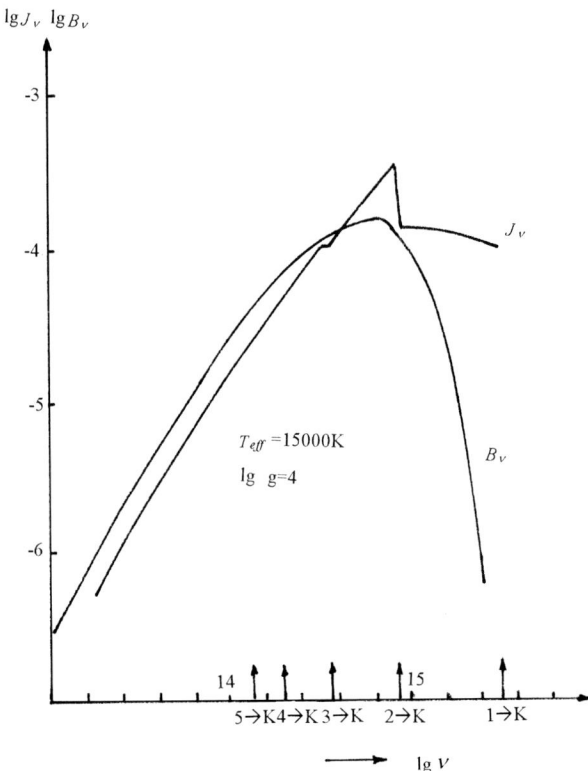

Fig. 13.10 The J_ν and Planck's function B_ν given by the LTE continuum model for an atmosphere composed of entirely hydrogen and $T_{eff} = 15000$ K and $\log g = 4$ (from Kudritzki 1978).

Eq. (13.273), we have

$$\bar\nu_{1K} = 3.76 \times 10^{15} s^{-1}$$
$$\bar\nu_{2K} = 1.24 \times 10^{15} s^{-1}$$

Based on the equation for photoelectric effect, the frequency for the photoelectron is $\Delta\bar\nu_i = \bar\nu_{iK} - \nu_i$, so

$$\Delta\bar\nu_1 = 4.72 \times 10^{14} s^{-1}$$
$$\Delta\bar\nu_2 = 4.15 \times 10^{14} s^{-1}$$

Therefore, we see that ionizations of hydrogen atoms in the region for formation of Lyman and Balmer continua will produce photoelectrons with an average kinetic energy higher than that for atmospheric particles in that region.

The condition for atomic ionizations in the atmosphere is the absorption of radiation energy by atoms. Therefore, the possibility of large amounts of atomic ionizations in the atmosphere can be checked by the variation of $\kappa_v J_v$. If $\kappa_v J_v$ increases in some regions, a large amount of atoms have been ionized.

For the LTE continuum model, the Lyman continuum forms in the region from $\log m = -2$ to -4. Here, the values of κ_v and thus $\kappa_v J_v$ become very large, so many atoms are ionized. This increases the temperature in the region and forms the first temperature plateau. In the region with $\log m < -4$ which is an outer region, the blanketing effects of Lyman spectral lines decrease J_v and thus $\kappa_v J_v$ and the temperature decreases to the second plateau.

For the non-LTE continuum model, the Lyman continuum forms in the region from $\log m = -2.5$ to -4.0. Here, the ground state occupancy number n_1 of hydrogen atoms is larger than n_1^* obtained under LTE, but not significantly, so κ_v increases only slightly. However, the formation of Lyman spectral lines decreases J_v. The decrease in J_v is more profound than the increase in κ_{1K}, so $\kappa_{1K} J_v$ decreases in the region from $\log m = -2.5$ to -4.0. For the energy state $n = 2$, κ_{2K} decreases since $n_2 < n_2^*$, which together with the decrease in J_v have caused the temperature in the region from $\log m = -2.5$ to -4.0 to become lower than the temperature under LTE. For regions with $\log m < -4.0$ or even outer regions, since $n_1 >> n_1^*$ or the ground state occupancy number for hydrogen atoms is far greater than that under LTE, κ_{1K} increases enormously. Although the formation of Lyman spectral lines will lower J_v, the increase in κ_{1K} is far greater than the decrease in J_v, so the value of $\kappa_{1K} J_v$ increases. The value of $\kappa_{2K} J_v$ decreases in the region with $\log m < -4$, but the value of $(\kappa_{1K} J_v + \kappa_{2K} J_v)$ increases, so the temperature for the non-LTE continuum model increases in the region with $\log m < -4$.

For the non-LTE line model, the occupancy number of hydrogen atoms in the two energy levels $n = 1$ and $n = 2$ increase more in the region with $\log m < -4$ (since the transition rates of atoms from high energy levels downwards are far greater than the transition rates upwards), so that κ_{1K} and κ_{2K} are much increased in this region. The increases in κ_{1K} and κ_{2K} are greater than the decrease of J_v so $(\kappa_{1K} J_v + \kappa_{2K} J_v)$ increases even faster, and the temperature is even higher in this region for the non-LTE line model.

Figure 13.11 shows the variation of temperature with depth obtained under LTE and non-LTE continuum models for an atmosphere composed of hydrogen and helium having $(N_{He}/N_H) = 0.1$, $T_{eff} = 45000$ K and $\log g = 4.5$ calculated by Kudritzki (1981). For the atmosphere composed of hydrogen and helium, we see that the temperature obtained for the non-LTE continuum model also increases in the region with $\log m < -2$.

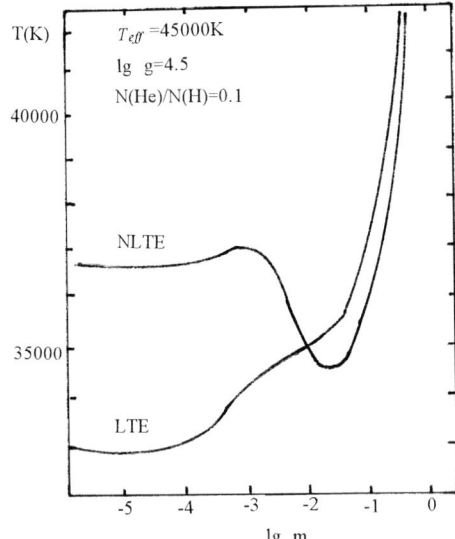

Fig. 13.11 The variation of the temperature with depth obtained under the LTE and non-LTE continuum models, for an atmosphere composed of hydrogen and helium having $N_{He}/N_H = 0.1$, $T_{eff} = 45000$ K and $\log g = 4.5$ m (from Kudritzki 1981).

(3) Spectral features

We now compare the color indices and the jump properties for the continuum flux obtained under LTE and non-LTE model atmospheres. Figure 13.12 gives the color indices obtained under LTE and non-LTE model atmospheres. From the figure, we see that for B type stars (15000 K $\leq T_{eff} \leq$ 30000 K), there is no difference between results in the visible range corresponding to main sequence stars with $\log g = 4$ obtained for non-LTE and LTE model atmospheres; there are differences for results corresponding to giant stars and super giants with $\log g = 3$ and $\log g = 2.5$. These show that the continua obtained from LTE and non-LTE models are the same for a main sequence star when the temperature is less than 30000 K; but are different for giant stars and super giants with $\log g = 3$ and $\log g = 2.5$ since the atmospheric density is extremely small. Figure 13.13 and Fig. 13.14 give the relation between the jump parameter $\phi = D_P/D_B$ and D_B for B type stars calculated using non-LTE and LTE model atmospheres, where D_P is the Paschen jump value and D_B is the Balmer jump value. From Fig. 13.13, we see that for late B type stars, the ϕ values obtained from the LTE model range between 0.16 to 0.17 for different $\log g$ values. However, the values obtained from non-LTE and LTE models are significantly different, and the difference increases with the decrease in $\log g$. Phenomena with similar properties also exist in intermediate B type stars (see Fig. 13.14). The above phenomena have already been proved through observations by Smith and Strom (1969) who found that the values of ϕ for super giants were always larger than

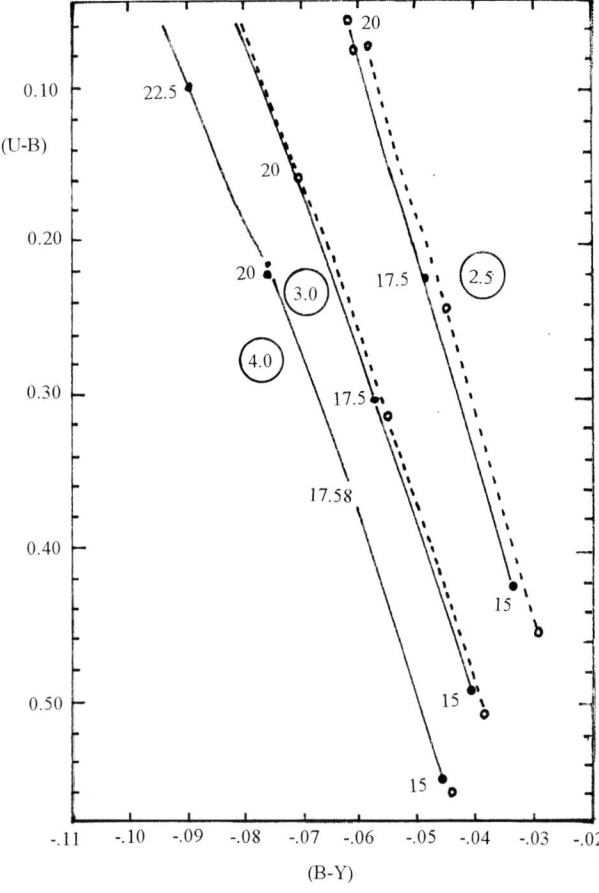

Fig. 13.12 The colour indices obtained under LTE and non-LTE model atmospheres. The outlined points and the broken lines are results for the LTE model atmosphere. The solid points and the solid lines are results for the non-LTE model. The numbers attached to the lines are effective temperatures in units of 10^3 K. The numbers in the circles represent the values of $\log g$.

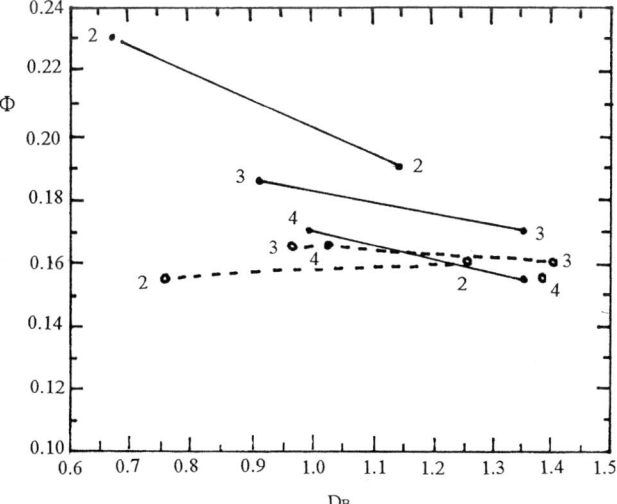

Fig. 13.13 The relation between the jump parameter $\phi = D_P/D_B$ and D_B for late B type stars calculated using non-LTE and LTE model atmospheres. The solid points and the solid lines are results for the non-LTE model. The outlined points and the broken lines are results for the LTE model atmosphere. The numbers represent the values of $\log g$.

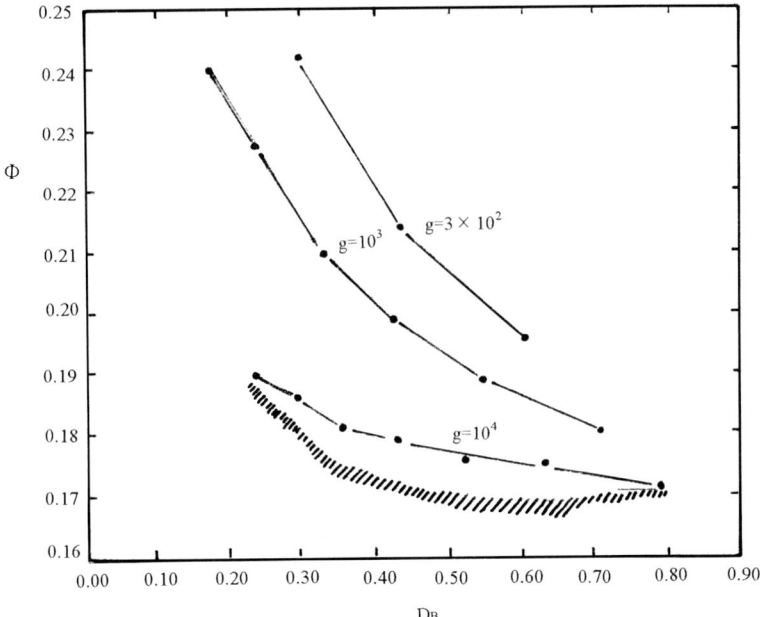

Fig. 13.14 The relation between the jump parameter $\phi = D_P/D_B$ and D_B for intermediate B type stars calculated for non-LTE and LTE model atmospheres. The solid points and the solid lines are results for the non-LTE model, while the hatched region are results for the LTE model atmosphere (corresponding to different g values).

those for main sequence stars for the same D_B value. For an O type star ($T_{eff} \geq 30000$ K), the continuum obtained by the non-LTE model is already significantly different from that obtained by the LTE model even if it is a main sequence star. Figure 13.15 gives the relation between the Balmer jump and the temperature for O type stars obtained by non-LTE and LTE model atmospheres. It can be seen that the results

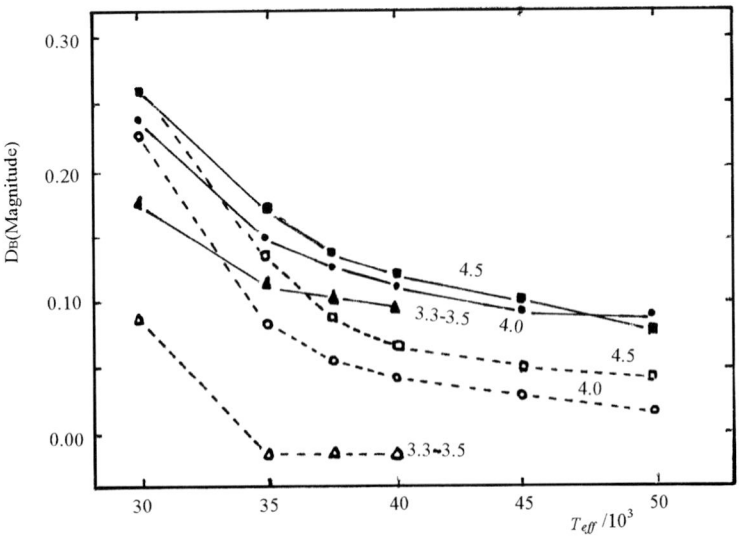

Fig. 13.15 The relation between the Balmer jump D_B and the temperature for O type stars obtained by non-LTE and LTE model atmospheres. The solid points and the solid lines are results for the non-LTE model, while the outlined points and the broken lines are results for the LTE model atmosphere. The numbers correspond to values of log$g$.

obtained by non-LTE and LTE models are already different for main sequence stars with $\log g = 4$. When $\log g$ decreases, i.e., the atmospheric density decreases, the difference grows larger. The above phenomena have also been proved by observations.

13.2 LINE FORMATION THEORY

A lot of spectral lines, absorption lines or emission lines, are embedded in the continuum spectra of stars. Different types of stars have spectral lines with different frequencies, intensities and shape. It is known that the continuum spectrum for a certain atom is formed at a greater depth inside the atmosphere, while the absorption lines are formed in the outer atmosphere through bound-bound or bound-free transitions of atoms or ions. The core parts of absorption lines are very opaque so they should be formed at the outermost layers of the atmosphere, and are formed when the atoms are at the lowest energy state and have large absorption coefficients. The wing parts of absorption lines are relatively transparent and are close to the continuum spectra, which shows that they are formed in regions close to where the continuum spectra are formed. The width of absorption lines are related to the motion of absorbing atoms in the atmosphere. We thus see that the frequencies, intensities and shape of spectral lines can provide rich information on the stellar atmosphere such as the physical state, motion and chemical composition etc., which can be revealed through theoretical analyses of spectral lines. The line formation theory is part of the theory for the entire stellar atmosphere.

13.2.1 Line Profile and Equivalent Width

The shape of a spectral line is usually described by the line profile or the line depth (or called the absorption depth). For a star, only the total radiation flux from the spherical surface of the star can be observed. If we denote the radiation flux from the stellar surface with frequency ν as F_ν^+, and the continuum radiation flux from the stellar surface as F_c^+ (see Fig. 13.16), the line profile is defined as

$$A_\nu \equiv F_\nu^+ / F_c^+ \tag{13.274}$$

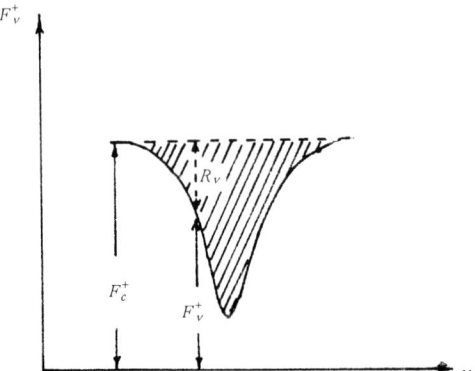

Fig. 13.16 Schematic sketch of a line profile.

The line profile A_ν represents the variation of the ratio of remnant energy flux after absorption to continuum radiation flux versus frequency.

The shape of the spectral line can also be described by the frequency distribution of the line depth which is defined as

$$R_\nu \equiv \frac{F_c^+ - F_\nu^+}{F_c^+} = 1 - A_\nu \qquad (13.275)$$

For the Sun, the radiation intensity $I_\nu(0, \mu)$ at different points on the surface of the Sun can be directly observed, so the line profile and the line depth can be defined as

$$\left.\begin{aligned} A_{\nu\odot} &\equiv \frac{I_\nu(0, \mu)}{I_c(0, \mu)} \\ R_{\nu\odot} &\equiv \frac{I_c(0, \mu) - I_\nu(0, \mu)}{I_c(0, \mu)} \end{aligned}\right\} \qquad (13.276)$$

The intensity of a spectral line is usually represented by the equivalent width. The intensity of a spectral line is the total energy absorbed by

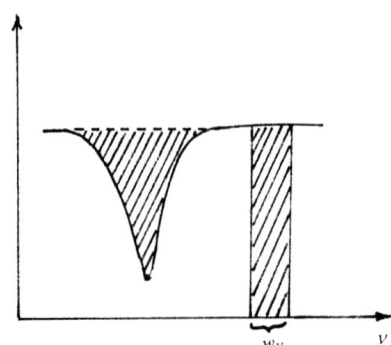

Fig. 13.17 The geometry for calculation of equivalent width.

the line with respect to the continuum spectrum, so the equivalent width is defined as

$$W_\nu \equiv \int_0^b R_\nu d\nu \tag{13.277}$$

If we adopt wavelength as the variable, the above equation can be written as

$$W_\lambda \equiv \int_0^b R_\lambda d\lambda \tag{13.278}$$

This is called equivalent width since the total area of the spectral line is numerically equal to the width of the same area corresponding to $R_\nu = 1$. For the sun, the equivalent width is defined as

$$W_{\lambda\odot} \equiv \int_0^b R_{\lambda\odot} d\lambda \tag{13.279}$$

If the radiation flux F_ν^+ and F_c^+ are known, the line profile and the equivalent width can be obtained from Eqs. (13.275) and (13.277). F_ν^+ and F_c^+ can be determined through

$$F_\nu^+ = 2 \int_0^1 I_\nu^+(0, \mu)\mu d\mu \tag{13.280}$$

$$F_c^+ = 2 \int_0^1 I_c^+(0, \mu)\mu d\mu \tag{13.281}$$

where I_ν^+ and I_c^+ can be obtained from Eq. (2.80) as

$$I_\nu^+(0, \mu) = \int_0^\infty S_\nu(\tau_\nu)e^{-\tau_\nu/\mu}\frac{d\tau_\nu}{\mu} \tag{13.282}$$

$$I_c^+(0, \mu) = \int_0^\infty S_c(\tau_c)e^{-\tau_c/\mu}\frac{d\tau_c}{\mu} \tag{13.283}$$

The optical depth for the above two equations can be obtained from

$$\tau_\nu = \int_Z^{Z_{max}} (\kappa_\nu + \sigma + l_\nu \phi_\nu) dZ' \tag{13.284}$$

$$\tau_c = \int_Z^{Z_{max}} (\kappa_\nu + \sigma) dZ' \tag{13.285}$$

where κ_ν and σ are the continuum absorption coefficient and the continuum scattering coefficient, and $l_\nu \phi_\nu$ is the line absorption coefficient.

To obtain the radiation flux F_ν^+ and F_c^+, we have to first solve the radiative transfer equation to get the variation of source function with optical depth, and then calculate F_ν^+ and F_c^+ from Eqs. (13.280) and (13.281).

13.2.2 Line Absorption Profile

The line absorption coefficient $l_\nu \phi_\nu$ is a critical factor in determining the line profile, in the same way the continuum absorption coefficient determines the distribution and shape of the continuum spectrum. The difference is that the line absorption coefficient is not only determined by atomic absorption, but is also related to other physical processes in the atmosphere, such as Doppler effect from thermal motion of absorbing atoms, and collisions between absorbing atoms and other particles. The effects are manifested in the function ϕ_ν of the line absorption coefficient, and ϕ_ν is called the line absorption profile.

The presence of the line absorption profile ϕ_ν in the line absorption coefficient shows that the line has a finite width. Within this width, the opacity is distributed according to a certain profile function. The finite width of the absorption line can be explained in terms of the finite width of bound states in atoms. If two bound states i and j have certain widths themselves, the absorption line resulting from the transition from bound state i to bound state j should also have a certain width (see Fig. 13.18). The finite widths of bound states in atoms are in turn determined by the uncertainty principle in quantum mechanics. From the uncertainty principle $\Delta E \Delta t \sim \hbar/2\pi$, the requirement of an infinitesimally narrow energy state or $\Delta E \to 0$ is equivalent to the requirement of $\Delta t \to \infty$, i.e. requirement of the atom to stay on that

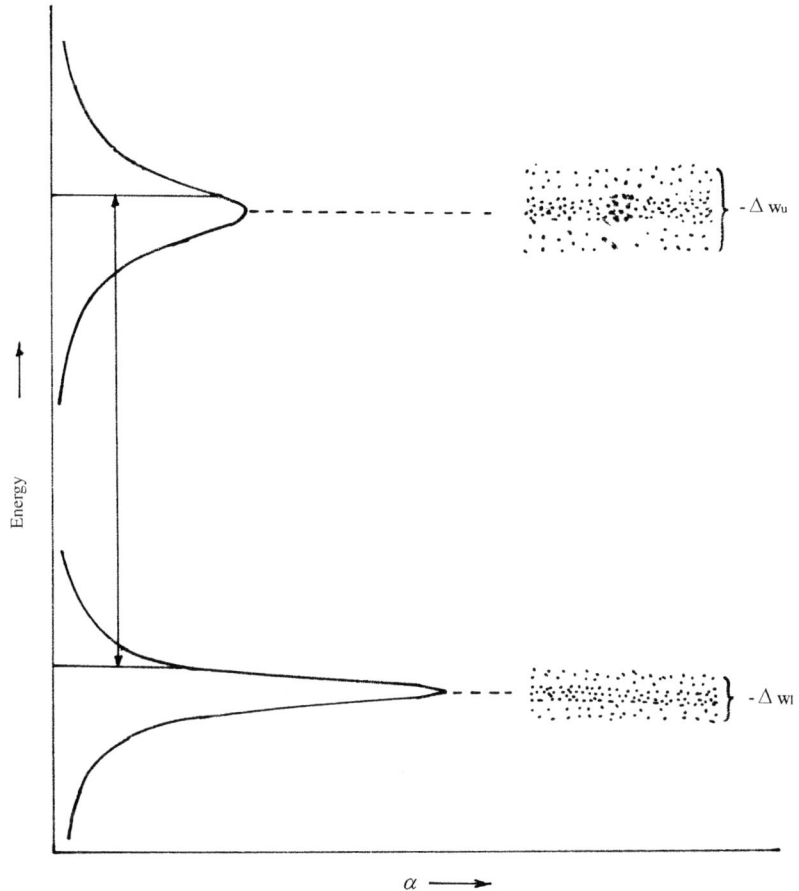

Fig. 13.18 Two bound states *i* and *j* have finite widths.

energy state for an infinitely long time without transition. Conversely, if there is a transition or Δt is finite, ΔE is also finite and the energy state should have a finite width. The uncertainty principle discussed above is in fact only one of the reasons for the finite line width, other reasons including Doppler effect of thermal motion of absorbing atoms, collisions between absorbing atoms and other particles, and macroscopic motion of the absorbing atoms etc. The function ϕ_ν in the absorption coefficient $l_\nu \phi_\nu$ is the resultant of profile functions produced by different line broadening processes, which will be discussed individually in the following.

(A) Radiative damping profile

Absorption lines and emission lines are produced through atomic bound-bound transitions, so the expression of the line absorption coefficient should be given through Eq. (5.42) as

$$\kappa_{ij} = n_i \frac{\pi e^2}{mc} f_{ij} \left(1 - \frac{g_i}{g_j} \frac{n_j}{n_i} \right) \varphi_v \qquad (13.286)$$

where φ_v is called the spontaneous absorption profile. According to Eq. (5.39), the expression for φ_v is

$$\varphi_v = \frac{\gamma}{(\omega_0 - \omega)^2 + \left(\frac{\gamma}{2}\right)^2} \qquad (13.287)$$

where $\gamma = (8\pi^2 v_0^2 e^2)/(3m_e c^3)$ is called the damping constant, ω_0 is the frequency of the line centre. Equation (13.287) can be written in terms of frequency. According to $\omega = 2\pi v$, and introducing $\delta = \gamma/4\pi$, Eq. (13.287) can be written as

$$\varphi_v = \frac{\delta/\pi}{(v - v_0)^2 + \delta^2} \qquad (13.288)$$

The profile functions expressed in Eqs. (13.287) and (13.288) are also called the "radiative damping profile" or "Lorentz profile". When we replace κ_{ij} in Eq. (13.286) by the line absorption coefficient $l_v \phi_v$, it is obvious that l_v is expressed as

$$l_v = n_i \frac{\pi e^2}{mc} f_{ij} \left(1 - \frac{g_i}{g_j} \frac{n_j}{n_i} \right) \qquad (13.289)$$

The radiative damping profile φ_v is only a component of the line absorption profile ϕ_v. Only when the line absorption profile is entirely determined by spontaneous atomic absorption will we have $\phi_v = \varphi_v$.

(B) Doppler profile

Absorbing atoms have different thermal motions. Due to Doppler effect, an absorbing atom with a radial velocity v_r relative to an observer will cause a frequency change of

$$\frac{\Delta\lambda}{\lambda} = -\frac{\Delta v}{v} = \frac{v_r}{c} \qquad (13.290)$$

where c is the speed of light. The distribution of $\Delta\lambda$ gives the profile function of Doppler effect, and $\Delta\lambda$ is proportional to v_r so the distribution of $\Delta\lambda$ is proportional to the distribution of v_r. The distribution of v_r fulfils Maxwellian distribution, i.e.,

$$\frac{dn}{n} = \frac{1}{\sqrt{\pi}} e^{-(v_r/v_0)^2} \frac{dv_r}{v_0} \tag{13.291}$$

where v_0 is the most probable speed. If there is turbulence in the atmosphere,

$$v_0^2 = \frac{2kT}{m} + v_t^2 \tag{13.292}$$

where v_t is the turbulent velocity.

From Eq. (13.291), the profile function for Doppler effect due to thermal motion of atoms is

$$\Psi_\nu = \frac{1}{\sqrt{\pi}\Delta\nu_D} e^{-(\Delta\nu/\Delta\nu_D)^2} \tag{13.293}$$

where $\Delta\nu_D = (v_0/c)\nu$ is called the Doppler frequency shift. The profile function Ψ_ν produced by Doppler effect is different from the profile function φ_ν for spontaneous atomic absorption. The function Ψ_ν follows the Gaussian distribution so it is called the "Gaussian profile".

(C) Collisional damping profile and linear Stark effect profile

The absorbing atoms collide continuously with other particles, such as electrons, ions, atoms of the same type or different types, or even molecules for late type cool stars, which lead to broadening of spectral lines. Since the probability of collisions is related to pressure, this line broadening mechanism is also called pressure broadening.

There are two classical theories for pressure broadening, namely the collision theory and the statistical theory, which have their own regimes of application. The features of spectral line cores are better explained by the collision theory, while the properties of the wings, in particular those of broader lines, are better explained by the statistical theory. A combination of the theories will constitute a more general pressure broadening theory.

(1) Collision theory
The atoms are treated as oscillators in the collision theory. When the oscillators are collided by other particles, the wave produced by the oscillators will be chopped into "wave sections". Each wave section can

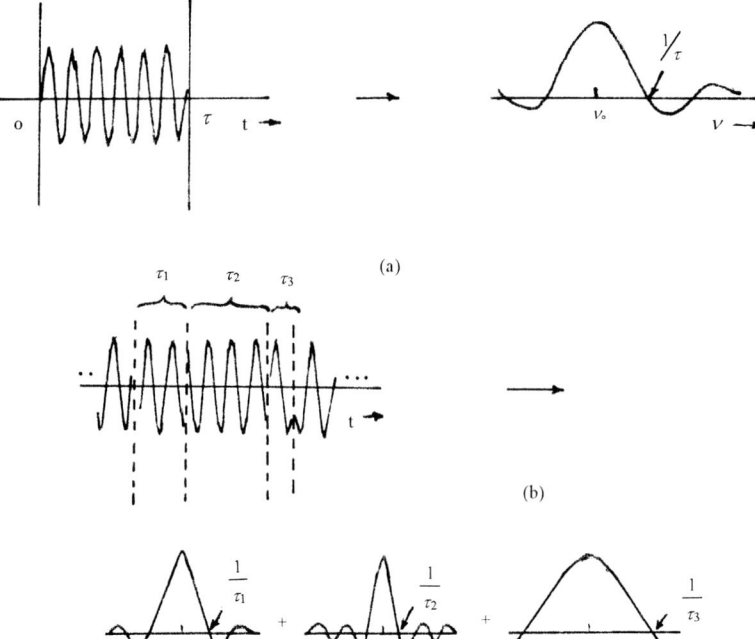

Fig. 13.19 The distribution of amplitudes in a wave section after superposition of waves with different frequencies.

be viewed as superposition of waves with different frequencies through Fourier analysis, and the amplitudes of these waves have a certain distribution [see Fig. 13.19(a)]. Since the time intervals between two collisions are different, different wave sections can be formed and the distributions of amplitudes are also different [see Fig. 13.19(b)]. Considering the statistical weightings of different wave sections, the collision damping profile can be obtained.

We now analyze the distribution of amplitudes in a wave section when this wave section is formed by superposition of waves with different frequencies. For clarity, we neglect the spontaneous atomic absorption (radiative damping). Suppose the two collisions take place at $-T/2$ and $+T/2$. When the oscillator radiates with frequency ω_0, the electromagnetic wave section can be expressed as

$$
\left.
\begin{aligned}
\varepsilon(t) &= const \cdot e^{i\omega_0 t} \quad \left(-\frac{T}{2} \le t \le \frac{T}{2}\right) \\
\varepsilon(t) &= 0 \qquad\qquad \left(t < -\frac{T}{2}, \frac{T}{2} < t\right)
\end{aligned}
\right\}
\tag{13.294}
$$

When we use the Fourier integral to represent $\varepsilon(t)$, i.e., when we treat $\varepsilon(t)$ as superposition of oscillations with different frequencies, we have

$$\varepsilon(t) = const \cdot \int_{-\infty}^{+\infty} E(\omega)e^{i\omega t} d\omega \qquad (13.295)$$

where $E(\omega)$ is the amplitude of the oscillation with frequency ω. From Fourier analysis, we know that

$$E(\omega) = const \cdot \int_{-T/2}^{+T/2} \varepsilon(t)e^{-i\omega t} dt \qquad (13.296)$$

that is, the amplitude of oscillations at each frequency can be treated as combinations of oscillations from $-T/2$ to $+T/2$. Substituting Eq. (13.294) into Eq. (13.296), we have

$$E(\omega) = const \cdot \int_{-T/2}^{+T/2} e^{i(\omega_0-\omega)t} dt = const \cdot \frac{\sin\frac{\omega_0-\omega}{2}T}{\frac{\omega_0-\omega}{2}} \qquad (13.297)$$

Since the radiation intensity is proportional to the square of the amplitude, we have

$$I(\omega) = const \cdot \left[\frac{\sin\frac{\omega_0-\omega}{2}T}{\frac{\omega_0-\omega}{2}} \right]^2 \qquad (13.298)$$

The intensity distribution given by Eq. (13.298) corresponds to the wave section for a time interval T between two collisions. We must also consider the statistical weightings of different wave sections. Suppose the probability for a particle having a collision in a unit time is p, so the probability of having a collision in a time interval Δt is $\Delta t \cdot p$ and the probability of not having a collision in the time interval is $q = (1 - \Delta t \cdot p)$. If we denote $Q(t)$ as the probability of a particle not having a collision in a time $t = n\Delta t$, we have

$$Q(t) = (1 - p\Delta t)^{t/\Delta t} \qquad (13.299)$$

When $\Delta t \to 0$, we get the limit

$$Q(t) = e^{-pt} \qquad (13.300)$$

The reciprocal of p is the period τ_c for one collision, so Eq. (13.300) can also be written as

$$Q(t) = e^{-t/\tau_c} \tag{13.301}$$

Substituting t with T in Eq. (13.301), we obtain the probability of not having a collision between two collisions. We use $Q(T) = e^{-T/\tau_c}$ as statistical weightings to calculate the statistical mean for Eq. (13.298), and obtain the intensity distribution inside the spectral line as

$$I_\nu(\omega) = const \int\limits_0^\infty \left[\frac{\sin(\nu - \nu_0)T}{\pi(\nu - \nu_0)} \right]^2 e^{-T/\tau_c} \frac{dT}{\tau_c}$$

$$= const \cdot \frac{\gamma_c/4\pi^2}{(\nu - \nu_0)^2 + (\gamma_c/4\pi)^2} \tag{13.302}$$

where $\gamma_c = 2/\tau_c$. In the derivation of the distribution in Eq. (13.302), we have treated the atoms as oscillators, and the radiation intensity of oscillators due to collisions will follow this distribution. Since the oscillators are set into forced oscillation with the same frequency with the incident radiation, the effect can be viewed as atomic absorption of the incident radiation. In this way, we can obtain the profile of the line absorption coefficient, which is called the collisional damping profile function

$$\theta_\nu = \frac{\gamma_c/4\pi^2}{(\nu - \nu_0)^2 + (\gamma_c/4\pi)^2} \tag{13.303}$$

On comparing the collisional damping profile in Eq. (13.303) with the radiative damping profile in Eq. (13.288), we will find that the collisional damping profile function θ_ν and the radiative damping profile function φ_ν have exactly the same form; both are Lorentz profiles and differ only in the damping constants. For collisional damping, the damping constant is

$$\gamma_c = \frac{2}{\tau_c} = 2N\bar{\nu}\sigma \tag{13.304}$$

where N is the number of perturbing particles in a unit volume, $\bar{\nu}$ is the mean relative velocity between the oscillator and the perturbing particles, and σ is the collision cross section. If we express the radiative damping constant as γ_r, and combine it with the collisional damping constant γ_c, the combined profile is still a Lorentz profile and the combined damping constant is $\gamma = \gamma_r + \gamma_c$.

To use Eq. (13.303), we still have to know the damping constant γ_c. The collisional damping constant γ_c is different from the radiative damping constant in that the former is a function of the depth and thus the density, and is also related to properties of interaction between the perturbing particles and the oscillators. There have been a lot of studies on interactions among different particles to calculate the value of γ_c, all of which have started from the following approximation. When the perturbing particle flies near the oscillator, the relation between the frequency change of the oscillator and their distance can be expressed as

$$\Delta\nu = \frac{c_n}{r^n} \tag{13.305}$$

where the coefficient c_n can be determined experimentally or calculated using quantum theories, and the constant n signifies interactions among particles of different types. In the following, we will introduce the expressions for γ_c obtained for different values of n by different authors (Lindholm 1945; Foley 1946; Unsöld 1958).

(i) $n = 2$
   Here,

$$\Delta\nu = \frac{c_2}{r^2} = \frac{c_2}{e}\cdot\frac{e}{r^2} = \frac{c_2}{e}\cdot F \tag{13.306}$$

where F is the electric field strength produced by ions or electrons. Equation (13.306) shows that the frequency change is linearly proportional to the electric field strength produced by the perturbing particles. The effect is commonly referred to as the linear Stark effect. In this situation, the collision theory does not have a solution, i.e., γ_c cannot be solved, and the linear Stark effect should be treated using statistical theories. The case $n = 2$ is valid for interactions between hydrogen atoms and other charged particles.

(ii) $n = 3$
This is the case for interactions among the same type of atoms. Here, the damping constant can be obtained as

$$\gamma_3 = 4\pi^3 C_3 N \tag{13.307}$$

where N is the number of perturbing particles in a unit volume, the constant C_3 is given by

$$C_3 = \frac{e^2}{8\pi^2 m_e \nu_0} f \tag{13.308}$$

and f is the oscillator strength, v_0 is the frequency of the line center. The case $n = 3$ may occur in the first few spectral lines in the Balmer series in some cooler stars. This is because neutral hydrogen atoms are more abundant in the atmosphere of cooler stars, so the probability of having collisions between hydrogen atoms is greater.

(iii) $n = 4$

The case for $n = 4$ is valid for collisions between non-hydrogen atoms and charged particles such as electrons and ions. Here

$$\Delta v = \frac{C_4}{r^4} = \frac{C_4}{e^2}\left(\frac{e}{r}\right)^2 = \frac{C_4}{e^2}F^2 \tag{13.309}$$

or the frequency change is proportional to the square of the strength of electric field produced by the perturbing particles, so it is called the second order Stark effect. Under such circumstances, the damping constant is

$$\gamma_4 = 38.8 C_4^{2/3}\bar{v}^{1/3}N \tag{13.310}$$

for which the relation between γ_4 and pressure and temperature can be obtained from

$$\log \gamma_4 = 19.4 + \frac{2}{3}\log C_4 + \log P_e - \frac{5}{6}\log T \tag{13.311}$$

where C_4 for some spectral lines have been experimentally determined and are shown in Table 13.6.

(iv) $n = 6$

The case for $n = 6$ is valid for collisions between atoms and other neutral atoms or molecules. Here, their interactions belong to the Van der Waals force type. The damping constant is

$$\gamma_6 = 17.0 C_6^{2/5}\bar{v}^{3/5}N \tag{13.312}$$

In general, neutral hydrogen atoms are the most abundant in stellar atmospheres with effective temperatures T_{eff} lower than 6000 K, so most of the spectral line broadening is due to collisions between absorbing atoms and neutral hydrogen atoms and belongs to the $n = 6$ case. However, for $T_{eff} > 8000$ K, the major collisions are those with electrons, that is, the above $n = 4$ case. Therefore, the $n = 4$ and $n = 6$ cases are very important.

Table 13.6 The experimentally determined values of $\log C_4$ for some spectral lines.

Spectral line		$\log C_4$
Na	λ5890	−15.17
	λ5896	−15.33
Mg	λ5172	−14.52
	λ5183	−14.52
	λ5528	−13.12

The above collision theory should fulfil the assumption that the collision time between oscillators and the perturbing particles is far shorter than the intervals between collisions. If spectral line broadening is large, i.e., the frequency change Δv of the oscillator is large, the above assumption is not fulfilled so the collision theory is invalid. This situation refers to the two wing parts of the broadened spectral line, which can only be explained using the statistical theory.

(2) Statistical theory

The core feature of the statistical theory is the absorbing atom surrounded by the electric field of perturbing atoms. According to the Stark effect, the spectral line will be broadened by an amount $\Delta\lambda$ which is related to the electric field strength of the perturbing particles. If $\Delta\lambda$ is linearly proportional to the electric field strength, this is called the linear Stark effect; if it is proportional to the square of the electric field strength, this is called the second order Stark effect. In general, for a hydrogen atom, the linear Stark effect is most important when the electric field strength F is not very large (less than 10^5V/cm); and for other atoms (except for some helium lines), there are in general only second order Stark effects.

The variation of electric field strength F of the surrounding perturbing particles with the change in relative positions of particles will follow a certain pattern, and the associated variation in the radiation intensity I will also follow a certain pattern. The variation of the absorption coefficient with $\Delta\lambda$ (i.e., the absorption profile) is proportional to the distribution of the radiation intensity $I(\Delta\lambda)$.

(i) The distribution function of electric field strength of perturbing particles

We choose the position of the absorbing atom as the origin of our coordinate system. We denote $p(r)dr$ as the probability of finding the nearest perturbing particle in the range $(r, r + dr)$, $w^*(r)$ as the probability of no perturbing particles in the sphere with a radius r and $w(r, r + dr)$ as the probability of a perturbing particle in the range $(r, r + dr)$, so it is obvious that

$$p(r)dr = w^*(r)w(r, r + dr) \qquad (13.313)$$

Since the volume of the spherical shell $(r, r + dr)$ is $4\pi r^2 dr$, we have

$$w(r, r + dr) = N \cdot 4\pi r^2 dr \qquad (13.314)$$

where N is the number of perturbing particles in a unit volume. Furthermore, if we denote $w^*(r + dr)$ as the probability of no perturbing particles in the sphere with radius $r + dr$,

$$w^*(r + dr) = w^*(r)w^*(r, r + dr)$$

or

$$w^*(r) + \frac{dw^*(r)}{dr} dr = w^*(r)[1 - w(r, r + dr)]$$
$$= w^*(r)[1 - 4\pi r^2 N dr]$$

so that

$$\frac{dw^*(r)}{dr} = -4\pi r^2 N w^*(r) \qquad (13.315)$$

Integrating the above equation, we get

$$w^*(r) = e^{-\frac{4\pi}{3}Nr^3} \qquad (13.316)$$

Substituting Eqs. (13.316) and (13.314) into Eq. (13.313), we obtain

$$p(r)dr = e^{-\frac{4\pi}{3}Nr^3} \cdot 4\pi N r^2 dr \qquad (13.317)$$

If r_0 represents half the average distance between perturbing particles, we have

$$\frac{4\pi}{3} r_0^3 \cdot N = 1$$

or

$$N = \frac{3}{4\pi r_0^3} \qquad (13.318)$$

Substituting Eq. (13.318) into Eq. (13.317), we have

$$p(r)dr = e^{-\left(\frac{r}{r_0}\right)^3} d\left(\frac{r}{r_0}\right)^3 \qquad (13.319)$$

Supposing the electric charge of a perturbing particle is e, the electric field at the absorbing atom from the perturbing particle is

$$F = \frac{e}{r^2} \qquad (13.320)$$

We denote

$$F_0 \equiv \frac{e}{r_0^2} = \frac{e}{\left(\frac{4}{3}\pi N\right)^{-2/3}} = 2.5985 \cdot N^{2/3} \qquad (13.321)$$

and

$$\beta \equiv F/F_0 \qquad (13.322)$$

Substituting Eqs. (13.320) to (13.322) into Eq. (13.319), we obtain

$$p(\beta)d\beta = \frac{3}{2}\beta^{-5/2}\exp(-\beta^{-3/2})d\beta \qquad (13.323)$$

which is the distribution function of the electric field strength determined by the nearest perturbing particle. It is easy to prove that Eq. (13.323) fulfils the normalization condition.

In the above discussions, we have only considered the electric field strength of the nearest perturbing particle, and have neglected the contribution from other particles. If all perturbing particles are considered, the derivation will be more complicated but a more accurate $p(\beta)$ can be obtained which is called the Holtsmark distribution. For example, the Holtsmark distribution given by Underhill and Waddell (1959) was

$$p(\beta)d\beta = \frac{4}{3\pi}\sum_{j=0}^{\infty}(-1)^{j}\frac{\Gamma[(4j+6)/3]}{(2j+1)}\beta^{2j+2} \qquad (13.324)$$

where Γ is the Γ function. Figure 13.20 gives the function $p(\beta)$ from Eqs. (13.323) and (13.324).

(ii) Profile function

The line broadening under the electric field of perturbing particles can be treated as splitting of the line into n sub-lines, where the relative intensity of the kth sub-line is I_k, and the intensity contribution from the kth line to the region within $(\Delta\lambda, \Delta\lambda + d(\Delta\lambda))$ from the line center is

$$I_K p(F/F_0)\frac{dF}{F_0} \qquad (13.325)$$

Summing the above equation over all sub-lines, we obtain the distribution of radiation intensity with the wavelength

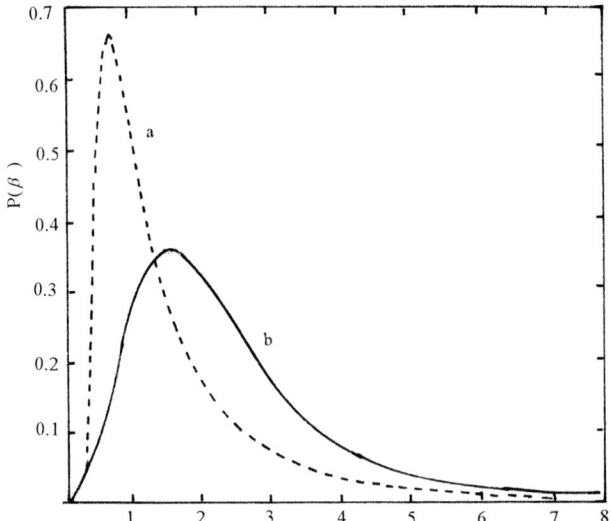

Fig. 13.20 The distribution function of electric field strength determined for perturbing particles. Curve a: the distribution function of electric field strength of the nearest perturbing particle; curve b: the distribution function of electric field strength of all perturbing particles.

inside the line as

$$I(\Delta\lambda)d(\Delta\lambda) = \sum_{K} I_K p(F/F_0)\frac{dF}{F_0}$$

$$= \sum_{K} I_K p\left(\frac{\Delta\lambda}{C_K F_0}\right)\frac{d(\Delta\lambda)}{C_K F_0} \qquad (13.326)$$

Introducing $\alpha \equiv \Delta\lambda/F_0$, and representing the result of the above equation as $S(\alpha)d\alpha$, we have

$$S(\alpha)d\alpha = \sum_{K} I_K p\left(\frac{\alpha}{C_K}\right)\frac{d\alpha}{C_K} \qquad (13.327)$$

Since the variation of absorption coefficient with wavelength is proportional to $I(\Delta\lambda)$, the absorption profile function should have the form of Eq. (13.327). It is easy to prove that it also fulfils the normalization condition, i.e.,

$$\int_{-\infty}^{+\infty} S(\alpha)d\alpha = 1 \qquad (13.328)$$

For the wing part of the line, $\Delta\lambda$ is large so β is also large, and from Eq. (13.324), we have

$$p(\beta) \cong 1.496 \cdot \beta^{-5/2} \qquad (13.329)$$

Substituting Eq. (13.329) into Eq. (13.327), we have

$$S(\alpha) = C \cdot (\Delta \lambda)^{-5/2} \qquad (13.330)$$

which is the wing profile distribution for the linear Stark effect.

(D) Total absorption profile

In the above, we have discussed different line broadening mechanisms and have given the corresponding absorption profile functions. The total absorption profile is the resultant of all absorption profiles. For spectral lines with insignificant linear Stark effects, the total absorption profile ϕ_ν is generally composed of the damping profile φ_ν (including radiative damping and collisional damping) and the Doppler profile Ψ_ν. From Eqs. (13.288), (13.293) and (13.303), we obtain

$$\Phi_\nu = \int_{-\infty}^{+\infty} \left(\frac{\gamma/4\pi^2}{(\Delta\nu - \Delta\nu_1)^2 + (\gamma/4\pi)^2} \right) \left(\frac{1}{\sqrt{\pi}\Delta\nu_D} e^{-\left(\frac{\Delta\nu_1}{\Delta\nu_D}\right)^2} \right) d(\Delta\nu_1)$$

$$(13.331)$$

which is called the Voigt profile. If we further introduce

$$u = \frac{\Delta\nu}{\Delta\nu_D}, \quad u_1 = \frac{\Delta\nu_1}{\Delta\nu_D}, \quad a = \frac{\gamma}{4\pi\Delta\nu_D} \qquad (13.332)$$

Eq. (13.331) can be written as

$$\Phi_\nu = \frac{1}{\sqrt{\pi}\Delta\nu_D} \cdot \frac{a}{\pi} \int_{-\infty}^{+\infty} \frac{e^{-u_1^2}}{(u - u_1)^2 + a^2} du_1$$

$$= \frac{1}{\sqrt{\pi}\Delta\nu_D} H(u, a) \qquad (13.333)$$

in which the function $H(u, a)$ is

$$H(u, a) = \frac{a}{\pi} \int_{-\infty}^{+\infty} \frac{e^{-u_1^2}}{(u - u_1)^2 + a^2} du_1 \qquad (13.334)$$

The function $H(u, a)$ is called the Hjerting function whose shape is mainly determined by the damping parameter a which normally ranges from 10^{-3} to 10^{-1}, i.e., $a << 1$. For $a << 1$, we can study the function $H(u, a)$ for different values of u.

(i) When u is very small, i.e., at the core part of the spectral line, we have

$$H(u, a) \approx \frac{a}{\pi} e^{-u^2} \int_{-\infty}^{+\infty} \frac{du_1}{(u - u_1)^2 + a^2} = e^{-u^2} \tag{13.335}$$

that is, there is only the shape for the Doppler profile.

(ii) When u is very large, i.e., at the wing part of the spectral line, we have $e^{-u_1^2} \approx e^{-u^2} << 1$ so

$$H(u, a) = \frac{a}{\pi} \int_{-\infty}^{+\infty} \frac{e^{-u_1^2} du_1}{(u - u_1)^2 + a^2} \cong \frac{a}{\pi} \frac{1}{u^2} \int_{-\infty}^{+\infty} e^{-u_1^2} du_1$$

$$= \frac{a}{\pi} \frac{1}{u^2} \tag{13.336}$$

which is the Lorentz profile for damping.

(iii) From (i) and (ii) above, we can have the approximation

$$H(u, a) \approx e^{-u^2} + \frac{a}{\pi u^2} \tag{13.337}$$

where the first term is the core part ($u \leq u^*$) and the second term is the wing part ($u \geq u^*$).

13.2.3 Calculations of Non-LTE Multilevel Lines

From non-LTE line calculations, we can get the profiles or the equivalent widths of some absorption (or emission) lines in stellar spectra. The comparison between theoretical profiles or equivalent widths with those from spectral observations is an important step in learning about the structure of stellar atmosphere.

The calculations of non-LTE multilevel lines are similar to these for the non-LTE model atmosphere, but with the following distinct differences.

(1) The spectral lines are calculated assuming the model atmosphere is known, i.e., the temperature T, density ρ, total particle number n_g, electron number density n_e and the total number of different atoms except the one we are calculating, are known for all layers

inside the atmosphere. Therefore, in the basic equation set of the line model, there are no hydrostatic equilibrium equation, energy equation, equation for conservation of total particles, equation for conservation of electrons and equation for the conservation of number of other atoms.

(2) The radiative transfer equation in line calculations is the line radiative transfer equation. The number of line radiative transfer equations is determined by the number of lines to be calculated and the frequency mesh points adopted for each spectral line. The absorption coefficients or emission coefficients in the line radiative transfer equations are made up of line absorption coefficients or line emission coefficients and the continuum absorption coefficients or continuum emission coefficients. The line absorption coefficients or line emission coefficients are determined from bound-bound transitions and bound-free transitions for the atoms we are studying, and the continuum absorption coefficients or continuum emission coefficients are determined by free-free transitions and electron scattering for all atoms in the atmosphere.

(3) In the calculations for spectral lines, the choice of the frequency mesh points are limited to the profile regions of the spectral lines being calculated.

(1) Basic equation set

1. The line radiative transfer equation is

$$\mu \frac{dI_v}{d\tau_v} = I_v - \eta_v/\kappa_v \qquad (13.338)$$

If the spectral line being calculated is generated from the bound-bound transition of a certain atom, the absorption coefficient κ_v and the emission coefficient η_v can be written as

$$\kappa_v = a_{ij}^A(v) \left[n_i^A - \left(\frac{g_i^A}{g_j^A} \right) n_j^A \right] + \kappa_c \qquad (13.339)$$

$$\eta_v = \left(\frac{2hv^3}{c^2} \right) a_{ij}^A(v) \left(\frac{g_i^A}{g_j^A} \right) n_j^A + \eta_c \qquad (13.340)$$

where κ_c and η_c are the continuum absorption coefficient and the continuum emission coefficient, which are determined by free-free transitions and electron scattering for all atoms in the atmosphere.

From Eqs. (13.3) and (13.4), we have

$$\kappa_c = \sum_a \sum_{\beta=0}^{\beta_\alpha} n_e a_{KK}^{\alpha\beta}(v) \left(\sum_{i=1}^{l_{\alpha,\beta+1}} n_i^{\alpha\beta+1} \right) (1 - e)^{-hv/kT} + n_e a_e \qquad (13.341)$$

$$\eta_c = \frac{2hv^3}{c^2} \sum_\alpha \sum_{\beta=0}^{\beta_\alpha} e^{-hv/kT} n_e a_{KK}^{\alpha\beta}(v) \sum_{i=1}^{l_{\alpha,\beta}} n_i^{\alpha\beta+1} + n_e a_e J_v \qquad (13.342)$$

For a known model atmosphere, κ_c and η_c can be calculated and can be treated as known constants. In Eqs. (13.339) and (13.340), $a_{ij}^A(v)$ is the cross section for bound-bound transitions for the atom being calculated. From Eq. (5.44), we know

$$a_{ij}^A(v) = \frac{\pi e^2}{mc} f_{ij} \phi_v \qquad (13.343)$$

where f_{ij} is the oscillator strength and ϕ_v is the absorption profile of that atom (see §13.2.2).

If the calculated spectral line is produced by bound-free transitions of a certain atom, the absorption coefficient and the emission coefficient in Eq. (13.338) can be written as

$$\kappa_v = a_{iK}^A(v) \left[n_i^A - n_K^A \left(\frac{n_i^A}{n_K^A} \right)^* e^{-hv/kT} \right] + \kappa_c \qquad (13.344)$$

$$\eta_v = \frac{2hv^3}{c^2} a_{iK}^A(v) n_K^A \left(\frac{n_i^A}{n_K^A} \right)^* e^{-hv/kT} + \eta_c \qquad (13.345)$$

where $a_{iK}^A(v)$ is the cross section for bound-free transitions for that atom and can be calculated from Eq. (5.58).

The boundary conditions for the line radiative transfer equation are as follows. At the stellar surface where $\tau_v = 0$, the incoming radiation is zero, i.e.,

$$\tau_v = 0: \quad I_v(\tau_v = 0, \mu) = 0 \quad , \quad 0 > \mu > -1 \qquad (13.346)$$

At the bottom of the atmosphere, we have

$$\tau_v = \tau_{vN}: \quad I_v(\tau_{vN}, \mu) = I_v^+ \qquad (13.347)$$

2. The statistical equilibrium equation is

$$n_i^A \left(\sum_{i \neq j}^{NF} P_{ij}^A + P_{iK}^A \right) = \sum_{j \neq i}^{NF} n_j^A P_{ji}^A + n_K^A P_{Ki}^A \qquad (13.348)$$

where n_i^A and n_j^A are the occupancy numbers on energy levels i and j for the atoms associated with the spectral lines being calculated, and the transition rates P_{ij}^A and P_{iK}^A are

$$P_{ij}^A = R_{ij}^A + C_{ij}^A \qquad (13.349)$$

$$P_{iK}^A = R_{iK}^A + C_{iK}^A \qquad (13.350)$$

where the radiative transition rates R_{ij}^A and R_{iK}^A are expressed in Eq. (13.19) and the collisional transition rates C_{ij}^A and C_{iK}^A are expressed in Eq. (13.20).

3. The equation for conservation of particle numbers for atoms associated with the spectral lines being calculated is

$$n^A = \sum_{\beta=0}^{\beta_A} \sum_{i=1}^{NL} n_i^{A\beta} \qquad (13.351)$$

(2) The solution for the basic equation set
The basic equation set is solved by the linearized separate solution method described in §13.1.4. The radiative transfer equation (13.338) is first written in the form similar to Eq. (13.164) according to the method in §13.1.4, and then expressed in the general form

$$f^d(\vec{J}^{d-1}, \vec{J}^d, \vec{J}^{d+1}, \vec{n}^{d-1}, \vec{n}^d, \vec{n}^{d+1}) = 0, \quad d = 2, ..., N-1 \qquad (13.352)$$

where the vectors

$$\vec{J} = \{ J_1, J_2, ..., J_K, ...J_{NL} \} \qquad (13.353)$$

$$\vec{n} = \{ n_1^A, n_2^A, ..., n_i^A, ...n_{NF}^A \} \qquad (13.354)$$

The surface boundary condition for the radiative transfer equation in Eq. (13.346) can be rewritten in a form similar to Eq. (13.165) which can then be expressed as

$$f^1(\vec{J}^1, \vec{J}^2, \vec{n}^1, \vec{n}^2) = 0 \qquad (13.355)$$

The bottom boundary condition for the radiative transfer equation in Eq. (13.347) can be rewritten in a form similar to Eq. (13.166) which can then be expressed as

$$f^N(\vec{J}^{N-1}, \vec{J}^N, \vec{n}^{N-1}, \vec{n}^N) = 0 \tag{13.356}$$

The linearization of Eq. (13.353) gives

$$\frac{\partial f^d}{\partial \vec{J}^{d-1}} \delta \vec{J}^{d-1} + \frac{\partial f^d}{\partial \vec{J}^d} \delta \vec{J}^d + \frac{\partial f^d}{\partial \vec{J}^{d+1}} \delta \vec{J}^{d+1}$$

$$= -f^d(\vec{J}_0^{d-1}, \vec{J}_0^d, \vec{J}_0^{d+1}, \vec{n}_0^{d-1} + \delta \vec{n}^{d-1}, \vec{n}_0^d + \delta \vec{n}^d, \vec{n}_0^{d+1} + \delta \vec{n}^{d+1}) \tag{13.357}$$

The linearization of Eqs. (13.355) and (13.356) give

$$\frac{\partial f^1}{\partial \vec{J}^1} \delta \vec{J}^1 + \frac{\partial f^1}{\partial \vec{J}^2} \delta \vec{J}^2 = -f^1(\vec{J}_0^1, \vec{J}_0^2, \vec{n}_0^1 + \delta \vec{n}^1, \vec{n}_0^2 + \delta \vec{n}^2) \tag{13.358}$$

$$\frac{\partial f^N}{\partial \vec{J}^{N-1}} \delta \vec{J}^{N-1} + \frac{\partial f^N}{\partial \vec{J}^N} \delta \vec{J}^N = -f^N(\vec{J}_0^{N-1}, \vec{J}_0^N, \vec{n}_0^{N-1} + \delta \vec{n}^{N-1}, \vec{n}_0^N + \delta \vec{n}^N)$$

$$\tag{13.359}$$

Combining the statistical equilibrium equation (13.348) and the equation for conservation of particles in Eq. (13.351), we get the general expression

$$g^d(\vec{J}^d, \vec{n}^d) = 0 \tag{13.360}$$

The linearization of Eq. (13.360) gives

$$\frac{\partial g^d}{\partial \vec{n}^d} \delta \vec{n}^d = -g^d(\vec{J}_0^d + \delta \vec{J}^d, \vec{n}_0^d) \tag{13.361}$$

The solutions of Eqs. (13.357) to (13.359) and Eq. (13.361) using the linearized separate solution method follow the steps below. We first give the initial solution $\vec{J}_0^d$ and $\vec{n}_0^d$ and assume that $\delta \vec{n}^d = 0$. Using these to solve the radiative transfer equations (13.357) to (13.359), we obtain $\delta \vec{J}^d$. We then use $\vec{J}^d = \vec{J}_0^d + \delta \vec{J}^d$ and $\vec{n}_0^d$ to solve Eq. (13.361) to get a new $\delta \vec{n}^d$, and then use $\vec{n}^d = \vec{n}_0^d + \delta \vec{n}^d$ and $\vec{J}^d$ to solve the radiative transfer equation. The above procedures are repeated until both the correction terms $\delta \vec{J}^d$ and $\delta \vec{n}^d$ are smaller than a certain precision.

13.3 RADIATIVE TRANSFER IN THE MOVING ATMOSPHERE

13.3.1 Introduction

Observations have shown that the atmosphere of some stars, in particular some giant stars and Wolf–Rayet stars, are expanding out-wards with high velocities, sometimes reaching 10000 to 20000 km/s. The study of the structure and radiative transfer characteristics for atmospheres having conspicuous macroscopic outward motions and production of consistent results with the observed spectra are important topics for the theory of stellar atmospheres. The study of radiative transfer problems in these atmospheres will encounter difficulties different from those for stable atmospheres without macroscopic motion.

(1) Since the atmosphere has macroscopic motion, the hydrostatic equilibrium equation and the energy equation are not valid. Thus, the method introduced in §13.1 cannot be employed for calculations of the model atmosphere. Besides, the spectral line calculations described in §13.2 have assumed that the model atmosphere is already known, so they cannot be used for atmospheres with macroscopic motions. To calculate the spectral lines in atmospheres with macroscopic motions, we have to first make special assumptions for the atmospheric structure (such as the density distribution and temperature distribution in the atmosphere) and the velocity field.

(2) The thickness of most outward expanding atmospheres are not negligible compared to the stellar radius, so the plane-parallel structure can no longer be used as the approximate geometric structure. The atmospheres should be treated as spherically symmetric structures in the study of radiative transfer.

(3) In the study of radiative transfer problems in atmospheres without macroscopic motion, we choose the reference frame which is stationary with respect to the ground observer and adopt the complete redistribution assumption (i.e., assuming the absorption profile and the emission profile of atoms for each energy level are the same). However, for atmospheres with macroscopic motion, if we study radiative transfer problems in the reference frame which is stationary with respect to the ground observer, we will find that the complete redistribution assumption does not hold (Vardavas 1974). This makes line calculations very complicated. Only when we choose a reference frame co-moving with the motion will the

complete redistribution assumption hold (Hamman and Kudritzki 1977). Therefore, in the study of radiative transfer in an atmosphere with macroscopic motion, it is common to choose a reference frame co-moving with the motion, and one should take into account the frequency change term appearing in the radiative transfer equation due to Doppler effect.

13.3.2 Radiative Transfer Equation in the Co-moving Frame

Consider a point in the atmosphere which has an outward velocity $v(r)$ with respect to the observer's frame, and a volume element co-moving with the atmosphere. The volume element emits a radiation with frequency ν_L which is measured in a frame stationary relative to that volume element (co-moving frame) and is called the local frequency. Because of Doppler effect, the frequency observed in the observer's frame is

$$\nu = \nu_L + \mu v(r) \frac{\nu_o}{c} \tag{13.362}$$

where ν_o is frequency of the line center in the observer's frame and c is the speed of light.

When the beam has travelled a distance ds along the radiation direction, the local frequency ν has also changed according to

$$\frac{d\nu_L}{ds} = -\frac{\nu_o}{c} \left[\mu^2 \frac{dv(r)}{dr} + v(r) \frac{d\mu}{ds} \right] \tag{13.363}$$

where $(d\mu/ds)$ is observed from Fig. 13.21 to be

$$\frac{d\mu}{ds} = \frac{d\mu}{d\theta} \cdot \frac{d\theta}{ds} = -\sin\theta \frac{d\theta}{ds}$$

$$= \frac{\sin^2\theta}{r} = \frac{1 - \mu^2}{r} \tag{13.364}$$

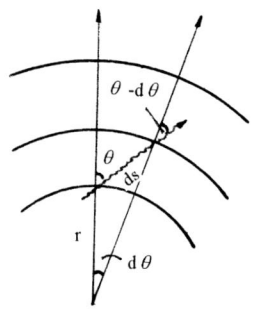

Fig. 13.21 Radiative transfer in a spherically symmetrical atmosphere.

so

$$\frac{d\nu_L}{ds} = -\frac{\nu_o}{c} \left[\mu^2 \frac{dv(r)}{dr} + \frac{v(r)}{r} (1 - \mu^2) \right] \tag{13.365}$$

The first term on the right-hand side is the frequency change caused by change in velocity from start to end of the beam, and the second term is the Doppler frequency change caused by the two positions moving away from each other.

Thus, when we employ the plane-polar coordinate (r, θ) to describe the radiative transfer equation in the co-moving frame, from Eqs. (2.63) and (13.365), we have

$$\mu \frac{\partial I(r, \mu, \nu_L)}{\partial r} + \frac{1 - \mu^2}{r} \frac{\partial I(r, \mu, \nu_L)}{\partial \mu} - \frac{\nu_0}{c} \left[\mu^2 \frac{d\nu(r)}{dr} + \frac{1 - \mu^2}{r} \nu(r) \right] \frac{\partial I(r, \mu, \nu_L)}{\partial \nu_L}$$

$$= \kappa(r, \mu, \nu_L)[S(r, \mu, \nu_L) - I(r, \mu, \nu_L)] \tag{13.366}$$

In the co-moving frame, when we employ the plane-cylindrical reference frame (Z, p) for the radiative transfer equations, from Eqs. (2.66) and (13.365), we have

$$\pm \frac{\partial I^\pm(Z, p, \nu_L)}{\partial Z} - \frac{\nu_0}{c} \left[\mu^2 \frac{d\nu(r)}{dr} + \frac{1 - \mu^2}{r} \nu(r) \right] \frac{\partial I^\pm(Z, p, \nu_L)}{\partial \nu_L}$$

$$= \kappa(Z, p, \nu_L) \left[S(Z, p, \nu_L) - I^\pm(Z, p, \nu_L) \right] \tag{13.367}$$

13.3.3 Treatment of Radiative Transfer Problems in the Co-moving Frame

(A) Methods and steps in solving radiative transfer problems

The basic difference between solving the radiative transfer problem in a moving atmosphere and solving it in a stationary atmosphere is the requirement of a solution in the co-moving frame and the treatment of the atmosphere as spherically symmetric in the former case.

In the co-moving frame, the line radiative transfer equation corresponding to the transition from an energy level i to an energy level j can be written in the plane-cylindrical frame as

$$\pm \frac{\partial I^\pm(Z, p, x)}{\partial Z} - P(Z, p) \frac{\partial I^\pm(Z, p, x)}{\partial x} = l_{ij} \left[S_{ij} - I^\pm(Z, p, x) \right] \tag{13.368}$$

where $P(Z, p)$ is

$$P(Z, p) = \mu^2 \frac{d\nu}{dr} + \frac{1 - \mu^2}{r} \nu \tag{13.369}$$

x is the dimensionless frequency, i.e., the frequency divided by the Doppler width $\Delta\nu_D$ and the measurement starts from the line center. If the dimensionless frequency in the observer's frame is represented by $\bar{x}$, we have

$$\bar{x} \equiv \frac{\nu - \nu_0}{\Delta\nu_D} \tag{13.370}$$

$$\Delta\nu_D \equiv \nu_0 \frac{\nu_s}{c} \tag{13.371}$$

$$\nu_s = (2kT/m)^{1/2} \tag{13.372}$$

where ν_s is the speed of thermal motion. In the co-moving frame, the dimensionless frequency x should be

$$x = \bar{x} - \mu\nu(r)/\nu_s \tag{13.373}$$

The line source function S_{ij} in Eq. (13.368) is

$$S_{ij} = \frac{2h\nu^3}{c^2} \left(\frac{n_i \, g_j}{n_j \, g_i} - 1 \right)^{-1} \tag{13.374}$$

and the line absorption coefficient l_{ij} is

$$l_{ij} = \frac{\pi e^2}{mc} f_{ij} \frac{\Phi_\nu}{\Delta\nu_D} \left(n_i - n_j \frac{g_i}{g_j} \right) \tag{13.375}$$

In the above expressions, we have made the complete redistribution assumption, i.e., we have assumed that the absorption profile is exactly the same as the emission profile.

Assume that the studied atom has M energy levels and can produce many spectral lines and correspondingly involves many line radiative transfer equations, line source functions and line absorption coefficients. The occupancy numbers n_1, n_2, ..., n_M on the energy levels are determined by a unified statistical equilibrium equation and the equation for conservation of total particles for that atom. The two equations can be combined to form a vector equation

$$M \cdot \vec{n} = \vec{U} \tag{13.376}$$

where the vector $\vec{n}$ is

$$\vec{n} = (n_1, n_2, ..., n_M) \tag{13.377}$$

Table 13.6 The components for the matrix M and the vector $\vec{U}$, where A_{ij} and B_{ij} are the Einstein coefficients.

M_{ij}	$-\left(A_{ii} + B_{ij}\int \Phi_\nu J_\nu d\nu + n_e\Omega_{ij}\right)$
M_{ii}	$\displaystyle\sum_{i\neq j}^{M}\left(A_{ij} + B_{ij}\int \Phi_\nu J_\nu d\nu + n_e\Omega_{ij}\right) + n_e\Omega_{iK} + R_{iK}$
U_i	$n_i^* n_e\Omega_{iK} + n_K R_{Ki}$

The components for the matrix M and the vector $\vec{U}$ can be found in Table 13.6.

We now introduce an iteration method to solve the radiative transfer equation (13.368) and the rate equation (13.376). Figure 13.22 gives the flow chart for this method and the detailed steps are as follows:

Step 1: A set of occupancy numbers for each atomic energy level (n_1, n_2, ..., n_M) is chosen as the initial approximation for each depth point. The choice of the initial approximation will be critical for the solution.

Step 2: Using the known (n_1, n_2, ..., n_M) values, and according to Eqs. (13.374) and (13.375), the line source function S_{ij} and the line absorption coefficient l_{ij} are calculated.

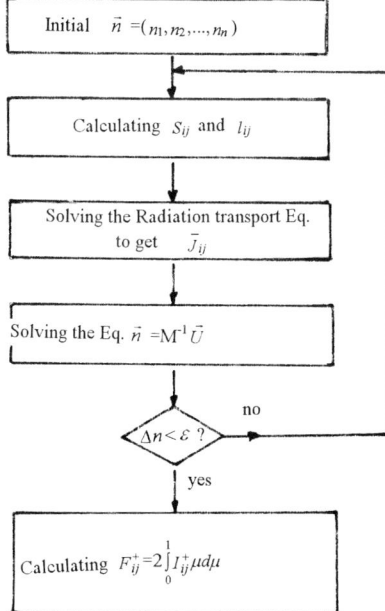

Fig. 13.22 Flow chart for the calculations for radiative transfer problems in a moving atmosphere.

Step 3: The obtained S_{ij} and l_{ij} are substituted into the line radiative transfer equation (13.368). After solving the line radiative transfer equation, we can calculate

$$\bar{J}_{ij} = \int_0^\infty \phi_\nu J_\nu d\nu \tag{13.378}$$

where J_ν can be calculated from

$$J_\nu = \int_0^1 u(x, \mu, Z) d\mu \tag{13.379}$$

and the integrand $u(x, \mu, Z)$ can be obtained from solving the line radiative transfer equation, which will be discussed in detail in the following.

Step 4: After obtaining all $\bar{J}_{ij}$, they are substituted into the rate equation (13.376) to obtain the occupancy numbers $(n_1, n_2, ..., n_M)$ for each energy levels:

$$\vec{n} = M^1 \vec{U} \tag{13.380}$$

Step 5: Check whether the differences between the atomic occupancy numbers obtained in step 4 and the initial approximations are within the acceptable precision, i.e., whether the calculations have converged. If the differences are larger than the acceptable precision, the atomic occupancy numbers obtained in step 4 are used as the initial approximation and the above calculations are repeated. The iterations will continue until the calculations converge.

Step 6: When the calculations finally converge, we obtain the line source functions and radiation fields at the above depth points. According to Eqs. (13.280) to (13.283), we can calculate the radiative flux and finally obtain the line profile or the equivalent width.

In the following, we will discuss in detail how we solve the line radiative transfer equation in step 2 and how we obtain the initial atomic occupancy numbers $(n_1, n_2, ..., n_M)$ in step 1.

(B) Solution for the line radiative transfer equation

Without loss of generality, the line radiative transfer equation in the co-moving frame expressed in the planar cylindrical system can be written as

$$\pm\frac{\partial I^{\pm}(Z, p, x)}{\partial Z} - P(Z, p)\frac{\partial I^{\pm}(Z, p, x)}{\partial x} = \kappa(r, x)\big[S(Z, p, x) - I^{\pm}(Z, p, x)\big]$$

$$(13.381)$$

which is a simplified equation considering only the Doppler effect but not the temporal variation of the radiative transfer or the aberration of light.

To solve Eq. (13.381), we introduce two variables u and v to replace the radiation intensity I, which are defined as

$$u(Z, p, x) \equiv \frac{1}{2}\big[I^{+}(Z, p, x) + I^{-}(Z, p, x)\big] \qquad (13.382)$$

$$v(Z, p, x) \equiv \frac{1}{2}\big[I^{+}(Z, p, x) - I^{-}(Z, p, x)\big] \qquad (13.383)$$

Here,

$$\begin{matrix} I^{+} = I, & \mu \geq 0 \\ I^{-} = I, & \mu < 0 \end{matrix} \quad (\mu = \cos\theta) \qquad (13.384)$$

u has properties of the radiation intensity while v has properties of the radiation flux. The sum and the difference of the two radiative transfer equations in Eq. (13.381) (equations on outward and inward radiation) give two equations in u and v as

$$\frac{\partial u(Z, p, x)}{\partial Z} - P(Z, p)\frac{\partial v(Z, p, x)}{\partial x} = -\kappa(r, x)v(Z, p, x) \qquad (13.385)$$

$$\frac{\partial v(Z, p, x)}{\partial Z} - P(Z, p)\frac{\partial u(Z, p, x)}{\partial x} = \kappa(r, x)\big[S(r) - u(Z, p, x)\big] \quad (13.386)$$

which are valid in a certain spatial region as shown in Fig. 13.23. The bottom layer of the atmosphere is at $r = r_c$ (called the core), and surface is at $r = R$. At the same time, we require $r = r_c = 1$ at the core, so Eqs. (13.385) and (13.386) are valid in the spatial region from the core to the surface, i.e., for the region $1 \leq r \leq R$. If $0 \leq p \leq 1$, the radiation can cut the core; if $1 < p \leq R$, it cannot cut the core. For a certain

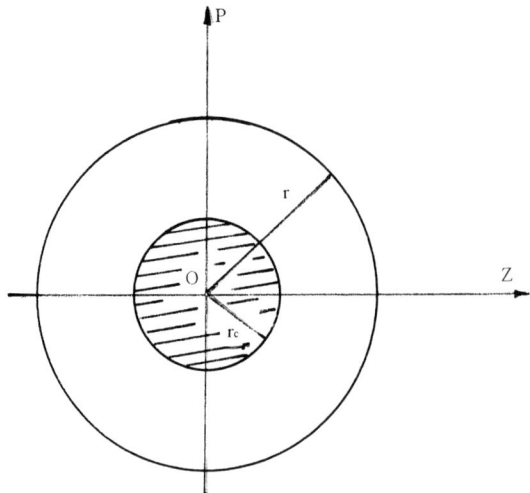

Fig. 13.23 Schematic illustration of a spherical stellar atmosphere. The bottom layer is at $r = r_c$, and the surface is at $r = R$.

p value, $Z_{min} \leq Z \leq Z_{max}$, where

$$Z_{min} = \sqrt{1 - p^2}, \; p \leq 1; \; Z_{min} = -Z_{max}, \; p > 1$$
$$Z_{max} = \sqrt{R^2 - p^2}$$

In solving Eqs. (13.385) and (13.386), we first give the boundary conditions for these two equations.

(1) Surface boundary condition
At the surface of the atmosphere, there is no incoming radiation, so $I^-(z_{max}, p, x) = 0$. From Eqs. (13.382) and (13.383), we know

$$u\big(Z_{max}, p, x\big) = v(Z_{max}, p, x)$$

and from Eq. (13.385) we get

$$\frac{\partial u(Z_{max}, p, x)}{\partial Z} - P(Z_{max}, p)\frac{\partial u(Z_{max}, p, x)}{\partial x} = -\kappa(R, x)u(Z_{max}, p, x)$$

$$(13.387)$$

(2) Internal boundary condition
The internal boundary condition will be discussed for two cases:

(i) *For the radiation not cutting the core, i.e., $p > 1$.* Since the radiation field is isotropic at $Z = 0$, i.e., $I^+(0, p, x) = I^-(0, p, x)$, we know $v(0, p, x) = 0$ from Eq. (13.383), and we get

$$\frac{\partial u(0, p, x)}{\partial Z} = 0, \quad > 1 \tag{13.388}$$

from Eq. (13.385).

(ii) *For the radiation cutting the core, i.e., $p \leq 1$.* The outward radiation intensity $I_c(p, x)$ at the core can be given. According to Eqs. (13.382) and (13.383), we know $v = I^+ - u$, so from Eq. (13.385), we have

$$\frac{\partial u(Z_{min}, p, x)}{\partial Z} + P(Z_{min}, p)\frac{\partial u(Z_{min}, p, x)}{\partial x} = \kappa(r=1, x)\big[u(Z_{min}, p, x) - I_c(p, x)\big]$$

$$+ P(Z_{min}, p)\frac{\partial I_c(p, x)}{\partial x} \tag{13.389}$$

(3) Frequency boundary condition
In previous discussions on the boundary conditions for the radiative transfer equation for an atmosphere without macroscopic motion, only the surface and internal boundary conditions are required. However, for the differential equations (13.385) and (13.386), we also need the boundary value for frequency in addition to the spatial boundary conditions, which is a special feature for solving the atmospheric radiative transfer equation in the co-moving frame.

During the outward expansion of the atmosphere, the speed of the expanding gas increases outwards, i.e., $dv/dr > 0$, $v(r) \geq 0$. Therefore, any two volume elements co-moving with the atmosphere are separating from each other. In other words, for any volume element, the radiations from all other volume elements are red-shifted. Thus, for the spectral line produced by that volume element, the radiation intensity of the bluest end of the wing (i.e., the maximum frequency x_{max}) will not be interfered by radiation of the same frequency from the atmosphere, and is determined by the continuous radiation field in that volume element alone, i.e.,

$$u(Z, p, x_{max}) = u_K(Z, p, x_{max})$$

$$v(Z, p, x_{max}) = v_K(Z, p, x_{max}) \tag{13.390}$$

We call this boundary value of frequency the bluest frequency boundary value.

Since there are mixed terms in Eqs. (13.385) and (13.386), we cannot eliminate v from them and build a second order differential equation in

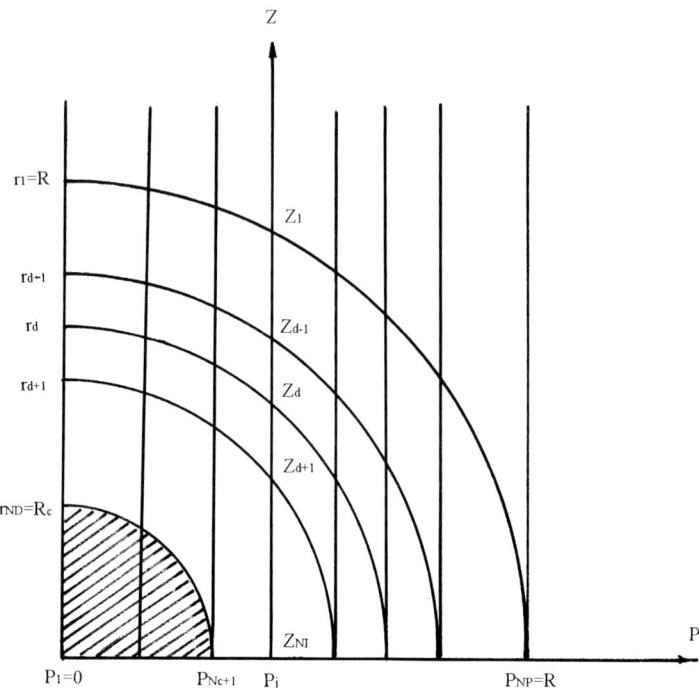

Fig. 13.24 Schematic sketch for giving the discrete values of the variables Z, p and x.

u. To numerically solve Eqs. (13.385) and (13.386), we have to rewrite the differential equations as difference equations. We first change the region in which the variables Z, p and x vary continuously to a series of discrete values as shown in Fig. 13.24.

(i) Choice of radius r_d: We take the surface to be $r_1 = R$ and the innermost layer to be $r_{ND} = 1$, so $r_1 = R > r_d > r_{ND} = 1$

(ii) Choice of variable p_j: For the radiation cutting the core, we have

$$p_1 = 0 < p_j < p_{NC} < 1$$

For the radiation not cutting the core, we choose a p value same as r, i.e.,

$$p_{NC+1} = r_{ND} = 1 < p_j = r_{NP+1-j} < p_{NP} = r_1 = R$$

(iii) Choice of variable Z_d: The value of Z_d is determined by r_d and p_j:

$$Z_d = \sqrt{r_d^2 - p_j^2}, \quad Z_1 = \sqrt{R^2 - p_j^2} > Z_d > Z_{NI}$$

for radiation cutting the core, $NI = ND$;
for radiation not cutting the core, $NI = NP + 1 - j$.

(iv) Choice of frequency x_K: The frequency points can be chosen using the equi-distance method, and the maximum value is x_{max}, so

$$x_K = x_{max} - (K - 1)\Delta x \quad , \quad K = 1, 2, ..., NF$$

We rewrite the differential equations as difference equations. From Eq. (13.385), when p is kept constant, we have

$$\frac{u(x_{K+1/2}, Z_{d+1}) - u(x_{K+1/2}, Z_d)}{\kappa(x_{K+1/2}, Z_{d+1/2})(Z_d - Z_{d+1})} + \frac{P(Z_{d+1/2})[v(x_K, Z_{d+1/2}) - v(x_{K+1}, Z_{d+1/2})]}{\kappa(x_{K+1/2}, Z_{d+1/2})(x_{K+1} - x_K)}$$

$$= v(x_{K+1/2}, Z_{d+1/2}) \tag{13.391}$$

From Eq. (13.386), we have

$$\frac{v(x_{K+1/2}, Z_{d+1/2}) - v(x_{K+1/2}, Z_{d-1/2})}{\kappa(x_{K+1/2}, Z_d)(Z_{d-1/2} - Z_{d+1/2})} + \frac{P(Z_d)[u(x_K, Z_d) - u(x_{K+1}, Z_d)]}{\kappa(x_{K+1/2}, Z_d)(x_K - x_{K+1})}$$

$$= u(x_{K+1/2}, Z_d) - S(Z_d) \tag{13.392}$$

For choice of discrete values for the variables, the subscript d is always an integer in the function u and is always half integers in the function v, i.e., $d = \pm(1/2)$. For intermediate values for the frequency, we define

$$u(x_{K+1/2}, Z_d) = \frac{1}{2}[u(x_{K+1}, Z_d) + u(x_K, Z_d)] \tag{13.393}$$

$$v(x_{K+1/2}, Z_{d+1/2}) = \frac{1}{2}[v(x_{K+1}, Z_{d+1/2}) + v(x_K, Z_{d+1/2})] \tag{13.394}$$

We now introduce the following abbreviations:

$$TAUZ(d) \equiv Z_d - Z_{d+1} \quad , \quad TAU(d) \equiv Z_{d-1/2} - Z_{d+1/2}$$
$$PPZ(d) \equiv P(Z_{d+1/2})/\Delta x \quad , \quad PP(d) \equiv P(Z_d)/\Delta x$$
$$AK \equiv \kappa(x_{K+1/2}, Z_d) \quad , \quad AZ \equiv \kappa(x_{K+1/2}, Z_{d+1/2})$$
$$DTZ \equiv 0.5/[AZ \cdot TAUZ(d)] \quad , \quad DXZ \equiv 2 \cdot PPZ(d)/AZ$$
$$DTZM \equiv DTZ(d-1) \quad , \quad DXZM \equiv DXZ(d-1)$$
$$DT \equiv 1/[AK \cdot TAU(d)] \quad , \quad DX \equiv PP(d)/AK$$

and

$$u(x_K, Z_d) \equiv u_{0,0}$$
$$u(x_{K+1}, Z_{d\pm1}) \equiv u_{1\pm1}$$

$$v(x_K, Z_{d+1/2}) \equiv v_{0,0}$$

$$v(x_K, Z_{d-1/2}) \equiv v_{0,-1}$$

According to Eq. (13.394), Eq. (13.391) can be written as

$$\frac{1}{AZ \cdot TAUZ(d)} \left[v(x_{K+1/2}, Z_{d+1}) - u(x_{K+1/2}, Z_d) \right] + \frac{2P(Z_{d+1/2})}{\kappa(x_{K+1/2}, Z_{d+1/2}) \Delta x}$$

$$\left[u(x_K, Z_{d+1/2}) - v(x_{K+1/2}, Z_{d+1/2}) \right] = v(x_{K+1/2}, Z_{d+1/2}) \qquad (13.395)$$

which can also be written as

$$DTZ[u(x_{K+1}, Z_{d+1}) + u(x_K, Z_{d+1}) - u(x_{K+1}, Z_d) - u(x_K, Z_d)]$$

$$+ DXZ \cdot v(x_K, Z_{d+1/2}) = (1 + DXZ) \cdot v(x_{K+1/2}, Z_{d+1/2}) \qquad (13.396)$$

Replacing d by $(d - 1)$ in the above equation, we have

$$DTZM[u(x_{K+1}, Z_d) + u(x_K, Z_d) - u(x_{K+1}, Z_{d-1}) - u(x_K, Z_{d-1})]$$

$$+ DXZM \cdot v(x_K, Z_{d-1/2}) = (1 + DXZM) \cdot v(x_{K+1/2}, Z_{d-1/2}) \qquad (13.397)$$

Using abbreviations, Eq. (13.392) can be written as

$$DT\left[v(x_{K+1/2}, Z_{d+1/2}) - v(x_{K+1/2}, Z_{d-1/2}) \right] + DX[u(x_K, Z_d) - u(x_{K+1}, Z_d)]$$

$$\frac{1}{2}[u(x_{K+1}, Z_d) + u(x_K, Z_d)] - S(x_{K+1/2}, Z_d) \qquad (13.398)$$

Substituting $v(x_{K+1/2}, Z_{d+1/2})$ by Eq. (13.396), and $v(x_{K+1/2}, Z_{d-1/2})$ by Eq. (13.397), and using the abbreviations, Eq. (13.398) becomes

$$\frac{DT \cdot DTZ}{1 + DXZ} \left(u_{1,1} + u_{0,1} - u_{1,0} - u_{0,0} \right) + \frac{DT \cdot DXZ}{1 + DXZ} v_{0,0}$$

$$- \frac{DT \cdot DTZM}{1 + DXZM} \left(u_{1,0} + u_{0,0} - u_{1,-1} - u_{0,-1} \right) - \frac{DT \cdot DXZM}{1 + DXZM} v_{0,-1}$$

$$+ DX\left(u_{0,0} - u_{1,0} \right) - \frac{1}{2} \left(u_{1,0} - u_{0,0} \right) = -S(x_{K+1/2}, Z_d) \qquad (13.399)$$

Rearranging,

$$TA \cdot u_{1,-1} + TB \cdot u_{1,0} + TC \cdot u_{1,1} + UA \cdot u_{0,-1} + UB \cdot u_{0,0} + UC \cdot u_{0,1}$$

$$+ VA \cdot v_{0,-1} + VB \cdot v_{0,0} = -S_{1/2,0} \qquad (13.400)$$

We write the above equation in the form of a vector equation,

$$\vec{T}_K \cdot \vec{u}_{K+1} + \vec{U}_K \cdot \vec{u}_K + \vec{V}_K \cdot \vec{v}_K = -\vec{S}_K \qquad (13.401)$$

where

$$\left.\begin{aligned}
(\vec{T}_K)_d &= (TA(K), TB(K), TC(K))_d \\
(\vec{U}_K)_d &= (UA(K), UB(K), UC(K))_d \\
(\vec{V}_K)_d &= (VA(K), VB(K))_d \\
(\vec{u}_K)_d &= u_{0,0} \\
(\vec{v}_K)_d &= v_{0,0} \\
(\vec{S}_K)_d &= S_{1/2,0}
\end{aligned}\right\} \qquad (13.402)$$

The matrix T and U with elements $(\vec{T}_K)_d$ and $(\vec{U}_K)_d$ are tri-diagonal matrices and the matrix V with elements $(\vec{V}_K)_d$ is a bi-diagonal matrix, i.e.,

$$U = \begin{pmatrix}
UB_1 & UC_1 & & & & \\
UA_2 & UB_2 & UC_2 & & o & \\
& \ddots & \ddots & \ddots & & \\
& & UA_d & UB_d & UC_d & \\
& o & & \ddots & \ddots & \\
& & & & UA_{NI} & UB_{NI}
\end{pmatrix}$$

$$T = \begin{pmatrix}
TB_1 & TC_1 & & & & \\
TA_2 & TB_2 & TC_2 & & o & \\
& \ddots & \ddots & \ddots & & \\
& & TA_d & TB_d & TC_d & \\
& o & & \ddots & \ddots & \\
& & & & TA_{NI} & TB_{NI}
\end{pmatrix} \qquad (13.403)$$

$$V = \begin{pmatrix}
VB_1 & & & & & \\
VA_2 & VB_2 & & & o & \\
& VA_3 & VB_3 & & & \\
& & \ddots & \ddots & & \\
& & & VA_d & VB_d & \\
& o & & & \ddots & \ddots \\
& & & & & VA_{NI}
\end{pmatrix}$$

The matrix elements for the matrices T, U and V are listed in Table 13.7. Again from Eq. (13.395), we obtain

$$2DTZ(u_{1,1} + u_{0,1} - u_{1,0} - u_{0,0}) + (DXZ - 1)v_{0,0} = (DXZ + 1)v_{1,0}$$

$$(13.404)$$

Table 13.7 Matrix elements for the matrices T, U, V, G and H.

	A (TA, UA, VA, GA, HA)	B (TB, UB, VB, GB)	C (TC, UC)
T	0, $d = 1$ $DT \cdot DTZM/(1 + DXZM)$, $2 \le d \le NI{-}1$ $2 \cdot DT \cdot DTZ/(1 + DXZ)$, $d = NI$ not cut the core DT, $d = NI$ cut the core	$-TC_1 - DX - 0.5$, $d = 1$ $-TA_d - TC_d - DX - 0.5$, $2 \le d \le NI{-}1$ $-TA_{NI} - DX - 0.5$, $d = NI$ not cut the core $-DT - DX - 0.5$, $d = NI$ cut the core	$0.5/[AK \cdot TAUZ(1)]$, $d = 1$ $DT \cdot DTZ/(1 + DXZ)$, $2 \le d \le NI{-}1$ 0, $d = NI$, not cut the core 0, $d = NI$, cut the core
U	0, $d = 1$ TA_d, $2 \le d \le NI - 1$ TA_d, $d = NI$ not cut the core DT, $d = NI$ cut the core	$-TC_1 + DX - 0.5$, $d = 1$ $-TA_d - TC_d + DX$, $2 \le d \le NI - 1$ $2 \cdot DX - TB_d$, $d = NI$ not cut the core $-DT + DX - 0.5$, $d = NI$ cut the core	TC_1, $d = 1$ TC_d, $2 \le d \le NI - 1$ 0, $d = NI$ not cut the core 0, $d = NI$ cut the core
V	0, $d = 1$ $-DT \cdot DXZM/(1 + DXZM)$, $2 \le d \le NI - 1$ $2 \cdot DT \cdot DXZ/(1 + DXZ)$, $d = NI$ not cut the core 0, $d = NI$ cut the core	0, $d = 1$ $DT \cdot DXZ/(1 + DXZ)$, $2 \le d \le NI - 1$ 0, $d = NI$ not cut the core 0, $d = NI$ cut the core	
G	$-2 \cdot DTZM/(1 + DXZM)$, $d = 1$ $-2 \cdot DTZ/(1 + DXZ)$, $2 \le d \le NI$	$-GA$, $1 \le d \le NI$	
H	$(DXZM - 1)/(DXZM + 1)$, $d = 1$ $(DXZ - 1)/(DXZ + 1)$, $2 \le d \le NI$		

or in the form of vector equation

$$\vec{v}_{K+1} = \vec{G}_K(\vec{u}_K + \vec{u}_{K+1}) + \vec{H}_K \cdot \vec{v}_K \qquad (13.405)$$

where

$$\left.\begin{aligned}
(\vec{G}_K)_d &= (GA(K), GB(K))_d \\
(\vec{H}_K)_d &= (HA(K))_d
\end{aligned}\right\} \qquad (13.406)$$

The matrix G with matrix elements $(\vec{G}_K)_d$ is a bi-diagonal matrix, and the matrix H with matrix elements $(\vec{H}_K)_d$ is a diagonal matrix. The matrix elements of the matrices G and H can be found in Table 13.7. Eq. (13.401) should be supplemented by the boundary conditions in Eqs. (13.387) to (13.389). From Eq. (13.387), we can get

$$\frac{1}{\kappa(x_{K+1/2}, Z_1)TAUZ(1)}\left[u(x_{K+1/2}, Z_2) - u(x_{K+1/2}, Z_1)\right]$$

$$+ \frac{P(Z_1)}{\kappa(x_{K+1/2}, Z_1)\Delta x}[u(x_K, Z_1) - u(x_{K+1}, Z_1)] = u(x_{K+1/2}, Z_1) \qquad (13.407)$$

which can be written as

$$TB \cdot u_{1,0} + TC \cdot u_{1,1} + UB \cdot u_{0,0} + UC \cdot u_{0,1} = 0 \qquad (13.408)$$

which supplements the first row ($d = 1$) in Eq. (13.401). For the radiation cutting the core, from Eq. (13.389), we obtain

$$DT(u_{0,0} + u_{1,0} - u_{0,-1} - u_{1,-1}) + DX(u_{0,0} - u_{1,0}) - \frac{1}{2}(u_{1,0} + u_{0,0}) = -I^+ \qquad (13.409)$$

which supplements the last row ($d = NI$) in Eq. (13.401). The above equation has employed the relation $v = I^+ - u$. For the radiation not cutting the core, the boundary condition should be written as a second order equation for precision (see Mihalas *et al.* 1975). Equations (13.396) and (13.397) can first be applied for $d = NI$, and then substituted into the symmetric conditions $u(x_K, Z_{NI+1}) = u(x_K, Z_{NI-1})$ and $v(x_K, Z_{NI+1}) = -v(x_K, Z_{NI-1})$ to get

$$DTZ(u_{0,1} + u_{1,1} + u_{0,0} - u_{0,1}) + DXZ \cdot v_{0,0} = (1 + DXZ) \cdot v(x_{K+1/2}, Z_{NI+1/2}) \qquad (13.410)$$

$$DTZ(u_{0,0} + u_{0,1} - u_{0,-1} - u_{1,-1}) + DXZ \cdot v_{0,-1} = (1 + DXZ) \cdot v(x_{K+1/2}, Z_{NI-1/2})$$

$$(13.411)$$

Equation (13.392) is further applied for $d = NI$, so

$$\frac{1}{\kappa(Z_{NI})TAUZ(NI-1)}\left[v(x_{K+1/2}, Z_{NI+1/2}) - v(x_{K+1/2}, Z_{NI-1/2})\right]$$

$$+ \frac{P(Z_{NI})}{\kappa(Z_{NI})\Delta x}[u(x_K, Z_{NI}) - u(x_{K+1}, Z_{NI})] = u(x_{K+1/2}, Z_{NI}) - S(Z_{NI})$$
$$(13.412)$$

Substituting $v(x_{K+1/2}, z_{NI+1/2})$ and $v(x_{K+1/2}, z_{NI-1/2})$ into Eqs. (13.410) to Eq. (13.412), and using the abbreviations, we get

$$\frac{DT \cdot DTZ}{1 + DXZ}(u_{0,1} + u_{1,1} - u_{0,0} - u_{1,0} - u_{0,0} - u_{1,0} + u_{0,-1} + u_{1,-1})$$

$$+ \frac{DT \cdot DXZ}{1 + DXZ}(v_{0,0} - v_{0,-1}) + DX(u_{0,0} - u_{1,0}) - \frac{1}{2}(u_{0,0} + u_{1,0}) = -S(Z_{NI})$$

Further substituting the conditions $u(x_K, Z_{NI+1}) = u(x_K, Z_{NI-1})$ and $v(x_K, Z_{NI+1}) = -v(x_K, Z_{NI-1})$, and rearranging, we get

$$TA \cdot u_{1,-1} + TB \cdot u_{1,0} + TC \cdot u_{1,1} + UA \cdot u_{0,-1} + UB \cdot u_{0,0} + UC \cdot u_{0,1}$$

$$+ VA \cdot v_{0,-1} + VB \cdot v_{0,0} = -S(Z_{NI}) \qquad (13.413)$$

which supplements the last row ($d = NI$) in Eq. (13.401). The matrix elements for the vector equations for Eqs. (13.408), (13.409) and (13.413) can be referred from Table 13.7.

Up to now, we have rewritten the radiative equations (13.385) and (13.386) and the boundary conditions in Eqs. (13.387) to (13.389) as two vector equations (13.401) and (13.405). In the following, we will discuss the steps in solving Eqs. (13.401) and (13.405).

We write Eqs. (13.401) and (13.405) in the form

$$\vec{u}_{K+1} = -\vec{T}_K^{-1}(\vec{U}_K\vec{u}_K + \vec{V}_K\vec{v}_K + \vec{S}) \qquad (13.414)$$

$$\vec{v}_{K+1} = G(\vec{u}_{K+1} + \vec{u}_K) + \vec{H}_K\vec{v}_K \qquad (13.415)$$

When the bluest frequency boundary value is given according to Eq. (13.390), it can be substituted into Eq. (13.414) to give the value of $\vec{u}$ for the next frequency point, which can be inserted into Eq. (13.415) to calculate the value of $\bar{v}$. Equations (13.414) and (13.415) are then used to calculate the values of $\bar{u}$ and $\bar{v}$ for the next frequency point. The procedures are continued from the bluest frequency to the reddest frequency boundary. It should be noted that Eqs. (13.414) and (13.415) are obtained for a certain d value. Therefore, the above should be calculated for different d values to get all $u(Z, p, x)$ and $v(Z, p, x)$.

When all $u(Z, p, x)$ and $v(Z, p, x)$ are obtained, we can use the equations

$$\bar{J}_{ij} = \int_0^\infty \Phi_v J_v dv$$

$$J_v = \int_0^1 u(Z, p, x) d\mu$$

to evaluate $\bar{J}_{ij}$.

13.3.4 Extended Sobolev Method

In the flow chart shown in Fig. 13.22, we have pointed out that the first step in solving the line radiative transfer equation and the constraint equation is to choose a set of occupancy numbers of atomic energy levels ($n_1, n_2, ..., n_M$) as the initial approximations. The choice of these initial approximations is critical for convergence of calculations. If the velocity of the moving atmosphere is very large, the extended Sobolev method can be used to obtain the initial approximations.

The properties of the radiation field in a certain region in a moving atmosphere with a very large velocity gradient are determined completely by local physical parameters (temperature and density etc.). This is easy to visualize. We know that two arbitrary volume elements co-moving with the motion of the outward expanding atmosphere are always separating from each other. Thus, for a volume element being examined, all the radiation coming from other volume elements will be Doppler shifted. The Doppler shift will be more significant for larger relative motion (i.e., when the velocity gradient in the expanding atmosphere is larger). If we study a certain spectral line with the profile region $\pm x_{max}$, the frequency of photons in this region emitted by any volume elements at a certain distance d from the reference volume element will not interfere with the radiation in this region emitted by the reference volume element due to the Doppler

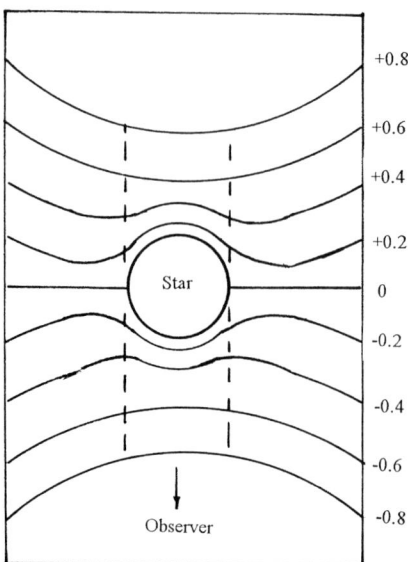

+0.8

+0.6

+0.4

+0.2

0

Star

-0.2

-0.4

-0.6

Observer

-0.8

Fig. 13.25 Surfaces of iso-radial velocities for an expanding atmosphere.

shift. In other words, the photons in the region of that spectral line emitted by other volume elements will escape from being absorbed by the reference volume element. Apparently, the value of d mentioned above will be smaller for a larger velocity gradient in the expanding atmosphere. When the velocity gradient reaches a certain value, d can be so small that the properties of the radiation field in a small region can be determined by local physical properties.

The radial velocities for different positions in the expanding atmosphere are different as viewed by a ground observer. Figure 13.25 gives the surfaces for iso-radial velocities, the shape of which are determined by the velocity field. If we look at the stellar atmosphere at a certain p value along the Z direction (see Figs. 13.24 and 13.25), and examine the radiation at a distance Δx from the line center, we can see that most regions of the atmosphere along the line of sight are optically transparent, since the radiation produced in these regions will have a Doppler shift greater than Δx. There is only a certain point in the atmosphere which is optically opaque: this point corresponds to a Doppler shift of Δx. According to Δx, we can obtain the intersection between the surface of iso-radial velocities and the line of sight, which is the optically opaque point. For the reference radiation at a distance Δx from the line center, the majority of the atmosphere is optically transparent. This implies that the photons from these regions have escaped.

For an expanding atmosphere with a large velocity gradient, according to Rybicki and Hummer (1978), the average radiation intensity at a certain point in the atmosphere can be calculated through the equation

$$\bar{J}_{ij} = (1 - \beta_{ij})S_{ij}(r) + \beta_c I_c \qquad (13.416)$$

where $S_{ij}(r)$ is the line source function, I_c is the core continuum radiation intensity, β_{ij} and β_c are escape integrals.

We first consider β_{ij}. The photon escape probability along an arbitrary direction is $\exp(-\tau_\infty)$ where τ_∞ is the optical depth from the reference point to infinity. Integrating over all angles and frequencies, we obtain

$$\beta_{ij}(r) = \frac{1}{2} \int\limits_{-1}^{+1} d\mu \int\limits_{-\infty}^{+\infty} dx' \phi(x') \exp\left[-\tau_{ij}(Z, p, x')\right] \qquad (13.417)$$

where the optical depth $\tau(Z, p, x)$ can be obtained from

$$\tau_{ij}(Z, p, x) = \int\limits_{Z}^{\infty} \kappa_{ij}(Z', p, x)dZ' = \int\limits_{Z}^{\infty} l_\nu(r')\phi(x')dZ' \qquad (13.418)$$

where

$$
\begin{aligned}
r' &= (Z'^2 + p^2)^{1/2} \\
\mu &= (Z'/r') \\
x' &= x - \mu V(r') \\
V(r) &= v(r)/v_s
\end{aligned}
\qquad (13.419)
$$

We wish to transform the variable Z' in Eq. (13.418) to x', so we have the transformation

$$-\left(\frac{\partial x'}{\partial Z}\right)_p = \left[\frac{\partial(\mu V)}{\partial Z}\right]_p = \mu^2 \left(\frac{\partial V}{\partial r}\right) + \frac{1 - \mu^2}{r} V$$

$$\equiv Q(r, \mu) \qquad (13.420)$$

We further define a function $y(x)$ as

$$y(x) \equiv \int\limits_{-\infty}^{+\infty} \phi(\xi')d\xi' \qquad (13.421)$$

Since the profile function ϕ should fulfil the normalization condition, we know

$$y(-\infty) = 0$$
$$y(\infty) = 1 \tag{13.422}$$

Inserting the relations in Eqs. (13.417) to (13.421) into Eq. (13.417), we obtain

$$\beta_{ij}(r) = \frac{1}{2} \int_{-1}^{+1} d\mu \int_{0}^{1} dy \cdot \exp\left[-l_\nu(r)y/Q(r,\mu)\right]$$

$$= \int_{0}^{1} d\mu \left[1 - \frac{\exp[-l_\nu(r)/Q]}{l_\nu/Q}\right]$$

$$= \int_{0}^{1} d\mu \left[1 - \frac{\exp\left(\frac{-\tau_{ij}^0(r)}{1+\mu^2(d\ln V/d\ln r - 1)}\right)}{\tau_{ij}^0(r)/[1 + \mu^2(d\ln V/d\ln r - 1)]}\right] \tag{13.423}$$

where

$$\tau_{ij}^0(r) = \frac{\pi e^2}{mc} f_{ij}\lambda_{ij}\left(n_i - n_j\frac{g_i}{g_j}\right)\frac{r}{V} \tag{13.424}$$

We now determine β_c. From the physical meaning of β_c, we get

$$\beta_c(r) = \frac{1}{2} \int_{\sqrt{1-(r_c/r)^2}}^{1} d\mu \int_{0}^{1} dy \cdot \exp\left[-l_\nu(r)y/Q(r,\mu)\right]$$

$$= \frac{1}{2} \int_{\sqrt{1-(r_c/r)^2}}^{1} d\mu \left[1 - \frac{\exp\left(\frac{-\tau_{ij}^0(r)}{1+\mu^2(d\ln V/d\ln r - 1)}\right)}{\tau_{ij}^0(r)/[1 + \mu^2(d\ln V/d\ln r - 1)]}\right] \tag{13.425}$$

which can be written as

$$\beta_c = \frac{1}{2}\int_{0}^{1} d\mu\left[1 - \exp(-\tau_{ij})/\tau_{ij}\right] - \frac{1}{2}\int_{0}^{\sqrt{1-(r_c/r)^2}} d\mu\left[1 - \exp(-\tau_{ij})/\tau_{ij}\right]$$

so

$$\beta_c = \frac{1}{2}\beta_{ij}(1 - \sqrt{1 - (r_c/r)^2} \cdot W_1/\beta_{ij}) \qquad (13.426)$$

where

$$W_1 = \int_0^1 dx \left[1 - \frac{\exp\left(\frac{-\tau_{ij}^o}{1+\mu^2(d\ln V/d\ln r - 1)}\right)}{\tau_{ij}^o/[1 + \mu^2(d\ln V/d\ln r - 1)]} \right] \qquad (13.427)$$

When $d\ln V/d\ln r = 1$, $V \sim r$, $\beta_c = \frac{1}{2}\left[1 - \sqrt{1 - \left(\frac{r_c}{r}\right)^2}\right] \cdot \beta_{ij}$, so

$$W_1 \equiv \left[1 - \sqrt{1 - \left(\frac{r_c}{r}\right)^2} \right] \qquad (13.428)$$

which is known as the "sparse factor".

Finally, we summarize the extended Sobolev method for finding the initial approximations (n_1, n_2, ..., n_M). We start by assuming that LTE is fulfilled at the innermost layer of the atmosphere, so the Boltzmann equation and the Saha equation can be employed to calculate the atomic energy level occupancy numbers (n_1, n_2, ..., n_M) there, from which the corresponding values of β_{ij}, β_c and S_{ij} can be evaluated. These values are then inserted into Eq. (13.416) to calculate $\bar{J}_{ij}$ which is then substituted into the rate equation (13.37) to solve for a new set of occupancy numbers (n_1, n_2, ..., n_M). If the new values are not consistent with the initial ones obtained from LTE, the new values are used as the initial values and the above calculations are repeated. The above steps are repeated until the newly obtained occupancy numbers are consistent with the initial values in that their differences are smaller than a precision value. When the occupancy numbers for the innermost layer are known, the numbers for the second layer can be calculated from the assumed density distribution for the atmosphere. The above procedures can be repeated to get the occupancy numbers for the second layer. Using the same method, the occupancy numbers for each layer can be obtained. It is worth remarking that the extended Sobolev method does not only provide the atomic energy level occupancy numbers for each layer, but also calculates S_{ij} for each layer, so that the spectral line profile and the equivalent widths can be directly evaluated, since

$$I^+(Z_{max}, p, x) = I(o, p, x)\exp\left[-\tau(o, p, x)\right]$$

$$+ \int_0^{\tau_{max}} S_{ij}(Z, p, x)\exp\left[-\tau(Z, p, x)\right]d\tau \qquad (13.429)$$

where

$$\tau(Z, p, x) = \int_{Z_{max}}^{Z} \kappa(Z', p, x)dZ' \qquad (13.430)$$

$$I(o, p, x) = \begin{cases} I_c = 1 & p \leq 1 \\ o & p > o \end{cases} \qquad (13.431)$$

When $I^+(Z_{max}, p, x)$ is known, the spectral line profile can be computed from

$$F_x = \frac{1}{2}\int_0^R\int_0^1 I^+(Z_{max}, p, x)\mu d\mu dp \qquad (13.432)$$

The line profile thus obtained is sometimes very satisfactory. Of course, more rigorous calculations should involve the use of atomic energy level occupancy numbers (n_1, n_2, ..., n_M) provided by the extended Sobolev method as the initial values and calculations described in §13.3.2.

13.4 APPLICATIONS OF THEORIES OF STELLAR ATMOSPHERE

The comparison between results from model atmosphere calculations and results from observation can yield many important physical parameters such as effective temperature T_{eff}, surface gravitational acceleration g, chemical composition X_i, turbulence velocity v_t, stellar radius R, spin velocity $v \sin i$, angular diameter of the star θ, stellar mass M and luminosity L. Han (1988) summarized various methods in comparing theoretical and observational results in the last 10 to 20 years, which are described in the following.

13.4.1 Determination of Effective Temperature, Surface Gravitational Acceleration and Chemical Composition

(A) Determination of the effective temperature through theoretical calculations and observations of the flux in a certain wavelength range (Underhill 1982)

The observed energy flux f_λ from a star is related to the radiation flux at the stellar surface through Eq. (2.10) as

$$f_\lambda = \frac{\theta^2}{4}\pi F_\lambda \qquad (13.433)$$

where θ is the angular diameter of the star ($\theta = 2R/l$) and πF_λ can be obtained from theoretical model calculations, so from Eq. (13.433) we get

$$\theta = 2\left[f_\lambda / \pi F_\lambda(\text{model})\right]^{1/2} \qquad (13.434)$$

We define a function G which represents the ratio between the total energy flux $\pi F = \int_0^\infty \pi F_\lambda d\lambda$ and the energy flux $\int_{\lambda_1}^{\lambda_2} \pi F_\lambda d\lambda$ between λ_1 and λ_2, i.e.,

$$G \equiv \frac{\int_0^\infty \pi F_\lambda d\lambda}{\int_{\lambda_1}^{\lambda_2} \pi F_\lambda d\lambda} = \frac{\pi F}{\pi F_{(\lambda_1 - \lambda_2)}} \qquad (13.435)$$

The function G can be evaluated through theoretical model calculations, so from Eq. (13.435),

$$\pi F = G(\text{model}) \int_{\lambda_1}^{\lambda_2} \pi F_\lambda d\lambda \qquad (13.436)$$

Since $\pi F = \sigma T_{\text{eff}}^4$, by substituting Eq. (13.434), we get

$$\sigma T_{\text{eff}}^4 = \frac{4G(\text{model})}{\theta^2} \int_{\lambda_1}^{\lambda_2} f_\lambda d\lambda \qquad (13.437)$$

where $\int_{\lambda_1}^{\lambda_2} f_\lambda d\lambda$ is the observed total energy flux between λ_1 and λ_2, which is denoted by

$$F_{obs} \equiv \int_{\lambda_1}^{\lambda_2} f_\lambda d\lambda \qquad (13.438)$$

so Eq. (13.437) can be expressed as

$$\sigma T_{eff}^4 = \frac{4G(\text{model})}{\theta^2} F_{obs} \qquad (13.439)$$

Replacing θ in Eq. (13.439) by Eq. (13.434), and considering that the function G is only sensitive to the effective temperature but not to the surface gravitational acceleration, we get

$$\log T_{eff} = \frac{1}{4}\left[\log F_{obs} - \log f_\lambda + \log G(\text{model}) + \log F_\lambda(\text{model})\right] + C$$

$$(13.440)$$

Through observations, we can get f_λ and the energy flux F_{obs} for a certain wavelength range. We first assume an effective temperature $T_{eff}^{(1)}$ for the star and calculate the model atmosphere to get $G(\text{model})$ and $F_\lambda(\text{model})$. These are input into Eq. (13.436) to acquire another T_{eff} denoted as $T_{eff}^{(2)}$. If $T_{eff}^{(2)}$ does not equal $T_{eff}^{(1)}$, $T_{eff}^{(2)}$ is used to recalculate the model atmosphere. These steps will be repeated until the values of T_{eff} input for and output from Eq. (13.440) are equal. At this time, we obtain the effective temperature of the star.

In realistic calculations, we can base on a series of T_{eff} values to calculate the model atmosphere to evaluate the corresponding $G(\text{model})$ and $F_\lambda(\text{model})$ and work out a table for T_{eff}, $G(\text{model})$ and $F_\lambda(\text{model})$, which simplifies the entire calculation process.

Table 13.8 The variation of the Balmer jump D_B and color index U–V with the effective temperature for a main sequence star obtained from model atmosphere calculations.

$T_{eff}(K)$	D_B	$U - V$
37450	0.02	−1.47
28640	0.075	−1.34
21910	0.130	−1.17
16800	0.241	−0.82

(B) Determination of the effective temperature through the Balmer jump D_B and the color index U–V (Morton and Adams 1968)

The model atmosphere calculations for a main sequence star with $\log g = 4$ and with normal helium abundance show that the Balmer jump D_B and the color index U–V are strongly dependent on the effective temperature, which is demonstrated in Table 13.8. This table can be used in interpolations to give effective temperatures of stars.

This method is valid for determination of the effective temperature for B type stars, but is not accurate for O type stars because the

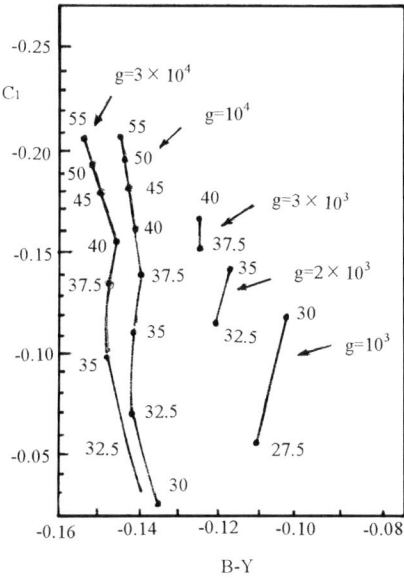

Fig. 13.26 Double-colour diagrams for different types of stars. The numbers represent effective temperatures ($T_{eff}/10^3$ K).

Balmer jumps are too small for the latter to allow accurate measurements.

(C) Determination of the effective temperature and the gravitational acceleration from the double color diagram (Mihalas 1972; Shipman and Sass 1980)

The color indices for stars can be obtained from model atmosphere calculations (Buser and Kurucz 1978; Matsushiman 1969; Matthews and Sandage 1963), and model atmosphere calculations using different surface gravitational accelerations or different effective temperatures will give different color indices and correspond to different positions ($\log g$, T_{eff}) in double-color diagrams. Figure 13.26 is a double-color diagram (C_1, $B - Y$) [$C_1 = (U - V) - (V - B)$] showing positions of stars with different effective temperatures (shown in numbers next to the points in units of $T_{eff}/10^3$ K) and for $g = 10^3$, 2×10^3, 3×10^3, 10^4, 3×10^4 cm.s^{-2} for normal helium abundance ($X_{He} = 0.1$) (points with same g are linked together).

The photoelectric method can be used to obtain color indices of stars which can then be used to obtain the corresponding position of the star in the double-color diagram. From this position, we can derive the effective temperature and the surface gravitational acceleration of the star.

(D) Determination of surface gravitational acceleration from equivalent widths of hydrogen lines (Dufton 1979)

The equivalent widths of hydrogen lines (e.g., H_γ and H_δ) will vary with the surface gravitational acceleration when the effective temperature T_{eff} is fixed. Therefore, for a fixed effective temperature, the equivalent widths of hydrogen lines calculated from different surface gravitational accelerations g can be tabulated with g, which can be used for interpolations to give the corresponding g values from observed equivalent widths of the hydrogen lines.

(E) Determination of the effective temperature from the ionization equilibrium method (Conti 1973)

Theoretical calculations have shown that the ratios of intensities of spectral lines for different degrees of ionization are not sensitive to the abundance of that element, but are sensitive to the effective temperature and surface gravitational acceleration of the atmosphere. For a given set of ($\log g$, T_{eff}) values, a model atmosphere can be calculated from which the ratios of intensities of spectral lines for different degrees of ionization for a certain element can be obtained. For different values of $\log g$, the curve between the intensity of spectral line and T_{eff} can be obtained. In this way, once the value of $\log g$ is known and the ratios of intensities of spectral lines are obtained from observations, the effective temperature can be inferred from the above curve.

Figure 13.27 gives the curves of the ratio between the intensities of the line $\lambda 4471$(HeI) and $\lambda 4541$(HeII) against the effective temperature T_{eff}. For a main sequence star ($\log g = 4$), once the ratio between the intensities of the line $\lambda 4471$(HeI) and $\lambda 4541$(HeII) is known, the effective temperature can be inferred from the curve for $\log g = 4$.

(F) Determination of the effective temperature and the surface gravitational acceleration from color-indices and equivalent widths of hydrogen lines (Morison 1975)

The color indices and equivalent widths of hydrogen lines can be obtained from theoretical calculations. For O type stars, the color index C_1 is very sensitive to the effective temperature T_{eff}, and the equivalent widths W of hydrogen lines are very sensitive to the effective temperature and the surface gravitational acceleration. The relation between ($\log g$, T_{eff}) and (C_1, $\log W$) can be obtained through theoretical calculations. For example, Fig. 13.28 is obtained from the model

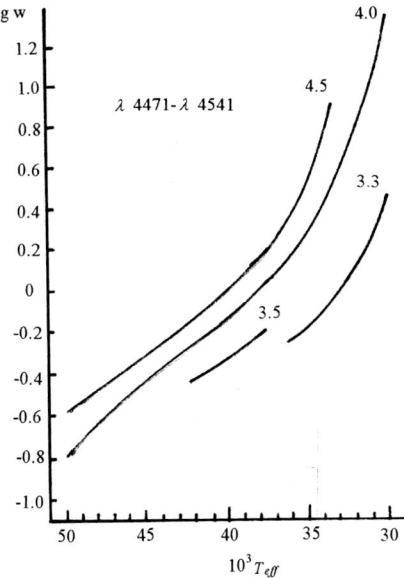

Fig. 13.27 The relations between the ratio W of the intensities of two spectral lines and the effective temperature T_{eff}.

atmosphere mesh given by Mihalas (1972). Figure 13.28 gives the points for different values of $(\log g,\ T_{eff})$ on the $(\log W(H_{\gamma}),\ C_1)$ plane. The points corresponding to the same T_{eff} are linked together, and the numbers attached to the curves are effective temperatures $(T_{eff}/10^3\ K)$. The numbers 3.0, 3.5, 4.0, 4.5 in the figure are values of $\log g$. In this way, once the color index C_1 and the equivalent width of the hydrogen line H_{γ} are obtained through observations, a point $(\log W(H_{\gamma}),\ C_1)$ can be identified on Fig. 13.28, from which the corresponding values of $(\log g,\ T_{eff})$ or the surface gravitational acceleration and the effective temperature for the star can be inferred.

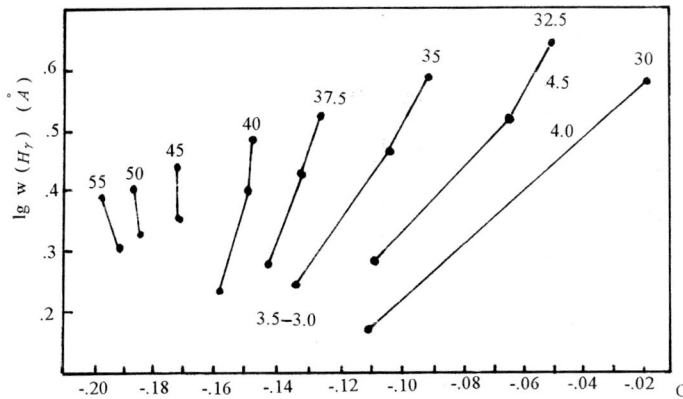

Fig. 13.28 The relation between equivalent widths of hydrogen lines and colour indices. The numbers attached to the curves are effective temperatures $(T_{eff}/10^3\ K)$. The numbers 3.0, 3.5, 4.0, 4.5 are values of $\log g$.

(G) Determination of the effective temperature and the surface gravitational acceleration from fitting diagrams

When the helium abundance is given for a star, the effective temperature T_{eff} and the surface gravitational acceleration $\log g$ will determine the Balmer jump D_B, color index and equivalent widths of the hydrogen lines (or profile widths). In other words, for a given helium abundance, values of (T_{eff}, $\log g$) will correspond to certain values of Balmer jump, color index and equivalent widths of hydrogen lines (or profile widths). Kudritzki (1976) calculated a mesh of non-LTE model atmospheres (40000 K $\leq T_{eff} \leq$ 50000 K, $4 \leq \log g \leq 8$, $0.1 \leq N(He)/N(H) \leq 1.0$), and used these results to determine Balmer jumps D_B, color indices and equivalent widths (or profile widths), and constructed fitting diagrams (Figs. 13.29 and 13.30). All curves in the fitting diagrams represent constant values of a variable. $D(0.1)$ are profile widths at the line depth of 0.1. In this way, for a given helium abundance [$N(He)/N((H)$], once the Balmer jump D_B, color index $U-B$, and the profile width $D(0.1)$ or the equivalent width W_λ of the hydrogen line are measured, the values of T_{eff} and $\log g$ can be obtained from Figs. 13.29 and 13.30. For example, if the helium abundance is 0.1, and from observations, $D_B = 0.10$ and $W_\lambda(H_\gamma) = 0.70$ Å, from the intersection between the $D_B = 0.10$ line and the $W_\lambda(H_\gamma) = 0.70$ line in Fig. 13.30, we obtain $\log T_{eff} = 4.66$ and $\log g = 6.0$.

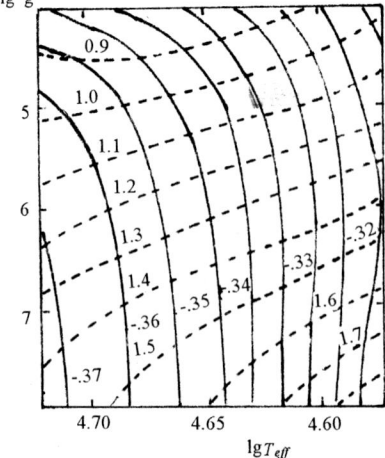

Fig. 13.29 Fitting diagram. The solid lines represent constant values of $U–B$ while the broken lines represent constant values of $\bar{D}$ (0.1) where $\bar{D}$ (0.1) is the average value of D(0.1) for H_γ and H_δ. $N(He)/N(H) = 0.1$ (according to Kudritzki 1976).

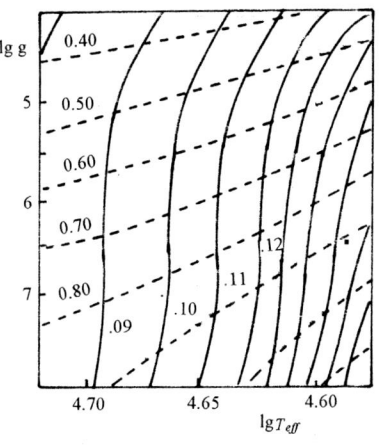

Fig. 13.30 Fitting diagram. The solid lines represent constant values of the Balmer jump D_B while the broken lines represent constant values of equivalent widths $W_\lambda(H_\gamma)$ of the H_γ line. $N(He)/N(H) = 0.1$ (according to Kudritzki 1976).

(H) Determination of the effective temperature, surface gravitational acceleration and helium abundance from intersection of fitting lines (Hunger *et al.* 1981)

The determination of the effective temperature, surface gravitational acceleration and helium abundance from intersection of fitting lines is the best method up to now because of its high precision. The fitting lines refer to those lines fitting theoretical values and observational values (such as equivalent widths) on the $\log g$-$\log T_{eff}$ plane (see Fig. 13.31). To obtain the fitting lines, we need calculated values of the model atmosphere mesh, and use them to tabulate the surface gravitational accelerations $\log g$, effective temperatures T_{eff} and equivalent widths for a certain line (such as $W_\lambda(H_\gamma)$ for the H_γ line). On the other hand, $W_\lambda(H_\gamma)$ can be obtained from observations. From the observed $W_\lambda(H_\gamma)$, we can get a series of ($\log g$, $\log T_{eff}$) values from interpolation within the table between $W_\lambda(H_\gamma)$ and ($\log g$, $\log T_{eff}$). From this series of values, the fitting line for the equivalent width of the H_γ line which fits the observed data can be plotted on the $\log g$-$\log T_{eff}$ plane for a helium abundance. The similar method can also be employed to obtain fitting lines for Balmer jump D_B, color index and spectral line profile on the $\log g$-$\log T_{eff}$ plane. When the main parameters (T_{eff}, g, X_i) of the model atmosphere are correct, the theoretical results such as the Balmer jump, color index, spectral index profile and equivalent width should be consistent with the observed values. Therefore, all fitting lines should intersect at one single point in the ideal case. In reality, however, this does not happen because of the presence of observational uncertainties and the simplified assumptions in the model atmosphere; instead, they intersect within a small area.

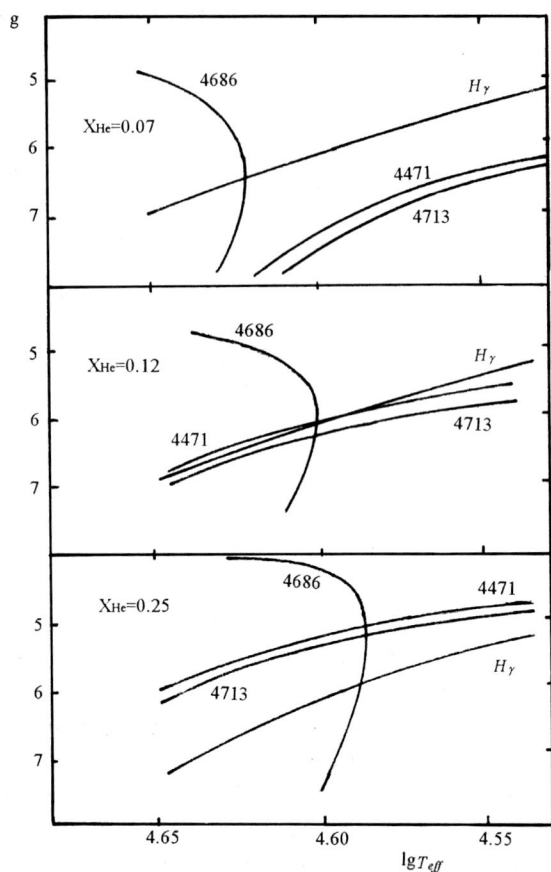

Fig. 13.31 The schematic diagram showing the determination of effective temperature T_{eff}, surface gravitational acceleration g and helium abundance from the intersection of fitting lines (from Hunger *et al.* 1981).

For a given helium abundance, different fitting (such as those for the color index, equivalent width etc.) can be plotted on the $\log g$-$\log T_{eff}$ plane. On changing the helium abundance to minimize the area of intersections among different fitting lines, we can obtain the best helium abundance, and the corresponding value of $(\log g, \log T_{eff})$ which are the effective temperature T_{eff} and the surface gravitational acceleration g of the star. For example, for the star SB884, Fig. 13.31 gives the fitting lines for the equivalent widths of spectral lines $\lambda 4686$, $\lambda 4471$, $\lambda 4713$ and H_γ for different helium abundance $X_{He} = 0.07$, 0.12 and 0.25 on the $\log g$-$\log T_{eff}$ plane. From Fig. 13.31, it can be seen that when $X_{He} = 0.12$, the fitting lines intersect within the smallest area so this is the required helium abundance. Furthermore, according to the intersection area, we infer that the effective temperature is $\log T_{eff} = 4.60$ and the surface gravitational acceleration is $\log g = 6.1$ for the star.

13.4.2 Determination of Angular Diameter, Radius, Luminosity, Mass and Rotational Velocity

After obtaining the main parameters of the model atmosphere from method (H) including the effective temperature, surface gravitational acceleration and chemical composition, the model atmosphere is fixed and the radiation flux $\pi F_\lambda(\text{model})$ can be theoretically calculated at the stellar surface. Then, according to the observed energy flux f_λ, we can obtain the angular diameter of the star from Eq. (13.434) as (Underhill 1982)

$$\theta = 2\left[f_\lambda / \pi f_\lambda(\text{model})\right]^{1/2} \qquad (13.441)$$

Further using other methods (such as according to the intensity of absorption lines of the interstellar medium), we can obtain the distance of the star, so the stellar radius is given by

$$R = \theta \cdot \frac{l}{2} \qquad (13.442)$$

The relation among the absolute magnitude M_V, radius R and radiation flux πF_λ is

$$M_V = -2.5 \log \int_0^\infty 4\pi R^2 \cdot \pi F_\lambda \cdot S_V^0(\lambda) d\lambda + c \qquad (13.443)$$

where S_V^0 is the V response function in the UBV photoelectric measurement system for a zero atmospheric mass. The Schulz relation can be obtained from Eq. (13.443) as

$$5 \log \frac{R}{R_\odot} = 29.57 - (M_V - V) \qquad (13.444)$$

where $V = -2.5\log \int_0^\infty F_\lambda S_V^0(\lambda) d\lambda$, and $F_\lambda (\text{erg/cm}^2.\text{s}.\text{Å})$ can be obtained from model atmosphere calculations, so the stellar radius R can be obtained from V, the absolute magnitude M_V and Eq. (13.444).

From the relation among the luminosity, effective temperature and radius

$$L = 4\pi R^2 \cdot \sigma T_{eff}^4 \qquad (13.445)$$

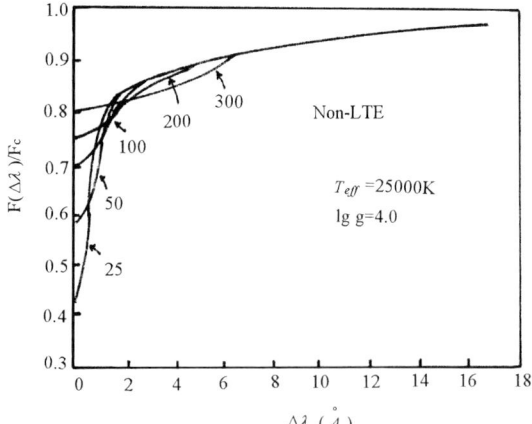

Fig. 13.32 The relation between the depth of the center of the H_α line and the rotational velocity (in km/s).

or

$$\log \frac{L}{L_\odot} = 4 \log T_{eff} + 2 \log\left(\frac{R}{R_\odot}\right) - 15.045 \qquad (13.446)$$

the stellar luminosity can be obtained from T_{eff} and R.

From the surface gravitational acceleration g and the radius R, we can obtain the stellar mass M by

$$\frac{M}{M_\odot} = \left(\frac{g}{g_\odot}\right)\left(\frac{R}{R_\odot}\right)^2 \qquad (13.447)$$

Mihalas and Auer (1970) calculated the model atmosphere and the H_α profile for no rotation, and discovered that the depth of the center of the H_α line is very sensitive to the rotational velocity. Figure 13.32 shows the relation between the depth of the center of the H_α line and the rotational velocity. Through comparisons with the observed depth of the center of the H_α line, the rotational velocity of the star can be obtained.

REFERENCES

Abbott, D.C., 1982, in "Wolf Rayet Stars: Observations, Physics and Evolution," *IAU Symposium No. 99*, p. 185, eds. C.de Loore, A.J. Willis, Dordrecht, Reidel.

Andreasen, G.K., 1988, Astronomy and Astrophysics, Vol. 201, p. 72

Andreasen, G.K., Petersen, J.O., 1988, Astronomy and Astrophysics, Vol. 192, L4.

Arnett, W.D., 1969, Astrophysics and Space Science, Vol. 5, p. 180.

Arnett, W.D., Baheall, J.N., Kirshner, R.P., Woosley, S.E., 1989, Annual Review of Astronomy and Astrophysics, Vol. 27, p. 627.

Auer, L.H., Mihalas, D., 1969, Astrophysical Journal, Vol. 158, p. 641.

Baad, W., 1994, Astrophysical Journal, Vol. 100, p. 137.

Baker, N., 1963, "Tables of Convective Stellar Envelope Models", Institute of Space Studies, New York.

Baker, N., 1966, in *Stellar Evolution*, eds. R.F. Stein, A.G.W. Cameron, Plenum Press, New York, p. 333.

Baker, N., and Kippenhahn, R., 1962, Zeitschrift für Astrophysik, Vol. 54, p. 114.

Baker, N., and Kippenhahn, R., 1965, Astrophysical Journal, Vol. 142, p. 868.

Bethe, H.A., and Wilson, J.R., 1985, Astrophysical Journal, Vol. 295, p. 11.

Biermann, P., and Kippenhahn, R., 1971, Astronomy and Astrophysics, Vol. 14, p. 32.

Böhn-Vitense, E., 1958, Zeitschrift für Astrophysik, Vol. 46, p. 108.

Bonsack, W.K., 1961, Astrophysical Journal, Vol. 133, p. 340.

Bonsack, W.K., and Greenstein, J.L., 1960, Astrophysical Journal, Vol. 131, p. 83.

Bressan, G., Bertelli, G., Chiosi, C., 1981, Astronomy and Astrophysics, Vol. 102, p. 25.

Brown, G.E., Bether, H.A., Baym, G., 1982, Nuclear Physics, A375, p. 481.

Brunish, W.M., and Truran, J.W., 1982, Astrophysical Journal Supplement, Vol. 49, p. 447

Buser, R., and Kurucz, R.L., 1978, Astronomy and Astrophysics, Vol. 70, p. 555.

Cannizzo, J.K., Pudritz, R.E., 1988, Astrophysical Journal, Vol. 327, p. 840.

Castor, J.I., 1971, Astrophysical Journal, Vol. 166, p. 109.

Castor, J.I., Abbott, D.C., Klein, R.I., 1975, Astrophysical Journal, Vol. 195, p. 157.

Chandrasekha, S., 1939, *An Introduction to the Study of Stellar Structure*, University of Chicago Press, p. 388.

Chevalier, C., 1971, Astronomy and Astrophysics, Vol. 14, p. 24.

Chiosi, C., Maeder, A., 1986, Annual Review of Astronomy and Astrophysics, Vol. 24, p. 329.

Clayton, D., 1968, *Principles of Stellar Evolution and Nucleosynthesis*, McGraw-Hill Book Company, p. 289, Table 4-1.

Conti, P.S., 1973, Astrophysical Journal, Vol. 179, p. 181.

Counselman, C.C., 1973, Astrophysical Journal, Vol. 179, p. 181.

Counselman, C.C., 1973, Astrophysical Journal, Vol. 180, p. 307.

Cowling, T.G., 1941, Monthly Notices of the Royal Astronomical Society, Vol. 101, p. 367.

Cox, A.N., King, D.S., Tabor, J.E., 1973, Astrophysical Journal, Vol. 184, p. 201.

Cox, A.N., Starrfield, S.G., Kidman, R.B., Pesnell, W.D., 1987, Astrophysical Journal, Vol. 317, p. 303.

Cox, J.P., 1963, Astrophysical Journal, Vol. 138, p. 487.

Cox, J.P., 1974, Reports on the Progress inPhysics, Vol. 37, p. 563.

Cox, J.P., 1980, *Theory of Stellar Pulsation*, Princeton University Press, Princeton, New Jersey.

Cox, J.P., and Giuli, R.T., 1968, *Stellar Structure*, Gordan and Breach, Science Publishers Inc.

Dalgarno, A., Willians, D.A., 1962, Astrophysical Journal, Vol. 136, p. 690.

Däppen, W., Mihalas, D., Hummer, D.G., and Mihalas, B.W., 1988, Astrophysical Journal, Vol. 332, p. 261.

Darwin, C.H., 1908, Scientific Papers, Vol. 2, Cambridge University Press.

Delgado, A.J., Thomas, H.-C., 1981, Astronomy and Astrophysics, Vol. 96, p. 142.

Doughty, N.A. et al., 1966, Monthly Notices of the Royal Astronomical Society, Vol. 132, p. 255.

Drew, J.E., 1987, Monthly Notices of the Royal Astronomical Society, Vol. 224, p.595.

Dufton, P.L., 1977, Astronomy and Astrophysics, Vol. 73, p. 203.

Dziembowski, W., Kozlowski, M., 1974, Acta Astronomica, Vol. 24, p. 245.

Dziembowski, W., Koester, D., 1981, Astronomy and Astrophysics, Vol. 97, p. 16.

Eggleton, P.P., 1983, Astrophysical Journal, Vol. 268, p. 368.

Ezer, D., Cameron, A.G.W., 1967, Canadian Journal of Physics, Vol. 45, p. 3429.

Faulkner, J., 1966, Astrophysical Journal, Vol. 144, p. 978.

Faulkner, J., 1971, Astrophysical Journal Letters, Vol. 170, L99.

Feautrier, P.C.R., 1964, Academy of Sciences of Paris, Vol. 258, p. 3189.

Fernie, J.D., 1990a, Astrophysical Journal, Vol. 354, p. 295.

Fernie, J.D., 1990b, Astrophysical Journal Supplement, Vol. 70, p. 153.

Flower, P.J., 1977, Astronomy and Astrophysics, Vol. 54, p. 31.

Foley, H.M., 1946, Physics Review, Vol. 69, p. 616.

Fowler, W.A., 1960, *Memories de la Sociètè Royale des Sciences de Liège*, Series 5, Vol. III.

Fowler, W.A., Caughlan, G.R., Zimmerman, B.A., 1975, Annual Review of Astronomy and Astrophysics, Vol. 13, p. 69.

Friend, D.B., Castor, J.I., 1982, Astrophysical Journal, Vol. 261, p. 293.

Gabriel, M., and Ledoux, P., 1967, Annual Review of Astronomy and Astrophysics, Vol. 31, p. 167.

Gaunt, J.A., 1930, Philosophical Transactions of the Royal Society of London Series, A229, 163.

Gaustadt, J.E., 1963, Astrophysical Journal, Vol. 138, p. 1050.

Geltman, S., 1962, Astrophysical Journal, Vol. 136, p. 944.

Geltman, S., 1965, Astrophysical Journal, Vol. 141, p. 376.

Georgeanne, R.C., and Fowler, W.A., 1988, Atomic Date and Nuclear Date Tables, Vol. 40, p. 283.

Gibson, D.M., 1978, Bulletin of the American Astronomy Society, Vol. 10, p. 631.

Gieren, W.P., Barnes, T.G., Moffelt, T.J., 1989, Astrophysical Journal, Vol. 342, p. 467.

Goldberg, L., Müller, E.A., Aller, L.H., 1960, Astrophysical Journal Supplement, Vol. 5.

Greenstein, J.L., and Richardson, R.S., 1951, Astrophysical Journal, Vol. 113, p. 536.

Hamada, T., Salpeter, E.E., 1961, Astrophysical Journal, Vol. 134, p. 683.

Hameury, J.M., King, A.R., Lasota, J.P., Ritter, H., 1988, Monthly Notices of Royal Astronomical Society, Vol. 231, p. 535.

Hamman, W.R., and Kudritzki, R.P., 1977, Astronomy and Astrophysics, Vol. 54, p. 524.

Han Z.W., 1987, Acta Astrophysica Sinica, Vol. 7, No. l, p. 33.

Han Z.W., 1988, Publications of Yunnan Observatory, No. 3, p. 77.

Harris, III, D.L., Strand, K.Aa., Worley, C.E., 1963, in *Basic Astronomical Date*, Chapter 15, ed. K.Aa. Strand, University of Chicago Press.

Harris, M.J., Fowler, W.A., Caughlan, G.R., Zimmerman, B.A., 1983, Annual Review of Astronomy and Astrophysics, Vol. 21, p. 165.

Hayashi, C., Hoshi, R., Sugimoto, D., 1962, Progress of Theoretical Physics Supplement, Vol. 22, p. 1.

Henyey, L.G., Vardya, M.S., Bodenheimer, P.L., 1965, Astrophysical Journal, Vol. 142, p. 841.

Van den Heuvel, E.P.J., 1992, in *Interacting Binaries*, eds. H. Nussbaumer, A. Orr, Springer-Verlag, p. 392.

Van den Heuvel, E.P.J., Habets, G.M.H.J., 1984, *Nature*, Vol. 309, p. 598.

Hillebrandt, W.D., 1985, in *High-Energy Phenomena Around Collapsed Stars*, Dordrecht, Reidel, p. 73.

Hofmeister, E., 1967, Zeitschrift für Astrophysik, Vol. 65, p. 164.

Hofmeister, E., Kippenhahn, R., Weigert, A., 1964, Zeitschrift für Astrophysik, 59, 215.

Huang, R.Q., 1997, Acta Astrophysical Sinica,Vol. 17, No. 2, p. 157.

Huang, R.Q., Li, Y., 1990, *Theory of Stellar Oscillations*, Science Press, Beijing.

Huang, R.Q., and Taam, R.E., 1990, Astronomy and Astrophysics, Vol. 230, p. 107.

Huang, R.Q., and Xie, X., 1986, Astronomy and Astrophysics, Vol. 161, p. 142.

Huang, R.Q., Yu, K.N., 1996, Acta Astrophysical Sinica, Vol. 16, No. 1, p. 41.

Huang, R.Q., Yu, K.N., Han, Z.W., 1992, Astronomy and Astrophysics, Vol. 256, p. 438.

Huang, R.Q., and Weigert, A., 1982, Astronomy and Astrophysics, Vol. 112, p. 281.

Huang, R.Q., and Weigert, A., 1983, Astronomy and Astrophysics, Vol. 127, p. 309.

Hummer, D.G., Mihalas, D., 1988, Astrophysical Journal, Vol. 331, p. 794.

Humphreys, R.M., and Davidson, K., 1979, Astrophysical Journal, Vol. 232, p. 409.

Hunger, K., Grushinski, J., Kudritzki, R.P., Simon, K.P., 1981, Astronomy and Astrophysics, Vol. 95, p. 244.

Iben, I. Jr., 1965, Astrophysical Journal, Vol. 141, p. 993.

Iben, I. Jr., 1971, Astrophysical Journal, Vol. 166, p. 131.

Iben, I. Jr., 1974, in *Stellar Instability and Evolution*, eds. P. Ledoux, A. Noels, A.W. Rodgers, IAU Symposium No. 59, Dordrecht, Reidel, p. 3.

Iben, I. Jr., Renzini, A., 1983, Annual Review of Astronomy and Astrophysics, Vol. 21, p. 271.

Iben, I. Jr., Tuggle, R.S., 1975, Astronomy and Astrophysics, Vol. 197, p. 39.

De Jager, C., Nieuwenhuijzen, H., van der Hucht, K.A., 1988, Astronomy and Astrophysics Supplement, Vol. 72, p. 259.

De Jager, C., 1980, *The Brightest Stars*, Dordrecht, Reidel.

Jeans, J., 1928, *Astronomy and Cosmogony* (Cambridge University Press, Cambridge), republished 1961 (Dover, New York).

Jiang, S.Y., Huang, R.Q., 1997a, Astronomy and Astrophysics, Vol. 317, p. 114.

Jiang, S.Y., Huang, R.Q., 1997b, Astronomy and Astrophysics, Vol. 317, p. 121.

John, T.L., 1966, Monthly Notices of the Royal Astronomical Society, Vol. 131, p. 315.

Kähler, H., 1972, Astronomy and Astrophysics, Vol. 20, p. 105.

King, D.S., Cox, A.N., 1987, in *Stellar Pulsation*, Lecture Notes in Physics, No. 274, eds. A.N. Cox, W.M. Sparks, S.G. Starrfield, Springer-Verlag, p. 124.

Kippenhahn, R., Thomas, H.-C., 1964, Zeitschrift, für Astrophysik, Vol. 60, p. 19.

Kippenhahn, R., Thomas, H.-C., Weigert, A., 1965, Zeitschrift für Astrophysik, Vol. 61, p. 241.

Kippenhahn, R., Thomas, H.-C., Weigert, A., 1968, Zeitschrift für Astrophysik, Vol. 69, p. 265.

Kippenhahn, R., Weigert, A., Hofmeister, E., 1967, *Method in Computational Physics*, Vol. 7, p. 129, Academic Press, New York.

Kippenhahn, R., Weigert, A., 1991, *Stellar Structure and Evolution*, Springer-Verlag.

Kippenhahn, R.,Weigert, A., 1967, Zeitschrift für Astrophysik, Vol. 65, p. 251.

Kopal, Z., 1959, *Close Binary System*, Chapman and Hall Ltd, London.

Kraft, R.P., Mathews, J., Greenstein, J.L., 1962, Astrophysical Journal, Vol. 136, p. 312.

Kramers, H.A., 1923, Philosophical Magazine, Vol. 46, p. 836.

Krodgahl, W.S., Miller, J.E., 1967, Astrophysical Journal, Vol. 150, p. 273.

Kudritzki, P.P., 1976, Astronomy and Astrophysics, Vol. 52, p. 11.

Kudritzki, R.P., 1978, Dissertation, Christian-Albrecht University, Kiel.

Kudritzki, R.P., 1981, ESO Workshop "The Most Massive Stars", eds. S. Dodorico, B. Baade, K. Kjar, Garching.

Kudritzki, R.P., Reimers, D., 1978, Astronomy and Astrophysics, Vol. 70, p. 227.

Kurucz, R.L., 1979, *A Computer Program for Calculating Model Stellar Atmospheres*, Smithsonian Astrophysical Observatory.

Lamers, H.L.G.L.M., 1985, in *Luminous Stars and Stellar Associations*, eds. P.S. Conti, C. de Loore, E. Kontizas, Dordrecht, Reidel.

Landau, L., Lifshitz, E., 1972, *The Classical Theory of Fields*, Addison-Wesley, Reading, Massachusetts.

Langer, N., El Eid, M.F., Fricke, K.J., 1985, Astronomy and Astrophysics, Vol. 145, p. 169.

Larson, R.B., 1969, Monthly Notices of the Royal Astronomical Society, Vol. 145, p. 271.

Lauterborn, D., 1970, Astronomy and Astrophysics, Vol. 7, p. 150.

Ledoux, P., Walraven, Th., 1958, in *Handbuch der Physik*, ed. S.Flugge, Springer-Verlag.

Li, Y., 1992a, Ph.D. Thesis, Yunnan Observatory Chinese Academy of Sciences.

Li, Y., 1992b, Astronomy and Astrophysics, Vol. 257, p. 133.

Li, Y., 1992c, Astronomy and Astrophysics, Vol. 257, p. 145.

Li, Y., 1993, Astronomy and Astrophysics, Vol. 276, p. 357.

Li, Y., Gong, Z.G., 1994, Astronomy and Astrophysics, Vol. 289, p. 446.

Li, Y., Stix, M., 1994, Astronomy and Astrophysics, Vol. 286, p. 815.

Lindholm, E., 1945, Arkiv. f. Math. Astron. Fysik., 32A, No. 17.

Livio, M., Soker, N., 1988, Astrophysical Journal, Vol. 329, p. 764.

De Loore, C., de Greve, J.P., de Guyper, J.P., 1975, Astrophysics and Space Science, Vol. 36, p. 219.

De Loore, C., de greve, J.P., Lamers, H.J.G.L.M., 1977, Astronomy and Astrophysics, Vol. 61, p. 251.

Low, C., Lynden-Bell, D., 1976, Monthly Notices of the Royal Astronomical Society, Vol. 176, p. 367.

Luo, G., 1994a, Astronomy and Astrophysics, Vol. 281, p. 460.

Luo, G., 1994b, Astronomy and Astrophysics, Vol. 284, p. 679.

Luo, G., 1994c, Astronomy and Astrophysics, Vol. 284, p. 684.

Maeder, A., 1975, Astronomy and Astrophysics, Vol. 40, p. 303.

Maeder, A., Conti, P.S., 1994, Annual Review of Astronomy and Astrophysics, Vol. 32, p. 227.

Martraka, B., Wassermann, C., Weigert, A., 1981, Astronomy and Astrophysics, Vol. 107, p. 283.

Mathews, T.A., Sandage, A.R., 1963, Astrophysical Journal, Vol. 138, p. 30.

Matsushiman, S., 1969, Astrophysical Journal, Vol. 158, p. 1137.

McDowell, M.R.C., Williamson, J.M., Myerscough, V.P., 1966, Astrophysical Journal, Vol. 144, p. 827.

Mihalas, D., 1972, Astrophysical Journal, Vol. 176, p. 139.

Mihalas, D., Auer, L.H., 1970, Astrophysical Journal, Vol. 161, p. 1129.

Mihalas, D., Däppen, W., Hummer, D.G., 1988, Astrophysical Journal, Vol. 331, p. 815.

Mihalas, D., Kunanz, P.B., Hummer, D.G., 1975, Astrophysical Journal, Vol. 202, p. 465.

Millis, R.L., 1966, Information Bulletin on Variable Stars, No. 137.

Miyaji, S., Nomoto, K., Yoko, K., Sugimoto, D., 1980, Publications of the Astronomical Society of Japan, 32, 303.

Morison, N.D., 1975, Astrophysical Journal, Vol. 200, p. 113.

Morton, D.C., Adams, T.F., 1968, Astrophysical Journal, Vol. 151, p. 611.

Mullan, D.J., Doyle, J.D., Redman, R.O., Mathioudakis, M., 1992, Astrophysical Journal, Vol. 397, p. 225.

Nieuwenhuijzen, H., de Jager, C., 1990, Astronomy and Astrophysics, Vol. 231, p. 134.

Nomoto, K., Sugimoto, D., Neo, S., 1976, Astrophysics and Space Science, Vol. 39, L37.

Nomoto, K., Thielemann, F.-K., Miyaji, S., 1985, Astronomy and Astrophysics, Vol. 149, p. 239.

Paczynski, B., 1967, Acta Astronomica, Vol. 17, p. 287.

Paczynski, B., 1972, Acta Astronomica, Vol. 22, p. 163.

Patterson, J., 1984, Astrophysical Journal Supplement, Vol. 54, p. 443.

Pesnell, W.D., 1987, Astrophysical Journal, Vol. 314, p. 598.

Petit, M., 1987, *Variable Stars*, John Wiley & Sons, Chichester.

Petrosian, V., Beaudet, G., Salpeter, E.E., 1967, Physics Review, Vol. 154, p. 1445.

Plavec, M., Kratochvil, P., 1964, Bulletin of the Astronomic Institute of Czechoslovakia., Vol. 15, p. 165.

Popper, D.M., 1980, Annual Review of Astronomy and Astrophysics, Vol. 18, p. 115.

Rapparport, S., Joss, P.C., Webbink, R.F., 1982, Astrophysical Journal, Vol. 254, p. 616.

Refdal, S., Weigert, A., 1969, Astronomy and Astrophysics, Vol. 1, p. 167.

Refsdal, S., Weigert, A., 1970, Astronomy and Astrophysics, Vol. 6, p. 426.

Reimers, D., 1975, Memories de la Societe Royale des Sciences de Liège, Series 6, Vol. 8, p. 369.

Robinson, E.L., 1979, in *Nonradial and Nonlinear Stellar Pulsation*, ed. H.A. Hill, W.A. Dziembowski, Springer-Verlag, p. 446.

Roth, M.L., Weigert, A., 1972, Astronomy and Astrophysics, Vol. 20, p. 13.

Rybicki, G.R., Hummer, D.G., 1978, Astrophysical Journal, Vol. 219, p. 654.

Sahade, J., Brandi, E., 1991, Astrophysical Journal, Vol. 379, p. 706.

Saio, H., Kato, M., Nomoto, K., 1988, Astrophysical Journal, Vol. 331, p. 388.

Salpeter, E.E., 1954, Australian Journal of Physics, Vol. 7, p. 373.

Sandage, A.R., 1962, Astrophysical Journal, Vol. 135, p. 333.

Saslaw, W.C., Schwarzshild, M., 1965, Astrophysical Journal, Vol. 142, p. 1468.

Sengbusch, V., 1968, Zeitschrift für Astrophysik, 69, 79.

Shavir, G., Salpeter, E.E., 1973, Astrophysical Journal, Vol. 184, p. 191.

Shipman, H.L., Sass, C.A., 1980, Astrophysical Journal, Vol. 235, p. 177.

Shore, S.N., Brown, D.N., 1988, Astrophysical Journal, Vol. 334, p. 1021.

Shore, S.N., Corcoran, M.F., 1992, in IAU Symposium No. 151, ed. Y. Kondo, R. Siotero, R. Polidar, Dordrecht, Kluwer, p. 359.

Skumanich, A., 1972, Astrophysical Journal, Vol. 171, p. 565.

Smith, M., Strom, A., 1969, Astrophysical Journal, Vol. 158, p. 1161.

Spitzer, L. Jr., 1968, *Diffuse Matter in Space*, Wiley, New York.

Starrfield, S.G., Cox, A.N., Hodson, S.W., Pesnell, W.D., 1983, Astrophysical Journal Letters, Vol. 268, L27.

Starrfield, S.G., Cox, A.N., Kidman, R.B., Pesnell, W.D., 1985, Astrophysics Journal Letters, Vol. 293, L23.

Stellingwerf, R.F., 1979, Astrophysical Journal, Vol. 227, p. 935.

Stilley, J.L., Callaway, J., 1970, Astrophysical Journal, Vol. 160, p. 245.

Strom, S.E., Strom, K.M., Rood, R.T., Iben, I.Jr., 1970, Astronomy and Astrophysics, Vol. 8, p. 243.

Sugimoto, D., Nomoto, K., 1980, Space Science Review, Vol. 25, p. 155.

Sutantyo, W., 1992, in *X-ray Binaries and Recycled Pulsars*, eds. E.P.J. Van den Heuvel, S.A. Rappaport, Kluwer Academic Publications, Dordrecht, p. 293.

Taam, R.E., 1983, Astrophysical Journal, Vol. 268, p. 361.

Taam, R.E., Flannery, B.P., Faulkner, J., 1980, Astrophysical Journal, Vol. 239, p. 1017.

Thomas, H.-C., 1967, Zeitschrift für Astrophysik, Vol. 67, p. 420.

Underhill, A.B., 1982, Astrophysical Journal, Vol. 263, p. 741.

Unno, W., Osaki, Y., Ando, H., Saio, H., Shibahashi, H., 1989, *Nonradial Oscillations of Stars*, 2nd ed., University of Tokyo Press.

Unsöld, A., 1958, *Physik der Sternatmospharen*, 2. Auflage, Springer-Verlag, p. 167.

Underhill, A.B., Waddell, J.H., 1959, National Bureau of Standards, Circular No. 603.

Vanderlinden, 1982, Ph.D. Thesis, University of Amsterdam.

Vardavas, I.M., 1974, Journal of Quantitative Spectroscopy Radiative Transfer, Vol. 14, p. 909.

Verbunt, F., Zwaan, C., 1981, Astrophysical Journal, Vol. 100, L7.

Warner, B., Nather, R.E., 1972, Monthly Notice of the Royal Astronomics Society, Vol. 156, p. 1.

Weaver, T.A., Zimmerman, G.B., Woosley, S.E., 1978, Astrophysical Journal, Vol. 225, p. 1021.

Weigert, A., 1966, Zeitschrift für Astrophysik, Vol. 64, p. 396.

Weiss, A., 1989, Astrophysics Journal, Vol. 339, p. 365.

Werner, K., 1986, Astronomy and Astrophysics, Vol. 161, p. 177.

Wheeler, J.C., Harkness, R.P., 1992, in *Stellar Astrophysics*, ed. R.J. Tayler, IOP Publishing Ltd, p. 160.

Wilson, J.R., 1985, Numerical Astrophysics, ed. J. Centrella et al. (Boston: Jones and Bartlett), p. 422.

Winget, D.E., Van Horn, H.M., Tassoul, M., Hansen, C.J., Fontaine, G., 1983, Astrophysical Journal Letters, Vol. 268, L33.

Winget, D.E., Van Horn, H.M., Tassoul, M., Hansen, C.J., Fontaine, G., Carroll, B.W., 1982, ApJ Letters, Vol. 252, L65.

Woosley, S.E., Weaver, T.A., 1986a, Annual Review of Astronomics and Astrophysics, Vol. 24, p. 205.

Woosley, S.E., Weaver, T.A., 1986b, in *Radiation Hydrodynamics in Stars and Compact Object*, eds. D. Mihalas, K.-H.A. Winkler, Springer-Verlag, p.91.

Woosley, S.E., Weaver, T.A., 1986c, in *Nucleosynthesis and Its Implications on Nuclear and Particle Physics*, eds. J. Audouze, N. Mathieu, Dordrecht, Reidel, p. 145.

Wu, G.Q., 1992, Apstrophysics and Space Science, Vol. 189, p. 171.

Yu, K.N., Huang, R.Q., 1995, Acta Astrophysical Sinica, Vol. 15, No. 4, p. 337.

Zaidi, M., 1966, Nuovo Cimento, Vol. 40A, p. 502.

Zhao, D.F., 1995, Ph.D. Thesis, Yunnan Observatory Chinese Academy of Sciences.

SUBJECT INDEX